AF358675

Corn Rootworm: Biology, Ecology, Behavior and Integrated Management

Corn Rootworm: Biology, Ecology, Behavior and Integrated Management

Guest Editors

Lance J. Meinke
Joseph L. Spencer

Basel • Beijing • Wuhan • Barcelona • Belgrade • Novi Sad • Cluj • Manchester

Guest Editors

Lance J. Meinke
Department of Entomology
University of Nebraska
Lincoln
United States

Joseph L. Spencer
Prairie Research Institute
University of Illinois
Champaign
United States

Editorial Office
MDPI AG
Grosspeteranlage 5
4052 Basel, Switzerland

This is a reprint of articles from the Special Issue published online in the open access journal *Insects* (ISSN 2075-4450) (available at: www.mdpi.com/journal/insects/special_issues/corn_rootworm).

For citation purposes, cite each article independently as indicated on the article page online and using the guide below:

Lastname, A.A.; Lastname, B.B. Article Title. *Journal Name* **Year**, *Volume Number*, Page Range.

ISBN 978-3-7258-1804-4 (Hbk)
ISBN 978-3-7258-1803-7 (PDF)
https://doi.org/10.3390/books978-3-7258-1803-7

Contents

About the Editors

Lance J. Meinke

Dr Lance J. Meinke has been a faculty member in the Department of Entomology at the University of Nebraska, Lincoln, Nebraska, USA, since 1984. He recently retired and, currently, he is an emeritus professor. He received his Ph.D in Entomology with a minor in Crop Science from North Carolina State University. His general areas of expertise include insect ecology and behavior, insect/plant interactions, integrated pest management, and resistance management. A large portion of his career has been dedicated to the study of the insect genus *Diabrotica* with special emphasis on the biology, ecology, behavior, and management of the western corn rootworm, (*Diabrotica virgifera virgifera* LeConte), an important pest insect in field maize. A key focus has been the characterization of field-evolved resistance by the western corn rootworm to commonly used insecticides and rootworm-active Bt traits in the Western USA Corn Belt.

Joseph L. Spencer

Dr. Joseph Lee Spencer is the Principal Insect Behaviorist at Illinois Natural History Survey in the Prairie Research Institute at the University of Illinois, Urbana-Champaign, Illinois, USA. He received his Ph.D. in Entomology from Michigan State University in 1994. He has worked at the University of Illinois since 1996. His research interests are focused on insect behavior and reproduction, with a special interest in western corn rootworm (*Diabrotica virgifera virgifera* LeConte) ecology, biology, behavior, and management in rotated maize and soybean systems. Dr. Spencer has been involved in insect resistance management research and the study of rootworm resistance to annual crop rotation and Bt maize for over 25 years. The study of western corn rootworm intra-/interfield movement and migration has been a persistent theme throughout his professional career.

Editorial

Corn Rootworm: Biology, Ecology, Behavior, and Integrated Management

Lance J. Meinke [1],* and Joseph L. Spencer [2]

[1] Department of Entomology, University of Nebraska, Lincoln, NE 68583, USA
[2] Illinois Natural History Survey, Prairie Research Institute, University of Illinois, Champaign, IL 61820, USA; spencer1@illinois.edu
* Correspondence: lmeinke1@unl.edu

Citation: Meinke, L.J.; Spencer, J.L. Corn Rootworm: Biology, Ecology, Behavior, and Integrated Management. *Insects* **2024**, *15*, 235. https://doi.org/10.3390/insects15040235

Received: 21 March 2024
Accepted: 22 March 2024
Published: 28 March 2024

Species of the beetle genus *Diabrotica* (Coleoptera: Chrysomelidae) are native to North and South America, with their greatest diversity occurring in neotropical areas [1]. Little is known about the biology and ecology of many of the 400 described species, with current knowledge primarily limited to the small number of species that are pests in agricultural systems [1,2]. This Special Issue focuses primarily on key economically important pest species, i.e., *D. virgifera virgifera* LeConte, *D. speciosa* (Germar), *D. balteata* (LeConte), and *D. viridula* (F.).

The western corn rootworm, *D. v. virgifera*, is a key pest of grain maize in North America and the specific focus of many contributions to this Special Issue. The costs of managing this pest and the value of lost production annually exceed USD 2 billion in the U.S.A. This species was also accidentally introduced into Europe, where it is now an established maize pest [3]. The highly adaptable nature of this species has made management an ongoing challenge. Over time, this species has evolved resistance to active ingredients in four insecticide classes, annual crop rotation, and all commercially available rootworm-active Cry toxins (derived from the soil microbe *Bacillus thuringiensis*) expressed in *Bt*–maize hybrids in the U.S.A. [4–6].

The future success of *Diabrotica* pest management may depend on a more holistic view of management than that implemented in the past. This requires movement away from single-tactic approaches to a combination of tactics deployed within an integrated pest management framework [5]. This will include the conceptualization and development of new tactics that are based on an increased understanding of *Diabrotica* biology, physiology, ecology, and population dynamics [6–10].

This Special Issue provides original research and comprehensive reviews that summarize the current knowledge in key areas of *Diabrotica* biology, ecology, behavior, and management. The contributions include the following:

- An overview of the evolutionary history and host relationships of *Diabrotica* species, plus natural enemies of *Diabrotica* [1];
- The biology and management of *D. v. virgifera* in Europe [3] and *D. speciosa* (Germar), *D. balteata* (LeConte), and *D. viridula* (F.) in South America [2];
- Host–microbe relationships and aspects of chemical ecology that influence *D. v. virgifera* behavior and host plant resistance [9];
- The movement ecology of *D. v. virgifera* and its relation to management [6];
- The potential of RNAi technologies as components of *D. v. virgifera* management strategies [10];
- An overview of *D. v. virgifera* resistance to insecticides and plant-incorporated *Bt* traits in maize [4,5];
- Advances in *D. v. virgifera* monitoring/sampling technologies [7];
- An overview of the available computer models and modeling approaches to support research on the biology and management of *Diabrotica* species [8].

The history of field-evolved resistance is marked by repeated failures to appreciate the capabilities of pest insects. The western corn rootworm is a good example of this. Therefore, it is our hope that these publications, which refresh our current understanding of *Diabrotica*, will inform future research and efforts to develop novel and sustainable management tactics/strategies.

Conflicts of Interest: The authors declare no conflicts of interest.

References

1. Eben, A. Ecology and evolutionary history of Diabrotica beetles—Overview and update. *Insects* **2022**, *13*, 156. [CrossRef] [PubMed]
2. Cabrera Walsh, G.; Ávila, C.; Cabrera, N.; Nava, D.; de Sene Pinto, A.; Weber, D. Biology and management of pest Diabrotica species in South America. *Insects* **2020**, *11*, 421. [CrossRef] [PubMed]
3. Bažok, R.; Lemić, D.; Chiarini, F.; Furlan, L. Western corn rootworm (Diabrotica virgifera virgifera LeConte) in Europe: Current status and sustainable pest management. *Insects* **2021**, *12*, 195. [CrossRef] [PubMed]
4. Gassmann, A. Resistance to Bt maize by western corn rootworm: Effects of pest biology, the pest–crop interaction and the agricultural landscape on resistance. *Insects* **2021**, *12*, 136. [CrossRef] [PubMed]
5. Meinke, L.; Souza, D.; Siegfried, B. The use of insecticides to manage the western corn rootworm, Diabrotica virgifera virgifera, LeConte: History, field-evolved resistance, and associated mechanisms. *Insects* **2021**, *12*, 112. [CrossRef]
6. Sappington, T.; Spencer, J. Movement ecology of adult western corn rootworm: Implications for management. *Insects* **2023**, *14*, 922. [CrossRef]
7. Tóth, Z.; Tóth, M.; Jósvai, J.; Tóth, F.; Flórián, N.; Gergócs, V.; Dombos, M. Automatic field detection of western corn rootworm (Diabrotica virgifera virgifera; Coleoptera: Chrysomelidae) with a new probe. *Insects* **2020**, *11*, 486. [CrossRef] [PubMed]
8. Onstad, D.; Caprio, M.; Pan, Z. Models of Diabrotica populations: Demography, population genetics, geographic spread, and management. *Insects* **2020**, *11*, 712. [CrossRef] [PubMed]
9. Paddock, K.; Robert, C.; Erb, M.; Hibbard, B. Western corn rootworm, plant and microbe interactions: A review and prospects for new management tools. *Insects* **2021**, *12*, 171. [CrossRef] [PubMed]
10. Darlington, M.; Reinders, J.; Sethi, A.; Lu, A.; Ramaseshadri, P.; Fischer, J.; Boeckman, C.; Petrick, J.; Roper, J.; Narva, K.; et al. RNAi for western corn rootworm management: Lessons learned, challenges, and future directions. *Insects* **2022**, *13*, 57. [CrossRef] [PubMed]

Review

Ecology and Evolutionary History of *Diabrotica* Beetles—Overview and Update

Astrid Eben

Julius Kühn–Institut (JKI), Federal Research Centre for Cultivated Plants, Institute for Plant Protection in Fruit Crops and Viticulture, Schwabenheimer Straße 101, 69221 Dossenheim, Germany; astrid.eben@julius-kuehn.de; Tel.: +49-(0)-394647-4760; Fax: +49-(0)-394647-4805

Simple Summary: This review provides an overview on selected aspects of the ecology and biology of *Diabrotica* leaf beetles. A special focus is on the western corn rootworm as a major pest insect on corn. Furthermore, general information on host plant relationships and natural enemies of this beetle group is presented. Current knowledge of these leaf beetles is mostly focused on a limited number of economically important species. For the majority of the species in the group little to nothing is known about their biology and their host plants. This information, however, could be useful for future plant breeding programs and pest control strategies.

Abstract: An overview is given on several aspects of evolutionary history, ecology, host plant use, and pharmacophagy of *Diabrotica* spp. with a focus on the evolution of host plant breadth and effects of plant compounds on natural enemies used for biocontrol of pest species in the group. Recent studies on each aspect are discussed, latest publications on taxonomic grouping of *Diabrotica* spp., and new findings on variations in the susceptibility of corn varieties to root feeding beetle larvae are presented. The further need for in-depth research on biology and ecology of the large number of non-pest species in the genus is pointed out.

Keywords: antibiosis; cucurbitacins; *Diabrotica*; corn; coevolution; entomopathogenic nematodes; teosinte

Citation: Eben, A. Ecology and Evolutionary History of *Diabrotica* Beetles—Overview and Update. *Insects* **2022**, *13*, 156. https://doi.org/10.3390/insects13020156

Academic Editors: Lance J. Meinke and Joseph L. Spencer

Received: 23 December 2021
Accepted: 29 January 2022
Published: 31 January 2022

Publisher's Note: MDPI stays neutral with regard to jurisdictional claims in published maps and institutional affiliations.

1. Introduction

Leaf beetles of the section Diabroticites (Chrysomelidae: Galerucinae: Luperini) comprise 823 species. Among these, the most species rich genus is *Diabrotica*, with about 400 described species [1–3]. The majority of species ($n = 354$) are grouped in the *fucata* group, and these species are supposed to be polyphagous on a number of plant families and multivoltine with more than one generation per year, while fewer of the *Diabrotica* spp. are oligophagous and univoltine: 24 species in the *virgifera* group and 11 in the *signifera* group [4]. Species of the genus *Diabrotica* are native to North and South America [2], with the greatest diversity found in Mexico and in Brazil. The highest number of species is found in Central America, where the group has its evolutionary origin [5,6]. The species of the *signifera* group are distributed only in South America [2,6].

Reliable information on biology, ecology, and above all host plant relationships is scarce or unavailable for many species [7]. To date, most data were published on the few species that are economically important as pest species—namely, *Diabrotica balteata* LeConte, *D. barberi* Smith & Lawrence, *D. undecimpunctata howardi* Barber, *D. virgifera* LeConte, and *D. speciosa* Germar [2,8,9].

In North America, *D. virgifera*, or western corn rootworm, is the most important pest insect on corn in the USA [10,11]. In the 1990s, this species was accidentally brought to Serbia, in Eastern Europe, where it quickly spread over large parts of eastern and central Europe and became a threat to the corn-growing regions foremost in Hungary and Germany [12–16]. Nevertheless, in those newly invaded regions, cultivating corn in a two- or three-year rotation with non-monocotyledon crops made it possible to maintain the populations of *D. v. virgifera* below economic threshold level [17,18]. *D. v. virgifera* is found in most European countries and was eradicated in Belgium, Netherlands, and United Kingdom [19], whereas in France and Germany, for example, populations are maintained below levels of economic damage due to intensive monitoring efforts and corn rotation strategies with particular regulations for each susceptible region and country [18].

2. Life Cycle

Species in the genus *Diabrotica* feed as adults on leaves and pollen of their host plants; one female can deposit 300–400 eggs in the soil at a depth of about 5–15 cm close to the roots of host plants, and larvae develop while feeding and tunnelling in host plant roots. Larval development encompasses three instars and mature larvae pupate in a soil chamber constructed by the late third instar larva. The pupal state lasts 7–10 days [2]. Newly hatched adults are softbodied and light coloured with scarcely visible elytral patterns (Eben, pers. observ). The entire life cycle of multivoltine species is completed within about 30 days. Moreover, univoltine species overwinter as eggs, and the multivoltine species spend the dry and cold season often as diapausing adults [2,20]. In *D. barberi*, an important pest species on corn, it is known that the egg diapause can be extended to two years due to availability of corn host plants in a biannual corn–soybean crop rotation system in Nebraska [21], whereas in semi-tropical and tropical climates, such as in Brazil, a multivoltine species such as *D. speciosa* can develop up to six generations per year [9].

3. Host Plant Use—Plant Resistance

Diabrotica species feed on plants from about 50 different families [6]. Some species are polyphagous—for example, a number of species in the *fucata* group—while others show narrow host use—i.e., most species in the *virgifera* group [8]. Information on host plants, however, remains scarce and difficult to find for most of the Diabroticites that are not of economic importance [7,22]. *Diabrotica* spp. were described to feed on at least 60 different crop plant species, but only a small number of species can be considered pests of these host plants [23,24]. Because some species distributed in the USA are of high economic importance as pest insects on corn, soybean, squash, and beans, their life cycle, their host plants, and their phylogenetic relationships have been well-studied [8,25,26].

Polyphagous species in the *fucata* group of the genus *Diabrotica* and their respective host plants are mainly found in the tropics and sub-tropics. Their distribution and abundance is thus often patchy and restricted to small areas. These habitats are very much in contrast with those used by prairie inhabiting and corn feeding species of the *virgifera* group, where host plants grow in large, landscape scale monocultures [7,27–29]. For many species, however, the knowledge of non-cultivated host plants remains restricted or unknown, even more so for larval hosts. The evolutionary switch of host plant families and the broadening or reduction in the host range in some species probably occurred in parallel with speciation events within the genus *Diabrotica* since several hosts were independently lost and gained throughout the evolution of the genus.

The intimate relationship of some species in the *virgifera* group with monocotyledonous plants in North American prairie habitats, mainly with corn, was long supposed to have originated in southern Mexico or further south in Guatemala [2,30]. Recently, it was confirmed that wild Mexican Balsas teosinte (*Zea mays* ssp. *Parviglumis diploperennis* Iltis and Doebley) is the ancestor of the modern, domesticated maize [31]. Domestication of maize in Mexico occurred about 9000 years ago [31,32], whereas the beginning of the diversification of cucurbits was dated back to at least 50 M years ago [33], with Asia being the area of origin of the family Cucurbitacae. The domestication of the genus *Cucurbita* was dated to about 10,000 years ago and is supposed to have taken place in northeastern Mexico [5,34,35].

Unfortunately, teosinte species in their native habitat are threatened by extinction before being investigated in detail for a number of potentially important characteristics, i.e., resistance towards biotic and abiotic stress. For example, one study of resistance mechanisms in these wild corn ancestors provided evidence for greater resistance against herbivores in teosinte than in domesticated corn [36]. Like a number of other herbivores, *Diabrotica* spp. feed on shoots and on roots of corn. The damage caused on the belowground root system may cause plants to partially fall to the ground and possibly wither or dry out. In contrast with corn, teosinte is a perennial plant and provides food for more than one generation of rootworms. Nevertheless, during comparative field investigations in Mexico, rootworms were found more abundant on corn than on teosinte [37]. This observation might be due to the generally much higher biomass in both, below- and aboveground plant parts of cultivated corn plants when compared with less sumptuous teosinte plants. Maize is well studied with regard to its inducible and constitutive defences such as hormones, volatile compounds, peptides, enzymes, and physical defences. Much less is known about the defence system of teosinte. Erb et al. [38] found that the response of corn and teosinte to simultaneous root herbivory by *D. v. virgifera* and *Spodoptera frugiperda* (J.E. Smith) (Lepidoptera: Noctuidae) was more intense for teosinte. In general, wild Mexican land races and teosinte seem to be more tolerant of herbivores than domesticated corn. Furthermore, since corn rootworms are associated with several symbiontic gut microbes, this interaction might affect plant defences as well as insect physiology [36,39–41].

Volatile compounds emitted by plants after herbivore attack are also broadly studied for corn [42,43]. These compounds are induced through feeding by rootworm larvae and reported to be highly variable in the different maize varieties [44]. The volatiles interact with above and below ground natural enemies of herbivores, such as parasitic wasps and entomopathogenic nematodes, respectively [45]. The capacity for production of induced defences was reported to be reduced or partially lost in North American domesticated maize varieties but was still broadly found in teosinte [46]. In a laboratory study, maize resistance to root herbivory by *D. v. virgifera* was compared for several biotypes of Balsas teosinte, Mexican landraces, US landraces, and US inbred corn [47]. The results showed that larval weight and survival was highest on US landraces and lowest on teosinte. The authors interpreted their findings as evidence for decreased corn resistance to feeding damage by rootworm larvae through domestication (landraces). Nevertheless, this negative correlation was only partially mediated through breeding since US inbred lines of maize were more resistant than US landrace ancestors. Such a negative correlation for domestication and resistance was also found for Mexican landraces that were less resistant to feeding damage than their teosinte ancestors. A similar study with Brazilian landraces and cultivars of maize found higher level of resistance towards feeding by *D. speciosa*, the most important Diabroticite pest species in Brazil, in two genotypes [48]. Larval development was longer and mortality was increased on one landrace 1 of 17 tested, and on one of two cultivars of maize. In laboratory experiments, those varieties were found to produce higher levels of lignin, cellulose, and fibres in the roots for a prolonged period after larval feeding. The identification of maize-genotypes with such so-called "antibiosis" effects on rootworms are of importance for future breeding programmes of rootworm-resistant corn varieties or for their use in crop rotation strategies.

4. Evolutionary History

The large diversity of species in this beetle group has been interpreted in different ways based on assumptions of adaptive radiation, coevolution [49], and competitive exclusion plus geomorphological patterns [7,50]. The origin of the Neotropical genus *Diabrotica* can be dated back to the Cretaceous, and its diversification began about 60 Mya with a peak between 30 and 45 Mya [7]. This is much earlier than previously proposed by Metcalf and Lampman [51], who related the diversification in Diabroticite beetles to the agricultural practice of intercropping corn with beans and cucurbits all over the Americas (inter-cropping theory) [34]. In Mexico, the cultivation of maize from its ancestor, teosinte (*Zea mexicana*), began about 9000 years before the present in the southern regions of Mexico [31], from where its use in agriculture spread north to Arizona, New Mexico, Utah, and Colorado (for more details see [10]). This is much later than the speciation events in *Diabrotica*.

Moreover, the diversification of *Diabrotica* spp. cannot be interpreted without taking into account its intricate relationship with wild species of the family Cucurbitaceae [2,5]. This plant family is characterized by the production of bitter, toxic secondary compounds—the cucurbitacins that render the plants unpalatable for most herbivores, including humans [35,52,53]. To date over 40 cucurbitacins and cucurbitacin-metabolites are known from Cucurbitaceae, and many of them have important medicinal properties [54]. The tetracyclic triterpenoids are typically found in wild species of cucurbits [55] and act as compulsive feeding arrestants for *Diabrotica* species [56,57]. When Diabroticite beetles encounter plant tissues with cucurbitacins they are arrested and continue to feed compulsively on the bitter tissues (Figure 1) [58–62].

Figure 1. Diabroticite beetle of the species Diabrotica *dissimilis* Jacoby feeding on the seeds of an open fruit from wild, bitter *Cucurbita okeechobeensis* ssp. *martinezii* (L.H. Bailey). This plant species of the family Cucurbitaceae is native to Veracruz, Mexico (Photo: A. Eben).

In a number of cucurbits, the highest concentrations of cucurbitacins are found in cotyledons, roots, and seeds of cucurbit species [63]. The beetles even abandon suitable host plants and prefer cucurbitacin containing tissues [60]. This compulsive feeding behaviour results from contact of receptors on the maxillae with the bitter cucurbitacins [64,65]. This trait might indicate an ancestral relationship with Cucurbitaceae that is maintained even in species no longer associated with cucurbits or in species where larvae develop on non-cucurbit hosts [2,60,66,67], probably due to benefits obtained through sequestration of the bitter compounds [68]. By feeding on cucurbitacin containing plant tissues, the beetles sequester these compounds in their haemolymph [51]. This was interpreted as an example of chemically mediated coevolution, as well as for the so-called feeding behaviour named pharmacophagy [69,70]. The latter is based on the fact that the beetles not only prefer to consume the bitter cucurbit tissues but also sequester the imbittering secondary compounds. Consequently, adult and larval *Diabrotica* spp. are protected from natural enemies by those sequestered plant compounds [68,71–73]. However, results on chemical defence are contradictory and depend on which beetle species and natural enemy are studied (see details in [74]). Furthermore, cucurbitacins are transferred into eggs and larvae through female beetles [68] and can act as antibiotics against the entomopathogenic fungus *Metarhizium anisopliae* [72].

Under field conditions in Veracruz, Mexico, mostly male beetles of Diabroticite species accumulate in the flowers and on tissues of wild cucurbit species (Figure 2) [73].

Figure 2. Flower of *Cucurbita pepo* L. with several Diabroticite beetles of the species *Acalymma blomorum* Munroe & Smith 1980 in the surroundings of Coatepec, Ver., Mexico (Photo: A. Eben).

Such a behaviour seemingly, what is called "lek formation", might contribute to sequestration of cucurbitacins in male tissues as well as in sperm transferred to females during copulation. In this way the cucurbitacins that are transferred to the females might have been already detoxified in the males and formed conjugates which reduce the metabolic costs for the receiving females. Such possible mating advantage or female choice preference for more "bitter" males was tested by Tallamy et al. [75]. In their study, however, this hypothesis could not be proven. Moreover, the compulsive feeding upon encounter of cucurbitacins was successfully incorporated in pest control strategies against the western corn rootworm. There, cucurbitacins were offered in combination with insecticides [76,77].

Cucurbitacin metabolism and detoxification in *Diabrotica* spp. differ in species with more restricted host plant breadth, such as *D. virgifera virgifera* compared with polyphagous *D. balteata* or *D. undecimpunctata howardi*. The least metabolic costs were reported from specialized cucurbit feeders such as closely related *Acalymma* spp. [53]. Among Diabroticite beetles, all 73 known species in the genus *Acalymma* [1,78] are specialized on cucurbits as adults and larvae. One species, *A. vittatum* (F.), is an important pest species on squash in the USA [79,80]. Like most investigated Diabroticites, these species show compulsive feeding on cucurbitacin-laden plant tissues and prefer them over non-cucurbitacin-containing host plants [61,62,81]. In a new study [82], the importance of cucurbitacin type and concentration in *Cucurbita pepo* cultivars was investigated. This research found that adult *A. vittatum* host preference was positively correlated with cucurbitacin content of cotyledons, but preference was also affected by the presence of other chemical compounds in the first true leaves that contained very low cucurbitacin levels. Moreover, cucurbitacin content in the cotyledons was not increased, i.e., induced, through feeding. The results of the genetic mapping and gene expression analyses were interpreted as an indication that genes responsible for the production of cucurbitacins in cotyledons and roots might contribute to intraspecific differences in plant susceptibility to herbivores. Such a finding is important for future breeding programs of edible Cucurbitaceae species.

5. Taxonomy and Phylogenetic Relationships

The taxonomic positioning and the relationships among species in the genus *Diabrotica* have been examined recently [3,83]. The authors described 18 new species for North and Central America and extended the taxonomic key including more morphological characters to distinguish species of the genus *Diabrotica* from other Diabroticite species. However, the revision was restricted to the North and Central American *fucata* and *virgifera* species groups with previously 123 reported species [1]. Only 107 species remained as valid species in the genus with a distribution in North and Central America. Unfortunately, only scarce data on host plants were provided in this new revision, probably due to the lack of information on sampling locations of the museum specimens. Especially, the species-rich and less-studied South American *fucata* group and *signifera* group still warrant exhaustive research on biology, ecology, host plant associations, and phylogeny.

An interesting morphological parameter used for species comparisons and taxonomic grouping is the sexual dimorphism of the basi-tarsus on the pro- and mesothoraxic legs found in *D. collicola* Cabrera & Cabrera Walsh [84], *D. speciosa*, *D. viridula* (Fabricius) [85], and other species in the tribes Diabroticites and Phyllectrites [86,87]. The adhesive setae found in males of some species were supposed to play a role in copulatory behaviour and male–male competition [88]. This morphological character can distinguish between male and female specimens in a non-destructive way and can also be used with living individuals.

Among the species comprising the genus *Diabrotica*, some are specialists on cucurbits (i.e., *D. scutellata*), whereas others are extremely polyphagous (*D. balteata*, *D. speciosa*). Since larvae are hidden from sight as below ground root-feeding and pollen-feeding adults are often found on hosts from diverse plant families, the evolution of host plant use in the genus cannot yet be fully understood. Efforts are further complicated by the lack of a complete phylogeny for this Neotropical genus. A phylogenetic reconstruction including gene sequences of 40 species found monophagous feeding behaviour as the ancestral trait [89]. The evolutionary diversification in the genus *Diabrotica* was initiated with the inclusion of host plant families that led to polyphagy. Secondary regressions to oligophagy and monophagy on cucurbits later occurred in some species. In the cited study, data on host plant use were obtained based on field observations and published host records from the literature. Overall, a high plasticity in the genus *Diabrotica* with regard to the evolution of host plant breadth was found. One characteristic that facilitated such an evolutionary pathway might be the subsequent optimization of cucurbitacin metabolism and sensory reception [59,90]. Further studies of cucurbitacin detoxification in non-pest species as well as their cucurbitacin receptor physiology and biochemistry are needed.

6. Natural Enemies of Diabroticites—Old and New Findings

Long common evolutionary histories in the native habitat made natural enemies adapted to the behavioural and chemical defences of their hosts. Parasitic flies of the species *Celatoria compressa* Wulp (Diptera: Tachinidae) were reared from several species of Diabroticite beetles in Veracruz, Mexico [71]. Based on extensive field surveys in central Mexico [73], this fly species was later considered for classical biological control of *D. v. virgifera* in Europe [91,92]. The fly, however, cannot be considered a specialist on species in the genus *Diabrotica*. Therefore, mass releases in the newly invaded areas can be undertaken only after intensive investigation taking into account potential negative effects on native Chrysomelid species. In Mexico, *C. compressa* is not abundant—like most parasitic insects, we found only around 5–10% of adult Diabroticites parasitized by the flies during several years of intensive field surveys [71,73].

With a focus on the most damaging life stage, the root-feeding larvae of *Diabrotica* spp., soil-inhabiting entomopathogenic nematodes of the families Heterorhabditidae and Steinernematidae were broadly investigated [39]. A significant effect of larval host plants on the efficiency and propagation of nematodes was found for *D. undecimpunctata howardi* and its main hosts. Feeding on the roots of corn and peanuts was positively correlated with greater rootworm weight and increased survival, whereas larvae fed on cucurbit roots were smaller and showed higher mortality [40]. Most importantly, it was the symbiontic bacteria in the nematodes that were negatively affected by cucurbitacins in roots of domesticated *C. pepo* [41,93]. In this sense, not only cucurbitacins can affect these symbionts but also secondary compounds produced by maize. Research on metabolites sequestered by *D. v. virgifera* larvae when feeding on maize roots found benzoxazinoid glucosides. Rootworm larvae stabilize benzoxazinoid derived from corn roots by glucosylation. When infested by entomopathogenic nematodes, the larvae hydrolyse the compound and release toxic

MBOA for their defence [94]. In this tri-trophic level system, the toxic compounds act against the 4th trophic level, the symbiontic bacteria in the nematodes. Recently, Bruno et al. [95] detected that most of 40 Mexican isolates of *Heterorhabditis bacteriophora* were resistant to benzoxazinoids, secondary compounds in the tested maize varieties. These compounds were sequestered by *D. v. virgifera* larvae and thus acted as chemical defence against some of the nematode isolates. Interestingly, all nematode isolates successfully infested polyphagous *D. balteata* larvae that did not sequester these metabolites from corn roots. Nevertheless, most of the nematodes obtained from the soil of Mexican cornfields were resistant to maize metabolites encountered in their insect hosts and effectively killed oligophagous *D. v. virgifera* larvae. Based on those results, it is unknown if symbiontic bacteria in entomopathogenic nematodes are able to detoxify the plant metabolites encountered in their host and make larval hosts suitable for their own propagation. Furthermore, the highly variable nutritional quality of the different maize genotypes tested might also affect susceptibility of both larvae and nematodes to each other.

7. Conclusions

Despite the economic importance of several *Diabrotica* spp. and their well-studied feeding relationship with corn and bitter plants of the Cucurbitaceae, knowledge of their biology related to topics such as diapause requirements or voltinism, is still unavailable for most species. Little information on host plant use of adults and larvae is published for the majority of the Central and South American species that are not of economic interest. Taxonomic studies, further phylogenetic reconstructions, and much broader knowledge of host plant associations and the distributions of the many species in Mexico and Central and South America are needed. That information will help explain why some species develop into such successful pests, such as the western corn rootworm, and how pest species might be controlled with environmentally sustainable means in the near future.

Funding: This research received no external funding.

Acknowledgments: Many thanks to L. Meinke (University of Nebraska, Lincoln) and J. Spencer (University of Illinois, Urbana-Champaign) for inviting me to contribute to this Special Issue.

Conflicts of Interest: The author declares no conflict of interests.

References

1. Smith, B.D.; Lawrence, J.F. *Clarification of the Status of Type Specimens of Diabroticites (Coleoptera: Chrysomelidae: Galerucini)*; Univ. Calif. Press: Berkeley, CA, USA, 1967.
2. Branson, T.F.; Krysan, J.L. Feeding and oviposition behaviour and live cycle strategies of *Diabrotica*: An evolutionary view with implications for pest management. *Environ. Entomol.* **1981**, *10*, 826–831. [CrossRef]
3. Derunkov, A.; Konstantinov, A. Taxonomic changes in the genus *Diabrotica* Chevrolat (Coleoptera: Chrysomelidae: Galerucinae): Results of a synopsis of North and Central America *Diabrotica* species. *Zootaxa* **2013**, *3686*, 301–325. [CrossRef] [PubMed]
4. Webster, R.M. On the probable origin, development and diffusion of North American species of the genus *Diabrotica*. *J. New York Entomol. Soc.* **1895**, *3*, 158–166.
5. Whitaker, R.W.; Bemis, W.P. Origins and evolution of the cultivated *Cucurbita*. *Bull. Torey Bot. Club* **1975**, *102*, 362–368. [CrossRef]
6. Krysan, J.L.; Branson, T.F. Biology, ecology and distribution of *Diabrotica*. In Proceedings of the International Maize Virus Disease Colloquium and Workshop, Wooster, OH, USA, 2–6 August 1982; Gordon, D.R., Knoke, J.K., Nault, L.R., Ritter, R.M., Eds.; Agricultural Research and Development Center (OARDC): Wooster, OH, USA, 1983; pp. 144–150.
7. Eben, A.; Espinosa de los Monteros, A. Trophic interaction network and the evolutionary history of Diabroticina beetles (Chrysomelidae: Galerucinae). *J. Appl. Entomol.* **2015**, *139*, 468–477. [CrossRef]
8. Krysan, J.L.; Smith, R.F. Systematics of the virgifera species group of *Diabrotica* (Coleoptera: Chrysomelidae: Galerucinae). *Entomography* **1987**, *5*, 375–484.
9. Walsh, G.C. Host range and reproductive traits of *Diabrotica speciosa* (Germar) and *Diabrotica viridula* (F.) (Coleoptera: Chsysomelidae), two species of South American pest rootworms, with notes on other species of Diabroticina. *Env. Entomol.* **2003**, *32*, 276–285. [CrossRef]
10. Lombaert, E.; Ciosi, M.; Miller, N.J.; Sappington, T.W.; Blin, A.; Guillemaud, T. Colonization history of the western corn rootworm (*Diabrotica virgifera virgifera*) in North America: Insight from random forest ABC using microsatellite data. *Biol. Invasions* **2018**, *20*, 665–677. [CrossRef]

11. Onstad, D.W.; Caprio, M.A.; Pan, Z. Models of *Diabrotica* populations: Demography, population genetics, geographic spread, and management. *Insects* **2020**, *11*, 712. [CrossRef]

12. Camprag, G.; Baca, F. *Diabrotica virgifera* (Coleoptera: Chrysomelidae) a new pest of maize in Yugoslawia. *Pest. Sci.* **1996**, *45*, 291–292. [CrossRef]

13. Baufeld, P.; Enzian, S. *Diabrotica virgifera virgifera*—Einschleppungsszenarien und Konsequenzen für Deutschland. Mitt. BBA Land-Und Forstwirtsch. *Berl.-Dahl.* **2000**, *376*, 221.

14. Baufeld, P. IWGO-Workshop und EPPO ad hoc Panel zum Westlichen Maiswurzelbohrer (*Diabrotica* virgifera) in Stuttgart. *Nachr. Dt. Pflanzenschutzd.* **2001**, *53*, 69–70.

15. Miller, N.; Estoup, A.; Toepfer, S.; Borguet, B.; Lapchin, L.; Derridj, S.; Kim, K.S.; Reynaus, P.; Furlan, L.; Guillemaud, T. Multiple transatlantic introductions of the western corn rootworm. *Science* **2005**, *310*, 992. [CrossRef]

16. Ciosi, M.; Miller, N.J.; Kim, K.S.; Giordano, R.; Estoup, A.; Guillemaud, T. Invasion of Europe by the western corn rootworm, *Diabrotica* virgifera virgifera: Multiple transatlantic introductions with various reductions of genetic diversity. *Mol. Ecol.* **2008**, *17*, 3614–3627. [CrossRef]

17. Gray, M.E.; Sappington, T.W.; Miller, N.J.; Moeser, J.; Bohn, M.O. Adaptation and invasiveness of Western Corn Rootworm: Intensifying research on a worsening pest. *Ann. Rev. Entomol.* **2008**, *54*, 303–321. [CrossRef]

18. Bažok, R.; Lemić, D.; Chiarini, F.; Furlan, L. Western Corn Rootworm (*Diabrotica* virgifera virgifera LeConte) in Europe: Current Status and Sustainable Pest Management. *Insects* **2021**, *12*, 195. [CrossRef]

19. EPPO Global Database. 2021. Available online: https://gd.eppo.int/ (accessed on 22 December 2021).

20. Krysan, J.L.; Miller, T.A. *Methods for the Study of Pest* Diabrotica; Springer: New York, NY, USA, 1986.

21. Geisert, R.W.; Meinke, L.J. Frequency and distribution of extended diapause in Nebraska populations of *Diabrotica* barberi (Coleoptera: Chrysomelidae). *J. Econ. Entomol.* **2013**, *106*, 1619–1627. [CrossRef]

22. Eben, A.; Espinosa de los Monteros, A. Ideas on the systematics of the genus *Diabrotica* Wilcox and other related leaf beetles. In *New Developments in the Biology of Chrysomelidae*; Jolivet, P., Santiago-Blay, J.A., Schmitt, M., Eds.; SPB Academic Publishing: The Hague, The Netherlands, 2004; pp. 59–74.

23. Cabrera Walsh, G.; Cabrera, N. Distribution and hosts of the pestiferous and other common Diabroticites from Argentina and southern South America: A geographical and systematic view. In *New Developments in the Biology of Chrysomelidae*; Jolivet, P., Santiago-Blay, J.A., Schmitt, M., Eds.; SSP Academic Publishing: The Hague, The Netherlands, 2004; pp. 333–350.

24. Cabrera Walsh, G.; Avila, C.J.; Cabrera, N.; Nava, D.E.; de Sene Pinto, A.; Weber, D.C. Biology and Management of pest *Diabrotica* species in South America. *Insects* **2020**, *11*, 421. [CrossRef]

25. Szalanski, A.L.; Roehrdanz, R.L.; Taylor, D.B. Genetic relationship among *Diabrotica* species (Coleoptera: Chrysomelidae) based on rDNA and mtDNA sequences. *Fla. Entomol.* **2000**, *83*, 262–267. [CrossRef]

26. Clark, T.L.; Meinke, L.J.; Foster, J.E. Molecular phylogeny of *Diabrotica* beetles (Coleoptera: Chrysomelidae) inferred from analysis of combined mitochondrial and nuclear DNA sequences. *Insect Mol. Biol.* **2001**, *10*, 303–314. [CrossRef]

27. Onstad, D.W.; Crowder, D.W.; Isard, S.A.; Levine, E.; Spencer, J.L.; O´Neal, M.; Ratcliffe, S.; Gray, M.E.; Bledsoe, L.W.; DiFonzo, C.D.; et al. Does landscape diversity slow the spread of rotation-resistant western corn rootworm (Coleoptera: Chrysomelidae)? *Environ. Entomol.* **2003**, *32*, 992–1001. [CrossRef]

28. Campbell, L.A.; Meinke, L.J. Seasonality and adult habitat use by four *Diabrotica* species at prairie-corn interfaces. *Env. Ent.* **2004**, *35*, 922–936. [CrossRef]

29. Clark, T.L.; Hibbard, B.E. Comparison of nonmaize host to support Western Corn Rootworm (Coleoptera: Chrysomelidae) larval biology. *Environ. Entomol.* **2004**, *33*, 681–689. [CrossRef]

30. Smith, R.F. Distributional patterns of selected western North American insects: The distribution of diabroticites in western North America. *Bull. Entomol. Soc. Am.* **1966**, *12*, 108–110.

31. Matsuoka, Y.; Vigouroux, Y.; Goodman, M.M.; Sanchez, G.J.; Buckler, E.; Doebley, J. A single domestication for maize shown by multilocus microsatellite genotyping. *Proc. Natl. Acad. Sci. USA* **2002**, *99*, 6080–6084.

32. Maag, D.; Erb, M.; Bernal, J.S.; Wolfender, J.L.; Turlings, T.C.J.; Glauser, G. Maize domestication and anti-herbivore defences: Leaf-specific dynamics during early ontogeny of maize and its wild ancestors. *PLoS ONE* **2015**, *10*, e0135722. [CrossRef]

33. Schaefer, H.; Heibl, C.; Renner, S.S. Gourds afloat: A dated phylogeny reveals an Asian origin of the gourd family (Cucurbitaceae) and numerous oversea dispersal events. *Proc. R. Soc. B* **2009**, *276*, 843–851. [CrossRef]

34. Smith, B.D. The initial domestication of *Cucurbita pepo* in the Americas10,000 years ago. *Science* **1997**, *276*, 932–934. [CrossRef]

35. Kistler, L.; Newsom, L.A.; Ryan, T.M.; Clarke, A.C.; Smith, B.D.; Perry, G.H. Gourds and squashes (*Cucurbita* spp.) adapted to megafaunal extinction and ecological anachronism through domestication. *Proc. Natl. Acad. Sci. USA* **2015**, *112*, 15107–15112. [CrossRef]

36. Lange de, E.S.; Balmer, D.; Mauch-Mani, B.; Turlings, T.C.J. Insect and pathogen attack and resistance in maize and its wild ancestors, the teosintes. *New Phytol.* **2014**, *204*, 329–341. [CrossRef]

37. De la Paz Gutiérrez, S.; Sánchez González, J.J.; Ruiz Corral, J.A.; Ron Parra, J.; Miranda Medrano, R.; De la Cruz Larios, L.; Lépiz Ildefonso, R. Diversidad de especies insectiles en maíz y teocinte en México. *Folia Entomol. Mex.* **2010**, *48*, 16.

38. Erb, M.; Flors, V.; Karalen, D.; de Lange, E.; Planchamp, C.; D´ Alessandro, M.; Turlings, T.C.J.; Ton, J. Signal signature of aboveground-induced resistance upon belowground herbivory in maize. *Plant. J.* **2009**, *59*, 292–302. [CrossRef] [PubMed]

39. Barbercheck, M.E.; Wang, J.; Hirsh, I.A. Host plant effects on entomopathogenic nematodes. *J. Invertebr. Pathol.* **1995**, *68*, 141–145. [CrossRef] [PubMed]

40. Eben, A.; Barbercheck, M.E. Host plant and substrate effects on mortality of southern corn rootworm from entomopathogenic nematodes. *Biol. Contr.* **1997**, *8*, 89–96. [CrossRef]

41. Barbercheck, M.E.; Wang, J. Effects of bacteria on entomopathogenic nematodes. *J. Invertebr. Pathol.* **1996**, *68*, 141–145. [CrossRef] [PubMed]

42. Degen, T.; Dillmann, C.; Marion-Poll, F.; Turlings, T.C.J. High genetic variability of herbivore-induced volatile emission within a broad range of maize inbred lines. *Plant. Physiol.* **2004**, *135*, 1928–1938. [CrossRef] [PubMed]

43. Gouinguené, S.; Degen, T.; Turlings, T.C.J. Variability in herbivore—Induced odour emissions among maize cultivars and their wild ancestors (teosinte). *Chemoecol.* **2001**, *11*, 9–16. [CrossRef]

44. Köllner, T.G.; Held, M.; Lenk, C.; Hiltpold, I.; Turlings, T.C.J.; Gershenzon, J.; Degenhardt, J. A maize (E)-ß-caryophyllene synthase implicated in indirect defense responses against herbivores is not expresse in most American maize varieties. *Plant. Cell* **2008**, *20*, 482–494. [CrossRef]

45. Tamiru, A.; Bruce, T.J.A.; Woodcock, C.M.; Caulfield, J.C.; Midega, C.A.O.; Ogol, C.K.P.O.; Mayon, P.; Birkett, M.A.; Pickett, J.A.; Khan, Z.R. Maize landraces recruit egg and larval parasitoids in response to egg deposition by a herbivore. *Ecol. Lett.* **2011**, *14*, 1075–1083. [CrossRef]

46. Rasmann, S.; Köllner, T.G.; Degenhardt, J.; Hiltpold, I.; Toepfer, S.; Kuhlmann, U.; Gershenzon, J.; Turlings, T.C.J. Recruitment of entomopathogenic nematodes by insect-damaged maize roots. *Nature* **2005**, *434*, 732–737. [CrossRef]

47. Fontes-Puebla, A.A.; Bernal, J.S. Resistance and tolerance to root herbivory in maize were mediated by domestication, spread, and breeding. *Frontiers* **2020**, *11*, 223. [CrossRef]

48. Neves Costa, E.; Nogueira, L.; Sardinha de Souza, B.H.; Ribeiro, Z.A.; Couvandini, H.; Zukoff, S.N.; Leal Bioca, A., Jr. Characterization of antibiosis to *Diabrotica speciosa* (Coleoptera: Chrysomelidae) in Brazilian maize landraces. *J. Econ. Ent.* **2018**, *111*, 454–462. [CrossRef]

49. Metcalf, R.L. Coevolutionary adaptations of rootworm beetles (Coleoptera: Chrysomelidae) to cucurbitacins. *J. Chem. Ecol.* **1986**, *12*, 1109–1124. [CrossRef]

50. Eben, A.; Espinosa de los Monteros, A. Tempo and mode of evolutionary radiation in Diabroticina beetles (genera *Acalymma, Cerotoma,* and *Diabrotica. ZooKeys* **2013**, *332*, 207–231. [CrossRef]

51. Metcalf, R.L.; Lampman, R.L. The chemical ecology of Diabroticites and Cucurbitaceae. *Experientia* **1989**, *45*, 240–247. [CrossRef]

52. Metcalf, R.L.; Metcalf, R.A.; Rhodes, A.M. Cucurbitacins as kairomones for Diabroticite beetles. *Proc. Natl. Acad. Sci. USA* **1980**, *17*, 3769–3772. [CrossRef]

53. Ferguson, J.E.; Metcalf, R.L.; Fischer, D.C. Disposition and fate of cucurbitacin B in five species of Diabroticites. *J. Chem. Ecol.* **1985**, *11*, 1307–13421. [CrossRef]

54. Blanco Díaz, M.T.; Font, R.; Gómez, P.; Celestino, M. Summer squash. In *Nutritional Composition and Antioxidant Properties of Fruits and Vegetables*; Jaiswal, A.K., Ed.; Academic Press: London, UK, 2020; pp. 239–254.

55. Rehm, S.; Enslin, P.A.; Meeuse, A.D.J.; Wessels, J.H. Bitter principles of the Cucurbitaceae VII. The distribution of bitter principles in the plant family. *J. Sci. Food Agric.* **1957**, *8*, 679–686.

56. Tallamy, D.W.; Hibbard, B.E.; Clark, T.L.; Gillespie, J.J. *Western corn rootworm, cucuurbits, and cucurbitacins. In Western Corn Rootworm: Ecology and Management*; Vidal, S., Kuhlmann, U., Edwards, R., Eds.; CABI Publishers: Wallingford, UK; pp. 67–93.

57. Andersen, J.F.; Metcalf, M.R. Factors influencing the distribution of *Diabrotica* spp. in the blossoms of cultivated *Cucurbita* sp. *J. Chem. Ecol.* **1987**, *14*, 681–689. [CrossRef]

58. Chambliss, O.L.; Jones, C.M. Cucurbitacins: Specific insect attractants in Cucurbitraceae. *Science* **1966**, *153*, 1392–1393. [CrossRef]

59. Andersen, J.F.; Plattner, R.D.; Weisleder, D. Metabolic transformations of cucurbitacins by *Diabrotica virgifera virgifera* LeConte and *D. undecimpunctata howardi* Barber. *Insect Biochem.* **1988**, *18*, 71–77. [CrossRef]

60. Lampman, R.L.; Metcalf, R.L. Multicmponent kairomone lures for southern and western corn rootworms (Coleoptera: Chrysomelidae: *Diabrotica* spp.). *J. Econ. Entomol.* **1987**, *76*, 1049–1051.

61. Eben, A.; Barbercheck, M.E.; Aluja, M. Mexican Diabroticite beetles: I. *Laboratory tests on host breadth of* Acalymma *and* Diabrotica *spp. Entomol. Exp. Appl.* **1997**, *82*, 53–62.

62. Eben, A.; Barbercheck, M.E.; Aluja, M. Mexican Diabroticite beetles: II. *Tests for preference of cucurbit hosts by* Acalymma *and* Diabrotica *spp. Entomol. Exp. Appl.* **1997**, *82*, 63–72.

63. Metcalf, R.L.; Rhodes, A.M.; Metcalf, R.A.; Ferguson, J.E.; Metcalf, E.R.; Lu, P.Y. Cucurbitacin contents and Diabroticites (Coleoptera: Chrysomelidae) feeding upon *Cucurbita* spp. *Environ. Entomol.* **1982**, *11*, 931–937. [CrossRef]

64. Chyb, S.; Eichenseer, H.; Hollister, B.; Mullin, C.A.; Frazier, J. Identification of sensilla involved in taste mediation of adult western corn rootworm (*Diabrotica virgifera virgifera* LeConte). *J. Chem. Ecol.* **1995**, *21*, 313–329. [CrossRef] [PubMed]

65. Eichenseer, H.; Mullin, C.A. Maxillary appendages used by western corn rootworms, *Diabrotica virgifera virgifera*, to discriminate between a phagostimulatory and deterrent. *Entomol. Exp. Appl.* **1996**, *78*, 237–242. [CrossRef]

66. Levine, E.; Oloumi-Sadeghi, H. Management of diabroticite rootworms in corn. *Annual Rev. Entomol.* **1991**, *36*, 229–255. [CrossRef]

67. Gillespie, J.J.; Kjer, K.M.; Ducket, C.N.; Tallamy, D.W. Convergent evolution of cucurbitacin feeding in spatially isolated rootworm taxa (Coleoptera: Chrysomelidae, Galerucinae, Luperini). *Mol. Phylogen. Evol.* **2003**, *29*, 161–175. [CrossRef]

68. Ferguson, J.E.; Metcalf, R.L. Cucurbitacins: Plant derived defense compounds for Diabroticites (Coleoptera: Chrysomelidae). *J. Chem. Ecol.* **1985**, *11*, 311–318. [CrossRef]

69. Boppré, M. Redefining „pharmacophagy". *J. Chem. Ecol.* **1984**, *10*, 1151–1154. [CrossRef]

70. Nishida, R.; Fukami, H. Sequestration of distasteful compounds by some pharmacophagous insects. *J. Chem. Ecol.* **1990**, *16*, 151–164. [CrossRef]

71. Eben, A.; Barbercheck, M.E. Field observations on host associations and natural enemies of Diabroticite beetles (Chrysomelidae: Luperini) in Veracruz, Mexico. *Acta Zool. Mex.* **1996**, *67*, 47–55.

72. Tallamy, D.W.; Whittington, D.P.; Defurio, F.; Fontaine, D.A.; Gorski, P.M.; Gothro, T.W. The effect of sequestered cucurbitacins on the pathogenicity of *Metarhizium anisopliae* (Monilialeas Moniliaceae) on spotted cucumber beetle eggs and larvae (Coleoptera: Chrysomelidae). *Environ. Entomol.* **1998**, *27*, 366–372. [CrossRef]

73. Gamez-Virues, S.; Eben, A. Comparison of beetle diversity and incidence of parasitism in Diabroticina (Coleoptera: Chrysomelidae) species collected on cucurbits. *Fla. Entomol.* **2005**, *88*, 72–76. [CrossRef]

74. Eben, A. Por qué amargarse la vida? La asociación de los escarabajos Diabroticina (Coleoptera: Chrysomelidae) con plantas de la familia Cucurbitaceae. In *Temas Selectos en Ecología Química de Insectos*; Rojas, J.C., Malo, E.A., Eds.; El Colegio de la Frontera Sur: Tapachula Chiapas, Mexico, 2012; pp. 193–216.

75. Tallamy, D.W.; Gorski, P.M.; Burzon, J.K. Fate of male-derived cucurbitacins in spotted cucurmber beetle females. *J. Chem. Ecol.* **2000**, *26*, 413–427. [CrossRef]

76. Metcalf, R.L.; Ferguson, J.E.; Lampman, R.L.; Andersen, J.F. Dry cucurbitacin containing baits for controlling Diabroticite beetles (Coleoptera: Chrysomelidae). *J. Econ. Entomol.* **1987**, *80*, 870–875. [CrossRef]

77. Gerber, C.K.; Edwards, C.R.; Bledsoe, L.W.; Gray, M.E.; Steffey, K.L.; Chandler, L.D. Application of the areawide concept using semiochemical-based insecticide baits for managing the Western Corn Rootworm (*Diabrotica virgifera virgifera* LeConte) variant in the Eastern Midwest. In *Western Corn Rootworm: Ecology and Management*; Vidal, S., Kuhlmann, U., Edwards, C.R., Eds.; CABI publishing: Wallingford, UK, 2005; pp. 221–238.

78. Cabrera, N.C.; Durante, S.P. Comparative morphology of mouthparts in species of the genus *Acalymma* Barber (Coleoptera, Chrysomelidae, Galerucinae). *Coleopt. Bull.* **2003**, *57*, 5–16. [CrossRef]

79. Cavanagh, A.; Hazzard, R.; Adler, L.S.; Coucher, J. Using trap crops for control of *Acalymma vittatum* (Coleoptera: Chrysomelidae) reduces insecticide use in butternut squash. *J. Econ. Ent.* **2009**, *102*, 1101–1107. [CrossRef]

80. Lewis, P.A.; Lampman, R.L.; Metcalf, R.L. Kaironomal attractants for *Acalymma vittatum* (Coeloptera: Chrysomelidae). *Environ. Entomol.* **1990**, *19*, 8–14. [CrossRef]

81. Tallamy, D.W.; Gorski, P.M. Long- and short-term effect of cucurbitacin consumption on *Acalymma vittatum* (Coleoptera: Chrysoemlidae) fitness. *Environ. Entomol.* **1997**, *26*, 672–677. [CrossRef]

82. Brzozowski, L.J.; Gore, M.A.; Agrawal, A.A.; Mazourek, M. Divergence of defensive cucurbitacins in independent *Cucurbita pepo* domestication events leads to differences in specialist herbivore preference. *Plant. Cell Environ.* **2020**, *43*, 2812–2825. [CrossRef] [PubMed]

83. Derunkov, A.; Prado, L.R.; Tishechkin, A.K.; Konstantinov, A.S. New species of *Diabrotica* Chevrolat (Coeloptera: Chrysomelidae: Galerucinae) and a key to *Diabrotica* and related genera: Resuts of a synopsis of North and Central American *Diabrotica* species. *J. Insect Biodiv.* **2015**, *3*, 1–55. [CrossRef]

84. Cabrera, N.; Cabrera Walsh, G.C. *Diabrotica collicola* (Coleoptera: Chrysomelidae), a new species of leaf beetle from Argentina and key to species of the *Diabrotica virgifera* group and relatives. *Zootaxa* **2010**, *2683*, 45–55. [CrossRef]

85. De Carli, M.; Rech, C.; Simões Bento, J.M.; Nardi, C. Sexual dimorphism in *Diabrotica speciosa* and *Diabrotica viridula* (Coleoptera: Chysomelidae). *Rev. Bras. Entomol.* **2018**, *62*, 172–175. [CrossRef]

86. Hammack, L.; French, B.W. Sexual dimorphism of basitarsi in pest species of *Diabrotica* and *Cerotoma* (Coleoptera: Chrysomelidae). *Ann. Entomol. Soc. Am.* **2007**, *100*, 59–63. [CrossRef]

87. Prado, L.R. Review on the use of sexually dimorphic characters in the taxonomy of Diabroticites (Galerucinae, Luperini, Diabroticina). *ZooKeys* **2013**, *332*, 33–54. [CrossRef]

88. Lew, A.C.; Ball, H.J. The mating behavior of the western corn rootworm *Diabrotica virgifera* (Coleoptera: Chrysomelidae). *Ann. Entomol. Soc. Am.* **1979**, *72*, 15. [CrossRef]

89. Eben, A.; Espinosa de los Monteros, A. Specialization is not a dead end: Further evidence from Diabroticina beetles. In *Research on Chrysomelidae I*; Jolivet, P., Santiago-Blay, J.A., Schmitt, M., Eds.; Brill Academic Publishers: Leiden, The Netherlands, 2005; pp. 42–58.

90. Metcalf, R.L.; Lampman, R.L. Evolution of Diabroticite (Coleoptera: Chrysomelidae) receptors for *Cucurbita* blossom volatiles. *Proc. Natl. Acad. Sci. USA* **1991**, *88*, 1869–1872. [CrossRef]

91. Kuhlmann, U.; Toepfer, S.; Zhang, F. Is classical biological control against Western Corn Rootworm in Europe a potential sustainable management strategy. In *Western Corn Rootworm. Ecology and Management*; Vidal, S., Kuhlmann, U., Edwards, C.R., Eds.; CABI Publishing: Wallingford, UK, 2005; pp. 263–284.

92. Toepfer, S.; Cabrera Walsh, G.; Eben, A.; Alvarez-Zagoya, R.; Haye, T.; Zhang, F.; Kuhlmann, U. A critical evaluation of host ranges of parasitoids of the subtribe Diabroticina (coeloptera: Chrysomelidae: Glaerucinae: Luperini) using field and laboratory host records. *Biocontrol Sci. Technol.* **2008**, *18*, 483–504. [CrossRef]

93. Hirsh, I.S.; Barbercheck, M.E. Effects of host plant and cucurbitacins on larval *Diabrotica undecimpuntata howardi* (Coleoptera: Chrysomelidae). *Entomol. Exp. Appl.* **1996**, *64*, 187–193.

94. Robert, C.A.M.; Zhang, X.; Machado, R.A.R.; Schirmer, S.; Lori, M.; Mateo, P.; Erb, M.; Gershenzon, J. Sequestration and activation of plant toxins protect the western corn rootworm from enemies at multiple trophic levels. *eLife* **2017**, *6*, e29307. [CrossRef]

95. Bruno, P.; Machado, R.A.R.; Glauser, G.; Köhler, A.; Campos-Herrera, R.; Bernal, J.; Toepfer, S.; Erb, M.; Robert, C.A.M.; Arce, C.C.M.; et al. Entomopathogenic nematodes from Mexico that can overcome the resistance mechanisms of the western corn rootworm. *Nature Sci. Rep.* **2020**, *10*, 8257. [CrossRef]

Review

Movement Ecology of Adult Western Corn Rootworm: Implications for Management

Thomas W. Sappington [1,2,*] **and Joseph L. Spencer** [3,*]

1   Corn Insects and Crop Genetics Research Unit, United States Department of Agriculture, Agricultural Research Service, Ames, IA 50011, USA
2   Department of Plant Pathology, Entomology and Microbiology, Iowa State University, Ames, IA 50011, USA
3   Illinois Natural History Survey, Prairie Research Institute, University of Illinois, Champaign, IL 61820, USA
*   Correspondence: tom.sappington@usda.gov (T.W.S.); spencer1@illinois.edu (J.L.S.)

Simple Summary: The western corn rootworm is a destructive and mobile insect pest of corn in North America and Europe. It is difficult to manage, in part because resistance has evolved to many forms of control. Understanding spatial patterns and distances of adult flight is critical to improving pest and resistance management strategies. However, a holistic understanding of adult rootworm movement has remained elusive because of conflicting observations of short- and long-distance lifetime dispersal, a type of dilemma in ecology called Reid's paradox. Estimates of gene exchange between populations provide indirect estimates of dispersal distances, suggesting movement that is much farther than that measured using direct field observations, a similar type of dilemma called Slatkin's paradox. Taken together, the evidence is clear that many individual rootworms do not travel very far in their lifetime, often laying eggs in the same field in which they emerged. However, a substantial number of others take long flights of many kilometers before leaving offspring. We conclude that western corn rootworm is a partially migratory species consisting of two distinct behavioral types, *residents* and *migrants*. This interpretation will be useful in improving models of rootworm population dynamics and devising better rootworm pest management methods.

Abstract: Movement of adult western corn rootworm, *Diabrotica virgifera virgifera* LeConte, is of fundamental importance to this species' population dynamics, ecology, evolution, and interactions with its environment, including cultivated cornfields. Realistic parameterization of dispersal components of models is needed to predict rates of range expansion, development, and spread of resistance to control measures and improve pest and resistance management strategies. However, a coherent understanding of western corn rootworm movement ecology has remained elusive because of conflicting evidence for both short- and long-distance lifetime dispersal, a type of dilemma observed in many species called Reid's paradox. Attempts to resolve this paradox using population genetic strategies to estimate rates of gene flow over space likewise imply greater dispersal distances than direct observations of short-range movement suggest, a dilemma called Slatkin's paradox. Based on the wide-array of available evidence, we present a conceptual model of adult western corn rootworm movement ecology under the premise it is a partially migratory species. We propose that rootworm populations consist of two behavioral phenotypes, *resident* and *migrant*. Both engage in local, appetitive flights, but only the migrant phenotype also makes non-appetitive migratory flights, resulting in observed patterns of bimodal dispersal distances and resolution of Reid's and Slatkin's paradoxes.

Keywords: *Diabrotica virgifera virgifera*; resistance; dispersal; partial migration; ranging; station keeping; flight; behavior; Reid's paradox; Slatkin's paradox

Citation: Sappington, T.W.; Spencer, J.L. Movement Ecology of Adult Western Corn Rootworm: Implications for Management. *Insects* 2023, 14, 922. https://doi.org/10.3390/insects14120922

Academic Editor: David A.J. Teulon

Received: 25 October 2023
Revised: 23 November 2023
Accepted: 27 November 2023
Published: 3 December 2023

1. Introduction

The western corn rootworm, *Diabrotica virgifera virgifera* LeConte (Coleoptera: Chrysomelidae), is the most significant pest of corn in North America and Europe [1], responsible for over USD 2 billion in combined management costs and yield losses annually in the U.S. alone [2]. Injury is caused mainly by larvae feeding on roots, disrupting water and nutrient uptake, and structurally weakening the plant, which may lodge in storms. Western corn rootworm biology and ecology evolved to exploit the annual availability of its host plant corn (or maize), *Zea mays*. This species has only one generation per year, overwintering in the soil as a diapausing egg [3,4]. Although the larvae can develop to adulthood on the roots of several wild grasses [5–8], for all practical purposes, they survive in significant numbers only on corn roots [1,9,10]. Thus, rotation of a field from corn, where eggs are laid one year, to another crop such as soybean (*Glycine max*) the following year effectively purges that field of larvae [11]. Annual crop rotation was recognized as an effective management tool since the earliest days of the western corn rootworm's emergence as a pest [12] and remains an important option today [13–17]. In non-rotated corn in North America, larval control with soil insecticides was the most common management tool in the last half of the 20th century [18], until the advent of transgenic Bt corn in 2003 [19]. Control of adults with foliar insecticides was a common tactic in parts of the Great Plains as well [18].

However, nearly every management tactic deployed against western corn rootworm has been compromised by evolution of resistance. For example, adult females in most populations oviposit predominantly in cornfields; however, important exceptions are rotation-resistant populations found in parts of Illinois and surrounding states, characterized by females with relaxed oviposition fidelity to cornfields [1,11,13,20,21]. A capacity for adult movement at multiple spatial and temporal scales has facilitated the rapid evolution and spread of resistant populations. Having a solid, comprehensive framework of rootworm dispersal is critical for efforts to understand and predict the ecological and demographic consequences of adult movement under different scenarios of biotic and abiotic conditions, landscape matrices, and human interventions. The goal of this paper is to review and synthesize the wide array of available evidence illuminating adult western corn rootworm dispersal activity as the basis for a proposed conceptual model of this species' adult movement ecology.

1.1. The Paradox of Western Corn Rootworm Movement

Many, perhaps most of the puzzle pieces needed to reveal the full picture of western corn rootworm adult movement have been painstakingly gathered over decades of research by numerous scientists. But how to fit all those pieces together into a coherent whole has remained frustratingly elusive. The challenge lies in reconciling incongruities between direct observations of readily measured, short-distance movement with indirect but compelling evidence of long-distance displacement of rootworm adults. Such a dilemma is called *Reid's paradox* [22]. Reid [23] pointed out that as the glaciers retreated at the end of the Pleistocene, trees expanded northward in the British Isles at a much faster rate than seemed possible based on observations of seed dispersal mechanisms. In addition to plant range expansions (e.g., [24]), Reid's paradox has been encountered in other taxa and ecological contexts as well [25–28].

Short-distance lifetime dispersal by western corn rootworm adults is well supported by both direct and indirect evidence. Based on capture of adults self-marked with Bt-corn tissue in the gut, Spencer et al. [29] determined 85% of adults in corn moved 4.6–9.1 m/d on average. Larval population density builds over consecutive generations in fields of continuously planted corn [4,30–36]. For density to build in cornfields over time, a large proportion of the adults emerging in a field must also lay many of their eggs in the same field. Furthermore, resistance to Bt toxins such as Cry3Bb1, expressed in transgenic Bt corn specifically targeting corn rootworm larvae, appears to have emerged independently in multiple locations across the Corn Belt, creating resistance hotspots in the landscape [37–40]. Resistance to a single-toxin Bt-corn hybrid can evolve in as few as three generations under artificial selection in the laboratory or greenhouse [41,42], or naturally in continuous Bt cornfields [34,37,38,43,44]. Fast evolution of resistance in the field is facilitated by assortative mating brought about in part by incomplete dispersal of adults from the field before mating and oviposition [19,45,46]. Shrestha and Gassmann [47] found that a western corn rootworm population in a field with a long history of Cry3Bb1 resistance had higher survival than rootworms from a field with a more recent history of resistance. This finding is consistent with a build-up of resistance in a field over time, indicating emigration of resistant beetles from that field is not complete and immigration of susceptible beetles is not sufficient to counteract the increasing frequency of resistance alleles in situ.

At the same time, evidence for emigration of western corn rootworm adults out of their natal field is also robust. Except in regions where rotation resistance is common [1,13,48], virtually no western corn rootworm adults emerge from first-year cornfields. Thus, adults found in a first-year cornfield are immigrants from elsewhere. Colonization of and oviposition in first-year cornfields is evidenced by accumulation of adults during that first year and by larval injury and adult emergence in second-year cornfields [30,49–52]. Similarly, an aerial insecticide application to suppress ovipositing western corn rootworm adult populations can protect a cornfield from economic injury by larvae the following year [18,35,53–56]. Pruess et al. [53] concluded that fields in central Nebraska treated for ovipositing adults the previous year were repopulated the next year in part by immigrant adults. Although the origin of immigrants recolonizing a depopulated field is primarily from nearby fields [50,52,57], there is evidence for dispersal of western corn rootworms well beyond the immediate surroundings of the natal field. This evidence includes rates of range expansion and spread of adaptive traits [1,4], ascent of freshly mated females into the atmosphere [58], wash-ups of adults on the shores of Lake Michigan [59,60], tethered-flight experiments [61], and population genetic estimates of gene flow [62], all of which will be examined later in this paper.

Long-distance flight by at least some western corn rootworms is not in doubt. Thus, in one respect, the solution to Reid's paradox is simplified—it seems obvious that dispersal in this species is bimodal, in the sense that some individuals remain throughout their lifetime to reproduce in or near the natal field, while others fly long distances and reproduce elsewhere. However, recognition of a bimodal pattern of flight distances raises important questions. What constitutes a long-distance flight (how "long" is "long")? What proportion of a rootworm population engages in long-distance flight? What motivations trigger a long-distance flight—i.e., what determines whether an individual takes only short-distance flights during its lifetime versus taking one or more long-distance flights? Is the timing of a long flight related to age or reproductive development? How many bouts of long-distance flight can be expected of an individual? Is migratory behavior involved? What external factors impact propensity and capacity to engage in a long-distance flight? We have clues and even answers to many of these and other questions for western corn rootworm, but before reviewing them, it will be helpful to briefly present some relevant concepts and terms used to describe insect movement and how these apply in general to western corn rootworm.

1.2. Types of Movement—Scale and Motivation

In a landmark paper, Nathan et al. [63] presented a unifying conceptual framework termed "movement ecology" for studying and understanding organismal movement. They describe four mechanistic components that interact to generate an observed movement path of an individual over a defined timeframe (from a few seconds to a lifetime), namely (1) internal state—the motivation to move; (2) motion capacity—the ability and modality of movement; (3) navigation capacity—the ability to direct movement toward a goal or target; and (4) external factors, which include any biotic or abiotic environmental conditions that affect the movement path directly or indirectly through their influence on the other three components. For many applications of this information, such as predicting population dynamics in a pest management context, or modeling evolution of resistance to a control tactic, the goal is to understand population-level patterns of movement and their impacts on the phenomenon of interest. Such population-level phenomena are emergent properties of individual behaviors and individual responses to selection [46,64–67]. Understanding the effects of dispersal on ecological and evolutionary dynamics depends on a robust understanding of the underlying mechanisms of individual movement [63].

All four mechanistic components of western corn rootworm movement by flight vary depending on the spatial and temporal scales of focus, as do the manner and consequences of their interactions at a given time and place. For example, the internal motivation for flight activity by an individual changes with intrinsic variables like age and physiological status (nutrition, mating, past flight activity, and reproductive development), and with extrinsic variables (biotic and abiotic) such as population density, crop maturity, and ambient weather conditions [68]. Fixed interindividual differences in characteristics like sex, genotype, and morphology can affect internal motivation or capacity for flight. Targets of navigation differ depending on motivation, such as the search for resources. The convergence or divergence of individual movement paths [63] at the population level can even affect extrinsic factors, which in turn can influence future flight behavior of individuals through density-dependent mechanisms.

Motivations underlying insect flight can be categorized as either *appetitive* or *non-appetitive*. Most flight activity is appetitive in nature, meaning that the individual is foraging or searching for a resource such as a mate, food, oviposition site, or shelter. Appetitive flight behavior is triggered by immediate conditions experienced by the individual and is arrested upon encountering the resource being sought [64]. Appetitive flight can be subcategorized based on the resulting degree of net displacement relative to the individual's *home range*, the area in which day-to-day maintenance and/or reproductive activities occur. *Station-keeping* behaviors involve localized movement within the home range, particularly foraging activities [64,69]. For western corn rootworm, station-keeping behavior often applies to movement within a single field and fields in the immediate vicinity, together constituting the individual's home range. *Ranging* is appetitive flight behavior that results in a permanent displacement out of the previous home range [69–71] and can be thought of as extended foraging for whatever resource is motivating flight at the moment.

For example, an adult western corn rootworm leaving a senescing cornfield in appetitive flight to search for later-planted, more attractive corn [49,72,73] may disperse across the road in a station-keeping foraging flight or across a considerable distance in ranging flight before finding such a habitat. Whether the individual leaves its current field because it is pushed by deteriorating conditions [21] or pulled by detection of relatively more-stimulatory volatiles emanating from a less mature cornfield [74] is unresolved, but the nature of the flight is appetitive regardless. Even if displacement occurs over a longer distance, such a ranging flight ends as soon as the sought-after resource is encountered. The beetle then resumes normal day-to-day station-keeping behaviors in its new home range, which now includes the location where its resource-seeking ranging flight terminated.

At a behavioral level, station-keeping and ranging flights are fundamentally the same in the sense that they both reflect appetitive behavior. The categories are descriptive, in that they are distinguished by whether the distance traversed is great enough to preclude re-encounter with the field of origin during subsequent station-keeping flights; if so, the individual has left its home range, and if not, it remains within its home range. Thus, only the distances traversed distinguish these two categories, and even then, displacement distances via station-keeping and ranging behaviors may grade into one another [64]. Nevertheless, these categories are useful, because the distances traveled are determined by the insects' responses to their proximate environment, which are important to understand for predicting movement patterns within and between fields in the local landscape. The consequences of the distances traveled are also of practical importance for population management, crop management, and insect resistance management (IRM). Even in seemingly homogenous agricultural landscapes, like those in the Corn Belt, ranging and some station-keeping movement likely result in displacement exposing individuals to fields—i.e., local environments—that vary in the particulars of seasonal and even day-to-day management. Such variables include which Bt/RNAi traits are expressed; crop phenology; local population density; the likelihood that a field may be sprayed or not at pollination; and weed, disease, and other pest pressures. The western corn rootworm "experience" can be quite different on the other side of the road or fence line.

In contrast, *migratory* flight behavior is non-appetitive. Migratory flight is not initiated as an immediate response to lack of a resource, and a migrating individual is not distracted by encounters with resources. It is characteristically persistent, straightened-out (i.e., non-meandering, non-searching) flight, which is terminated in a systematic way by environmental or internal cues unrelated to resource cues [64,66,75–78]. Migratory flight in insects that utilize winds at altitude for transport is characterized by three phases [79,80]: (1) *ascent* into the atmosphere above the flight boundary layer; (2) *transmigration* during horizontal displacement, characterized by maintenance of straight-line flight, altitude, and suppression of response to resource cues; and (3) *termination* of migratory flight accompanied by descent and landing. After termination of migratory flight, the insect finds itself in (or over, if the switch occurs while still airborne) a new environment, which may fortuitously be a habitat suitable for a new home range where it can begin station-keeping activities. However, if the habitat is unsuitable, the individual may, in some species, recommence with another bout of migratory flight, or it may begin appetitive ranging behavior in the local landscape in search of favorable habitat.

Migration is a phenomenon associated with a *migratory syndrome*, a suite of developmental, physiological, morphological, behavioral, and life-history traits that together direct and accomplish successful migration [64,77,81]. The migratory syndrome is underlain by genetically controlled mechanisms shaped by natural selection. Although migratory flight can be directional and long-distance, it does not have to be either. Western corn rootworm is not commonly thought of as a migratory insect because this species' overwintering and breeding ranges are the same. This differs from more familiar insect migrants like fall armyworm (*Spodoptera frugiperda*) [82–85] and monarch butterfly (*Danaus plexippus*) [86,87], which engage in long-distance migration between geographically disjunct overwintering and summer breeding ranges. However, insect migration is not a phenomenon limited to directional movement between seasonal ranges [75,88,89], and there is considerable evidence for western corn rootworm migratory flight behavior, which we present later in this review.

With this background in mind, we can better describe the overall nature of the bimodal dispersal pattern observed in western corn rootworm populations (Figure 1). Some, but not all, western corn rootworm adults migrate, the defining characteristic of a *partially migratory species* [65,67,90,91]. Partial migration is the most common type among migratory animal taxa [65,90,92]. We infer two behavioral phenotypes of western corn rootworm adults, *resident* and *migrant*. During its lifetime, a *resident* engages in flight only in the context of station-keeping and ranging behaviors, that is, only in appetitive flight. Rootworm females that oviposit all their lifetime eggs in the natal field and its immediate vicinity (*natal home range*) are residents. Individuals that travel far enough via appetitive ranging behavior to establish a new home range outside their natal home range are also residents. The distinguishing characteristic of a western corn rootworm *migrant* is that during its lifetime, it engages in at least one non-appetitive flight consistent with this species' migratory syndrome. Not all aspects of the rootworm migratory syndrome are understood, but it includes behaviors such as purposeful ascent into the atmosphere above the flight boundary layer (altitude at which wind speed exceeds unaided flight speed of the insect [93]), and developmental timing in which young females migrate after mating but before egg maturation. Both are attributes of the migratory syndromes of many other migratory insect species as well [66,69,94].

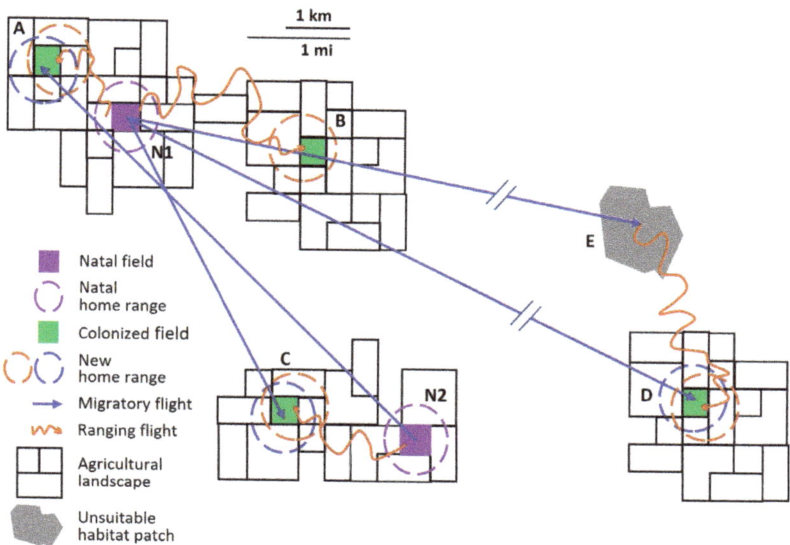

Figure 1. Schematic of types of flight behavior by *resident* and *migrant* phenotypes of western corn rootworm, and resulting spatial displacement across local and regional landscapes relative to natal and colonized cornfields. The schematic crop fields in the figure represent typical dimensions of rectangular fields in the U.S. Corn Belt, where the smallest squares shown are 0.25 mi (0.40 km) per side, or 40 acres (16.2 ha). In this depiction, western corn rootworm adults eclose in two natal fields (**N1** and **N2**). A *resident* may ultimately engage only in appetitive station-keeping behaviors throughout its lifetime, resulting in net displacement of only short distances within the natal field or after emigration into habitat in the natal field's immediate vicinity. This area of station-keeping activity constitutes the natal home range (purple circle) and is observed as diffusive spread of the field's population. Some residents engage in an appetitive ranging flight (meandering brown arrow) in search of a needed resource or better habitat, emigrating from the natal field and beyond the natal home range. Ranging flight is facultative and triggered by proximate conditions such as lack of a needed resource or deteriorating habitat. A ranging resident stops when it encounters the sought-after resource, which may be a relatively short distance away (e.g., **N1** → **A**, **N2** → **C**), or a longer distance across the local landscape (e.g., **N1** → **B**). In either case, after immigration into a suitable habitat, station-keeping behaviors resume in the colonized field, and a new home range (brown circle) emerges. A *migrant* adult emigrates from the natal field via a non-appetitive migratory flight (straight blue arrow) over relatively long distances, not only beyond the natal home range, but often beyond the local landscape (e.g., **N1** → **C**, **N1** → **D**, **N2** → **A**). Migration is not initiated in direct response to proximate conditions. Thus, migrants may emigrate from a highly suitable natal field in which a large number of residents remain to reproduce. Migration is innate to the migrant phenotype and is initiated in females during a narrow developmental window after mating but before egg maturation. Migratory flight is straight-line (non-meandering), and in western corn rootworm is not directed toward a geographic goal or in a preferred direction (e.g., beetles on opposite trajectories, such as **N1** → **D** and **N2** → **A**, can issue from the same field). The migrating insect does not respond to resource cues or cease flight when encountering suitable habitat (e.g., blue arrow passing over field **B**). Instead, migration is terminated in response to global environmental cues like sunset, or intrinsic cues such as an internal clock or physiological status, which have not yet been elucidated for western corn rootworm. If the insect fortuitously terminates its migratory flight in suitable habitat (e.g., fields **A**, **C**, and **D**), it then resumes station-keeping behavior, and a new home range emerges (blue circle). If migration terminates in unsuitable habitat (e.g., **N1** → **E**), the migrant initiates appetitive ranging behavior through the local landscape in search of appropriate habitat (e.g., **E** → **D**). Once suitable habitat is encountered, ranging ends, station-keeping behaviors resume, and a new home range emerges.

The mixture of migrants and residents in a western corn rootworm population may have arisen as a bet-hedging strategy. A female risks a catastrophic loss of long-term fitness when ovipositing all its eggs in a single field ("putting all its eggs in one basket") destined for rotation to a non-host crop, whereas a female that oviposits in more than one field, only some of which may be destined for crop rotation, virtually ensures some of its offspring will emerge in a cornfield the next year and thus survive to reproduce. Bet hedging is an evolutionary strategy to spread risk in unpredictable environments by producing alternative phenotypes. The strategy involves accepting reduced mean within-generation fitness to increase geometric mean fitness over generations [95]. Producing some offspring that migrate from the natal field could be an adaptation in western corn rootworm to spread the risk of crop rotation of the natal field, or to a progenitorial host with a patchy distribution. It seems likely that selection will favor females of either phenotype that produce a mixture of resident and migrant offspring. The proportions among the offspring of an individual female most likely will depend on inherited environmental threshold responses ([91]; see Section 7.2). Among residents, a female that oviposits on the isolated progenitorial host plant on which it (and some number of siblings) developed may also doom the offspring to crowding on a locally limited food supply. Larval overcrowding has negative effects on mortality, development time, and size [96–100]. The female would do well to move before ovipositing, but such movement only needs to be local and at the within-field scale.

While bimodal dispersal is probably an adaptation, in part, to the unpredictability of crop rotation of the natal field, the rotation-resistant phenotype of western corn rootworm arose in eastern Illinois in response to a highly predictable crop rotation of nearly all fields of corn in the natal landscape. It is interesting that rotation-resistant western corn rootworm populations are also partially migratory, as evidenced by rates of range expansion correlated with prevailing wind direction, including stratified dispersal, and ascent of individuals into the atmosphere, as will be discussed (see especially Sections 5.3 and 5.5).

Western corn rootworm movement ecology is complicated (Figure 1), but recognizing the distinction between residents and migrants provides a framework for assessing the wide array of observational and experimental data on rootworm dispersal and placing them in proper perspective. Characterizing the many aspects of the movement of both residents and migrants is a challenge. Adequate description and understanding of long-distance movement are particularly lacking [4]. Much of the difficulty in understanding long-distance movement arises from conflation of appetitive ranging flight with non-appetitive migratory flight, both of which displace an adult rootworm out of its natal home range (Figure 1) but which derive from fundamentally different motivations and behaviors. The consequences of bimodal intrapopulation variability in dispersal behavior of individual western corn rootworms both determine and introduce variation in patterns of net generational displacement and gene flow at the population level. Knowledge of such patterns can help us predict population changes in a particular field or set of adjacent fields, even if we do not fully understand the motivations and determinative dynamics at the level of individuals. In the context of short-term management of populations infesting cornfields, and of designing and implementing strategies to delay, contain, or otherwise mitigate resistance to control measures, we can benefit from recognizing movement patterns within fields, between fields in the local landscape, and across larger expanses of space.

1.3. Incorporating Short- and Long-Distance Dispersal in Population Models

Short-range daily movement distances are used to parameterize the dispersal component of most models exploring western corn rootworm resistance evolution [101–106] and resistance mitigation [107]. The values used are generally in the range of a few tens of meters per day, although local ranging among fields in the modeling landscape is often assumed (see [108] for a review). Short-distance movement is associated with station-keeping activities within fields and ranging activities between nearby fields. Heavy reliance on the short-range, diffusive movements of the resident portion of the population to parameterize flight distances in models of western corn rootworm population dynamics, ecology, and evolution means that the role of concomitant long-distance movement by migrants remains mostly unknown and unaccounted for. Modelers are not unaware of the occurrence and potential importance of long-distance flight by at least some adults, but such movement is ill-defined, and including it as a realistic parameter can be challenging. Nevertheless, a few attempts have been made, yielding insightful results.

Caprio et al. [109] incorporated long-distance dispersal indirectly in their model for evolution of methyl parathion resistance. Instead of assigning distances traveled during long-distance flights, they simply included the daily rate of adults of different developmental maturities dispersing out of the natal field to any one of the other 24 fields in the modeling landscape, based on the percentage of sustained fliers, presumably migrants, observed in tethered-flight experiments [61,110]. Most commonly, long-distance flight is included in models focused on geographic spread of a trait or range expansion of the species itself. Onstad et al. [111,112] modeled predictions of geographic spread of rotation resistance through the western corn rootworm metapopulation from its point-source origin in east-central Illinois. They incorporated a maximum distance of wind-aided flight of 33 km calculated from speed of long-duration (>30 min) flights on flight mills [61] and typical characteristics of summer wind and storm events in the central Corn Belt.

The invasion of Europe by western corn rootworm generated great interest in developing models to predict expansion into new areas [113,114], and to assess efficacy of potential containment strategies and tactics to slow the spread [115–117]. As in the case of IRM models, dispersal is a key parameter that helps drive these models' output. But because the context of most European models relates to predicting or slowing range expansion, long-distance dispersal plays a prominent role in parameterization. With some exceptions (e.g., [117]), the approach taken to identify realistic values with which to parameterize long-distance dispersal components has been empirical in nature, using rates of range expansion already observed for this species during the ongoing invasion process. This approach is practical and does not require detailed understanding of underlying processes governing the expansion. Though conceptually straightforward, acquiring good estimates of long-distance dispersal based on observed rates of range expansion is not easy, as will become evident later. Nevertheless, the effort to obtain such estimates for both the North American and European expansions has helped clarify the movement ecology of western corn rootworm, including the interconnected contributions of residents and migrants to net rates of expansion.

Many details of western corn rootworm adult movement ecology, especially understanding motivations (and the mechanisms controlling motivations) for movement and dispersal over different scales and involving different behaviors, will take much focused research to sort out. At the same time, we already know a great deal about the spatial patterns of adult movement of this pest species (Figure 1). In this review, we focus on describing these movement patterns, and their implications for pest management, IRM, and predicted range expansions. Notably, the experimentation and observations that have helped elucidate patterns of movement and dispersal also give us much insight into the mechanisms and fundamental drivers underlying the movement ecology of individuals, as well as exposing critical gaps in our understanding that would benefit from future research.

2. Pre-Mating Movement

Western corn rootworm adult emergence begins in late June or early July. Adult males emerge ca. 5–6 d before females (protandry) [32,118,119] due to faster pre- and post-eclosion development rates [119,120]. Newly emerged males are not sexually mature, requiring 5–9 d to become responsive to female pheromone [121]. Delayed sexual maturity results in time for pre-mating male activity, including feeding and dispersal. In contrast, adult females are sexually mature upon emergence; 54% and 96% engaged in pheromone calling behavior during the 1st or 2nd day after emergence, respectively [122].

Post-emergence, pre-mating movement by western corn rootworm adults varies by sex. Newly emerged females rest on corn leaves an average of 0.78 m above the soil surface and move little before mating [3,123–125]. Most females release pheromone from a position near their natal plant [11,40]. Much male activity, including responses to pheromone, occurs below 2.0 m in the corn canopy [126,127]. Marquardt and Krupke [125] suggested that pre-mating males move more than females.

When directly observing individual behavior is not feasible, the movement patterns of many beetles can be tracked using a variety of markers. Spencer et al. [128] and Hughson [129] used two Cry proteins differentially expressed in rootworm Bt-corn or non-Bt-corn ("refuge") hybrids as ingestible markers [29] acquired during normal feeding. Cry protein-containing tissue remained detectable in beetle gut contents for up to ca. 24 h post-ingestion [29,129]. The presence of hybrid-specific Cry proteins in the gut contents of beetles collected from Bt or refuge corn rows reveals intrafield movement between field areas. Marker and labeling studies of western corn rootworm recovered in mating pairs [128–130] revealed there can be substantial movement of both sexes before mating.

Western corn rootworm intrafield movement is greatest during the vegetative period of corn phenology [128,129]. Among individual beetles collected in cornfields, the proportion engaged in intrafield movement drops significantly from 0.22 during the vegetative period to 0.06 and 0.07 during the pollination and post-pollination periods [129]. The wide availability of nutritious pollen and silks are likely factors that reduce hunger-related intrafield movement during pollination and the early portion of the post-pollination period. Individual gut-content analyses of the partners from a mating pair reveal details of pre-mating movement. Like the overall field population, among males collected in mating pairs, the proportion that engaged in pre-mating intrafield movement (0.38) was greatest during the vegetative period of corn plant phenology [128]. Mating females also engage in intrafield movement; however, the proportion of moving females collected in mating pairs (0.09) was significantly less than that of their male partners (0.25) [128]. Hughson [129] reported similar proportions of moving males (0.246) and females (0.030) in mating pairs.

Among females, the likelihood of pre-mating intrafield movement may be influenced by female age and/or access to mating opportunities. While Spencer et al. [128] identified females that moved across Bt/refuge boundaries before mating, these intrafield moving individuals were almost exclusively (93.1%) mature and non-teneral (i.e., they emerged >24 h before mating). Scarcity of teneral individuals among moving females suggests that pre-mating intrafield movement is a phenomenon largely limited to females that remain unmated more than one day after emergence [128]. The implication is that in areas where mate-seeking males are not abundant, female mating may be delayed. Eventually, as unmated females move and enter areas of higher male abundance (e.g., refuges), they are intercepted by mate-seeking males and are mated as older, non-teneral adults.

Most western corn rootworm beetles in these studies did not cross between Bt and refuge areas of cornfields before finding a mate. In fact, mating pairs that included a beetle that engaged in intrafield movement (between refuge and Bt corn) were mostly detected within a few rows of the interface between those areas [128,129]. Despite only modest proportions of males and females engaging in pre-mating intrafield movement, the average movement rate of those individuals was 29.5 m/d (based on Figures 3 and 4 in [128]).

Taylor and Krupke [130] studied western corn rootworm dispersal and mating interactions in Bt and non-Bt refuge corn. They labeled non-Bt refuge corn plants with ^{15}N, which was ingested by feeding larvae and was later detectable in the adults, enabling definitive identification of refuge beetles when present in mating pairs [130]. In their study, 41.5% of mating pairs included partners that originated from different areas of their study plots. However, like Spencer et al. [128] and Hughson [129], when mating pairs with beetles from two different areas of their fields were detected, they were found not far from boundaries between non-Bt refuge corn and Bt corn [130]. While some significant pre-mating movement occurred in males, and to a lesser extent in females, most mate-seeking western corn rootworm beetles did not move very far from their emergence location to find a mate [130].

3. Mating and Movement

Western corn rootworm mating behavior was described in detail by Lew and Ball [124]. Initial detection of sex pheromone prompts an excited response from males [131]. In a laboratory wind tunnel, rapid waving of antennae (antennation; also important in the mating sequence [124]) combined with orienting the body with respect to the pheromone source precede initiation of an upwind or lateral hovering flight toward the source [131]. A slow hovering flight while approaching a pheromone source is documented for the related species *Diabrotica balteata* [132] in which males responded to pheromone lures in soybean fields from as far as 49 m. A similar slow, hovering flight is commonly observed among western corn rootworm males approaching a calling female. Branson and Krysan [133] characterized western corn rootworm pheromone as "extremely efficient" but unnecessarily so for a species where the sexes are at high density in corn. They speculated such efficiency is evidence of rootworm evolution under conditions of much lower adult density. This suggestion is consistent with other life-history traits indicative of a species adapted to low-density adult populations, such as protandry, female calling from the natal host plant, and post-mating female migration. An environment with scattered perennial progenitorial hosts [133] would favor other characteristics present in western corn rootworm, including high adult mobility and sensitivity to host plant volatiles [74,134], oviposition near the host plant [135–138], and high sensitivity and responsiveness to root-produced volatiles among larvae with limited capacity for moving through soil [139,140].

Western corn rootworm mating may be observed throughout the period of adult emergence. During a 3–4 h mating [141], males transfer a large spermatophore to the female [142–144]. The western corn rootworm spermatophore may equal up to 9% of the male's mass [143] and constitutes a significant male investment in the female. Some of the spermatophore components are incorporated into the eggs [144]. Most females are thought to mate only once [3,123], though some mate a second time in the laboratory [39]. Dissection of females from extensive field collections by Hughson [129] revealed multiple spermatophores in some and other evidence indicating at least 4–5% likely mate multiply. Males are also capable of multiple matings [143,145,146], though acquiring resources to provision large spermatophores may limit a male's capacity in this regard, especially as cornfields mature. Kang and Krupke [146] found most males were capable of only 2–3 matings during an approximately 10–14-d reproductive period following an initial mating. Bermond et al. [147] found three-fold greater survival of females than males under starvation in the laboratory and hypothesized that female use of nutrients from the male spermatophore may explain this result. In contrast, Murphy and Krupke [144] found no difference in longevity of mated and unmated females under starvation conditions in the laboratory. The potential contribution of spermatophore components to the metabolic demands of female migratory flight has not been evaluated. Evidence of recent mating is strongly associated with female western corn rootworm engaged in migratory behavior [58, 148,149]. However, any potential role for the mating act itself or spermatophore components in stimulating some females to engage in migratory flight is unknown.

4. Post-Mating Movement

After mating, male and female movement patterns may diverge. Males presumably continue mate-seeking behavior in areas where female emergence is ongoing. Among females, mating stimulates egg development [141,150,151], necessitating that they locate and feed on nutritious host plant tissues. Newly mated females require 6–21 d of feeding to mature their first batch of eggs [141,152]. It is during this post-mating, pre-ovipositional period that females of the migrant phenotype engage in non-appetitive migratory flight (as will be described in detail in Section 5). Resident beetles of both sexes engage in appetitive station-keeping and ranging flight behavior after mating, which may keep an individual in its natal field or may lead to displacement over a relatively short distance within the local landscape.

Areas in Bt cornfields with the greatest adult abundance, along with significant mating activity, frequently coincide with nearby non-Bt refuge rows that are the areas of greatest adult emergence [32,129]. Adult intrafield movement between blocks of non-Bt refuge and Bt corn becomes less likely with advancing corn phenology [129]. Among free-moving adults in Bt cornfields with structured refuges, the probability of intrafield movement measured during the vegetative period (0.27) dropped significantly during pollination (0.01–0.07) and post-pollination (0.01–0.07) [129]. Among the minority of beetles that moved each day between refuge and Bt corn areas, intrafield (refuge to Bt corn) movement rates (males: 28.83 ± 3.62 m/d (mean $\pm$ SEM); females: 23.19 ± 0.19 m/d) were similar to pre-mating rates reported by Spencer et al. [128], though they were not significantly different between the sexes [129]. This contrasts with Ludwig and Hill [153], who suggested males are more mobile within fields than females. The declining probability of intrafield movement after the vegetative period of corn phenology may be attributed to the abundance of corn silks and pollen in the current field, to which western corn rootworm respond strongly [74] and which can concentrate populations [154].

Quantifying interfield movement is complicated by the scale of distances over which movement must be tracked and the difficulty of identifying specific beetles among large populations [148]. Mark-recapture experiments provide direct evidence of the net displacement and origin of individual immigrants. However, decreasing density of marked individuals with increasing distance from the source location usually limits efficacy of such a strategy to the local landscape. Various marking methods have been used to identify western corn rootworm movers, including fluorescent powder [148,155,156], ingested Bt protein [29], and N-isotope signatures [157]. For example, Toepfer et al. [158] mass-marked adults in cages with fluorescent powder and documented recapture of 0.03% in small corn plots 300 m from their release points in steppe habitat in Hungary. Using ingested Bt tissue as a marker, Hughson [129] examined the relationship of corn phenology in the local landscape to interfield movement. Among males, the proportion engaging in local interfield movement (measured between adjacent cornfields) was significantly higher during the vegetative period (0.09) than during pollination (0.02) or post-pollination (0.03) periods; female interfield movement was not measured. Failure to detect increasing interfield ranging movement as post-pollination cornfields matured may be attributed to near-identical phenology among the hybrids in all fields where movement was monitored [129].

Phenology has a role in various aspects of interfield western corn rootworm beetle movement. Interfield movement toward a less mature cornfield can occur when there is variation in crop phenology among local fields. At field interfaces, western corn rootworms express a short-range flight orientation preference for flowering corn vs. pre- or post-flowering stages and for corn vs. other crops (i.e., soybean, wheat, or sweet clover) regardless of corn developmental stage [134]. By extension, this tendency to orient toward corn should help beetles that have moved out of cornfields to relocate them. Campbell and Meinke [9] reported that western corn rootworm beetles moved from cornfields into adjacent non-corn habitats (native upland and lowland prairie) when corn had pollinated. McKone et al. [159] reported similar western corn rootworm usage of tallgrass prairie. Contrary to most studies suggesting that females predominate among western corn rootworms leaving corn, Campbell and Meinke [9] join Moeser and Vidal [160] to suggest there is significant movement out of corn by males. Western corn rootworm attraction to relatively less mature corn plants [74] is likely a factor in many instances of local interfield movement. This phenomenon is the basis for using late-planted corn as a "trap crop" to attract and concentrate egg-laying beetles into specific cornfields, creating infestations for study purposes or as a tactic to diminish infestations in the fields from which beetles were attracted [72,161]. A response among western corn rootworm beetles to a specific stage of corn phenology, R2 (post-silking blister stage [162]), was documented by Pierce and Gray [21]. They showed that increased interfield movement by rotation-resistant western corn rootworm adults out of corn into soybean or by rotation-susceptible adults into late-planted corn was associated with onset of the R2 stage [21]. Females (of both rotation-resistant and -susceptible populations) predominated significantly among the interfield movers from the R2 stage onward [163]. Notably, while differences in phenology of adjacent cornfields made the late corn a competitive sink for egg laying with soybean among the rotation-resistant population, extreme phenological differences did not lead to egg laying outside of corn where rotation resistance was absent [21]. While western corn rootworm responsiveness to corn phenology facilitates highly adaptive beetle movement, ovipositional fidelity to corn in the face of an extreme phenological difference suggests that its relaxation [164] is unlikely to account for the origins of rotation resistance.

Indirect evidence of interfield movement is frequently inferred from the proportions of western corn rootworm males and females detected outside of their natal fields. Female-biased sex ratios among beetles collected in rotated corn (and other rotated crops, especially where crop-rotation-resistant western corn rootworm populations are present [1]) testify to the different probabilities of male and female interfield dispersal from their natal fields [21,30,49,158,163,165,166]. Furthermore, these probabilities vary with flight altitude. While females outnumber males in most collections of movers, males are still present among the beetles flying between fields; nevertheless, females predominate at higher altitudes to a great extent [20,58,59,126,148,149,167]. Interfield movement (including initiation of migratory behavior and high-elevation flight) is strongly periodic and gated by permissive environmental conditions typical of mid-morning and late afternoon/early evening [58,167,168].

Use of indirect evidence to quantify interfield movement can provide valuable insights into western corn rootworm movement ecology, but interpretation of results can be more fraught with difficulty than often realized. Inferences about interfield movement drawn from sampling methods such as trapping or visual counts of adults often rely on the assumption that shifting patterns of spatial abundance emerge from movement patterns of individuals engaging in local appetitive ranging or station-keeping behavior. However, the partial migratory nature of western corn rootworm means that some portion of immigrants to a particular field are of the migrant phenotype that originated from somewhere beyond the local landscape. Likewise, of those individuals that emigrate from a field, not all will disperse to nearby fields—some will exit from the local landscape entirely via non-appetitive migratory flight. Levay et al. [52] concluded from trapping studies of semi-isolated pairs of fields separated by distances of 1–1400 m (one field of first-year corn and one field of continuous-planted corn) that ~38% of the adults that emerged in the continuous-planted field emigrated to colonize the paired first-year cornfield as immigrants. The experimental design and interpretation of data relied on important assumptions: (1) Emigrants from the continuous cornfield all colonized the first-year cornfield. This implies movement trajectories of emigrants all in the same direction, which in turn implies attraction from a distance during appetitive ranging flight. (2) Immigrants to the first-year field all originated from the (paired) nearest continuous cornfield and not from outside the 3 km semi-isolation zone. These are reasonable assumptions if all emigration and immigration resulted from local, appetitive station-keeping or ranging flight. However, they are potentially problematic under the proposition that emigrants and immigrants are a variable mixture of residents and migrants engaged in appetitive and non-appetitive flights of differing distances, directions, and motivations. This was a demanding and exceptionally well-conducted study, which illustrates the point that wringing insights about western corn rootworm movement ecology from experimental and observational data is especially challenging given this species' partially migratory nature.

Among the examples of interfield western corn rootworm movement, there are significant differences between populations. In the eastern Corn Belt, movement between corn and soybean fields by crop-rotation-resistant females happens in a very different context from the interfield movements of females from cornfield to cornfield. Interfield movement by rotation-resistant females that lay some of their eggs in soybean fields is the behavior that allows them to circumvent annual crop rotation. However, distinguishing between normal station-keeping or ranging flight that moves a beetle into a new cornfield and the interfield flight of rotation-resistant western corn rootworm beetles leading to egg laying in soybean is challenging and context-dependent.

Rotation resistance in this species is almost certainly a genetically determined trait, as evidenced by its geographic spread from a point source in Ford County, Illinois [1,13], and it is clear that loss of female fidelity to oviposition in cornfields is the basis of rotation resistance [11,20,21,169,170]. Although the mechanism underlying the loss of fidelity is still undetermined, it is likely related to differences in flight behavior. It seems plausible, for example, that the extremely high selection pressure imposed by high-frequency crop rotation across a vast landscape could act on heritable variation to generate the observed differences in flight propensity between rotation-resistant and susceptible populations [171], but this remains a hypothesis to be tested. Some oviposition in non-corn crops by both rotation-resistant and susceptible (wild-type) beetles is undoubtedly the result of station-keeping behavior along the edge of a cornfield. While investigating invasive western corn rootworm in Croatia, Barčić et al. [172] showed that suspicious larval damage in rotated corn could be explained by an interfield movement edge effect extending ca. 20 m into the previous wheat or soybean field from adjacent continuous corn. Damage in the rotated crop was a consequence of adults laying eggs along the edge of the rotated crop. Because the crop fields were small (50 m wide), oviposition around the perimeter generated a pattern of damage that could be incorrectly construed as the presence of a rotation-resistant population. There is evidence for movement and edge effects on a similar scale in other studies. Well before rotation resistance became evident, Shaw et al. [136] reported larval damage in rotated corn (after soybean) at locations within 12 rows (9.1 m) of the edge of the previously adjacent cornfield. Spencer et al. [29] reported that 85–93% of crop-rotation-resistant western corn rootworm beetle movement within cornfields and from cornfields into soybean fields occurred at rates of 4.6–9.1 m/d. In another study of rotation-resistant western corn rootworms, Schroeder et al. [173] suggested that when crop fields (corn and three rotated crops) were small (i.e., 0.06 ha or ca. 24.6 m × 24.6 m) and in close proximity, it becomes a trivial task for beetles to move short distances to enter and lay eggs in an adjacent cornfield plot. Thus, in spite of equally high western corn rootworm female abundance measured in plots of corn, soybean, and other crops, egg laying and subsequent root injury in the rotated cornfields (mostly from soybean) were significantly reduced compared to that in the corn plots. Rondon and Gray [174] also evaluated adult western corn rootworm abundance in corn and a variety of rotated crops using small (0.1 ha) fields in a 5 × 5 Latin square. While they found significantly more females in soybean, there were equal numbers of eggs laid in all crops. They interpreted these results as evidence of non-preference in egg laying.

Something beyond station-keeping edge effects must be responsible for oviposition throughout large commercial soybean fields, observed as widespread damage to first-year cornfields. Compelling evidence for robust attraction to soybean tissues is lacking. The initial flight from a cornfield into an adjacent soybean field may result from a simple expression of enhanced propensity for flight. Field and laboratory behavioral assays showed that western corn rootworm adults from rotation-resistant populations are more active and ready to initiate flight than those from wild-type populations under the same rearing and environmental conditions [171]. In addition, nutritional stress caused by feeding on soybean tissue increases activity and probability of female flight compared to females in adjacent cornfields [175,176]. Though beetles from crop-rotation-resistant populations have a variety of adaptations to blunt the negative nutritional consequences of soybean herbivory that are lacking in rotation-susceptible populations [177–179], consumption of poor food (soybean foliage) stimulates behavior leading to interfield flight. Once females are active in soybean, consequences of soybean herbivory increase the probability of egg laying and flight. Thus, movement may play a role in egg laying if a soybean field is sufficiently large that some females remain long enough for the negative consequences of soybean herbivory to stimulate oviposition [21,175]. Soybean herbivory-stimulated flight out of soybean fields may be the mechanism that eventually returns a female to a nearby cornfield where it can feed on host tissues, allowing maturation of eggs that may be laid in soybean on a subsequent return to the field [176]. While rotation-resistant western corn rootworm herbivory in soybean can sometimes be dramatic [13], the high soybean tolerance for defoliation [180] suggests rootworm activity is unlikely to affect soybean yield, without contributions by other resident soybean herbivores.

Unlike most interfield ranging flights by western corn rootworm, the conditions responsible for initiating interfield flight by crop-rotation-resistant females differ depending on whether they are moving from corn to soybean or *vice versa*. Interfield movement from soybean to corn may be interpreted as an expression of appetitive ranging behavior in response to urgent nutritional stress. In contrast, most western corn rootworm females collected flying into a soybean field were not capable of ovipositing immediately, and only 20% carried resources sufficient to eventually lay eggs [181]. Combined with the poor quality of soybean tissues as food for western corn rootworm, a simple appetitive argument for this type of interfield movement seems problematic. Perhaps the initial flight out of corn is a manifestation of "typical" station-keeping behavior in a more mobile western corn rootworm population, resulting in inadvertent arrestment in a soybean field where opportunistic herbivory leads to nutritional stress. Alternatively, increased flight activity of rotation-resistant beetles [171] may also represent non-appetitive behavior, where non-directional short-distance displacement is itself the goal. Such behavior would not be conventionally migratory because a migration syndrome promoting long-distance flight is not involved, but it also would not be conventional appetitive flight behavior. Though speculative, a similar type of short-distance non-appetitive flight behavior has been hypothesized in reproductive (non-migratory) generations of the monarch butterfly [182].

5. Long-Distance, Migratory Movement

5.1. Laboratory Tethered Flight Behavior

Coats et al. [61] conducted tethered-flight experiments with mated female western corn rootworm of different ages on rotary flight mills and observed a clear differentiation between beetles engaging in "sustained" flights of 42–230 min duration and "trivial" flights of 1–17 min. Only 15% (28) of the 183 tested females engaged in sustained flight during the 24-h test period, and these averaged 4.5 sustained flights, each of about 72 min on average. That percentage includes all ages tested (2–15 d post-eclosion), even though females engaged in no sustained flights after 9 d of age. When broken down, 21% of females aged 2–9 d made a sustained flight, and 31% of those tested at the age of peak sustained flight activity (5–6 d) engaged in such flights (Table 1). All beetles engaged in numerous trivial flights across all ages tested, averaging about 3.9 min each. Based on their results, Coats et al. [61] argued that sustained flights represent migratory, non-appetitive flights by migratory beetles, and conversely that trivial flights represent appetitive flights. There are several lines of evidence that can point to an individual insect's flight behavior being migratory, and Coats et al. [61] provided supporting observations that make a convincing case in this regard for western corn rootworm. Much additional evidence for western corn rootworm as a migratory species has accumulated in the subsequent 35+ years. Using Coats et al. [61] as a jumping-off point, those data merit a detailed summary.

Table 1. Mated female western corn rootworm of indicated age ranges undertaking a sustained flight of ≥20 min in laboratory tethered-flight studies.

Study	Age (d)	No. Tested	No. Making Sustained Flight	% Making Sustained Flight
Coats et al. [61]	2–15	183	28	15
	2–9 [a]	135	28	21
	2–7 [a]	95	23	24
	5–6 [a]	48	15	31
	10–15 [a]	48	0	0
Coats et al. [b] [151]	2–11	97	28	29
	3–7 [c]	54	28	52
	5–6 [c]	28	15	54
	9–11 [c]	27	0	0
Naranjo [110]	2–7	75 [d]	18	24
	10–17	34	5 [d]	15
	23–30	100 [d]	4	4
Naranjo [e] [183]	5–10	77	19	25
	20–25	61	4	7
Wilson [184]	5–6	204	65 [d]	32
Stebbing et al. [185]	3–20	54	15	28
Yu et al. [f] [100]	6	159	36	23

[a] Subset of data included in the 2–15-d age group; [b] untreated mated female controls; [c] subset of data included in the 2–11-d age group; [d] inferred from reported data; [e] two sets of untreated controls for each age group, summed; [f] from data in Supplemental Material of Yu et al. [100].

Coats et al. [61] designated a fight duration threshold of 30 min to delimit trivial and migratory flights of western corn rootworm, based on reported flight duration distributions in tethered-flight experiments with the large milkweed bug, *Oncopeltus fasciatus* (Hemiptera: Lygaeidae) [186], and the convergent lady beetle, *Hippodamia convergens* (Coleoptera: Coccinellidae) [187]. However, a flight duration threshold to distinguish migratory flights in the laboratory depends on the species and can be difficult to resolve. It is best determined based on observations of flight duration distributions in the test population, along with associations of categories of flight duration with behavioral, age, physiological, and developmental characters. There was a clear, broad break in flight durations (no flights of 18–41 min) among the female rootworms tested by Coats et al. [61], making the 30-min threshold applicable to their dataset. Coats et al. [151] used the same threshold to distinguish short and long flights in a follow-up study demonstrating the role of juvenile hormone in regulating sustained flight activity. Naranjo [110,183] measured flight duration of western corn rootworm on actographs, a device that allows a tethered beetle to fly upward in a vertical plane and return to the platform to end a flight by landing. Although flight distance cannot be measured on an actograph, flight duration and timing of flights can. He also found a distinct bimodal distribution of flight durations by mated females, with no flights occurring between 18 and 29 min [110]. He therefore adjusted the flight duration threshold distinguishing migratory flight to 20 min, which also was consistent with data reported by Coats et al. [61].

Stebbing et al. [185] used the same actographs to compare flight performance of methyl parathion-resistant and normal susceptible western corn rootworms. For the susceptible beetles, the authors observed a continuous distribution of flight durations instead of a distribution with a distinctive gap between trivial and sustained flights. They retained the 20-min threshold for sustained flight used by Naranjo [110,183] for purposes of group comparisons. The beetles were tested in two broad age categories of 3–10 d and 11–20 d, which did not differ significantly in duration of sustained flights, and the percentage of mated females making a sustained flight was reported only for both age groups combined (Table 1). Interestingly, although no females >9 d old made a sustained flight in the studies by Coats et al. [61,151], some did in the 11–20-d-old category of Stebbing et al. [185], as well as in the 10–17-d and 23–30-d age groups tested by Naranjo [110], and the 20–25-d age group in the untreated controls of Naranjo [183] (Table 1). Yu et al. [100] also did not observe a distinct bimodal distribution in trivial versus sustained flight durations when they compared flight behavior of mated, 6-d-old western corn rootworm females reared at different larval densities. Frequency of longest-flight duration had reached a low level by 10 min, so that was declared the duration threshold of sustained flight for treatment comparison purposes. However, data in the supplemental material indicate 23% of females engaged in a sustained flight ≥ 20 min. The lack of an obvious break in duration distributions in the studies by Stebbing et al. [185] and Yu et al. [100] suggests there may not be a strict duration threshold defining migratory flight under all conditions. Nevertheless, in these six studies and an unpublished dissertation [184], the percentage of mated females that engaged in sustained flights of at least 20 min were fairly similar and highest among those in age bins under 10 d, ranging from about 21% to 54% (Table 1). Also, the frequency distribution of flight duration in a test population was always positively skewed and leptokurtic, with many short bouts of flight, tapering off rapidly to a fat tail of progressively rarer, longer flights. In a laboratory study, Li et al. [188] found that 5-d-old adults had greater propensity to takeoff and a shorter time to takeoff than 30-d-old adults in free flight after release at the base of a vertical stick. While this is consistent with age differences in sustained tethered flight (Table 1), the data presumably reflect an unknown mixture of trivial and sustained flights.

5.2. Synchrony of Immature Ovaries and Sustained Flight by Females

Evidence that the sustained flights by tethered western corn rootworm were migratory and not simply long appetitive flights is that most of the females making sustained flights had immature ovaries [110,151]. Migratory behavior of many insects occurs during the pre-oviposition period, sometimes concomitant with delayed egg maturation, a phenomenon called the "oogenesis-flight syndrome" [94]. Although many migratory species do not display this syndrome [91,189–193], migratory behavior is strongly suggested when long flights by individuals of a species are characteristically restricted to early in the pre-oviposition period. Female western corn rootworm mate soon after adult emergence, many while still teneral [11,124]. Though sexually mature at emergence in terms of mating competence [122], females emerge with undeveloped, pre-vitellogenic oocytes [152] and have a long pre-oviposition period ranging from 6–21 d [141,150,152,194,195]. Age of peak flight activity in tethered insects is a useful indicator of the migratory window [196]. Western corn rootworm flight mill data (Table 1) support an age of peak sustained flight activity of 3–7 d after emergence, with sustained flight uncommon or rare before and after these ages [151]. Thus, the age of peak migratory activity in this species corresponds to very early stages of ovarian development [61,110,151,152,197] when no or only a few oocytes have left the germarium and entered the vitellarium, and are mostly previtellogenic. Frequency of sustained flight behavior in the lab began to decline as females entered the stage when many oocytes were present in the vitellarium and vitellogenesis had begun, and it ended as yolk deposition accelerated [110,151].

5.3. Ascent into Atmosphere for Transport by Wind

The flight speed of migratory insects is usually not great enough to account for observed distances covered in only a few days [66,198–200]. Distances of 1000–2000 km traveled by seasonal migrants are not uncommon [201]. For example, migratory flight speed of black cutworm (*Agrotis ipsilon*, Noctuidae), a strong-flying seasonal migrant, averaged about 3.8 km/h (1.06 m/s) on flight mills [202], but a marked male released in Texas took only 2–4 nights to fly 1266 km to Iowa, where it was recaptured in a trap [203]. To accomplish such long displacements in a short amount of time, most migratory insects take advantage of tailwinds [66,80,204]. Wind speeds generally increase with increasing altitude to a maximum at about 200–400 m above ground level (a.g.l.) as strength of the interaction of the atmosphere with the Earth's surface decreases [205,206]. Migratory flight can occur at low altitudes, especially if the migrant needs to maintain control of its flight direction; diurnal migrants are more likely to employ this strategy [66,200]. However, most migratory insects ascend to high altitudes, up to about 2 km, to facilitate fast downwind transport in winds of higher speed than those available near the ground [80,199,206–210]. Thus, although insect flight near ground level may or may not be migratory, ascent into the atmosphere above the flight boundary layer is a strong indicator of migratory behavior [80].

The reproductive maturity distinctions between migratory and non-migratory western corn rootworms in the study by Coats et al. [61] hold true among field-collected adults flying at different elevations above corn and soybean fields. Western corn rootworm flight activity was monitored [68,167] along with high-elevation flight and ascent around corn and soybean fields in east-central Illinois [58,148,149,211]. In addition to confirming a diel periodicity to canopy-level interfield flight and high-elevation (10 m) flight, analyses of adults collected at 1 m, 3 m, and 10 m a.g.l. reveal a significant increase in the proportion of females among fliers at increasing elevation (0.424, 0.725, and 0.878, respectively) [211]. There were also trends among high-flying (at 10 m) females ascending from field crops and those flying within or just above (at 1 m) the corn or soybean canopies [148]. Females flying at 10 m weighed the least (11.7×10^{-3} g), had the lowest percentage carrying any mature eggs (0.6%), and were significantly more likely to contain a spermatophore (84%) [148]. Compared to other females, the characteristics of those flying at 10 m elevation suggest they were mated within the previous 5–7 d. The females flying at 10 m were also likely of local origin as analysis of their gut contents detected the presence of Bt proteins (or other plant tissues) available in upwind source fields [58,211]. Furthermore, the flux (females/min) of beetles active at 10 m elevation rose and fell with variation in the conditions for flight as measured in the canopy of the cornfields surrounding the 10-m platforms used for collection of flying beetles [58]. Thus, most western corn rootworm beetles captured while flying at 10 m had likely just initiated flight from nearby fields and were ascending when they were captured. Their characteristics are similar to those of the migratory individuals identified by Coats et al. [61].

5.4. Range Expansion

The historical range expansion of western corn rootworm from its established distribution in the central Great Plains to the East Coast of the United States and southern Canada is well documented (reviewed by [1,4]). Expansion occurred via stratified dispersal [1,4], a process characterized by two scales of spread into previously uncolonized territory [212,213] (see also [214]). Slower spread of the main invasion front can be described as neighborhood diffusion [212], the result of short-range movement driven by station-keeping and local ranging behaviors of individuals, some of whom find themselves in virgin territory. Concurrently, outlying populations are founded by colonizers originating from somewhere at or behind the invasion front. The founder populations have the effect of accelerating the overall range expansion as they spread backward to coalesce with the main front while also spreading forward from their advanced position via diffusion and production of their own long-distance emigrants. Such founder populations were frequently observed during the eastward North American range expansion of western corn rootworm (e.g., [215–218] and reviewed in detail by Meinke et al. [4]).

Similarly, the western corn rootworm range expansion in Europe, which began after the first detection of an introduced population in 1992 near the Belgrade airport in present-day Serbia [219], has been well documented [16,31,220–225]. However, the dynamics of the rootworm's spread in Europe have a different, more complex flavor than that in North America for several reasons. Analyses of genetic markers indicate there were at least five independent introductions from the U.S. into Europe, which resulted in disjunct areas of infestation [226,227]. Range expansion in Europe has been characterized by spread in multiple directions from multiple disconnected infestations, compared to the more or less unidirectional spread along a broad front that occurred in North America. Numerous disconnected infestations, characteristic of stratified dispersal, have been detected at various, often long distances from the main fronts of diffusive spread in Europe [116,222,228,229].

Distances from the main front of expansion to disconnected outbreaks ahead of the front offer clues about dispersal distances traveled by colonizing western corn rootworm adults. The average rates of expansion, ranging from roughly 20 to 200 km/y in North America [4,215] and 40–100 km/y in Europe [113,224], are logically related to dispersal distances of rootworm adults per generation. The association is not entirely direct, however, making the interpretation of expansion data less than straightforward due to three primary complications.

First, it almost surely takes more than a single mated female to successfully seed a self-sustaining population in a new location [213,230–232]. Losses of genetic variation were observed in colonizing populations of western corn rootworm in Europe compared to the likely parent populations [222,224,227,233,234], as is expected from genetic bottlenecks associated with founder events [235–238]. However, the losses were not of a magnitude suggesting only one or a very few founder females. Regardless, founder events often involve a relatively small number of individuals [213,232,239], and it may take more than one generation for a new population to increase enough to be detectable [240], especially in the absence of systematic monitoring (e.g., rootworm monitoring in the UK pre-2003 [241]). During the time lag between founding and detection, nearby expansion fronts draw closer, so that the minimum distance between a potential parent and founder population at the time of the colonization event is underestimated. Furthermore, if a critical number of immigrants are required to successfully found a new population after settling in close proximity to one another, it seems likely that other individuals disperse beyond that distance but in numbers too few to establish a new population. This consideration is another reason why range expansion data may underestimate the distances western corn rootworms can disperse. Such long-distance movement in small numbers could nevertheless be important for gene flow if the long-distance dispersal event terminates within the already-established larger distribution of the species [89].

Second, the nearest possible source population may not be the actual source. Identification of the most probable source population can be attempted via comparison of genotype profiles between disjunct and potential source populations across an array of selectively neutral genetic markers using population assignment analyses [242–247]. This strategy has been employed to determine probable source populations for disjunct western corn rootworm infestations in Europe [222,226,227,234]. Importantly, the closest population is not always the most probable source population based on genetic profiles. For example, the geographically closest possible source population of the Friuli outbreak of western corn rootworm in northeastern Italy was the northwestern Italy infestation, and the nearest possible source of the Frickingen outbreak in Germany was an established population in the Alsace region of France. However, genetic population assignment tests identified the large but more distant Central and Southeastern European (CSE) infestation as the most likely source for both the Friuli and Frickingen disjunct populations [222].

Other types of natural markers can provide evidence for source populations of immigrant insects. For example, genetic and pollen fingerprint analyses were combined with backtrack wind trajectory analyses to identify the most likely source population of boll weevils (*Anthonomus grandis grandis*, Curculionidae) that reinvaded an eradication zone in Texas in large numbers after the passage of a tropical storm [245]. Comparisons of exotic pollen and morphometrics of immigrant black cutworm (*Agrotis ipsilon*, Noctuidae) adults captured near Maryville, Missouri, in the central U.S. in Spring 1985 to those from potential source populations established the Brownsville, Texas area, 1600 km to the south, as the likely source of migrants [248,249]. Similarly, morphometrics of adult western corn rootworm are affected by soil type and other environmental factors and thus differ depending on geographic area, or even field of origin [250–254]. Population genetics analyses could not detect differences between populations within Croatia [233,252,255], but environmentally influenced hindwing morphometrics distinguished four populations within a small 600 km^2 region in the southeast of the country [252]. Consequently, hindwing shape can potentially serve as a natural biomarker in studies of adult movement and determination of likely source areas of immigrants, including at geographic scales too small for genetic markers to be of much assistance [224,253]. Microbiome communities in western corn rootworm vary by location of sampling across both large and small geographic scales and demonstrate a decay of similarity with geographic distance along transects of populations separated by 12–50 km in northeastern Colorado [256]. The complex interactions between host insects, environment, and means of microbial dispersal outside the host are only beginning to be examined for this genus [257,258]. However, intragenerational comparisons of communities from potential source and receiving populations could perhaps be used as natural markers to identify immigrants, analogous to how species composition of pollen grains attached to an insect's surface can provide information on its geographic origin [245,259].

Third, it is not always clear whether the founders of a disjunct western corn rootworm population arrived via natural dispersal by flight or by human-mediated transport [222]. After all, at least five different introductions to Europe occurred over the Atlantic Ocean, presumably onboard transcontinental aircraft [226,227]. It is generally assumed that the eastward range expansion in North America was wholly through natural dispersal by flight, an assumption supported by the very long invasion front stretching through areas of the country often far from major air transportation hubs. On the other hand, the U.S. is crisscrossed by major highways, sustaining heavy commercial truck and recreational vehicle traffic, so the role of human-mediated transport in North America cannot be entirely discounted. In Europe, topographical features like the Alps argue against solely long-distance dispersal of rootworms by natural flight from the most logical and genetically supported source population in the CSE infestation region to the disconnected outbreaks of Friuli in northeastern Italy and Frickingen in southern Germany. Carrasco et al. [260] developed a model to help distinguish whether a disconnected outbreak of an invasive insect from an expanding population was most likely from natural flight or from human-mediated transport. They examined western corn rootworm expansion data from Austria to test their model and concluded that many of the disjunct populations ahead of the main front resulted from human transport to the west along the Danube River basin. However, this conclusion is based on assumptions about the ease and frequency of rootworm hitchhiking on boats, trains, and trucks, which so far lack empirical support and may or may not be warranted or more likely than long-distance dispersal by natural flight. Difficulty distinguishing between transport mechanisms (natural flight or human-mediated) in specific cases of founding distant disjunct populations risks underestimating or overestimating western corn rootworm dispersal capacity.

5.5. Geographic Spread of Resistance

The rate of spread of insect resistance to a control tactic from a focal population to a new area of previous susceptibility can also provide insight into dispersal distances of the insect. The principle is similar to that of a species range expansion but with the rate of resistance expansion through the existing metapopulation being the measure of the dispersal rate of individuals carrying the resistance allele(s) [40]. The same issues that can complicate inference of flight distances from the rate of the species range expansion also apply to rate of resistance expansion but with additional complicating factors. The most important is that resistance is a phenotype of an individual [46] controlled by one or more genes, each with its own set of alleles, and the resistance phenotype is not a selectively neutral marker.

A well-studied example of using resistance as a marker of western corn rootworm dispersal capacity is the spread of crop rotation resistance. Rotation resistance is manifested as a loss of strict fidelity to cornfields for oviposition. Normally, crop rotation from corn to soybean is a very effective way to control western corn rootworm because the larvae hatching from overwintered eggs deposited in a cornfield cannot survive on soybean roots the following year. In areas with a high frequency of corn–soybean rotation, laying eggs in a soybean field by rotation-resistant females dramatically improves the chance that offspring will hatch in corn the following year. Rotation resistance was first reported in six fields in a 3-km^2 area of Ford County in eastern Illinois in 1987 [13], from which it spread over the next 15–20 years through most of Illinois and into several neighboring states [170]. Maps of the spread [1,111,112,148,170,261] show a pattern of stratified dispersal with several disjunct pockets of rotation resistance in counties ahead of the main expansion front. Early in the spread of rotation resistance, the annual rate of expansion averaged 10–30 km/generation, with slower rates toward the west and south against prevailing winds than to the east [111,261]. Expansion slowed after 1997 as the front entered regions with higher landscape diversity and possibly higher rates of continuous corn planting [4,112,170]. After the introduction of rootworm-targeting Bt corn into the landscape beginning in 2003, the spread of rotation resistance fully stalled and even contracted [4,262]. Miller and Sappington [40] suggested this could be, in part, a result of widespread planting of rootworm Bt corn even after rotation from soybean, largely negating the selective advantage of rotation resistance. Wetter springtime weather patterns, which significantly reduce populations, have also been hypothesized to contribute to reduced western corn rootworm abundance in soybean fields across areas of Illinois [48]. Regardless, this history of changing rates of rotation resistance spread emphasizes the caution that must be exercised in using spread of resistance, or any other adaptive trait under selection, as a way to estimate dispersal capacity of the insect. Even so, maximum observed rates of resistance spread probably provide a decent estimate in most cases of minimum dispersal capacity.

The spread of cyclodiene insecticide resistance is also instructive. Western corn rootworm resistance to aldrin and heptachlor, used as soil insecticides at planting, was first noticed in 1959 in a small area of south-central Nebraska [263]. By 1963, resistance had spread from that focal area to the eastern front of the species range expansion in western Iowa, southwest Minnesota, and northwest Missouri. It was also spreading in other directions through the existing metapopulation, reaching the western limits of its established distribution in Colorado by 1964 [1,264]. After reaching the eastern boundary of the species distribution in 1963, the subsequent eastward spread of resistance and the species range expansion coincided. At the same time, the annual rate of species range expansion increased greatly compared to preceding decades.

To account for this striking coincidence of resistance development with the accelerated species range expansion, Metcalf [264] suggested that cyclodiene resistance likely resulted, in part, in a behavioral change that caused greater dispersal. While this possibility cannot be discounted, there are several reasons to suspect the increased rate of expansion was due to the phenomenon of stratified dispersal [1,4]. Evolution and spread of cyclodiene resistance occurring at about the same time as the acceleration of the species range expansion is not as surprising a coincidence as it may at first seem, and it does not necessarily signify direct cause and effect. The species' range expansion out of the western Great Plains was triggered in large part by revolutionary agronomic changes in corn production after World War II, including the use of soil insecticides which played a major role. In other words, the high selection pressure for cyclodiene resistance was intertwined with the same conditions promoting rootworm population growth and eastward range expansion into the Corn Belt: These included increased corn acreage and continuous corn production as a viable option for farmers on the Great Plains, all made possible by efficient irrigation, modern synthetic fertilizer, and use of soil insecticide (cyclodienes) [4]. Until demonstration of increased flight propensity or performance of cyclodiene-resistant rootworms compared to susceptible beetles, e.g., using tethered-flight experiments [185], the most parsimonious explanation for the increased rate of western corn rootworm range expansion starting in the early 1960s is stratified dispersal ahead of the broadening invasion front. Regardless, the rate of spread of cyclodiene resistance within the contemporary range of the species reflects dispersal distances of rootworm adults carrying those resistance alleles [1,264]. Based on distribution maps in Metcalf [264] and Hamilton [265], aldrin resistance spread outward from its origin in the Grand Island and Kearny area of Nebraska in 1959 reaching

Brookings, South Dakota, to the northeast by 1963 at a rate of about 80 km/y, and reaching Fort Collins, Colorado, to the west by 1964 at a rate of about 95 km/y.

5.6. Spatial Distribution of Resistance

Distributions of adaptive traits in the landscape, including resistance to conventional insecticides and Bt corn, are often quite heterogeneous [169], providing another way to gauge western corn rootworm dispersal patterns and distances. For example, many growers in the western Corn Belt switched to aerial applications of organophosphate insecticides like methyl parathion to control adult western corn rootworm soon after resistance rendered cyclodiene soil insecticides ineffective against the larvae [18,54,56,169]. Surveys of adult resistance to methyl parathion using vial bioassays across several counties in south-central Nebraska from 1995 to 2002 [56,169,266,267] show a patchwork of percentage resistance between and within counties. Most populations in 1996 along a roughly 100-km corridor of susceptibility between highly resistant populations in Phelps Co. and populations of building resistance in York Co. to the northeast had become highly resistant by 1998 [169]. When resistance appears to spread from concentrated sources like this, it suggests geographic spread of resistance alleles by dispersing adults. However, resistance to methyl parathion or other control measures does not behave like a selectively neutral marker; thus, local selection, while high in many areas, was low or non-existent in others. Reinders et al. [34] listed some factors influencing the level of Bt resistance of rootworms in and between fields at the local landscape scale. With modification, these principles apply to spatial variation in resistance to conventional insecticides and rotation resistance as well, including how long the control measure has been used, population density in the area, rates of gene flow with distance (a function of adult dispersal), and frequency of alternative control tactics including efforts to mitigate resistance. If resistance alleles are already present in a susceptible population, the rate at which the phenotype is manifested depends in part on local selection pressure, rate of introduction of resistance and susceptible alleles carried by immigrants, and population density [34].

As with the spread of rotation resistance, it is problematic to directly translate rates of spread of methyl parathion resistance phenotype into estimates of rootworm dispersal distances. Nevertheless, some inferences about western corn rootworm adult dispersal can be gleaned from such data. The presence of a susceptible population in the near vicinity of one or more resistant populations indicates that the immigration rate of the resistant insects was not sufficient to overcome susceptibility. For example, 8 of 11 populations surveyed in 1996 in Phelps Co., Nebraska, were highly resistant to methyl parathion, 3 of them located only 5–7 km to the west, south, and east of a susceptible population [169]. Populations in Gosper and Buffalo Counties susceptible to methyl parathion in 1995 [56] had become resistant by 2002, but this was probably facilitated by selection from this insecticide used commonly in those counties [267]. However, a formerly susceptible population in central Clay Co. (as of 1998 [169]) was also resistant by 2002 [267] despite lack of selection pressure from methyl parathion, indicating that the increase in resistance must have been spread by resistant immigrants. If a nearby population was the source of resistance alleles, the closest known resistant population in time and space (1999, northwest Clay Co. [266]) was ~30 km away, yielding a dispersal estimate of about 8 km/y. Because many more than one resistant beetle would be needed to shift the susceptible population to resistant, 8 km/y can only be considered a minimum estimate of dispersal distance by the resistant adults. Tethered-flight experiments [185] indicated flight activity of methyl parathion-resistant adults did not differ from that of susceptible adults, although exposure to the insecticide reduced flight activity of resistant beetles. Duration of sustained flights (i.e., >20 min) among unexposed resistant rootworm adults averaged 52.7 min. Given average speeds for sustained flights of western corn rootworm adults measured on rotary flight mills of 50 m/min (0.83 m/s) [61] and 28.8 m/min (0.48 m/s) (archived data in [100] https://figshare.com/s/8f7b424145e5ac9fb8e9?file=14362775, accessed on 26 November 2023), methyl parathion-resistant adults could be predicted to fly 1.5–2.6 km per sustained flight. Expected total distances covered by individuals would be greater if aided by wind, or if more than one sustained flight is taken by a migratory adult along the same trajectory.

Another interesting example involves development of resistance to carbaryl, the active ingredient in an insecticidal bait used in a pilot areawide management program from 1997 to 2001 [268,269]. In this pilot program, fields in management areas of ~41 km^2 located in four Corn Belt states were treated with a cucurbitacin bait mixed with the insecticide carbaryl to control adults [270]. Cucurbitacin is a semiochemical arrestant and feeding stimulant for western corn rootworm adults [271–273]. Pruess et al. [53] had previously shown that areawide adult control with malathion across a block of this size could successfully protect treated fields from economic injury by larvae the following year, but that cumulative suppression over years was not achieved because of recolonization by adult immigrants the year after treatment. Similar results were observed in the areawide cucurbitacin bait study [18,274]. In three of the four states, resistance to carbaryl increased significantly within the managed block during the period of the study compared to nearby untreated control fields [269]. There was also a trend of decreasing susceptibility in the nearby control fields, suggesting possible spread of resistance from the management block, but the change was subtle and not statistically significant [268,269]. In addition, there was a significant decrease in behavioral response (as arrestant, feeding stimulant, or both) to cucurbitacin in the managed areas compared to the control fields [269]. Together, these results imply that immigration of surrounding susceptible western corn rootworm adults into an area of 41 km^2 was not enough to overcome localized selection for resistance, and that emigration out of the managed area was not great enough to significantly increase resistance in nearby susceptible populations. At the same time, immigration into the managed area was enough to prevent cumulative population suppression.

Rootworm Bt corn was rapidly embraced by growers throughout most corn-growing regions east of the Rocky Mountains after its commercial release in 2003. It provided excellent control of rootworm larvae, allowing farmers to plant continuous corn while forgoing the use of chemical insecticides. Unfortunately, the western corn rootworm lived up to its reputation for quickly surmounting control tactics through evolution of resistance, in this case to the various Cry3 toxins expressed in the Bt-corn hybrids [37,38, 43,275–280]. Resistance to the different Cry3 toxins (Cry3Bb1, mCry3A, and eCry3.1Ab) is now widespread, thanks in part to extensive cross-resistance among the structurally similar toxins [37,38,276,277,281,282], but was initially patchy [43]. Reports of field-evolved resistance to Cry34Ab1/35Ab1 (hereafter "Cry34/35"; note this toxin was recently renamed Gpp34Ab1/Tpp35Ab1 [283]) Bt corn [275,278] emerged after an increase in its adoption in response to Cry3 resistance [284]. Like early resistance to Cry3 hybrids, the geographic distribution of Cry34/35 resistance is patchy [275,280].

For both Cry3 and Cry34/35 toxins, the initial appearance of resistance in widely separated locations, as well as cases of co-occurrence of resistant populations with nearby susceptible populations (e.g., [34,38,43]), suggest multiple instances of independent evolution of resistance in local selection "hotspots" [37,38,40,43,46]. The existence of resistance hotspots is made possible by the large proportion of resident beetles comprising an adult western corn rootworm population [19]. Adaptive alleles can build up over time in response to strong selection on a (mostly) sedentary population of residents that oviposit most or all of their eggs in the natal field or surrounding cluster of fields comprising the natal home range. As described earlier, however, not all residents remain in the natal home range. Depending on environmental conditions, some residents may range via appetitive flights in search of more attractive habitat, depositing a portion of their eggs in fields in the local landscape but outside the natal home range (Figure 1). The migrant portion of the population migrates after mating but before ovipositing, ultimately leaving their genes in fields potentially many kilometers away. The spatial distribution of resistance to Bt-corn in the landscape around a hotspot reflects the outward rate of resistance spread, or expansion through the surrounding metapopulation, and thus may be informative about adult dispersal rates and distances.

Gassmann et al. [37] concluded that despite the relative proximity of fields in northeastern Iowa that developed resistance to Cry3 Bt-corn hybrids by 2009 and 2010, field histories of at least 3 consecutive years of Bt-corn planting (the minimum necessary for resistance to evolve in a population [34,38,41,43,285,286]) were consistent with independent evolution of resistance. However, in one of these fields, the resistant rootworm population received only one year of Cry3 selection, indicating that the field must have been colonized by Cry3-resistant adults. Resistant populations were present in the same and adjacent counties, but the distance from which the resistant colonizers originated is not known.

St. Clair et al. [36] compared field histories, larval damage, and population sizes in "problem" counties in northeastern Iowa, where greater-than-expected injury (i.e., >1 on the Oleson et al. [287] node injury scale) to Bt cornfields had been reported previously, to "non-problem" counties in southeastern and central Iowa where there had been no such reports. Planting of continuous corn was common in the surveyed fields of all counties and years (2015–2017) examined. Surprisingly, larval damage, abundance, and resistance to Cry3Bb1 Bt-corn did not differ between these counties. The authors concluded this was because mitigation tactics quickly adopted in the problem counties, such as soil insecticide applications and switching to Cry34/35 Bt hybrids, successfully suppressed the Cry3Bb1-resistant rootworm populations. The temporal and spatial patterns of the widespread Cry3Bb1 Bt-corn resistance suggested independent selection for resistance dependent on differing selection pressures in the different regions of the state (e.g., less intensive maize production in central and southwestern Iowa than in the northeast) rather than spread of resistance from northeastern Iowa where field failures were first observed. In a companion study, St. Clair et al. [288] made similar comparisons at the local landscape scale, between focal fields with at least one year in the previous six of greater-than-expected damage to Cry3Bb1 or mCry3A Bt-maize and neighboring fields (within 2.2 km). Again there was no difference in injury or rootworm abundance, and populations from all fields were resistant to Cry3Bb1. The focal fields, which had suffered greater-than-expected damage to Cry3Bb1 maize in the past, had a field history of more years of planting Cry3 maize than surrounding fields. While it is clear that resistance alleles were spread from the focal fields to neighboring fields in the local landscape, the authors pointed out that field-level selection was also occurring in the surrounding fields. They suggested that immigration of resistant beetles from a focal field augments the effect of selection in receiving fields and thus helps homogenize the level of resistance in the landscape.

In a fine-grained geographic study of 17 fields in a 10 × 20 km area of Keith County, Nebraska, and 16 fields in an 11 × 13 km area of Buffalo County, Reinders et al. [34] documented a mosaic of Cry3Bb1 and mCry3A resistance and susceptibility in the landscape. They found multiple instances of fields harboring Cry3 Bt-resistant western corn rootworm populations within 1–3 km of fields with susceptible populations. The authors classified each field with a cumulative index score of selection pressure from Cry3 Bt corn derived from field history (e.g., number of years per trait, single or pyramided traits, number of resistant or susceptible populations within 1 mile, etc., with the index reset to 0 after crop rotation to soybean). By comparing selection index scores with observed levels of Bt resistance, inferences about gene flow of both susceptible and resistance alleles were possible in some cases. As in the Iowa case described by Gassmann et al. [37], Reinders et al. [34] documented substantial Bt resistance among larvae from second-year corn after rotating out of soybeans, indicating they must have been offspring of Bt-resistant immigrants arriving from elsewhere during the previous year. This field (Field 4) was less than 1 km from the border of the sampling area in Buffalo County, and the authors suspected the immigrants came from fields of continuous Cry3Bb1 Bt corn located "immediately adjacent" to Field 4 but outside the surveyed area, and thus they were not bioassayed for resistance.

5.7. Gene Flow

Because of its intimate relationship with dispersal, estimates of gene flow can, in principle, be used to infer distances and patterns of dispersal using a variety of population genetics approaches [246,247,289–291]. The most common methods use variation in allele frequency across panels of selectively neutral genetic markers to quantify genetic differentiation between sampled populations [292–295]. F_{ST} is a measure of genetic differentiation between populations or subpopulations, ranging from 0 (no difference in allele frequency) to 1 (fixed differences in alleles) [296,297]. Because genetic differentiation of two populations is inversely related to the rate of gene flow between them, a rough estimate of migrant exchange per generation, Nm, can be calculated directly from F_{ST} values using the equation $Nm = [(1/F_{ST}) - 1]/4$ [298]. In areas where individuals are distributed continuously across the landscape, one expects to see genetic differentiation increase with geographic distance, creating an isolation-by-distance (IBD) relationship [297,299,300]. IBD is assessed via regression of genetic distance, $F_{ST}/(1 - F_{ST})$, on geographic distance [297,301,302]. It reflects the balance between gene flow (determined by effective dispersal distances) and genetic drift (a function of effective population size).

While theoretically sound and intuitively appealing, use of F_{ST} to infer gene flow (and thus dispersal) relies on several critical assumptions: (1) The genetic markers are selectively neutral; (2) mutation rate of the markers is relatively low and constant; and (3) the sampled populations are at gene flow–genetic drift equilibrium [303–305]. Violation of any of these can affect F_{ST}, in which case geographic patterns in degree of differentiation cannot be ascribed unambiguously to gene flow [291,306].

A number of researchers have employed population genetics strategies to help address uncertainties related to the movement ecology of adult western corn rootworm and have reported pairwise F_{ST} values (Table 2). For example, Kim and Sappington [307] estimated genetic differentiation between 10 populations from New York to northwestern Texas using microsatellite markers. None of the pairwise F_{ST} values were significant except those involving Texas samples, and two involving the Illinois sample versus the Pennsylvania and Delaware samples. The IBD regression was significant but only when the Texas population was included, a region of probable hybridization with the partially reproductively isolated Mexican corn rootworm (*Diabrotica virgifera zeae*) [308,309]. Flagel et al. [310] used a large number of transcriptome-derived SNP markers to examine differentiation among 20 wild populations of western corn rootworm from northeastern Colorado to eastern Indiana. Pairwise F_{ST} values between populations were calculated for all (>10,000) unigenes identified from the transcriptome. The mean F_{ST} between all populations was 0.052 and indicated little differentiation even at great distances, up to 1500 km. IBD was detected across the sampled area but only when the easternmost population sampled, Indiana, was included. The reason for this outsized influence of the Indiana population is unclear [310].

Likewise, other studies consistently report low and mostly nonsignificant genetic differentiation among western corn rootworm populations through the vast U.S. Corn Belt (Table 2) (see [18] for a map of the Corn Belt). Although such findings can theoretically reflect dispersal and gene flow over very long distances, the various authors recognized the more likely explanation is that gene flow–genetic drift equilibrium has not yet been attained in the wake of the recent eastward range expansion [1]. This interpretation is supported by studies in which populations south of the Corn Belt, such as Mexico, Arizona, New Mexico, and Texas, were included in comparisons (e.g., [227,311,312]). The high and significant pairwise F_{ST} values reported in comparisons among themselves and with northern populations (Table 2) reflect the much longer period of western corn rootworm residency in southwestern North America than in the Corn Belt [312].

Table 2. Summary of pairwise F_{ST} estimates between western corn rootworm populations reported in the literature.

Region	No. Sites	Geographic Scale (km)		Genetic Markers	Signif. IBD?	Pairwise F_{ST} Range [a]	% Signif. F_{ST} Values	Notes	Citation
		Min	Max						
USA: NY to northwest TX	10 (9 states)	9	1400	Microsats	Yes (No when TX excluded)	0–0.002	24 (6 when TX excluded)	Texas sample from a probable hybrid zone with *D. v. zeae* (Mexican corn rootworm).	Kim and Sappington [307] (See also Kim et al. [244] and Coates et al. [313] for similar results using EST-derived microsat and SNP markers, respectively.),
USA: central IA east-central IL IA vs. IL	3 3 2	3 0.5 500	32 1 500	Microsats and AFLPs	—	IA: 0.007 IL: 0 IA vs. IL: 0.002_{AFLP} to $0.005_{Microsat}$	IA: 0 IL: 0 IA vs. IL: 100	Sampled pupae instead of adults. IL populations were rotation-resistant; IA populations were wild-type.	Miller et al. [314]
USA: PA, IA, eastern KS, western KS	4	254	1941	Microsats	—	0–0.004	0	Several lab colonies were also compared but not included here.	Kim et al. [315]
North America: PA, IL, TX, AZ, Mexico	5	763	3124	Microsats	—	0.009–0.118 0–0.023	100	For European comparisons, only populations within a single outbreak area are included here.	Ciosi et al. [227]
Europe: northwestern Italy	3	21	40				67		
Europe: CSE (total) East transect West transect	18 11 9	14 24 26	438 433 438	Microsats	Yes Yes No	0–0.039 0–0.017 0–0.039	13 2 13	Two locations were shared between the East and West transects.	Ciosi et al. [222]
Europe: northwestern Italy	3	73	151	Microsats	—	0–0.01	33	Only comparisons between populations within outbreak areas, and not admixed, are included here.	Bermond et al. [228]
CSE (Hungary, Serbia)	2	211	211			0	0		

Table 2. *Cont.*

Region	No. Sites	Geographic Scale (km)		Genetic Markers	Signif. IBD?	Pairwise F_{ST} Range [a]	% Signif. F_{ST} Values	Notes	Citation
		Min	Max						
USA: eastern NE, AZ	5	9	1400	Microsats	Yes (No when AZ excluded)	0.002–0.082	80	2 NE populations were insecticide-resistant (R), and 2 (+AZ) were susceptible (S). All comparisons were significant, except within R or S categories in NE.	Chen et al. [311]
Europe: Croatia 1996	11	7	182	Microsats	No	0–0.083	15	In 2009, 10 of 11 sites were within 80 km of each other in eastern Croatia, and all were ≥245 km from Ogulin at the western edge of the invasion front.	Lemic et al. [233] (See also Lemic et al. [234,252] for similar results including samples from 2011.)
2009	11	7	393		Yes (No when Ogulin excluded)	0–0.024	4		
USA: Corn Belt (CO to IN)	20 (8 states)	0.5	1528	SNPs (transcript-omic)	Yes (No when IN excluded)	0.029–0.054	—	Six laboratory colonies of differing years in culture and originating from various locations were also compared, but are not included here.	Flagel et al. [310]
USA: NE, IA, IL	7	110	800	Microsats	—	0–0.005	0	The 3 populations from IL were rotation-resistant.	Ivkosic et al. [255]
Europe: Croatia, Serbia	8	15	350			0–0.011	0		
North America: Mexico to Northeast USA	21	92	3456	Microsats	Yes	0–0.160	65 (31 from KS to north and east)	Neighbor-joining tree indicated tight clustering of sites north and east of KS.	Lombaert et al. [312]

IBD, isolation by distance. Microsats, microsatellites; AFLPs, amplified fragment length polymorphisms; SNPs, single-nucleotide polymorphisms. USA state abbreviations: AZ, Arizona; CO, Colorado; IA, Iowa; IL, Illinois; IN, Indiana; KS, Kansas; NE, Nebraska; NY, New York; PA, Pennsylvania; TX, Texas. CSE, Central and Southeastern European outbreak area. [a] Negative F_{ST} estimates in original papers are reported here as '0'.

In Europe, high and significant pairwise F_{ST} values are typically reported between geographically distinct regions of infestation. These disjunct populations either originated directly from independent introductions from North America or via secondary emigration from invasive bridgehead populations [316]; the observed genetic differentiation undoubtedly reflects founder effects associated with recent colonizations [222,226–228,234]. In contrast, differentiation within larger areas of infestation that originated from a single founder population is invariably low and mostly non-significant (Table 2) [222,226–228,233,234,255].

The lesson is that F_{ST}-based inference of western corn rootworm dispersal distances is unreliable for populations in areas of recent invasion, such as the U.S. Corn Belt and Europe because the assumption of gene flow–genetic drift equilibrium is violated. However, it should be possible to use patterns of genetic differentiation to infer dispersal distances in regions of long-term endemicity, such as in the western Great Plains. One such region is northeastern Colorado, which served as the source of the eastward range expansion. Based on molecular genetic analyses, Colorado was colonized in ~1838 CE by western corn rootworm originating from populations further south in New Mexico and Texas [312]. Pairwise F_{ST} values between northeastern Colorado and all locations further east in the Corn Belt indicated significant differentiation (Table S3 in Lombaert et al. [312]). Preliminary analyses among populations sampled along transects within northeastern Colorado indicate significant IBD as well (T.W.S. and J.L.S., unpublished data), as would be expected in an area at gene flow–genetic drift equilibrium.

Even in areas of invasion, however, other types of genetic analyses can be exploited to provide estimates of dispersal. For example, Bermond et al. [62] estimated dispersal distances of western corn rootworm by exploiting distinct genetic profiles at microsatellite marker loci of independently introduced and expanding CSE and NW populations in Europe as they came in contact in northeastern Italy. The principle behind their estimates is that transient clines of selectively neutral genetic markers will form through gene flow in the zone of contact (equivalent to a hybrid zone) when the expansion fronts of differentiated populations meet. The decay rate of the cline as genotypes homogenize via interbreeding, which is measured by change in the width of the contact zone, depends on the diffusion rate of dispersal [317]. In addition, Bermond et al. [62] calculated dispersal distance from the width of the linkage disequilibrium cline generated by gene flow between the two expanding populations, which is maximum in the center of the contact zone [318]. Estimates from both methods were in broad agreement, suggesting western corn rootworm dispersal of 13–21 km/generation. This is not only strong evidence for relatively long-distance movement by western corn rootworm adults, but it also presents a clear example of *Slatkin's paradox*, a dilemma in which estimates of gene flow suggest movement over much greater distances than the much shorter lifetime dispersal distances indicated by direct ecological studies [27,28,304]. As pointed out by Bermond et al. [62], the cline analysis estimate of per-generation dispersal is the same as the diffusion rate but only if long-distance dispersal is rare. It is likely, given ecological evidence of less-common, but not rare, long-distance movement by western corn rootworm, that the estimate of 13–21 km/generation from clinal decay analysis reflects both short-distance diffusion via station-keeping and ranging flights as well as long-distance leaps via migratory flight.

6. Conclusions and Synthesis

As summarized in this review, a wide array of experimental and observational data strongly supports short-distance lifetime displacement by western corn rootworm adults. This deduction starkly contrasts with evidence strongly supporting long-distance lifetime displacement from an equally imposing array of data from indirect ecological, behavioral, and population genetics studies. The geographic scale of estimated net lifetime displacement of western corn rootworm ranges from a few tens of meters to hundreds of kilometers (Figure 2). This lack of congruence in estimated dispersal distances across methodologies is a classic example of both Reid's and Slatkin's paradoxes. As profound and confounding as these contradictory estimates are, they are not illusory or generated by methodological shortcomings. The weight of evidence is clear that both short- and long-distance flights occur in this species, generating a population-level pattern of bimodal dispersal distances. However, the outstanding question concerns the mechanisms generating this pattern.

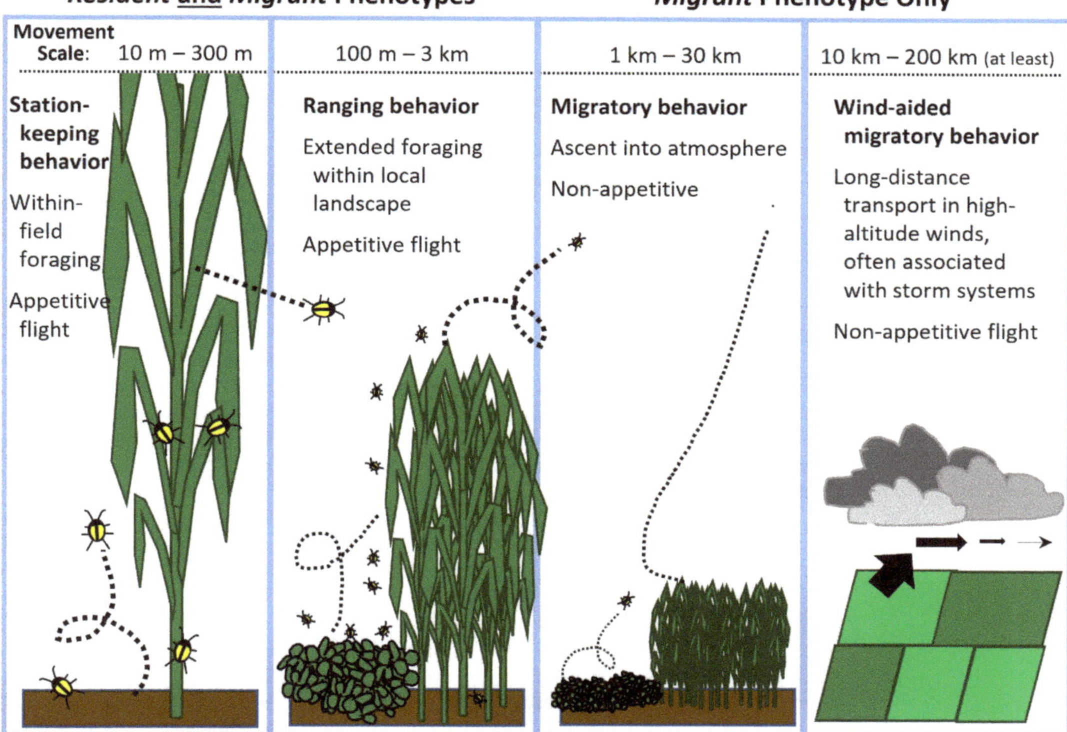

Figure 2. Approximate scales of net displacement during a bout of flight activity by western corn rootworm and assignment of behaviors to both *resident* and *migrant* phenotypes, or exclusively to the *migrant* phenotype. Movement behaviors [64] differ fundamentally as either appetitive or non-appetitive. Appetitive behaviors (station-keeping and ranging behaviors) are motivated by search for a needed resource. Non-appetitive behavior is migratory, motivated by the goal of spatial displacement itself. Migrant western corn rootworms begin migratory flight by purposely ascending high into the atmosphere where fast winds increase speed and distance of displacement.

In other species, both paradoxes are most commonly resolved by positing rare, unobserved, or underestimated long-distance dispersal events by a portion of the population [24,27]. In other words, it is assumed that essentially "invisible" long-distance movement (invisible because it is rare and difficult to observe) comprises the long tail of a leptokurtic frequency distribution of dispersal distances in a population [22,27,28,319]. As is true of frequency distributions for measures of dispersal of most organisms, including insects [196,214,320–322], the frequency distributions of flight duration of western corn rootworm observed in tethered-flight experiments are always positively skewed and leptokurtic, with many short bouts of flight tapering off to a long tail of progressively less-frequent longer flights [61,100,110,151,183,185]. In principle, the long leptokurtic tail could simply represent facultative, long-distance ranging flights by individuals in search of better habitat. If this were the case, however, one would expect a "thin"-tailed frequency distribution of distances typical of a normal or Gaussian distribution, where long-distance flights quickly become very rare [321,322]. This is the way stratified dispersal has been conventionally modeled [323]. Instead, the evidence for western corn rootworm indicates that very long flights >10 km, while certainly much less common than short-distance flights, are not exceptionally rare. The resulting pattern of a fat-tailed distribution of dispersal distances is evident in the tethered-flight experiments and is inferred from population genetics studies [62] and observations of disjunct founder populations far ahead of invasion fronts [4,222].

In this paper, we have reviewed the different lines of evidence accumulated over many decades for both short- and long-distance displacement of western corn rootworm, including what is known about this insect's flight behavior, such as temporal and developmental timing and the underlying motivations for bouts of movement. Based on these many studies and their findings, we conclude that two behavioral phenotypes, residents and migrants, exist in any western corn rootworm population. Individuals of both phenotypes engage overwhelmingly in appetitive flights throughout their life, but only one phenotype also engages in non-appetitive, migratory flight. We propose a conceptual model of western corn rootworm as a partially migratory species, where the resident portion of the population never engages in migratory behavior, while the migrant portion does (Figures 1 and 2). Thus, the observed "fat" leptokurtic tail of flight distance is shaped by a mixture of two dispersal kernels [22,25,322], one representing the frequency distribution of short-distance appetitive flights of both phenotypes and the other the distribution of long-distance non-appetitive flights by the migrant portion of the population. The presence of two behavioral phenotypes, manifesting as two dispersal kernels emerging from the same western corn rootworm populations, is the solution to both Reid's and Slatkin's paradoxes in this species.

Whether long-distance movement by western corn rootworm is via appetitive extended-ranging flight or non-appetitive migratory flight is not an exercise in semantics. It affects how experimental data on movement are interpreted and applied to predict population dynamics, efficacy of management strategies, and evolution of local adaptations like resistance. For example, emigrants from a field are often equated with the ~15% of tethered females across all ages tested by Coats et al. [61] that engaged in migratory flight. Consequently, this percentage is commonly used in models or interpretations of field data as the expected rate of emigration from a natal field. However, not all emigrants from the natal field are migrants. Additionally, not all immigrants entering a field are migrants. Individuals that leave a field and enter a different field may be residents engaging in simple appetitive station-keeping flights or ranging flights (Figure 1). The flight mill studies of western corn rootworm by Coats et al. [61] and Naranjo [110] only distinguished migrants from non-migrants. But "non-migrant" is a descriptive category, not a phenotype. A group of non-migrant rootworm adults observed on a particular day are a mixture of residents (which never migrate) and migrants that did not happen to engage in migratory flight on the day of observation. Residents emigrate from or immigrate to fields only via appetitive flight, even if they have traveled far enough to exit their previous home range. The point is that individuals of the migrant phenotype constitute only a subset of the emigrants from and immigrants to a given field, such as the natal field. The total number of adults emigrating from a field will be greater than the number of migrants emigrating because a collection of emigrants will include individuals of both migrant and resident phenotypes. Similarly, the number of all immigrants contributing to a field's total population will be greater than the number of migrants that entered and now comprise part of that field's total population. Thus, when a % value of female migrants observed in flight mill studies, like the 15% value of Coats et al. [61], is used to parameterize emigration rates from the natal field, it will underestimate actual emigration. This is because the 15% value only includes migrants and does not account for emigration via appetitive flight, like ranging behavior or even station-keeping behavior that spills over boundaries of adjacent fields. In addition, the percentage of migrant females in a population is probably closer to 25–30% (potentially up to 50%), based on tethered-flight data when restricted to females under

10 d old (Table 1), the ages encompassing the developmental window for pre-oviposition migratory behavior [58,148,149,211].

Immigrant western corn rootworm adults will include a subset of erstwhile migrants that henceforth behave like residents in their new home range (Figure 1). Any future emigration events the immigrants engage in probably will not involve migratory behavior but rather appetitive flights, which could include ranging over relatively long distances. An exception could be multi-leg migration flights taken in short succession by an individual still in a stage of early ovarian development, but whether an adult engages in more than one migratory flight on different days is unknown. Females in tethered-flight experiments that made a migratory flight, sometimes made more than one in the same 24-h test period [61,183]. However, direct application of laboratory-derived flight parameters may overlook real-world constraints on flight, a well-known issue with all tethered-flight studies [196,324–326]. For example, in the study by Coats et al. [61], some females engaged in multiple overnight flights. This is unlikely under field conditions. Isard et al. [58] showed that among individuals in field populations, ascending flight stops at or just before sunset. Thus, a long-distance migrant that stops flying and alights after dark will not be able to initiate another flight until the return of daylight and the presence of permissive atmospheric conditions [58,167].

7. Unanswered Questions and Future Directions
7.1. Rate of Emigration from the Natal Field

Characterizing the relative contributions of station-keeping, ranging, and migratory types of flight behavior to emigration and immigration rates of western corn rootworm will require different approaches to obtain the data needed to parameterize adult movement functions in models. Although the 15% value commonly used for lifetime emigration rate is almost certainly a substantial underestimation as explained above, determining a more realistic rate may not be simple. The true maximum rate of emigration must be considerably less than 100% because we know that enough females lay eggs in their natal field to cause the observed population increases over consecutive generations in continuously planted cornfields. At the same time, the true rate of emigration must be substantially greater than nil, because those of the migrant phenotype will always emigrate (unless prevented over an extended period by prohibitive abiotic conditions), and because we know that first-year cornfields are rapidly colonized by immigrants, presumably of both phenotypes.

A focus on measuring rate of emigration directly from an isolated field could be an effective strategy. However, determining the rate or ceiling rate of emigration may be difficult. One possibility would be to compare the observed rate of population increase in a focal field from one generation to another to the rate of predicted population increase in the absence of emigration and immigration, to calculate the net deficit of adults in the field at a given time. This is a question of population dynamics that lends itself to modeling. Contributors to the deficit will include adult mortality and the difference between immigration and emigration. The challenge lies in disentangling these three factors [327,328]. This approach is similar to that taken by Hein et al. [327] to quantify the relationship between population density of western corn rootworm adults and oviposition in cornfields. They obtained direct estimates of adult emergence with emergence cages and estimated adult population densities over the season via whole-plant visual counts [329]. Early in the period of adult emergence, there was good correspondence between rate of emergence and number of adults in the natal field. However, within three weeks after initial emergence, only a third of the expected beetles were present in the field, indicating either high mortality, high emigration, or some combination of the two. They recognized immigration could also have affected population density but assumed for their model that it was less than the rate of emigration; this was a reasonable assumption because the fields they studied were not late-planted and should not have been more attractive to would-be immigrants than other fields in the landscape.

7.2. Mechanisms Determining Resident and Migrant Phenotypes

In most partially migratory insect species [67], the underlying mechanisms determining the resident and migrant phenotypes are poorly understood. In general, there is a strong polygenic basis for migratory syndromes and component traits in insects [64]. It can also be presumed that even though the resident and migrant phenotypes in western corn rootworm are distinct dichotomous traits, it is likely that the final phenotype is not determined by a single major gene but by a threshold response [91,330–332] to a continuous distribution of many genes of small effect. This kind of situation manifests as a "developmental switch" mechanism: a migrant phenotype develops when enough genes with alleles associated with migration are present to exceed a threshold, while a resident phenotype develops when the threshold is not exceeded [67,333,334].

At the same time, the ultimate phenotype of partially migratory insects often depends on environmental interactions with the various genes associated with the behavioral phenotypes [67,91,333]. Environmental cues, often experienced by an immature stage during a sensitive period, influence the decision to develop as one or another phenotype [335]. The lag period between the environment-sensitive period in the immature stage and production of an adult migrant may be needed to allow time for development of the associated morphological and physiological components of the species' migratory syndrome. For example, in the partially migratory oriental armyworm (*Mythimna separata*; Noctuidae) and beet webworm (*Loxostege sticticalis*; Pyralidae), the decision to develop as a migrant or resident adult occurs in the larval stage. Environmental cues received as larvae, including poor nutrition, short photoperiod, cold stress, and larval crowding, lead to development of the migrant adult phenotype. Environmental conditions encountered during the first day of adulthood can shift an erstwhile migrant to a resident, but an adult resident cannot shift to become a migrant [81,336–339].

The western corn rootworm has not been studied as extensively in this regard, but there are indications of underlying genetic variation in flight behavior. For example, Li et al. [251] examined the effects of fields and regional populations of origin, and the number of generations of laboratory rearing on flight activity of offspring at 1–6 d of age under identical rearing and assay conditions. Significant differences in the means and variability of propensity to take off and seconds to takeoff from a vertical stick were common. Larvae reared at high density are more apt to fly and fly farther on flight mills than those reared under less crowded conditions [100], suggesting crowding may be an environmental cue affecting adult phenotype. Similarly, in the paired-field study by Levay et al. [52] described above, the percentage of immigration (and hence emigration under the assumptions of the experimental design) increased with increasing estimates of adult population density.

It is likely that most or all developmental control mechanisms underlying partial migration in insects include hormonal signaling [335,340], and juvenile hormone (JH) is commonly involved (e.g., [64,196,337,341–343]). Importantly, Coats et al. [151] demonstrated that JH applied to the adult western corn rootworm cuticle increased the likelihood of migratory flight on flight mills. This finding does not necessarily indicate the adult stage is the environment-sensitive period in rootworms for phenotype determination because environmental cues received by larvae can trigger effects on JH concentrations, timing of JH secretion or sensitive period, threshold sensitivity, or cellular responses much later in development [335]. There is great scope for future research focusing on the genetic and environmental factors and their interactions that determine the developmental trajectory of individual western corn rootworms into resident or migrant phenotypes.

7.3. Differential Migration of Males and Females

Evidence from atmospheric ascent and tethered-flight data suggests the percentage of males that migrate is much lower than that of females. The proportion of females among western corn rootworm captured 4.6–10.0 m a.g.l. consistently ranges between 72% and 89% [58,149,211,344]. Of the beetles washed up on the shores of Lake Michigan over a three-year period, an average of 89% were female. Although tethered flight studies have tended to focus on females, available data indicate only 1–11% of tethered males made a sustained flight [110,183–185], in sharp contrast to the 21–54% observed for females (Table 1). Although these same data support migratory flight by some males, the substantial differences suggest sex-dependent differences in costs and benefits of migration. Male genes migrate whether males do or not because most migrant females mate before departing. Thus, any advantage to the male of physically migrating may be reduced and its risks amplified. By migrating, a male could potentially improve its mating chances, but perhaps little more than by ranging in the local landscape. Conversely, leaving a landscape where receptive females are currently plentiful to alight where the presence of unmated females is not assured poses a higher level of risk than faced by a mated, migrant female that only needs to find a cornfield in which to feed and oviposit. This type of risk is elevated during range expansion, when a migrating male that leaps ahead of the invasion front will encounter a new landscape essentially devoid of young calling females. If, as seems likely, a quantitative genetic developmental switch is involved [67,91] in determining resident or migrant phenotypes, the observed female bias among migrating adults implies a higher "migrant" threshold response in males than in females. Regardless, observed sex-dependent differences in propensity to migrate potentially complicate parameterization of the movement components of IRM and population models, where male dispersal affects spatial encounters and mating with newly emerged females (e.g., [105]).

7.4. Migration Process

7.4.1. Ascent Phase

The three phases of migration (ascent, transmigration, and termination) are inherently difficult to observe and characterize for any migratory species [66,206,345], and much is unknown about the behavior of individual western corn rootworm during migratory flight. There are two diurnal peaks of western corn rootworm ascent to at least 10 m a.g.l. in east-central Illinois, one in the morning between about 06:45 and 11:00, and the other in the evening from 17:00 to 20:30, ending just before sunset [58]. Flight mill data generally support morning and evening peaks of sustained flight activity [61,100]. In the study by Coats et al. [61], sustained flight of females was most common from 16:00 to 22:00, with a smaller peak from 05:00 to 06:00, and a few instances between 23:00 and 03:00. However, the entire 24-h test period of this experiment was conducted under constant dim lighting to simulate twilight, making translation of timing in the lab to that in the field questionable. Flight mill trials in the study by Yu et al. [100] were conducted under a 14:10 L:D photoperiod with simulated dawn and dusk. The longest sustained flights ($\geq$10 min) were initiated by 75% of individual females between 07:00 and 15:00 (median 08:30). However, 40% of the longest flights were initiated at first-light at 07:00, suggesting a startle response, again making translation to field timing suspect. Regardless, none of the longest sustained flights started before the beginning of dawn, and all ended before sundown.

Dissections of female western corn rootworm captured at 10 m a.g.l. while ascending from corn revealed 99% were mated, 84% with spermatophores, and half of the spermatophores were large, indicating very recent mating [148]. It seems unlikely that this tight tie between mating and initiation of migration is coincidental. One possibility is that the spermatophore helps supplement energy requirements during migratory flight. Another is that mating stimulates sustained flight in the migrant phenotype.

If the potential for migratory flight is enabled by mating, perhaps flight mill trials using females from interrupted matings, like those studied by Sherwood and Levine [141], could yield insight into the role of mating (or mating quality) on initiation of migratory flight. Sherwood and Levine [141] used interrupted western corn rootworm mating treatments to show that a copulation duration (1 h) insufficient for sperm to be transferred into the spermatheca resulted in a lower percentage of egg-laying females and few eggs laid. Their study strongly suggests that the mechanical stimulation associated with copulation or fecundity-enhancing substances (FESs) produced by male accessory glands is responsible for eliciting the mated response in females. FESs are transferred to the female during mating, presumably as spermatophore components. Alternatively, in the many species in which FESs are implicated in the transition to mated-behavior patterns (i.e., cessation of mating receptivity, initiation of egg maturation and oviposition), including some beetles, hemocoelic injection of male reproductive tract homogenates alone is sufficient to change female behavior [346–351]. Hemocoel injection of homogenized WCR male accessory glands (which are the source of spermatophore components) or whole spermatophores themselves (recovered from freshly mated and incompletely mated females) into unmated females destined for flight mills would present another route to examine the male contribution on subsequent expression of migratory behavior—free from the potential influence of mechanical stimulation.

Though the western corn rootworm is clearly diurnal as an adult, the extent to which they may also fly at night is unclear. Collections on sticky traps in cornfields at heights up to 3.6 m showed morning and afternoon peaks of flight activity for both sexes, but some beetles were captured during the 8-h period of darkness [126,168]. Tóth et al. [352] found that male western corn rootworm responded to pheromone lures throughout the night. Laboratory tethered-flight studies under light–dark cycles show substantial short-duration flight activity after dark, although daytime activity is greater [100,110,185]. Isard et al. [58] interpreted cessation of western corn rootworm captures from 10-m towers at the approach of darkness as cessation of flight, but more properly, it meant ascent into the atmosphere during migration stopped. It is possible, even likely, that individuals ascending earlier in the evening maintained their flight at higher altitudes after dark during the transmigration phase, as is the case with a number of other migratory insect species [80,345]. In an interview, W.B. Showers described capturing western corn rootworm beetles on a television tower in light traps, which were not visible from the ground, suggesting migrants were flying at night [353]. Vörös et al. [354] reported captures of this species in light traps 2 km from the nearest corn.

7.4.2. Transmigration Phase

Behavior of western corn rootworm during the transmigration phase is largely unknown. Many migrating insects ascend to a layer of maximum winds just below the temperature inversion during stable atmospheric conditions, which is often around 200–400 m a.g.l., but cruising altitudes can be much higher. Laboratory studies of western corn rootworm flight lack manipulation of atmospheric and wind conditions or any understanding of how a migrating beetle behaves once it enters such fast-moving parcels of air. The use of entomological radar to observe ascent and transmigration of this species may help fill many of our knowledge gaps, as it has for numerous other migratory insects [80,198,355–357]. To account for wash-ups of western corn rootworm beetles on Lake Michigan shores, Isard et al. [60] postulated that migrating beetles encountered the daytime lake breeze at 2 km in elevation before being mixed downward over the water into onshore winds. Grant and Seevers [59] cite unpublished data that beetles can be "found readily at heights of at least 35 m" near cornfields. Spencer et al. [148] cited a newspaper report of western corn rootworm beetles accumulating on tall buildings in Chicago at about 130 m a.g.l. Adults were captured in light traps on the television tower in central Iowa mentioned earlier at 76, 152, and 275 m a.g.l. [353]. Systematic sampling of this insect at intervals above 10 m would help clarify behavior during the transmigration phase, such as flight altitude in relation to atmospheric and wind conditions. It would also help elucidate the demographic makeup of the migrating population. For example, although a relatively small proportion of ascending beetles captured at 10 m are males, the flight durations of males making sustained flights on flight mills are less than those of females. We predict that the proportionate number of males declines with altitude. Sampling at high altitudes can be attempted via balloon netting [199,210,345] or netting using remote-controlled aircraft (e.g., [358,359]). Employment of entomological radar in conjunction with 10 m tower netting would be valuable in determining altitude of transmigratory flight by ascending western corn rootworm.

7.4.3. Termination Phase

Little is known about the termination phase of western corn rootworm migratory flight. The lack of observational data for descent of migrating individuals leaves a troublesome hole in our knowledge. A cornfield out of which adults are captured at 10 m while ascending into the atmosphere represents a concentrated departure point, but spatial dilution of migrating beetle density inevitably occurs during the transmigration phase and during the termination phase itself as individuals descend after different durations of flight. On flight mills, sustained flights exhibit a fat-tailed distribution over a wide range of duration and distance [61,100,110,151,185], and vagaries in wind profiles at different flight altitudes and times will generate variation in descent locations as well. Because migratory flight is non-appetitive, differences in suitability of habitat will not serve to concentrate immigrants to particular cornfields in a landscape until after termination of migration and resumption of appetitive ranging flight. Whether this transition occurs before, during, or after descent is unknown, and undoubtedly affects the ultimate terminus of an immigrant.

7.4.4. Number of Migratory Flights

Many migratory insects engage in more than one bout of migratory flight. As discussed earlier, it is unlikely that western corn rootworm migrants engage in more than one migratory flight in a single day because of the need for permissive atmospheric conditions which change as part of a diurnal cycle of insolation [58,68,167]. Migratory flights may occur on more than one day, a hypothesis deserving investigation. Such multi-leg or "stopover" migratory flights are common in long-distance, seasonal insect migrants between overwintering and reproductive ranges (e.g., [182,203,360]). However, it seems unlikely to occur more than once per lifetime in western corn rootworm. This is an aseasonal, partial migratory species where the evolutionary impetus behind migration presumably is displacement itself rather than a destination, and displacement can be achieved in a single bout of non-appetitive flight.

7.5. Additional Knowledge Gaps

There are a number of other major knowledge gaps that can and should be addressed experimentally. The genetic and physiological mechanisms underlying development of a western corn rootworm larva into a resident or migrant adult behavioral phenotype are unknown. Crowding increases average distances flown by young mated females on flight mills [100], but is this a reflection of migrant flight behavior, resident flight behavior, or both? Or does it reflect a facultative increase in the proportion of migrant phenotypes in the population? To what extent does the proportion of residents and migrants within a population and across a landscape differ between years and spatially within years? Does resistance to Bt Cry toxins or crop rotation affect flight behavior, including the proportion of residents and migrants in a population? Performance of resistant beetles in tethered flight compared to wild-type beetles could be valuable in understanding and predicting evolution, maintenance, and spread of resistance. What is the physiology of migrants during preparation for, during, and after migratory flight? Juvenile hormone (JH) is involved in expression of sustained flight in western corn rootworm [151], but little else is known regarding endocrine control of migration. Advances have been made in a few other insects in identifying the genetic underpinnings of insect flight and movement [92,325,361,362]. The most powerful approach combines tethered-flight assays with genomic and transcriptomic tools. While not trivial, tethered-flight experimentation with western corn rootworm is within reach of most laboratories [363]. The recent publication of this species' transcriptome [364] and genome [365] will help make genetic studies of the biological components of western corn rootworm movement ecology far more tractable than they have been up to now.

Author Contributions: Both authors contributed to conceptualization, and contributed to all drafts of the manuscript. All authors have read and agreed to the published version of the manuscript.

Funding: This research received no external funding.

Data Availability Statement: No new data were created or analyzed in this study. Data sharing is not applicable to this article.

Acknowledgments: The findings and conclusions in this publication are those of the authors and should not be construed to represent any official USDA or U.S. Government determination or policy. Any mention of trade names or commercial products in this publication is solely for the purpose of providing specific information and does not imply a recommendation or endorsement by USDA. USDA ARS is an equal opportunity provider and employer.

Conflicts of Interest: The authors declare no conflict of interest.

References

1. Gray, M.E.; Sappington, T.W.; Miller, N.J.; Moeser, J.; Bohn, M.O. Adaptation and invasiveness of western corn rootworm: Intensifying research on a worsening pest. *Annu. Rev. Entomol.* **2009**, *54*, 303–321. [CrossRef] [PubMed]
2. Wechsler, S.; Smith, D. Has resistance taken root in US corn fields? Demand for insect control. *Am. J. Agric. Econ.* **2018**, *100*, 1136–1150. [CrossRef]
3. Ball, H.J. On the biology and egg-laying habits of the western corn rootworm. *J. Econ. Entomol.* **1957**, *50*, 126–128. [CrossRef]
4. Meinke, L.J.; Sappington, T.W.; Onstad, D.W.; Guillemaud, T.; Miller, N.J.; Komáromi, J.; Levay, N.; Furlan, L.; Kiss, J.; Toth, F. Western corn rootworm (*Diabrotica virgifera virgifera* LeConte) population dynamics. *Agric. For. Entomol.* **2009**, *11*, 29–46. [CrossRef]
5. Clark, T.L.; Hibbard, B.E. Comparison of non-maize hosts to support western corn rootworm (Coleoptera: Chrysomelidae) larval biology. *Environ. Entomol.* **2004**, *33*, 681–689. [CrossRef]
6. Moeser, J.; Vidal, S. Do alternative host plants enhance the invasion of the maize pest *Diabrotica virgifera virgifera* (Coleoptera: Chrysomelidae, Galerucinae) in Europe? *Environ. Entomol.* **2004**, *33*, 1169–1177. [CrossRef]
7. Oyediran, I.O.; Hibbard, B.E.; Clark, T.L. Prairie grasses as hosts of the western corn rootworm (Coleoptera: Chrysomelidae). *Environ. Entomol.* **2004**, *33*, 1497–1504. [CrossRef]
8. Wilson, T.A.; Hibbard, B.E. Host suitability of nonmaize agroecosystem grasses for the western corn rootworm (Coleoptera: Chrysomelidae). *Environ. Entomol.* **2004**, *33*, 1102–1108. [CrossRef]
9. Campbell, L.A.; Meinke, L.J. Seasonality and adult habitat use by four *Diabrotica* species at prairie-corn interfaces. *Environ. Entomol.* **2006**, *35*, 922–936. [CrossRef]
10. Bernal, J.S.; Medina, R.F. Agriculture sows pests: How crop domestication, host shifts, and agricultural intensification can create insect pests from herbivores. *Curr. Opin. Insect Sci.* **2018**, *26*, 76–81. [CrossRef]
11. Spencer, J.L.; Hibbard, B.E.; Moeser, J.; Onstad, D.W. Behaviour and ecology of the western corn rootworm (*Diabrotica virgifera virgifera* LeConte) (Coleoptera: Chrysomelidae). *Agric. For. Entomol.* **2009**, *11*, 9–27. [CrossRef]
12. Gillette, C.P. *Diabrotica virgifera* LeC as a corn root-worm. *J. Econ. Entomol.* **1912**, *5*, 364–366. [CrossRef]
13. Levine, E.; Spencer, J.L.; Isard, S.A.; Onstad, D.W.; Gray, M.E. Adaptation of the western corn rootworm to crop rotation: Evolution of a new strain in response to a management practice. *Am. Entomol.* **2002**, *48*, 94–107. [CrossRef]
14. Ludwick, D.C.; Hibbard, B.E. Rootworm management: Status of GM traits, insecticides and potential new tools. *CAB Rev.* **2016**, *11*, 48. [CrossRef]
15. Carrière, Y.; Brown, Z.; Aglasan, S.; Dutilleul, P.; Carroll, M.; Head, G.; Tabashnik, B.E.; Jørgensen, P.S.; Carroll, S.F. Crop rotation mitigates impacts of corn rootworm resistance to transgenic Bt corn. *Proc. Natl. Acad. Sci. USA* **2020**, *117*, 18385–18392. [CrossRef]
16. Bažok, R.; Lemić, D.; Chiarini, F.; Furlan, L. Western corn rootworm (*Diabrotica virgifera virgifera* LeConte) in Europe: Current status and sustainable pest management. *Insects* **2021**, *12*, 195. [CrossRef]
17. Furlan, L.; Chiarini, F.; Contiero, B.; Benvegnù, I.; Horgan, F.G.; Kos, T.; Lemić, D.; Bažok, R. Risk assessment and area-wide crop rotation to keep western corn rootworm below damage thresholds and avoid insecticide use in European maize production. *Insects* **2022**, *13*, 415. [CrossRef]
18. Meinke, L.J.; Souza, D.; Siegfried, B.D. The use of insecticides to manage the western corn rootworm, *Diabrotica virgifera virgifera*, LeConte: History, field-evolved resistance, and associated mechanisms. *Insects* **2021**, *12*, 112. [CrossRef]
19. Gassmann, A.J. Resistance to Bt maize by western corn rootworm: Effects of pest biology, the pest-crop interaction and the agricultural landscape on resistance. *Insects* **2021**, *12*, 136. [CrossRef]
20. Spencer, J.L.; Isard, S.A.; Levine, E. Free flight of western corn rootworm (Coleoptera: Chrysomelidae) to corn and soybean plants in a walk-in wind tunnel. *J. Econ. Entomol.* **1999**, *92*, 146–155. [CrossRef]
21. Pierce, C.M.F.; Gray, M.E. Western corn rootworm, *Diabrotica virgifera virgifera* LeConte (Coleoptera: Chrysomelidae), oviposition: A variant's response to maize phenology. *Environ. Entomol.* **2006**, *35*, 423–434. [CrossRef]

22. Clark, J.S.; Fastie, C.; Hurtt, G.; Jackson, S.T.; Johnson, C.; King, G.A.; Lewis, M.; Lynch, J.; Pacala, S.; Prentice, C.; et al. Reid's paradox of rapid plant migration: Dispersal theory and interpretation of paleoecological records. *BioScience* **1998**, *48*, 13–24. [CrossRef]

23. Reid, C. *The Origin of the British Flora*; Dulau & Company: London, UK, 1899.

24. Harnik, P.G.; Maherali, H.; Miller, J.H.; Manos, P.S. Geographic range velocity and its association with phylogeny and life history traits in North American woody plants. *Ecol. Evol.* **2018**, *8*, 2632–2644. [CrossRef] [PubMed]

25. Caswell, H.; Lensink, R.; Neubert, M.G. Demography and dispersal: Life table response experiments for invasion speed. *Ecology* **2003**, *84*, 1968–1978. [CrossRef]

26. Marko, P.B. 'What's larvae got to do with it?' Disparate patterns of post-glacial population structure in two benthic marine gastropods with identical dispersal potential. *Mol. Ecol.* **2004**, *13*, 597–611. [CrossRef] [PubMed]

27. Jones, A. Reconciling field observations of dispersal with estimates of gene flow. *Mol. Ecol.* **2010**, *19*, 4379–4382. [CrossRef] [PubMed]

28. Yu, H.; Nason, J.D.; Ge, X.; Zeng, J. Slatkin's Paradox: When direct observation and realized gene flow disagree. A case study in *Ficus*. *Mol. Ecol.* **2010**, *19*, 4441–4453. [CrossRef] [PubMed]

29. Spencer, J.L.; Mabry, T.R.; Vaughn, T. Use of transgenic plants to measure insect herbivore movement. *J. Econ. Entomol.* **2003**, *96*, 1738–1749. [CrossRef]

30. Hill, R.E.; Mayo, Z.B. Distribution and abundance of corn rootworm species as influenced by topography and crop rotation in eastern Nebraska. *Environ. Entomol.* **1980**, *9*, 122–127. [CrossRef]

31. Boriani, M.; Agosti, M.; Kiss, J.; Edwards, C.R. Sustainable management of the western corn rootworm (*Diabrotica virgifera virgifera* LeConte, Coleoptera: Chrysomelidae) in infested areas: Italian, Hungarian and USA experiences. *OEPP/EPPO Bull.* **2006**, *36*, 531–537. [CrossRef]

32. Hughson, S.A.; Spencer, J.L. Emergence and abundance of western corn rootworm (Coleoptera: Chrysomelidae) in Bt cornfields with structured and seed blend refuges. *J. Econ. Entomol.* **2015**, *108*, 114–125. [CrossRef] [PubMed]

33. Pereira, A.E.; Souza, D.; Zukoff, S.N.; Meinke, L.J.; Siegfried, B.D. Crossresistance and synergism bioassays suggest multiple mechanisms of pyrethroid resistance in western corn rootworm populations. *PLoS ONE* **2017**, *12*, e0179311. [CrossRef] [PubMed]

34. Reinders, J.D.; Hitt, B.D.; Stroup, W.W.; French, B.W.; Meinke, L.J. Spatial variation in western corn rootworm (Coleoptera: Chrysomelidae) susceptibility to Cry3 toxins in Nebraska. *PLoS ONE* **2018**, *13*, e0208266. [CrossRef] [PubMed]

35. Souza, D.; Peterson, J.A.; Wright, R.J.; Meinke, L.J. Field efficacy of soil insecticides on pyrethroid-resistant western corn rootworms (*Diabrotica virgifera virgifera* LeConte). *Pest Manag. Sci.* **2020**, *76*, 827–833. [CrossRef] [PubMed]

36. St. Clair, C.R.; Head, G.P.; Gassmann, A.J. Comparing populations of western corn rootworm (Coleoptera: Chrysomelidae) in regions with and without a history of injury to Cry3 corn. *J. Econ. Entomol.* **2020**, *113*, 1839–1849. [CrossRef]

37. Gassmann, A.J.; Petzold-Maxwell, J.L.; Clifton, E.H.; Dunbar, M.W.; Hoffmann, A.M.; Ingber, D.A.; Keweshan, R.S. Field-evolved resistance by western corn rootworm to multiple *Bacillus thuringiensis* toxins in transgenic maize. *Proc. Natl. Acad. Sci. USA* **2014**, *111*, 5141–5146. [CrossRef]

38. Wangila, D.S.; Gassmann, A.J.; Petzold-Maxwell, J.L.; French, B.W.; Meinke, L.J. Susceptibility of Nebraska western corn rootworm (Coleoptera: Chrysomelidae) populations to Bt corn events. *J. Econ. Entomol.* **2015**, *108*, 742–751. [CrossRef]

39. Gassmann, A.J. Resistance to Bt maize by western corn rootworm: Insights from the laboratory and the field. *Curr. Opin. Insect Sci.* **2016**, *15*, 111–115. [CrossRef]

40. Miller, N.J.; Sappington, T.W. Role of dispersal in resistance evolution and spread. *Curr. Opin. Insect Sci.* **2017**, *21*, 68–74. [CrossRef]

41. Meihls, L.N.; Higdon, M.L.; Siegfried, B.D.; Miller, N.J.; Sappington, T.W.; Ellersieck, M.R.; Spencer, T.A.; Hibbard, B.E. Increased survival of western corn rootworm on transgenic corn within three generations of on-plant greenhouse selection. *Proc. Natl. Acad. Sci. USA* **2008**, *105*, 19177–19182. [CrossRef]

42. Oswald, K.J.; French, B.W.; Nielson, C.; Bagley, M. Selection for Cry3Bb1 resistance in a genetically diverse population of nondiapausing western corn rootworm (Coleoptera: Chrysomelidae). *J. Econ. Entomol.* **2011**, *104*, 1038–1044. [CrossRef] [PubMed]

43. Gassmann, A.J.; Petzold-Maxwell, J.L.; Keweshan, R.S.; Dunbar, M.W. Field-evolved resistance to Bt maize by western corn rootworm. *PLoS ONE* **2011**, *6*, e22629. [CrossRef] [PubMed]

44. Shrestha, R.B.; Dunbar, M.W.; French, B.W.; Gassmann, A.J. Effects of field history on resistance to Bt maize by western corn rootworm, *Diabrotica virgifera virgifera* LeConte (Coleoptera: Chrysomelidae). *PLoS ONE* **2018**, *13*, e0200156. [CrossRef] [PubMed]

45. Deitloff, J.; Dunbar, M.W.; Ingber, D.A.; Hibbard, B.E.; Gassmann, A.J. Effects of refuges on the evolution of resistance to transgenic corn by the western corn rootworm, *Diabrotica virgifera virgifera* LeConte. *Pest Manag. Sci.* **2016**, *72*, 190–198. [CrossRef] [PubMed]

46. Andow, D.A.; Pueppke, S.G.; Schaafsma, A.W.; Gassmann, A.J.; Sappington, T.W.; Meinke, L.J.; Mitchell, P.D.; Hurley, T.M.; Hellmich, R.L.; Porter, R.P. Early detection and mitigation of resistance to Bt maize by western corn rootworm (Coleoptera: Chrysomelidae). *J. Econ. Entomol.* **2016**, *109*, 1–12. [CrossRef] [PubMed]

47. Shrestha, R.B.; Gassmann, A.J. Field and laboratory studies of resistance to Bt corn by western corn rootworm (Coleoptera: Chrysomelidae). *J. Econ. Entomol.* **2019**, *112*, 2324–2334. [CrossRef] [PubMed]

48. Tinsley, N.A.; Spencer, J.L.; Estes, R.E.; Estes, K.A.; Kaluf, A.L.; Isard, S.A.; Levine, E.; Gray, M.E. Multi-year surveys reveal significant decline in western corn rootworm densities in Illinois soybean fields. *Am. Entomol.* **2018**, *64*, 112–119. [CrossRef]

49. Godfrey, L.D.; Turpin, F.T. Comparison of western corn rootworm (Coleoptera: Chrysomelidae) adult populations and economic thresholds in first-year and continuous cornfields. *J. Econ. Entomol.* **1983**, *76*, 1028–1032. [CrossRef]
50. Beckler, A.B.; French, B.W.; Chandler, L.D. Characterization of western corn rootworm (Coleoptera: Chrysomelidae) population dynamics in relation to landscape attributes. *Agric. For. Entomol.* **2004**, *6*, 129–139. [CrossRef]
51. Sivcev, I.; Stankovic, S.; Kostic, M.; Lakic, N.; Popovic, Z. Population density of *Diabrotica virgifera virgifera* LeConte beetles in Serbian first year and continuous maize fields. *J. Appl. Entomol.* **2009**, *133*, 430–437. [CrossRef]
52. Levay, N.; Terpo, I.; Kiss, J.; Toepfer, S. Quantifying inter-field movements of the western corn rootworm (*Diabrotica virgifera virgifera* LeConte)—A central European field study. *Cereal Res. Commun.* **2014**, *43*, 155–165. [CrossRef]
53. Pruess, K.P.; Witkowski, J.F.; Raun, E.S. Population suppression of western corn rootworm by adult control with ULV malathion. *J. Econ. Entomol.* **1974**, *67*, 651–655. [CrossRef]
54. Meinke, L.J. Adult corn rootworm management. In *University of Nebraska Agricultural Research Division Miscellaneous Publication 63-C*; The University of Nebraska-Lincoln: Lincoln, NE, USA, 1995.
55. van Rozen, K.; Ester, A. Chemical control of *Diabrotica virgifera virgifera* LeConte. *J. Appl. Entomol.* **2010**, *134*, 376–384. [CrossRef]
56. Meinke, L.J.; Siegfried, B.; Wright, R.J.; Chandler, L. Adult susceptibility of Nebraska western corn rootworm (Coleoptera: Chrysomelidae) populations to selected insecticides. *J. Econ. Entomol.* **1998**, *91*, 594–600. [CrossRef]
57. Szalai, M.; Kőszegi, J.; Toepfer, S.; Kiss, J. Colonisation of first-year maize fields by western corn rootworm (*Diabrotica virgifera virgifera* LeConte) from adjacent infested maize fields. *Acta Phytopathol. Entomol. Hung.* **2011**, *46*, 213–223. [CrossRef]
58. Isard, S.A.; Spencer, J.L.; Mabry, T.R.; Levine, E. Influence of atmospheric conditions on high-elevation flight of western corn rootworm (Coleoptera: Chrysomelidae). *Environ. Entomol.* **2004**, *33*, 650–656. [CrossRef]
59. Grant, R.H.; Seevers, K.P. Local and long-range movement of adult western corn rootworm (Coleoptera: Chrysomelidae) as evidenced by washup along southern Lake Michigan shores. *Environ. Entomol.* **1989**, *18*, 266–272. [CrossRef]
60. Isard, S.A.; Kristovich, D.A.R.; Gage, S.H.; Jones, C.J.; Laird, N.F. Atmospheric motion systems that influence the redistribution and accumulation of insects on the beaches of the Great Lakes in North America. *Aerobiologia* **2001**, *17*, 275–291. [CrossRef]
61. Coats, S.A.; Tollefson, J.J.; Mutchmor, J.A. Study of migratory flight in the western corn rootworm (Coleoptera: Chrysomelidae). *Environ. Entomol.* **1986**, *15*, 620–625. [CrossRef]
62. Bermond, G.; Blin, A.; Vercken, E.; Ravigné, V.; Rieux, A.; Mallez, S.; Morel-Journel, T.; Guillemaud, T. Estimation of the dispersal of a major pest of maize by cline analysis of a temporary contact zone between two invasive outbreaks. *Mol. Ecol.* **2013**, *22*, 5368–5381. [CrossRef]
63. Nathan, R.; Getz, W.M.; Revilla, E.; Holyoak, M.; Kadmon, R.; Saltz, D.; Smouse, P.E. A movement ecology paradigm for unifying organismal movement research. *Proc. Natl. Acad. Sci. USA* **2008**, *105*, 19052–19059. [CrossRef] [PubMed]
64. Dingle, H. *Migration: The Biology of Life on the Move*, 2nd ed.; Oxford University Press: New York, NY, USA, 2014; ISBN 978-0-19-964039-3.
65. Chapman, B.B.; Brönmark, C.; Nilsson, J.-Å.; Hansson, L.-A. The ecology and evolution of partial migration. *Oikos* **2011**, *120*, 1764–1775. [CrossRef]
66. Chapman, J.W.; Reynolds, D.R.; Wilson, K. Long-range seasonal migration in insects: Mechanisms, evolutionary drivers and ecological consequences. *Ecol. Lett.* **2015**, *18*, 287–302. [CrossRef] [PubMed]
67. Menz, M.H.M.; Reynolds, D.R.; Gao, B.; Chapman, J.W.; Wotton, K.R. Mechanisms and consequences of partial migration in insects. *Front. Ecol. Evol.* **2019**, *7*, 403. [CrossRef]
68. Isard, S.A.; Nasser, M.A.; Spencer, J.L.; Levine, E. The influence of weather on western corn rootworm flight activity at the borders of a soybean field in east central Illinois. *Aerobiologia* **1999**, *15*, 95–104. [CrossRef]
69. Dingle, H.; Drake, V.A. What is migration? *BioScience* **2007**, *57*, 113–121. [CrossRef]
70. Jander, R. Ecological aspects of spatial orientation. *Annu. Rev. Ecol. Syst.* **1975**, *6*, 171–188. [CrossRef]
71. Bell, W.J. *Searching Behavior: The Behavioral Ecology of Finding Resources*; Chapman & Hall: London, UK, 1991.
72. Hill, R.E.; Mayo, Z.B. Trap-corn to control rootworms. *J. Econ. Entomol.* **1974**, *67*, 748–750. [CrossRef]
73. Naranjo, S.E. Movement of corn rootworm beetles, *Diabrotica* spp. (Coleoptera: Chrysomelidae), at cornfield boundaries in relation to sex, reproductive status, and crop phenology. *Entomol. Exp. Appl.* **1991**, *55*, 79–90. [CrossRef]
74. Prystupa, B.; Ellis, C.R.; Teal, P.E.A. Attraction of adult *Diabrotica* (Coleoptera: Chrysomelidae) to corn silks and analysis of the host-finding response. *J. Chem. Ecol.* **1988**, *14*, 635–651. [CrossRef]
75. Kennedy, J.S. A turning point in the study of insect migration. *Nature* **1961**, *189*, 785–791. [CrossRef]
76. Kennedy, J.S. Migration, behavioural and ecological. In *Migration: Mechanisms and Adaptive Significance*; Rankin, M.A., Ed.; University of Texas Marine Science Institute: Port Aransas, TX, USA, 1985; Volume 27, pp. 5–26.
77. Drake, V.A.; Gatehouse, A.G.; Farrow, R.A. Insect migration: A holistic conceptual model. In *Insect Migration: Tracking Resources Through Space and Time*; Drake, V.A., Gatehouse, A.G., Eds.; Cambridge University Press: Cambridge, UK, 1995; pp. 427–457.
78. Compton, S.G. Sailing with the wind: Dispersal by small flying insects. In *Dispersal Ecology*; Bullock, J.M., Kenward, R.E., Hails, R.S., Eds.; Blackwell Publishing: Oxford, UK, 2002; pp. 113–133.
79. Westbrook, J.K.; Isard, S.A. Atmospheric scales of biotic dispersal. *Agic. For. Meteorol.* **1999**, *97*, 263–274. [CrossRef]
80. Reynolds, D.R.; Chapman, J.W.; Drake, V.A. Riders on the wind: The aeroecology of insect migrants. In *Aeroecology*; Chilson, P.B., Frick, W.F., Kelly, J.F., Liechti, F., Eds.; Springer International Publishing AG part of Springer Nature: Cham, Switzerland, 2017; pp. 145–178, ISBN 978-3-319-68574-8.

81. Jiang, X.F.; Luo, L.Z.; Zhang, L.; Sappington, T.W.; Hu, Y. Regulation of migration in the oriental armyworm, *Mythimna separata* (Walker) in China: A review integrating environmental, physiological, hormonal, genetic, and molecular factors. *Environ. Entomol.* **2011**, *40*, 516–533. [CrossRef] [PubMed]

82. Nagoshi, R.N.; Meagher, R.L.; Hay-Roe, M. Inferring the annual migration patterns of fall armyworm (Lepidoptera: Noctuidae) in the United States from mitochondrial haplotypes. *Ecol. Evol.* **2012**, *2*, 1458–1467. [CrossRef] [PubMed]

83. Nagoshi, R.N.; Rosas-García, N.M.; Meagher, R.L.; Fleischer, S.J.; Westbrook, J.K.; Sappington, T.W.; Hay-Roe, M.; Thomas, J.M.G.; Murúa, G.M. Haplotype profile comparisons between *Spodoptera frugiperda* (Lepidoptera: Noctuidae) populations from Mexico with those from Puerto Rico, South America, and the United States and their implications to migratory behavior. *J. Econ. Entomol.* **2015**, *108*, 135–144. [CrossRef] [PubMed]

84. Westbrook, J.K.; Nagoshi, R.N.; Meagher, R.L.; Fleischer, S.J.; Jairam, S. Modeling seasonal migration of fall armyworm moths. *Int. J. Biometeorol.* **2016**, *60*, 255–267. [CrossRef]

85. Wu, M.-F.; Qi, G.-J.; Chen, H.; Ma, J.; Liu, J.; Jiang, Y.-Y.; Lee, G.-S.; Otuka, A.; Hu, G. Overseas immigration of fall armyworm, *Spodoptera frugiperda* (Lepidoptera: Noctuidae), invading Korea and Japan in 2019. *Insect Sci.* **2022**, *29*, 505–520. [CrossRef]

86. Reppert, S.M.; de Roode, J.C. Demystifying monarch butterfly migration. *Curr. Biol.* **2018**, *28*, R1009–R1022. [CrossRef]

87. Taylor, O.R., Jr.; Lovett, J.P.; Gibo, D.L.; Weiser, E.L.; Thogmartin, W.E.; Semmens, D.J.; Diffendorfer, J.E.; Pleasants, J.M.; Pecoraro, S.D.; Grundel, R. Is the timing, pace, and success of the monarch migration associated with sun angle? *Front. Ecol. Evol.* **2019**, *7*, 442. [CrossRef]

88. Tauber, M.J.; Tauber, C.A.; Masaki, S. *Seasonal Adaptations of Insects*; Oxford University Press: New York, NY, USA, 1986; ISBN 0-19-503635-2.

89. Sappington, T.W. Migratory flight of insect pests within a year-round distribution: European corn borer as a case study. *J. Integr. Agric.* **2018**, *17*, 1485–1505. [CrossRef]

90. Chapman, B.B.; Brönmark, C.; Nilsson, J.-Å.; Hansson, L.-A. Partial migration: An introduction. *Oikos* **2011**, *120*, 1761–1763. [CrossRef]

91. Asplen, M.K. Proximate drivers of migration and dispersal in wing-monomorphic insects. *Insects* **2020**, *11*, 61. [CrossRef] [PubMed]

92. Dällenbach, L.J.; Glauser, A.; Lim, K.S.; Chapman, J.W.; Menz, M.H.M. Higher flight activity in the offspring of migrants compared to residents in a migratory insect. *Proc. R. Soc. B* **2018**, *285*, 20172829. [CrossRef] [PubMed]

93. Taylor, L.R. Insect migration, flight periodicity and the boundary layer. *J. Anim. Ecol.* **1974**, *43*, 225–238. [CrossRef]

94. Johnson, C.G. *Migration and Dispersal of Insects by Flight*; Methuen: London, UK, 1969.

95. Philippi, T.; Seger, J. Hedging one's evolutionary bets, revisited. *Trends Evol. Ecol.* **1989**, *4*, 41–44. [CrossRef] [PubMed]

96. Branson, T.F.; Sutter, G.R. Influence of population density of immatures on size, longevity, and fecundity of adult *Diabrotica virgifera virgifera* (Coleoptera: Chrysomelidae). *Environ. Entomol.* **1985**, *14*, 687–690. [CrossRef]

97. Weiss, M.J.; Seevers, K.P.; Mayo, Z.B. Influence of Western Corn Rootworm Larval Densities and Damage on Corn Rootworm Survival Developmental Time, Size, and Sex Ratio (Coleoptera: Chrysomelidae). *J. Kans. Entomol. Soc.* **1985**, *58*, 397–402. Available online: https://www.jstor.org/stable/25084658 (accessed on 24 October 2023).

98. Elliott, N.C.; Sutter, G.R.; Branson, T.F.; Fisher, J.R. Effect of population density of immatures on survival and development of the western corn rootworm (Coleoptera: Chrysomelidae). *J. Entomol. Sci.* **1989**, *24*, 209–213. [CrossRef]

99. Hibbard, B.E.; Meihls, L.N.; Ellersieck, M.R.; Onstad, D.W. Density-dependent and density-independent mortality of the western corn rootworm: Impact on dose calculations of rootworm-resistant Bt corn. *J. Econ. Entomol.* **2010**, *103*, 77–84. [CrossRef]

100. Yu, E.Y.; Gassmann, A.J.; Sappington, T.W. Effects of larval density on dispersal and fecundity of western corn rootworm, *Diabrotica virgifera virgifera* LeConte (Coleoptera: Chrysomelidae). *PLoS ONE* **2019**, *14*, e0212696. [CrossRef]

101. Onstad, D.W.; Guse, C.A.; Spencer, J.L.; Levine, E.; Gray, M.E. Modeling the dynamics of adaptation to transgenic corn by western corn rootworm (Coleoptera: Chrysomelidae). *J. Econ. Entomol.* **2001**, *94*, 529–540. [CrossRef] [PubMed]

102. Onstad, D.W.; Crowder, D.W.; Mitchell, P.D.; Guse, C.A.; Spencer, J.L.; Levine, E.; Gray, M.E. Economics versus alleles: Balancing IPM and IRM for rotation-resistant western corn rootworm (Coleoptera: Chrysomelidae). *J. Econ. Entomol.* **2003**, *96*, 1872–1885. [CrossRef]

103. Storer, N.P. A spatially explicit model simulating western corn rootworm (Coleoptera: Chrysomelidae) adaptation to insect-resistant maize. *J. Econ. Entomol.* **2003**, *96*, 1530–1547. [CrossRef] [PubMed]

104. Crowder, D.W.; Onstad, D.W.; Gray, M.E.; Pierce, C.M.F.; Hager, A.G.; Ratcliffe, S.T.; Steffey, K.L. Analysis of the dynamics of adaptation to transgenic corn and crop rotation by western corn rootworm (Coleoptera: Chrysomelidae) using a daily timestep model. *J. Econ. Entomol.* **2005**, *98*, 534–551. [CrossRef] [PubMed]

105. Onstad, D.; Meinke, L.J. Modeling evolution of *Diabrotica virgifera virgifera* (Coleoptera: Chrysomelidae) to transgenic corn with two insecticidal traits. *J. Econ. Entomol.* **2010**, *103*, 849–860. [CrossRef] [PubMed]

106. Pan, Z.; Onstad, D.W.; Nowatzki, T.M.; Stanley, B.H.; Meinke, L.J.; Flexner, L. Western corn rootworm (Coleoptera: Chrysomelidae) dispersal and adaptation to single-toxin transgenic corn deployed with block and blended refuge. *Environ. Entomol.* **2011**, *40*, 964–978. [CrossRef]

107. Martinez, J.C.; Caprio, M.A. IPM use with the deployment of a non-high dose Bt pyramid and mitigation of resistance for western corn rootworm (*Diabrotica virgifera virgifera*). *Environ. Entomol.* **2016**, *45*, 747–761. [CrossRef]

108. Onstad, D.W.; Caprio, M.A.; Pan, Z. Models of *Diabrotica* populations: Demography, population genetics, geographic spread, and management. *Insects* **2020**, *11*, 712. [CrossRef]

109. Caprio, M.J.; Nowatzki, T.J.; Siegfried, B.D.; Meinke, L.J.; Wright, R.J.; Chandler, L.D. Assessing the risk of resistance to aerial applications of methyl-parathion in the western corn rootworm (Coleoptera: Chrysomelidae). *J. Econ. Entomol.* **2006**, *99*, 483–493. [CrossRef]

110. Naranjo, S.E. Comparative flight behavior of *Diabrotica virgifera virgifera* and *Diabrotica barberi* in the laboratory. *Entomol. Exp. Appl.* **1990**, *55*, 79–90. [CrossRef]

111. Onstad, D.W.; Joselyn, M.G.; Isard, S.A.; Levine, E.; Spencer, J.L.; Bledsoe, L.W.; Edwards, C.R.; Di Fonzo, C.D.; Willson, H. Modeling the spread of western corn rootworm (Coleoptera: Chrysomelidae) populations adapting to soybean-corn rotation. *Environ. Entomol.* **1999**, *28*, 188–194. [CrossRef]

112. Onstad, D.W.; Crowder, D.W.; Isard, S.A.; Levine, E.; Spencer, J.L.; O'Neal, M.E.; Ratcliffe, S.T.; Gray, M.E.; Bledsoe, L.W.; Di Fonzo, C.D.; et al. Does landscape diversity slow the spread of rotation-resistant western corn rootworm (Coleoptera: Chrysomelidae)? *Environ. Entomol.* **2003**, *32*, 992–1001. [CrossRef]

113. Baufeld, P.; Enzian, S. Simulation Model for Spreading Scenarios for Western Corn Rootworm (*Diabrotica virgifera virgifera*) in Case of Germany. In Proceedings of the XXI IWGO Conference and VIII *Diabrotica* Subgroup Meeting, Padova, Italy, 27 October–3 November 2001; pp. 63–67. Available online: http://www.iwgo.org/downloads/Proceedings2001_IWGO_Conference.pdf (accessed on 26 November 2023).

114. Hemerik, L.; Busstra, C.; Mols, P. Predicting the temperature-dependent natural population expansion of the western corn rootworm, *Diabrotica virgifera*. *Entomol. Exp. Appl.* **2004**, *111*, 59–69. [CrossRef]

115. Carrasco, L.R.; Harwood, T.D.; Toepfer, S.; MacLeod, A.; Levay, N.; Kiss, J.; Baker, R.H.A.; Mumford, J.D.; Knight, J.D. Dispersal kernels of the invasive alien western corn rootworm and the effectiveness of buffer zones in eradication programmes in Europe. *Ann. Appl. Biol.* **2010**, *156*, 63–77. [CrossRef]

116. Carrasco, L.R.; Cook, D.; Baker, R.; MacLeod, A.; Knight, J.D.; Mumford, J.D. Towards the integration of spread and economic impacts of biological invasions in a landscape of learning and imitating agents. *Ecol. Econ.* **2012**, *76*, 95–103. [CrossRef]

117. Szalai, M.; Kiss, J.; Kövér, S.; Toepfer, S. Simulating crop rotation strategies with a spatiotemporal lattice model to improve legislation for the management of the maize pest *Diabrotica virgifera virgifera*. *Agric. Sys.* **2014**, *124*, 39–50. [CrossRef]

118. Branson, T.F. The contribution of prehatch and posthatch development to protandry in the chrysomelid, *Diabrotica virgifera virgifera*. *Entomol. Exp. Appl.* **1987**, *43*, 205–208. [CrossRef]

119. Jackson, J.J.; Elliott, N.C. Temperature-dependent development of immature stages of the western corn rootworm, *Diabrotica virgifera virgifera* (Coleoptera: Chrysomelidae). *Environ. Entomol.* **1988**, *17*, 166–171. [CrossRef]

120. Ludwick, D.C.; Zukoff, A.; Higdon, M.; Hibbard, B.E. Protandry of the western corn rootworm (Coleoptera: Chrysomelidae) partially due to earlier egg hatch of males. *J. Kans. Entomol.* **2017**, *90*, 94–99. [CrossRef]

121. Guss, P.L. The sex pheromone of the western corn rootworm (*Diabrotica virgifera*). *Environ. Entomol.* **1976**, *5*, 219–223. [CrossRef]

122. Hammack, L. Calling behavior in female western corn rootworm beetles (Coleoptera: Chrysomelidae). *Ann. Entomol. Soc. Am.* **1995**, *88*, 562–569. [CrossRef]

123. Cates, M.D. Behavioral and Physiological Aspects of Mating and Oviposition by the Adult Western Corn Rootworm, *Diabrotica virgifera virgifera* LeConte. Ph.D. Thesis, University of Nebraska, Lincoln, Nebraska, 1968.

124. Lew, A.C.; Ball, H.J. The mating behavior of the western corn rootworm *Diabrotica virgifera virgifera* (Coleoptera: Chrysomelidae). *Ann. Entomol. Soc. Am.* **1979**, *72*, 391–393. [CrossRef]

125. Marquardt, P.T.; Krupke, C.H. Dispersal and mating behavior of *Diabrotica virgifera virgifera* (Coleoptera: Chrysomelidae) in Bt cornfields. *Environ. Entomol.* **2009**, *38*, 176–182. [CrossRef] [PubMed]

126. Witkowski, J.F.; Owens, J.C.; Tollefson, J.J. Diel activity and vertical flight distribution of adult western corn rootworms in Iowa cornfields. *J. Econ. Entomol.* **1975**, *68*, 351–352. [CrossRef]

127. Bartelt, R.J.; Chiang, H.C. Field studies involving the sex-attractant pheromones of the western and northern corn rootworm beetles. *Environ. Entomol.* **1977**, *6*, 853–861. [CrossRef]

128. Spencer, J.; Onstad, D.; Krupke, C.; Hughson, S.; Pan, Z.; Stanley, B.; Flexner, L. Isolated females and limited males: Evolution of insect resistance in structured landscapes. *Entomol. Exp. Appl.* **2013**, *146*, 38–49. [CrossRef]

129. Hughson, S.A. The Movement Behavior and Reproductive Ecology of Western Corn Rootworm Beetles (Coleoptera: Chrysomelidae) in Bt Cornfields with Structured and Seed Blend Refuges. Ph.D. Thesis, University of Illinois at Urbana-Champaign, Champaign, IL, USA, 2017. Available online: http://hdl.handle.net/2142/98361 (accessed on 26 November 2023).

130. Taylor, S.; Krupke, C. Measuring rootworm refuge function: *Diabrotica virgifera virgifera* emergence and mating in seed blend and strip refuges for *Bacillus thuringiensis* (Bt) maize. *Pest. Manag. Sci.* **2018**, *74*, 2195–2203. [CrossRef]

131. Dobson, I.D.; Teal, P.E.A. Analysis of long-range reproductive behavior of male *Diabrotica virgifera virgifera* LeConte and *D. barberi* Smith and Lawrence to stereoisomers of 8-methyl-2-decyl propanoate under laboratory conditions. *J. Chem. Ecol.* **1987**, *13*, 1331–1341. [CrossRef]

132. Cuthbert, F.P., Jr.; Reid, W.J., Jr. Studies of sex attractant of banded cucumber beetle. *J. Econ. Entomol.* **1964**, *57*, 247–250. [CrossRef]

133. Branson, T.F.; Krysan, J.L. Feeding and oviposition behavior and life cycle strategies of *Diabrotica*: An evolutionary view with implications for pest management. *Environ. Entomol.* **1981**, *10*, 826–831. [CrossRef]

134. Naranjo, S.E. Flight orientation of *Diabrotica virgifera virgifera* and *D. barberi* (Coleoptera: Chrysomelidae) at habitat interfaces. *Ann. Entomol. Soc. Am.* **1994**, *87*, 383–394. [CrossRef]
135. Kirk, V.M.; Calkins, C.O.; Post, F.J. Oviposition preferences of western corn rootworms for various soil surface conditions. *J. Econ. Entomol.* **1968**, *61*, 1322–1324. [CrossRef]
136. Shaw, J.T.; Paullus, J.H.; Luckmann, W.H. Corn rootworm oviposition in soybeans. *J. Econ. Entomol.* **1978**, *71*, 189–191. [CrossRef]
137. Kirk, V.M. Suitable oviposition site for corn rootworms (Coleoptera, Chrysomelidae) resulting from concentration of rainwater by corn plant. *Agric. Meteorol.* **1975**, *15*, 113–116. [CrossRef]
138. Kirk, V.M. Base of Corn Stalks as Oviposition Sites for Western and Northern Corn Rootworms (*Diabrotica*: Coleoptera). *J. Kansas Ent. Soc.* **1981**, *54*, 255–262. Available online: https://www.jstor.org/stable/25084157 (accessed on 24 October 2023).
139. Strnad, S.P.; Bergman, M.K. Movement of first-instar western corn rootworms (Coleoptera: Chrysomelidae) in soil. *Environ. Entomol.* **1987**, *16*, 975–978. [CrossRef]
140. Strnad, S.P.; Bergman, M.K. Distribution of western corn rootworm (Coleoptera: Chrysomelidae) larvae in corn roots. *Environ. Entomol.* **1987**, *16*, 1193–1198. [CrossRef]
141. Sherwood, D.R.; Levine, E. Copulation and its duration affects female weight, oviposition, hatching patterns, and ovarian development in the western corn rootworm (Coleoptera: Chrysomelidae). *J. Econ. Entomol.* **1993**, *86*, 1664–1671. [CrossRef]
142. Lew, A.C.; Ball, H.J. Effect of copulation time on spermatozoan transfer of *Diabrotica virgifera* (Coleoptera: Chrysomelidae). *Ann. Entomol. Soc. Am.* **1980**, *73*, 360–361. [CrossRef]
143. Quiring, D.T.; Timmins, P.R. Influence of reproductive ecology on feasibility of mass trapping *Diabrotica virgifera virgifera* (Coleoptera: Chrysomelidae). *J. Appl. Ecol.* **1990**, *27*, 965–982. [CrossRef]
144. Murphy, A.F.; Krupke, C.H. Mating success and spermatophore composition in western corn rootworm (Coleoptera: Chrysomelidae). *Environ. Entomol.* **2011**, *40*, 1585–1594. [CrossRef] [PubMed]
145. Branson, T.F.; Guss, P.L.; Jackson, J.J. Mating frequency of the western corn rootworm. *Ann. Entomol. Soc. Am.* **1977**, *70*, 506–508. [CrossRef]
146. Kang, J.; Krupke, C.H. Likelihood of multiple mating in *Diabrotica virgifera virgifera* (Coleoptera: Chrysomelidae). *J. Econ. Entomol.* **2009**, *102*, 2096–2100. [CrossRef] [PubMed]
147. Bermond, G.; Cavigliasso, F.; Mallez, S.; Spencer, J.; Guillemaud, T. No clear effect of admixture between two European invading outbreaks of *Diabrotica virgifera virgifera* in natura. *PLoS ONE* **2014**, *9*, e106139. [CrossRef]
148. Spencer, J.L.; Mabry, T.R.; Levine, E.; Isard, S.A. Movement, dispersal, and behavior of western corn rootworm adults in rotated corn and soybean fields. In *Western Corn Rootworm: Ecology and Management*; Vidal, S., Kuhlmann, U., Edwards, C.R., Eds.; CABI Publishing: Oxfordshire, UK, 2005; pp. 121–144, ISBN 0-85199-817-8.
149. Spencer, J.L. Getting high with the beetles. *Am. Entomol.* **2020**, *66*, 28–32. [CrossRef]
150. Hill, R.E. Mating, oviposition patterns, fecundity and longevity of the western corn rootworm. *J. Econ. Entomol.* **1975**, *68*, 311–315. [CrossRef]
151. Coats, S.A.; Mutchmor, J.A.; Tollefson, J.J. Regulation of migratory flight by juvenile hormone mimic and inhibitor in the western corn rootworm (Coleoptera: Chrysomelidae). *Ann. Entomol. Soc. Am.* **1987**, *80*, 697–708. [CrossRef]
152. Short, D.E.; Hill, R.E. Adult emergence, ovarian development, and oviposition sequence of the western corn rootworm in Nebraska. *J. Econ. Entomol.* **1972**, *65*, 685–689. [CrossRef]
153. Ludwig, K.A.; Hill, R.E. Comparison of gut contents of adult western and northern corn rootworms in northeast Nebraska. *Environ. Entomol.* **1975**, *4*, 435–438. [CrossRef]
154. Darnell, S.J.; Meinke, L.J.; Young, L.J. Influence of corn phenology on adult western corn rootworm (Coleoptera: Chrysomelidae) distribution. *Environ. Entomol.* **2000**, *29*, 587–595. [CrossRef]
155. Oloumi-Sadegi, H.; Levine, E. A simple, effective, and low-cost method for mass marking adult western corn rootworms (Coleoptera: Chrysomelidae). *J. Entomol. Sci.* **1990**, *25*, 170–175. [CrossRef]
156. Toepfer, S.; Levay, N.; Kiss, J. Suitability of different fluorescent powders for mass-marking the Chrysomelid, *Diabrotica virgifera virgifera* LeConte. *J. Appl. Entomol.* **2005**, *129*, 456–464. [CrossRef]
157. Taylor, S.V.; Smith, S.J.; Krupke, C.H. 2016. Quantifying rates of random mating in western corn rootworm emerging from Cry3Bb1-expressing and refuge maize in field cages. *Entomol. Exp. Appl.* **2016**, *161*, 203–212. [CrossRef]
158. Toepfer, S.; Levay, N.; Kiss, J. Adult movements of newly introduced alien *Diabrotica virgifera virgifera* (Coleoptera: Chrysomelidae) from non-host habitats. *Bull. Entomol. Res.* **2006**, *96*, 327–335. [CrossRef]
159. McKone, M.J.; McLauchlan, K.K.; Lebrun, E.G.; McCall, A.C. An edge effect caused by adult corn-rootworm beetles on sunflowers in tallgrass prairie remnants. *Conserv. Biol.* **2001**, *15*, 1315–1324. [CrossRef]
160. Moeser, J.; Vidal, S. Nutritional resources used by the maize pest *Diabrotica virgifera virgifera* in its new Southeast European distribution range. *Entomol. Exp. Appl.* **2005**, *114*, 55–63. [CrossRef]
161. Owens, J.C.; Peters, D.C.; Hallauer, A.R. Corn rootworm tolerance in maize. *Environ. Entomol.* **1974**, *3*, 767–772. [CrossRef]
162. Abendroth, L.J.; Elmore, R.J.; Boyer, M.J.; Marlay, S.K. 2011. Corn Growth and Development. Iowa State Univ. Extension Publication #PMR-1009. Available online: https://store.extension.iastate.edu/Product/Corn-Growth-and-Development (accessed on 26 November 2023).
163. Pierce, C.M.F.; Gray, M.E. Population dynamics of a western corn rootworm (Coleoptera: Chrysomelidae), variant in east central Illinois commercial maize and soybean fields. *J. Econ. Entomol.* **2007**, *100*, 1104–1115. [CrossRef]

164. O'Neal, M.E.; Landis, D.A.; Miller, J.R.; DiFonzo, C.D. Corn phenology influences *Diabrotica virgifera virgifera* emigration and visitation to soybean in laboratory assays. *Environ. Entomol.* **2004**, *33*, 35–44. [CrossRef]

165. Kuhar, T.P.; Youngman, R.R. Sex ratio and sexual dimorphism in western corn rootworm (Coleoptera: Chrysomelidae) adults on yellow sticky traps in corn. *Environ. Entomol.* **1995**, *24*, 1408–1413. [CrossRef]

166. O'Neal, M.E.; Gray, M.E.; Smyth, C.A. Population characteristics of a western corn rootworm (Coleoptera: Chrysomelidae) strain in East-Central Illinois corn and soybean fields. *J. Econ. Entomol.* **1999**, *92*, 1301–1310. [CrossRef]

167. Isard, S.A.; Spencer, J.L.; Nasser, M.A.; Levine, E. Aerial movement of western corn rootworm (Coleoptera: Chrysomelidae): Diel periodicity of flight activity in soybean fields. *Environ. Entomol.* **2000**, *29*, 226–234. [CrossRef]

168. Grant, R.H.; Seevers, K.P. The vertical movement of adult western corn rootworms (*Diabrotica virgifera virgifera*) relative to the transport of momentum and heat. *Agric. For. Meteorol.* **1990**, *49*, 191–203. [CrossRef]

169. Miller, N.J.; Guillemaud, T.; Giordano, R.; Siegfried, B.D.; Gray, M.E.; Meinke, L.J.; Sappington, T.W. Genes, gene flow and adaptation of *Diabrotica virgifera virgifera*. *Agric. For. Entomol.* **2009**, *11*, 47–60. [CrossRef]

170. Spencer, J.L.; Hughson, S.A. Resistance to crop rotation. In *Insect Resistance Management: Biology, Economics and Prediction*, 3rd ed.; Onstad, D.W., Knolhoff, L.M., Eds.; Academic Press: London, UK, 2023; pp. 191–244, ISBN 978-0-12-823878-8.

171. Knolhoff, L.M.; Onstad, D.W.; Spencer, J.L.; Levine, E. Behavioral differences between rotation-resistant and wild-type *Diabrotica virgifera virgifera* (Coleoptera: Chrysomelidae). *Environ. Entomol.* **2006**, *35*, 1049–1057. [CrossRef]

172. Barčić, J.I.; Bažok, R.; Edwards, C.R.; Kos, T. Western corn rootworm adult movement and possible egg laying in fields bordering maize. *J. Appl. Entomol.* **2007**, *131*, 400–405. [CrossRef]

173. Schroeder, J.B.; Ratcliffe, S.T.; Gray, M.E. Effect of four cropping systems on variant western corn rootworm (Coleoptera: Chrysomelidae) adult and egg densities and subsequent larval injury in rotated maize. *J. Econ. Entomol.* **2005**, *98*, 1587–1593. [CrossRef]

174. Rondon, S.I.; Gray, M.E. Captures of western corn rootworm (Coleoptera: Chrysomelidae) adults with Pherocon AM and vial traps in four crops in east central Illinois. *J. Econ. Entomol.* **2003**, *96*, 737–747. [CrossRef]

175. Mabry, T.R.; Spencer, J.L.; Levine, E.; Isard, S.A. Western corn rootworm (Coleoptera: Chrysomelidae) behavior is affected by alternating diets of corn and soybean. *Environ. Entomol.* **2004**, *33*, 860–871. [CrossRef]

176. Spencer, J.L.; Mabry, T.R.; Isard, S.A.; Levine, E. Soybean foliage consumption reduces adult western corn rootworm (*Diabrotica virgifera virgifera*) (Coleoptera: Chrysomelidae) survival and stimulates flight. *J. Econ. Entomol.* **2021**, *114*, 2390–2399. [CrossRef]

177. Curzi, M.; Zavala, J.; Spencer, J.L.; Seufferheld, M.J. Abnormally high digestive enzyme activity and gene expression explains the contemporary evolution of a *Diabrotica* biotype able to feed on soybeans. *Ecol. Evol.* **2012**, *2*, 2005–2017. [CrossRef] [PubMed]

178. Chu, C.-C.; Spencer, J.L.; Curzi, M.J.; Zavala, J.A.; Seufferheld, M.J. Gut bacteria facilitate adaptation to crop rotation in the western corn rootworm. *Proc. Natl. Acad. Sci. USA* **2013**, *110*, 11917–11922. [CrossRef] [PubMed]

179. Chu, C.-C.; Zavala, J.A.; Spencer, J.L.; Curzi, M.J.; Fields, C.J.; Drnevich, J.; Siegfried, B.D.; Seufferfeld, M.J. Patterns of differential gene expression in adult rotation-resistant and wild-type western corn rootworm digestive tracts. *Evol. Appl.* **2015**, *8*, 692–704. [CrossRef] [PubMed]

180. Basol, T. Keep An Eye Out for Soybean Defoliators. Integrated Crop Management News, 12 July 2023. Iowa State University Extension and Outreach. Ames, IA, USA. Available online: https://crops.extension.iastate.edu/cropnews/2023/07/keep-eye-out-soybean-defoliators (accessed on 26 November 2023).

181. Mabry, T.R.; Spencer, J.L. Survival and oviposition of a western corn rootworm variant feeding on soybean. *Entomol. Exp. Appl.* **2003**, *109*, 113–121. [CrossRef]

182. Grant, T.J.; Fisher, K.E.; Krishnan, N.; Mullins, A.N.; Hellmich, R.L.; Sappington, T.W.; Adelman, J.S.; Coats, J.R.; Hartzler, R.G.; Pleasants, J.M.; et al. Monarch butterfly ecology, behavior, and vulnerabilities in North Central United States agricultural landscapes. *BioScience* **2022**, *72*, 1176–1203. [CrossRef]

183. Naranjo, S.E. Influence of two mass-marking techniques on survival and flight behavior of *Diabrotica virgifera virgifera* (Coleoptera: Chrysomelidae). *J. Econ. Entomol.* **1990**, *83*, 1360–1364. [CrossRef]

184. Wilson, T.A. Fitness of the Western Corn Rootworm, *Diabrotica virgifera virgifera* LeConte, Exposed to Transgenic Plants and Farmer Perceptions of Transgenic corn. Ph.D. Dissertation, Iowa State University, Ames, IA, USA, 2003; p. 1401. Available online: https://lib.dr.iastate.edu/rtd/1401 (accessed on 26 November 2023).

185. Stebbing, J.A.; Meinke, L.J.; Naranjo, S.E.; Siegfried, B.D.; Wright, R.J.; Chandler, L.D. Flight behavior of methyl-parathion resistant and susceptible western corn rootworm (Coleoptera: Chrysomelidae) populations from Nebraska. *J. Econ. Entomol.* **2005**, *98*, 1294–1304. [CrossRef]

186. Dingle, H. The relation between age and flight activity in the milkweed bug, *Oncopeltus*. *J. Exp. Biol.* **1965**, *42*, 269–283. [CrossRef]

187. Rankin, M.A.; Rankin, S.M. The hormonal control of migratory flight behavior in the convergent lady beetle, *Hippodamia convergens*. *Physiol. Entomol.* **1980**, *5*, 175–182. [CrossRef]

188. Li, H.; Toepfer, S.; Kuhlmann, U. Flight and crawling activities of *Diabrotica virgifera virgifera* (Coleoptera: Chrysomelidae) in relation to morphometric traits. *J. Appl. Entomol.* **2010**, *134*, 449–461. [CrossRef]

189. Sappington, T.W.; Showers, W.B. Reproductive maturity, mating status, and long-duration flight behavior of *Agrotis ipsilon* (Lepidoptera: Noctuidae) and the conceptual misuse of the oogenesis-flight syndrome by entomologists. *Environ. Entomol.* **1992**, *21*, 676–688. [CrossRef]

190. Zhao, X.C.; Feng, H.Q.; Wu, B.; Wu, X.F.; Liu, Z.F.; Wu, K.M.; McNeil, J.N. Does the onset of sexual maturation terminate the expression of migratory behaviour in moths? A study of the oriental armyworm, *Mythimna separata*. *J. Insect Physiol.* **2009**, *55*, 1039–1043. [CrossRef] [PubMed]

191. Jiang, X.F.; Luo, L.Z.; Sappington, T.W. Relationship of flight and reproduction in beet armyworm, *Spodoptera exigua* (Lepidoptera: Noctuidae), a migrant lacking the oogenesis-flight syndrome. *J. Insect Physiol.* **2010**, *56*, 1631–1637. [CrossRef] [PubMed]

192. Tigreros, N.; Davidowitz, G. Flight-fecundity tradeoffs in wing-monomorphic insects. *Adv. Insect Physiol.* **2019**, *56*, 1–41. [CrossRef]

193. Guo, J.; Yang, F.; Zhang, H.; Lin, P.; Zhai, B.; Lu, Z.; Hu, G.; Liu, P. Reproduction does not impede the stopover departure to ensure a potent migration in *Cnaphalocrocis medinalis* moths. *Insect Sci.* **2022**, *29*, 1672–1684. [CrossRef] [PubMed]

194. Branson, T.F.; Johnson, R.D. Adult western corn rootworms: Oviposition, fecundity, and longevity in the laboratory. *J. Econ. Entomol.* **1973**, *66*, 417–418. [CrossRef]

195. Bayar, K.; Komaromi, J.; Kiss, J. Egg production of western corn rootworm (*Diabrotica virgifera virgifera* LeConte) in laboratory rearing. *Növényvédelem* **2003**, *38*, 543–545. (In Hungarian)

196. Minter, M.; Pearson, A.; Lim, K.S.; Wilson, K.; Chapman, J.W.; Jones, C.M. The tethered flight technique as a tool for studying life-history strategies associated with migration in insects. *Ecol. Entomol.* **2018**, *43*, 397–411. [CrossRef]

197. Cinereski, J.E.; Chiang, H.C. The pattern of movement of adults of the northern corn rootworm inside and outside of corn fields. *J. Econ. Entomol.* **1968**, *61*, 1531–1536. [CrossRef]

198. Chapman, J.W.; Klaassen, R.H.G.; Drake, A.V.; Fossette, S.; Hays, G.C.; Metcalfe, J.D.; Reynolds, A.M.; Reynolds, D.R.; Alerstam, T. Animal orientation strategies for movement in flows. *Curr. Biol.* **2011**, *21*, R861–R870. [CrossRef]

199. Florio, J.; Verú, L.M.; Dao, A.; Yaro, A.D.; Diallo, M.; Sanogo, Z.L.; Samaké, D.; Huestis, D.L.; Yossi, O.; Talamas, E.; et al. Diversity, dynamics, direction, and magnitude of high-altitude migrating insects in the Sahel. *Sci. Rep.* **2020**, *10*, 20523. [CrossRef] [PubMed]

200. Gao, B.; Hedlund, J.; Reynolds, D.R.; Zhai, D.R.; Hu, G.; Chapman, J.W. The 'migratory connectivity' concept, and its applicability to insect migrants. *Mov. Ecol.* **2020**, *8*, 48. [CrossRef] [PubMed]

201. Chapman, J.W.; Nesbit, R.L.; Burgin, L.E.; Reynolds, D.R.; Smith, A.D.; Middleton, D.R.; Hill, J.K. Flight orientation behaviors promote optimal migration trajectories in high-flying insects. *Science* **2010**, *327*, 682–685. [CrossRef] [PubMed]

202. Sappington, T.W.; Showers, W.B. Implications for migration of age-related variation in flight behavior of *Agrotis ipsilon* (Lepidoptera: Noctuidae). *Ann. Entomol. Soc. Am.* **1991**, *84*, 560–565. [CrossRef]

203. Showers, W.B.; Smelser, R.B.; Keaster, A.J.; Whitford, F.; Robinson, J.F.; Lopez, J.D.; Taylor, S.E. Recapture of marked black cutworm (Lepidoptera: Noctuidae) males after long-range transport. *Environ. Entomol.* **1989**, *18*, 447–458. [CrossRef]

204. Chapman, J.W.; Drake, V.A.; Reynolds, D.R. Recent insights from radar studies of insect flight. *Annu. Rev. Entomol.* **2011**, *56*, 337–356. [CrossRef]

205. Drake, V.A.; Farrow, R.A. The influence of atmospheric structure and motions on insect migration. *Annu. Rev. Entomol.* **1988**, *33*, 183–210. [CrossRef]

206. Reynolds, D.R.; Chapman, J.W.; Harrington, R. The migration of insect vectors of plant and animal viruses. *Adv. Virus Res.* **2006**, *67*, 453–517. [CrossRef]

207. Hu, G.; Lim, K.S.; Reynolds, D.R.; Reynolds, A.M.; Chapman, J.W. Wind-related orientation patterns in diurnal, crepuscular and nocturnal high-altitude insect migrants. *Front. Behav. Neurosci.* **2016**, *10*, 32. [CrossRef]

208. Hu, G.; Lim, K.S.; Horvitz, N.; Clark, S.J.; Reynolds, D.R.; Sapir, N.; Chapman, J.W. Mass seasonal bioflows of high-flying insect migrants. *Science* **2016**, *354*, 1584–1587. [CrossRef]

209. Reynolds, A.M.; Reynolds, D.R.; Sane, S.P.; Hu, G.; Chapman, J.W. Orientation in high-flying migrant insects in relation to flows: Mechanisms and strategies. *Phil. Trans. R. Soc. B* **2016**, *371*, 20150392. [CrossRef] [PubMed]

210. Satterfield, D.A.; Sillett, T.S.; Chapman, J.W.; Altizer, S.; Marra, P.P. Seasonal insect migrations: Massive, influential, and overlooked. *Front. Ecol. Environ.* **2020**, *18*, 335–344. [CrossRef]

211. Fontenot, L.E. The Role of High Elevation Flight in Western Corn Rootworm Dispersal. Master's Thesis, Department of Geography University of Illinois at Urbana-Champaign, Urbana, IL, USA, 2006; 55p.

212. Shigesada, N.; Kawasaki, K.; Takeda, Y. Modeling stratified diffusion in biological invasions. *Am. Nat.* **1995**, *146*, 229–251. [CrossRef]

213. Liebhold, A.M.; Tobin, P.C. Population ecology of insect invasions and their management. *Annu. Rev. Entomol.* **2008**, *53*, 387–408. [CrossRef] [PubMed]

214. Hallatschek, O.; Fisher, D.S. Acceleration of evolutionary spread by long-range dispersal. *Proc. Natl. Acad. Sci. USA* **2014**, *111*, E4911–E4919. [CrossRef] [PubMed]

215. Chiang, H.C.; Flaskerd, R.G. Northern and Western Corn Rootworms in Minnesota. *J. Minn. Acad. Sci.* **1969**, *36*, 48–51. Available online: https://digitalcommons.morris.umn.edu/jmas/vol36/iss1/16 (accessed on 26 November 2023).

216. Ruppel, R.F. Dispersal of western corn rootworm, *Diabrotica virgifera* Le Conte, in Michigan (Coleoptera: Chrysomelidae). *J. Kan. Entomol. Soc.* **1975**, *48*, 291–296. Available online: http://www.jstor.org/stable/25082756 (accessed on 26 November 2023).

217. Youngman, R.R.; Day, E.R. Incidence of western corn-rootworm beetles (Coleoptera, Chrysomelidae) on corn in Virginia from 1987 to 1992. *J. Entomol. Sci.* **1993**, *28*, 136–141. [CrossRef]

218. Meloche, F.; Rhainds, M.; Roy, M.; Brodeur, J. Distribution of western and northern corn rootworms (Coleoptera: Chrysomelidae) in Quebec, Canada. *Can. Entomol.* **2005**, *137*, 226–229. [CrossRef]

219. Bača, F. New member of the harmful entomofauna of Yugoslavia *Diabrotica virgifera virgifera* LeConte (Coleoptera, Chrysomelidae). *IWGO Newsletter* **1993**, *12*, 21.
220. Kiss, J.; Edwards, C.R.; Berger, H.K.; Cate, P.; Cean, M.; Cheek, S.; Derron, J.; Festic, H.; Furlan, L.; Igrc-Barčić, J.; et al. Monitoring of western corn rootworm (*Diabrotica virgifera virgifera* LeConte) in Europe 1992–2003. In *Western Corn Rootworm: Ecology and Management*; Vidal, S., Kuhlmann, U., Edwards, C.R., Eds.; CABI Publishing: Oxfordshire, UK, 2005; pp. 29–39, ISBN 0-85199-817-8.
221. Konefal, T.; Bereś, P.K. *Diabrotica virgifera* LeConte in Poland in 2005-2007 and regulations in the control of the pest in 2008. *J. Plant Prot. Res.* **2009**, *49*, 129–134. [CrossRef]
222. Ciosi, M.; Miller, N.J.; Toepfer, S.; Estoup, A.; Guillemaud, T. Stratified dispersal and increasing genetic variation during the invasion of Central Europe by the western corn rootworm, *Diabrotica virgifera virgifera*. *Evol. Appl.* **2011**, *4*, 54–70. [CrossRef] [PubMed]
223. De Luigi, V.; Furlan, L.; Palmieri, S.; Vettorazzo, M.; Zanini, G.; Edwards, C.R.; Burgio, G. Results of WCR monitoring plans and evaluation of an eradication programme using GIS and Indicator Kriging. *J. Appl. Entomol.* **2011**, *135*, 38–46. [CrossRef]
224. Mrganić, M.; Bažok, R.; Mikac, K.M.; Benítez, H.A.; Lemic, D. Two decades of invasive western corn rootworm population monitoring in Croatia. *Insects* **2018**, *9*, 160. [CrossRef] [PubMed]
225. Falkner, K.; Mitter, H.; Moltchanova, E.; Schmid, E. A zero-inflated Poisson mixture model to analyse spread and abundance of the western corn rootworm in Austria. *Agric. Sys.* **2019**, *174*, 105–116. [CrossRef]
226. Miller, N.; Estoup, A.; Toepfer, S.; Bourguet, D.; Lapchin, L.; Derridj, S.; Kim, K.S.; Reynaud, P.; Furlan, L.; Guillemaud, T. Multiple transatlantic introductions of the western corn rootworm. *Science* **2005**, *310*, 992. [CrossRef] [PubMed]
227. Ciosi, M.; Miller, N.J.; Kim, K.S.; Giordano, R.; Estoup, A.; Guillemaud, T. Invasion of Europe by the western corn rootworm, *Diabrotica virgifera virgifera*: Multiple transatlantic introductions with various reductions of genetic diversity. *Mol. Ecol.* **2008**, *17*, 3614–3627. [CrossRef]
228. Bermond, G.; Ciosi, M.; Lombaert, E.; Blin, A.; Boriani, M.; Boriani, M.; Furlan, L.; Toepfer, S.; Guillemaud, T. Secondary contact and admixture between independently invading populations of the western corn rootworm, *Diabrotica virgifera virgifera* in Europe. *PLoS ONE* **2012**, *7*, e50129. [CrossRef]
229. Kiss, J.; Edwards, C.R. *Diabrotica virgifera virgifera* LeConte in Europe 2012. Entomology Department at Purdue University. Available online: https://extension.entm.purdue.edu/wcr/images/pdf/2012/EuropeMap2012.pdf (accessed on 19 April 2023).
230. Mollison, D. Modeling biological invasions: Chance, explanation, prediction. *Philos. Trans. R. Soc. London Ser. B* **1986**, *314*, 675–693. [CrossRef]
231. Blackburn, T.M.; Lockwood, J.L.; Cassey, P. The influence of numbers on invasion success. *Mol. Ecol.* **2015**, *24*, 1942–1953. [CrossRef]
232. Saccaggi, D.L.; Wilson, J.R.U.; Terblanche, J.S. Propagule pressure helps overcome adverse environmental conditions during population establishment. *Curr. Res. Insect Sci.* **2021**, *1*, 100011. [CrossRef] [PubMed]
233. Lemic, D.; Mikac, K.M.; Bažok, R. Historical and contemporary population genetics of the invasive western corn rootworm (Coleoptera: Chrysomelidae) in Croatia. *Environ. Entomol.* **2013**, *42*, 811–819. [CrossRef] [PubMed]
234. Lemic, D.; Mikac, K.M.; Ivkosic, S.A.; Bažok, R. The temporal and spatial invasion genetics of the western corn rootworm (Coleoptera: Chrysomelidae) in Southern Europe. *PLoS ONE* **2015**, *10*, e0138796. [CrossRef] [PubMed]
235. Tsutsui, N.D.; Suarez, A.V.; Holway, D.A.; Case, T.J. Reduced genetic variation and the success of an invasive species. *Proc. Natl. Acad. Sci. USA* **2000**, *97*, 5948–5953. [CrossRef] [PubMed]
236. Sakai, A.K.; Allendorf, F.W.; Holt, J.S.; Lodge, D.M.; Molofsky, J.; With, K.A.; Baughman, S.; Cabin, R.J.; Cohen, J.E.; Ellstrand, N.C.; et al. The population biology of invasive species. *Annu. Rev. Ecol. Syst.* **2001**, *32*, 305–332. [CrossRef]
237. Renault, D.; Laparie, M.; McCauley, S.J.; Bonte, D. Environmental adaptations, ecological filtering, and dispersal central to insect invasions. *Annu. Rev. Entomol.* **2018**, *63*, 345–368. [CrossRef] [PubMed]
238. Bermond, G.B.; Li, H.; Guillemaud, T.; Toepfer, S. Genetic and phenotypic effects of hybridization in independently introduced populations of the invasive maize pest *Diabrotica virgifera virgifera* in Europe. *J. Entomol. Acarol. Res.* **2021**, *53*, 9559. [CrossRef]
239. Williamson, M.; Fitter, A. The varying success of invaders. *Ecology* **1996**, *77*, 1661–1666. [CrossRef]
240. Zenni, R.D.; Nuñez, M.A. The elephant in the room: The role of failed invasions in understanding invasion biology. *Oikos* **2013**, *122*, 801–815. [CrossRef]
241. Eyre, D.P.; Cannon, R.J.C.; Cheek, S.; MacLeod, A.; Baker, R.H.A. Surveying for *Diabrotica virgifera virgifera* (Coleoptera: Chrysomelidae) in England & Wales, 2003–2006. In *Aspects of Applied Biology 83, Crop Protection in Southern Britain*; Association of Applied Biologists: Wellesbourne, UK, 2007; pp. 141–145.
242. Paetkau, D.; Slade, R.; Burden, M.; Estoup, A. Genetic assignment methods for the direct, real-time estimation of migration rate: A simulation-based exploration of accuracy and power. *Mol. Ecol.* **2004**, *13*, 55–65. [CrossRef]
243. Kim, K.S.; Cano-Ríos, P.; Sappington, T.W. Using genetic markers and population assignment techniques to infer origin of boll weevils (Coleoptera: Curculionidae) unexpectedly captured near an eradication zone in Mexico. *Environ. Entomol.* **2006**, *35*, 813–826. [CrossRef]
244. Kim, K.S.; Ratcliffe, S.T.; French, B.W.; Liu, L.; Sappington, T.W. Utility of EST-derived SSRs as population genetics markers in a beetle. *J. Hered.* **2008**, *99*, 112–124. [CrossRef] [PubMed]

245. Kim, K.S.; Jones, G.D.; Westbrook, J.K.; Sappington, T.W. Multidisciplinary fingerprints: Forensic reconstruction of an insect reinvasion. *J. R. Soc. Interface* **2010**, *7*, 677–686. [CrossRef] [PubMed]

246. Kim, K.S.; Sappington, T.W. Population genetics strategies to characterize long-distance dispersal of insects. *J. Asia-Pac. Entomol.* **2013**, *16*, 87–97. [CrossRef]

247. Kim, K.S.; Sappington, T.W. Microsatellite data analysis for population genetics. In *Microsatellites: Methods and Protocols. Methods in Molecular Biology*; Kantartzi, S.K., Ed.; Humana Press: Totowa, NJ, USA; Springer Science+Business Media, LLC: New York, NY, USA, 2013; pp. 271–295.

248. Hendrix, W.H., III; Showers, W.B. Tracing black cutworm and armyworm (Lepidoptera: Noctuidae) northward migration using *Pithecellobium* and *Calliandra* pollen. *Environ. Entomol.* **1992**, *21*, 1092–1096. [CrossRef]

249. Sappington, T.W.; Showers, W.B.; McNutt, J.J.; Bernhardt, J.L.; Goodenough, J.L.; Keaster, A.J.; Levine, E.; McLeod, D.G.R.; Robinson, J.F.; Way, M.O. Morphological correlates of migratory behavior in the black cutworm (Lepidoptera: Noctuidae). *Environ. Entomol.* **1994**, *23*, 58–67. [CrossRef]

250. Benítez, H.A.; Lemic, D.; Bažok, R.; Gallardo-Araya, C.; Mikac, K.M. Evolutionary directional asymmetry and shape variation in *Diabrotica v. virgifera* (Coleoptera: Chrysomelidae): An example using hindwings. *Biol. J. Linn. Soc.* **2014**, *111*, 110–118. [CrossRef]

251. Li, H.; Guillemaud, T.; French, B.W.; Kuhlmann, U.; Toepfer, S. Phenotypic trait changes in laboratory-reared colonies of the maize herbivore, *Diabrotica virgifera virgifera*. *Bull. Entomol. Res.* **2014**, *104*, 97–115. [CrossRef]

252. Lemic, D.; Mikac, M.K.; Kozina, A.; Benitez, A.H.; McLean, M.C.; Bažok, R. Monitoring techniques of the western corn rootworm are the precursor to effective IPM strategies. *Pest Manag. Sci.* **2016**, *72*, 405–417. [CrossRef]

253. Mikac, K.M.; Lemic, D.; Bažok, R.; Benítez, H.A. Wing shape changes: A morphological view of the *Diabrotica virgifera virgifera* European invasion. *Biol. Invasions* **2016**, *18*, 3401–3407. [CrossRef]

254. Mikac, K.M.; Lemic, D.; Benítez, H.A.; Bažok, R. Changes in corn rootworm wing morphology are related to resistance development. *J. Pest Sci.* **2019**, *92*, 443–451. [CrossRef]

255. Ivkosic, S.A.; Gorman, J.; Lemic, D.; Mikac, K.M. Genetic monitoring of western corn rootworm (Coleoptera: Chrysomelidae) populations on a microgeographic scale. *Environ. Entomol.* **2014**, *43*, 804–818. [CrossRef] [PubMed]

256. Paddock, K.J.; Finke, D.L.; Kim, K.S.; Sappington, T.W.; Hibbard, B.E. Patterns of microbiome composition vary across spatial scales in a specialist insect. *Front. Microbiol.* **2022**, *13*, 898744. [CrossRef] [PubMed]

257. Ludwick, D.C.; Ericsson, A.C.; Meihls, L.N.; Gregory, M.L.J.; Finke, D.L.; Coudron, T.A.; Hibbard, B.E.; Shelby, K.S. Survey of bacteria associated with western corn rootworm life stages reveals no difference between insects reared in different soils. *Sci. Rep.* **2019**, *9*, 15332. [CrossRef] [PubMed]

258. Paddock, K.J.; Robert, C.A.M.; Erb, M.; Hibbard, B.E. Western corn rootworm, plant and microbe interactions: A review and prospects for new management tools. *Insects* **2021**, *12*, 171. [CrossRef] [PubMed]

259. Suchan, T.; Talavera, G.; Sáez, L.; Ronikier, M.; Vila, R. Pollen metabarcoding as a tool for tracking long-distance insect migrations. *Mol. Ecol. Resour.* **2019**, *19*, 149–162. [CrossRef]

260. Carrasco, L.R.; Mumford, J.D.; MacLeod, A.; Harwood, T.; Grabenweger, G.; Leach, A.W.; Knight, J.D.; Baker, R.H.A. Unveiling human-assisted dispersal mechanisms in invasive alien insects: Integration of spatial stochastic simulation and phenology models. *Ecol. Model.* **2010**, *221*, 2068–2075. [CrossRef]

261. Onstad, D.W.; Guse, C.A.; Crowder, D.W. Heterogeneous landscapes and variable behaviour: Modelling rootworm evolution and geographical spread. In *Western Corn Rootworm: Ecology and Management*; Vidal, S., Kuhlmann, U., Edwards, C.R., Eds.; CABI Publishing: Oxfordshire, UK, 2005; pp. 155–167, ISBN 0-85199-817-8.

262. Dunbar, M.W.; Gassmann, A.J. Abundance and distribution of western and northern corn rootworm (*Diabrotica* spp.) and prevalence of rotation resistance in eastern Iowa. *J. Econ. Entomol.* **2013**, *106*, 168–180. [CrossRef]

263. Ball, H.J.; Weekman, G.T. Insecticide resistance in the adult western corn rootworm in Nebraska. *J. Econ. Entomol.* **1962**, *55*, 439–441. [CrossRef]

264. Metcalf, R.L. Implications and prognosis of resistance to insecticides. In *Pest Resistance to Pesticides*; Georghiou, G.P., Saito, T., Eds.; Plenum Press: New York, NY, USA, 1983; pp. 703–733.

265. Hamilton, E.W. Aldrin resistance in corn rootworm beetles. *J. Econ. Entomol.* **1965**, *58*, 296–300. [CrossRef]

266. Zhou, X.; Scharf, M.E.; Parimi, S.; Meinke, L.J.; Wright, R.J.; Chandler, L.D.; Siegfried, B.D. Diagnostic assays based on esterase-mediated resistance mechanisms in western corn rootworms (Coleoptera: Chrysomelidae). *J. Econ. Entomol.* **2002**, *95*, 1261–1266. [CrossRef] [PubMed]

267. Parimi, S.; Meinke, L.J.; French, B.W.; Chandler, L.D.; Siegfried, B.D. Stability and persistence and of methyl-parathion and aldrin resistance in western corn rootworms. *Crop Prot.* **2006**, *25*, 269–274. [CrossRef]

268. Zhu, K.Y.; Wilde, G.E.; Higgins, R.A.; Sloderbeck, P.E.; Buschman, L.L.; Shufran, R.A.; Whitworth, R.J.; Starkey, S.R.; He, F. Evidence of evolving carbaryl resistance in western corn rootworm (Coleoptera: Chrysomelidae) in areawide-managed cornfields in north central Kansas. *J. Econ. Entomol.* **2001**, *94*, 929–934. [CrossRef] [PubMed]

269. Siegfried, B.D.; Meinke, L.J.; Parimi, S.; Scharf, M.E.; Nowatzki, T.J.; Zhou, X.; Chandler, L.D. Monitoring western corn rootworm (Coleoptera: Chrysomelidae) susceptibility to carbaryl and cucurbitacin baits in the areawide management pilot program. *J. Econ. Entomol.* **2004**, *97*, 1726–1733. [CrossRef] [PubMed]

270. Chandler, L.D. Corn rootworm areawide management program: United States Department of Agriculture-Agricultural Research Service. *Pest Manag. Sci.* **2003**, *59*, 605–608. [CrossRef] [PubMed]

271. Lampman, R.L.; Metcalf, R.L.; Andersen, J.F. Semiochemical attractants of *Diabrotica undecimpunctata howardi* Barber, southern corn rootworm, and *Diabrotica virgifera virgifera* LeConte, the western corn rootworm (Coleoptera: Chrysomelidae). *J. Chem. Ecol.* **1987**, *13*, 959–975. [CrossRef] [PubMed]

272. Metcalf, R.L.; Ferguson, J.E.; Lampman, R.; Andersen, J.F. Dry cucurbitacin-containing baits for controlling diabroticite beetles (Coleoptera: Chrysomelidae). *J. Econ. Entomol.* **1987**, *80*, 870–875. [CrossRef]

273. Weissling, T.J.; Meinke, L.J. Potential of starch encapsulated semiochemical-insecticide formulations for adult corn rootworm (Coleoptera: Chrysomelidae) control. *J. Econ. Entomol.* **1991**, *84*, 601–609. [CrossRef]

274. Gerber, C.K.; Edwards, C.R.; Bledsoe, L.W.; Gray, M.E.; Steffey, K.L.; Chandler, L.D. Application of the areawide concept using semiochemical-based insecticide baits for managing the western corn rootworm (*Diabrotica virgifera virgifera* LeCone) variant in the eastern Midwest. In *Western Corn Rootworm: Ecology and Management*; Vidal, S., Kuhlmann, U., Edwards, C.R., Eds.; CABI Publishing: Oxfordshire, UK, 2005; pp. 221–238, ISBN 0-85199-817-8.

275. Gassmann, A.J.; Shrestha, R.B.; Jakka, S.R.K.; Dunbar, M.W.; Clifton, E.H.; Paolino, A.R.; Ingber, D.A.; French, B.W.; Masloski, K.E.; Dounda, J.W.; et al. Evidence of resistance to Cry34/35 Ab1 corn by western corn rootworm (Coleoptera: Chrysomelidae): Root injury in the field and larval survival in plant-based bioassays. *J. Econ. Entomol.* **2016**, *109*, 1872–1880. [CrossRef]

276. Jakka, S.R.K.; Shrestha, R.B.; Gassmann, A.J. Broad-spectrum resistance to *Bacillus thuringiensis* toxins by western corn rootworm (*Diabrotica virgifera virgifera*). *Sci. Rep.* **2016**, *6*, 27860. [CrossRef]

277. Zukoff, S.N.; Ostlie, K.R.; Potter, B.; Meihls, L.N.; Zukoff, A.L.; French, L.; Ellersieck, M.R.; French, B.W.; Hibbard, B.E. Multiple assays indicate varying levels of cross resistance in Cry3Bb1-selected field populations of the western corn rootworm to mCry3A, eCry3.1Ab, and Cry34/35Ab1. *J. Econ. Entomol.* **2016**, *109*, 1387–1398. [CrossRef] [PubMed]

278. Ludwick, D.C.; Meihls, L.N.; Ostlie, K.R.; Potter, B.D.; French, L.; Hibbard, B.E. Minnesota field population of western corn rootworm (Coleoptera: Chrysomelidae) shows incomplete resistance to Cry34Ab1/Cry35Ab1 and Cry3Bb1. *J. Appl. Entomol.* **2017**, *141*, 28–40. [CrossRef]

279. Schrader, P.M.; Estes, R.E.; Tinsley, N.A.; Gassmann, A.J.; Gray, M.E. Evaluation of adult emergence and larval root injury for Cry3Bb1-resistant populations of the western corn rootworm. *J. Appl. Entomol.* **2017**, *141*, 41–52. [CrossRef]

280. Calles-Torrez, V.; Knodel, J.J.; Boetel, M.A.; French, B.W.; Fuller, B.W.; Ransom, J.K. Field-evolved resistance of northern and western corn rootworm (Coleoptera: Chrysomelidae) populations to corn hybrids expressing single and pyramided Cry3Bb1 and Cry34/35Ab1 Bt proteins in North Dakota. *J. Econ. Entomol.* **2019**, *112*, 1875–1886. [CrossRef]

281. Gassmann, A.J.; Shrestha, R.B.; Kropf, A.L.; St Clair, C.R.; Brenizer, B.D. Field-evolved resistance by western corn rootworm to Cry34/35Ab1 and other *Bacillus thuringiensis* traits in transgenic maize. *Pest Manag. Sci.* **2020**, *76*, 268–276. [CrossRef]

282. Shrestha, R.B.; Jakka, S.R.K.; Gassmann, A.J. Response of Cry3Bb1-resistant western corn rootworm (Coleoptera: Chrysomelidae) to Bt maize and soil insecticide. *J. Appl. Entomol.* **2018**, *142*, 937–946. [CrossRef]

283. Crickmore, N.; Berry, C.; Panneerselvam, S.; Mishra, R.; Connor, T.R.; Bonning, B.C. A structure-based nomenclature for *Bacillus thuringiensis* and other bacteria-derived pesticidal proteins. *J. Invert. Pathol.* **2021**, *186*, 107438. [CrossRef]

284. Dunbar, M.W.; O'Neal, M.E.; Gassmann, A.J. Effects of field history on corn root injury and adult abundance of northern and western corn rootworm (Coleoptera: Chrysomelidae). *J. Econ. Entomol.* **2016**, *109*, 2096–2104. [CrossRef]

285. Gassmann, A.J.; Petzold-Maxwell, J.L.; Keweshan, R.S.; Dunbar, M.W. Western corn rootworm and Bt maize: Challenges of pest resistance in the field. *GM Crops Food* **2012**, *3*, 235–244. [CrossRef]

286. Gassmann, A.J. Field-evolved resistance to Bt maize by western corn rootworm: Predictions from the laboratory and effects in the field. *J. Invert. Pathol.* **2012**, *110*, 287–293. [CrossRef]

287. Oleson, J.D.; Park, Y.-L.; Nowatski, T.M.; Tollefson, J.J. Node-injury scale to evaluate root injury by corn rootworms (Coleoptera: Chrysomelidae). *J. Econ. Entomol.* **2005**, *98*, 1–8. [CrossRef] [PubMed]

288. St. Clair, C.R.; Head, G.P.; Gassmann, A.J. Western corn rootworm abundance, injury to corn, and resistance to Cry3Bb1 in the local landscape of previous problem fields. *PLoS ONE* **2020**, *15*, e0237094. [CrossRef] [PubMed]

289. Broquet, T.; Petit, E.J. Molecular estimation of dispersal for ecology and population genetics. *Annu. Rev. Ecol. Evol. Syst.* **2009**, *40*, 193–216. [CrossRef]

290. Lowe, W.H.; Allendorf, F.W. What can genetics tell us about population connectivity? *Mol. Ecol.* **2010**, *19*, 3038–3051. [CrossRef] [PubMed]

291. Cayuela, H.; Rougemont, Q.; Prunier, J.G.; Moore, J.-S.; Clobert, J.; Besnard, A.; Bernatchez, L. Demographic and genetic approaches to study dispersal in wild animal populations: A methodological review. *Mol. Ecol.* **2018**, *27*, 3976–4010. [CrossRef]

292. Wright, S. Evolution in Mendelian populations. *Genetics* **1931**, *16*, 97–159. [CrossRef]

293. Weir, B.S.; Cockerham, C.C. Estimating F-statistics for the analysis of population structure. *Evolution* **1984**, *38*, 1358–1370. [CrossRef]

294. Holsinger, K.E.; Weir, B.S. Genetics in geographically structured populations: Defining, estimating and interpreting F_{ST}. *Nat. Rev. Genet.* **2009**, *10*, 639–650. [CrossRef]

295. Bradburd, G.S.; Ralph, P.L. Spatial population genetics: It's about time. *Annu. Rev. Ecol. Evol. Syst.* **2019**, *50*, 427–449. [CrossRef]

296. McCauley, D.E.; Goff, P.W. Intrademic genetic structure and natural selection in insects. In *Genetic Structure and Local Adaptation in Natural Insect Populations: Effects of Ecology, Life History, and Behavior*; Mopper, S., Straus, S.Y., Eds.; Chapman & Hall: New York, NY, USA, 1998; pp. 181–204.

297. De Meeûs, T.D.; Ravel, S.; Solano, P.; Bouyer, J. Negative density-dependent dispersal in tsetse flies: A risk for control campaigns? *Trends Parsitol.* **2019**, *35*, 615–621. [CrossRef]

298. Wright, S. *Evolution and Genetics of Populations*; University of Chicago Press: Chicago, IL, USA, 1969; Volume 2.

299. Wright, S. Isolation by distance. *Genetics* **1943**, *28*, 114–138. [CrossRef] [PubMed]

300. Séré, M.; Thévenon, S.; Belem, A.M.G.; De Meeûs, T. Comparison of different genetic distances to test isolation by distance between populations. *Heredity* **2017**, *119*, 55–63. [CrossRef] [PubMed]

301. Rousset, F. Genetic differentiation and estimation of gene flow from *F*-statistics under isolation by distance. *Genetics* **1997**, *145*, 1219–1228. [CrossRef] [PubMed]

302. Watts, P.C.; Rousset, F.; Saccheri, I.J.; Leblois, R.; Kem, S.J.; Thompson, D.J. Compatible genetic and ecological estimates of dispersal rates in insect (*Coenagrion mercuriale*: Odonata: Zygoptera) populations: Analysis of 'neighbourhood size' using a more precise estimator. *Mol. Ecol.* **2007**, *16*, 737–751. [CrossRef] [PubMed]

303. Whitlock, M.C.; McCauley, D.E. Indirect measures of gene flow and migration: $F_{ST} \neq 1/(4Nm + 1)$. *Heredity* **1999**, *82*, 117–125. [CrossRef] [PubMed]

304. Mallet, J.B. Gene flow. In *Insect Movement: Mechanisms and Consequences*; Woiwod, I.P., Reynolds, D.R., Thomas, C.D., Eds.; CABI Publishing: Wallingford, UK, 2001; pp. 337–360.

305. Marko, P.B.; Hart, M.W. The complex analytical landscape of gene flow inference. *Trends Ecol. Evol.* **2011**, *26*, 448–456. [CrossRef] [PubMed]

306. Slatkin, M. Gene flow and the geographic structure of natural populations. *Science* **1987**, *236*, 787–792. [CrossRef] [PubMed]

307. Kim, K.S.; Sappington, T.W. Genetic structuring of western corn rootworm (Coleoptera: Chrysomelidae) populations in the U.S. based on microsatellite loci analysis. *Environ. Entomol.* **2005**, *34*, 494–503. [CrossRef]

308. Krysan, J.L.; Smith, R.F.; Branson, T.F.; Guss, P.L. A new subspecies of *Diabrotica virgifera* (Coleoptera: Chrysomelidae): Description, distribution, and sexual compatibility. *Ann. Entomol. Soc. Am.* **1980**, *73*, 123–130. [CrossRef]

309. Giordano, R.; Jackson, J.J.; Robertson, H.M. The role of *Wolbachia* bacteria in reproductive incompatibilities and hybrid zones of *Diabrotica* beetles and *Gryllus* crickets. *Proc. Natl. Acad. Sci. USA* **1997**, *94*, 11439–11444. [CrossRef]

310. Flagel, L.E.; Bansal, R.; Kerstetter, R.A.; Chen, M.; Carroll, M.; Flannagan, R.; Clark, T.; Goldman, B.S.; Michel, A.P. Western corn rootworm (*Diabrotica virgifera virgifera*) transcriptome assembly and genomic analysis of population structure. *BMC Genom.* **2014**, *15*, 195. [CrossRef] [PubMed]

311. Chen, H.; Wang, H.; Siegfried, B.D. Genetic differentiation of western corn rootworm populations (Coleoptera: Chrysomelidae) relative to insecticide resistance. *Ann. Entomol. Soc. Am.* **2012**, *105*, 232–240. [CrossRef]

312. Lombaert, E.; Ciosi, M.; Miller, N.J.; Sappington, T.W.; Blin, A.; Guillemaud, T. Colonization history of the western corn rootworm (*Diabrotica virgifera virgifera*) in North America: Insights from random forest ABC using microsatellite data. *Biol. Invasions* **2018**, *20*, 665–677. [CrossRef]

313. Coates, B.S.; Sumerford, D.V.; Miller, N.J.; Kim, K.S.; Sappington, T.W.; Siegfried, B.D.; Lewis, L.C. Comparative performance of single nucleotide polymorphism and microsatellite markers for population genetic analysis. *J. Hered.* **2009**, *100*, 556–564. [CrossRef] [PubMed]

314. Miller, N.J.; Ciosi, M.; Sappington, T.W.; Ratcliffe, S.T.; Spencer, J.L.; Guillemaud, T. Genome scan of *Diabrotica virgifera virgifera* for genetic variation associated with crop rotation tolerance. *J. Appl. Entomol.* **2007**, *131*, 378–385. [CrossRef]

315. Kim, K.S.; French, B.W.; Sumerford, D.V.; Sappington, T.W. Genetic diversity in laboratory colonies of western corn rootworm (Coleoptera: Chrysomelidae), including a nondiapause colony. *Environ. Entomol.* **2007**, *36*, 637–645. [CrossRef]

316. Guillemaud, T.; Ciosi, M.; Lombaert, E.; Estoup, A. Biological invasions in agricultural settings: Insights from evolutionary biology and population genetics. *Comptes Rendus Biol.* **2011**, *334*, 237–246. [CrossRef]

317. Endler, J.A. *Geographic Variation, Speciation, and Clines*; Princeton University Press: Princeton, NJ, USA, 1977.

318. Barton, N.H. The structure of the hybrid zone in *Uroderma bilobatum* (Chiroptera, Phyllostomatidae). *Evolution* **1982**, *36*, 863–866. [CrossRef]

319. Nathan, R.; Perry, G.; Cronin, J.T.; Strand, A.E.; Cain, M.L. Methods for estimating long-distance dispersal. *Oikos* **2003**, *103*, 261–273. [CrossRef]

320. Steyn, V.M.; Mitchell, K.A.; Terblanche, J.S. Dispersal propensity, but not flight performance, explains variation in dispersal ability. *Proc. R. Soc. B* **2016**, *283*, 20160905. [CrossRef]

321. Caton, B.P.; Fang, H.; Manoukis, N.C.; Pallipparambil, G.R. Quantifying insect dispersal distances from trapping detections data to predict delimiting survey radii. *J. Appl. Entomol.* **2021**, *146*, 203–216. [CrossRef]

322. Liu, B.R. Biphasic range expansions with short-and long-distance dispersal. *Theor. Ecol.* **2021**, *14*, 409–427. [CrossRef]

323. Berthouly-Salazar, C.; Hui, C.; Blackburn, T.M.; Gaboriaud, C.; van Rensburg, B.J.; van Vuuren, B.J.; Le Roux, J.J. Long-distance dispersal maximizes evolutionary potential during rapid geographic range expansion. *Mol. Ecol.* **2013**, *22*, 5793–5804. [CrossRef] [PubMed]

324. Jones, H.B.; Lim, K.S.; Bell, J.R.; Hill, J.K.; Chapman, J.W. Quantifying interspecific variation in dispersal ability of noctuid moths using an advanced tethered flight technique. *Ecol. Evol.* **2016**, *6*, 181–190. [CrossRef] [PubMed]

325. Asplen, M.K. Dispersal strategies in terrestrial insects. *Curr. Opin. Insect Sci.* **2018**, *27*, 16–20. [CrossRef] [PubMed]

326. Naranjo, S.E. Assessing insect flight behavior in the laboratory: A primer on flight mill methodology and what can be learned. *Ann. Entomol. Soc. Am.* **2019**, *112*, 182–189. [CrossRef]

327. Hein, G.L.; Tollefson, J.J.; Foster, R.E. Adult Northern and Western Corn Rootworm (Coleoptera: Chrysomelidae) Population Dynamics and Oviposition. *J. Kan. Entomol. Soc.* **1988**, *61*, 214–223. Available online: https://www.jstor.org/stable/25084942 (accessed on 24 October 2023).

328. Elliott, N.C.; Hein, G.L. Population dynamics of the western corn rootworm: Formulation, validation, and analysis of a simulation model. *Ecol. Modelling* **1991**, *59*, 93–122. [CrossRef]

329. Steffey, K.L.; Tollefson, J.J. Spatial dispersion patterns of northern and western corn rootworm adults in Iowa cornfields. *Environ. Entomol.* **1982**, *11*, 283–286. [CrossRef]

330. Falconer, D.S.; Mackay, T.F.C. *Introduction to Quantitative Genetics*, 4th ed.; Longman: Harlow, UK, 1996.

331. Roff, D.A. The evolution of threshold traits in animals. *Q. Rev. Biol.* **1996**, *71*, 3–35. [CrossRef]

332. Pulido, F.; Berthold, P. Current selection for lower migratory activity will drive the evolution of residency in a migratory bird population. *Proc. Natl. Acad. Sci. USA* **2010**, *107*, 7341–7346. [CrossRef] [PubMed]

333. Pulido, F. Evolutionary genetics of partial migration–the threshold model of migration revis(it)ed. *Oikos* **2011**, *120*, 1776–1783. [CrossRef]

334. Fudickar, A.M.; Jahn, A.E.; Ketterson, E.D. Animal migration: An overview of one of nature's great spectacles. *Annu. Rev. Ecol. Evol. Syst.* **2021**, *52*, 479–497. [CrossRef]

335. Nijhout, H.F. Control mechanisms of polyphenic development in insects. *BioScience* **1999**, *49*, 181–192. [CrossRef]

336. Zhang, L.; Jiang, X.-F.; Luo, L.-Z. Determination of sensitive stage for switching migrant oriental armyworms into residents. *Environ. Entomol.* **2008**, *37*, 1389–1395. [CrossRef] [PubMed]

337. Zhang, L.; Cheng, L.; Chapman, J.W.; Sappington, T.W.; Liu, J.; Cheng, Y.; Jiang, X. Juvenile hormone regulates the shift from migrants to residents in adult oriental armyworm, *Mythimna separata*. *Sci. Rep.* **2020**, *10*, 11626. [CrossRef] [PubMed]

338. Cheng, Y.; Sappington, T.W.; Luo, L.; Zhang, L.; Jiang, X. Starvation on first or second day of adulthood reverses larval-stage decision to migrate in beet webworm (Lepidoptera: Pyralidae). *Environ. Entomol.* **2021**, *50*, 523–531. [CrossRef]

339. Wang, F.; Lv, W. Low temperature triggers physiological and behavioral shifts in adult oriental armyworm, *Mythimna separata*. *Bull. Entomol. Res.* **2022**, *112*, 546–556. [CrossRef] [PubMed]

340. Goossens, S.; Wybouw, N.; Van Leeuwen, T.; Bonte, D. The physiology of movement. *Mov. Ecol.* **2020**, *8*, 5. [CrossRef]

341. Rankin, M.A.; Riddiford, L.M. Significance of haemolymph juvenile hormone titer changes in timing of migration and reproduction in adult *Oncopeltus fasciatus*. *J. Insect Physiol.* **1978**, *24*, 31–38. [CrossRef]

342. McNeil, J.N.; Cusson, M.; Delisle, J.; Orchard, I.; Tobe, S.S. Physiological integration of migration in Lepidoptera. In *Insect Migration: Tracking Resources Through Space and Time*; Drake, V.A., Gatehouse, A.G., Eds.; Cambridge University Press: Cambridge, UK, 1995; pp. 279–302.

343. Zera, A.J.; Denno, R.F. Physiology and ecology of dispersal polymorphism in insects. *Annu. Rev. Entomol.* **1997**, *42*, 207–230. [CrossRef] [PubMed]

344. VanWoerkom, G.J.; Turpin, F.T.; Barret, J.R., Jr. Wind effect on western corn rootworm (Coleoptera: Chrysomelidae) flight behavior. *Environ. Entomol.* **1983**, *12*, 196–200. [CrossRef]

345. Chapman, J.W.; Reynolds, D.R.; Smith, A.D.; Smith, E.T.; Woiwod, I.P. An aerial netting study of insects migrating at high altitude over England. *Bull. Entomol. Res.* **2004**, *94*, 123–136. [CrossRef]

346. Leahy, M.G.; Craig, G.B., Jr. Accessory gland substance as a stimulant for oviposition in *Aedes aegypti* and *A. albopictus*. *Mosq. News* **1965**, *25*, 448–452.

347. Huignard, J. Influence de la copulation sur la fonction reproductrice female chez *Acanthoscelides obtectus* (Coleoptera: Bruchidae). I. Copulation et spermatophore. *Ann. Sci. Nat. Zool. Paris* **1974**, *16*, 361–434.

348. Chen, P.S.; Stumm-Zollinger, E.; Aigaki, T.; Balmer, J.; Bienz, M.; Böhlen, P. A male accessory gland peptide that regulates reproductive behavior of female *D. melanogaster*. *Cell* **1988**, *54*, 291–298. [CrossRef] [PubMed]

349. Kubli, E. My favorite molecule: The sex-peptide. *BioEssays* **1992**, *14*, 779–784. [CrossRef] [PubMed]

350. Spencer, J.L.; Bush, G.L., Jr.; Keller, J.E.; Miller, J.R. Modification of female onion fly, *Delia antiqua* (Meigen), reproductive behavior by male paragonial gland extracts (Diptera: Anthomyiidae). *J. Insect Behav.* **1992**, *5*, 689–697. [CrossRef]

351. Yamane, T.; Miyatake, T. Induction of oviposition by injection of male-derived extracts in two *Callosobruchus* species. *J. Insect Physiol.* **2010**, *56*, 1783–1788. [CrossRef]

352. Tóth, M.; Törőcsik, G.Y.; Imrei, Z.; Vörös, G. Diel rhythmicity of field responses to synthetic pheromonal or floral lures in the western corn rootworm *Diabrotica v. virgifera*. *Acta Phytopathol. Entomol. Hung.* **2010**, *45*, 323–328. [CrossRef]

353. Rice, M.E. William B. Showers, Jr.: *Semper Fi* and cutworms that fly. *Am. Entomol.* **2018**, *64*, 8–14. [CrossRef]

354. Vörös, G.; Szentkirályi, F.; Takács, J. Detection of flight of the western corn rootworm (*Diabrotica virgifera virgifera* LeConte) with light traps. *Növényvédelem* **2002**, *38*, 539–541. (In Hungarian with English Summary).

355. Chen, R.L.; Bao, X.Z.; Drake, V.A.; Farrow, R.A.; Wang, S.Y.; Sun, Y.J.; Zhai, B.P. Radar observations of the spring migration into northeastern China of the oriental armyworm, *Mythimna separata* and other insects. *Ecol. Entomol.* **1989**, *14*, 149–162. [CrossRef]

356. Drake, V.A.; Reynolds, D.R. *Radar Entomology: Observing Insect Flight and Migration*; CABI: Wallingford, UK, 2012; ISBN 978-1-84593-556-6.

357. Bauer, S.; Shamoun-Baranes, J.; Nilsson, C.; Farnsworth, A.; Kelly, J.F.; Reynolds, D.R.; Dokter, A.M.; Krauel, J.F.; Petterson, L.B.; Horton, K.G.; et al. The grand challenges of migration ecology that radar aeroecology can help answer. *Ecography* **2019**, *42*, 861–875. [CrossRef]

358. Shields, E.J.; Testa, A. Fall migratory flight initiation of the potato leafhopper, *Empoasca fabae* (Harris) (Homoptera: Cicadellidae): Observations in the lower atmosphere using remote piloted vehicles. *Agric. For. Meteorol.* **1999**, *97*, 317–330. [CrossRef]
359. Taylor, R.A.J.; Shields, E.J. Revisiting potato leafhopper, *Empoasca fabae* (Harris), migration: Implications in a world where invasive insects are all too common. *Am. Entomol.* **2018**, *64*, 44–51. [CrossRef]
360. Chapman, J.W.; Bell, J.R.; Burgin, L.E.; Reynolds, D.R.; Pettersson, L.B.; Hill, J.K.; Bonsall, M.B.; Thomas, J.A. Seasonal migration to high latitudes results in major reproductive benefits in an insect. *Proc. Natl. Acad. Sci. USA* **2012**, *109*, 14924–14929. [CrossRef]
361. Jones, C.M.; Papanicolaou, A.; Mironidis, G.K.; Vontas, J.; Yang, Y.; Lim, K.S.; Oakeshott, J.G.; Bass, C.; Chapman, J.W. Genomewide transcriptional signatures of migratory flight activity in a globally invasive insect pest. *Mol. Ecol.* **2015**, *24*, 4901–4911. [CrossRef]
362. Merlin, C.; Iiams, S.E.; Lugena, A.B. Monarch butterfly migration moving into the genetic era. *Trends Gen.* **2020**, *36*, 689–701. [CrossRef]
363. Yu, E.Y.; Gassmann, A.J.; Sappington, T.W. Using flight mills to measure flight propensity and performance of western corn rootworm, *Diabrotica virgifera virgifera* (LeConte). *J. Vis. Exp.* **2019**, *152*, e59196. [CrossRef]
364. Coates, B.S.; Deleury, E.; Gassmann, A.J.; Hibbard, B.E.; Meinke, L.J.; Miller, N.J.; Petzold-Maxwell, J.; French, B.W.; Sappington, T.W.; Siegfried, B.D.; et al. Up-regulation of apoptotic- and cell survival-related gene pathways following exposures of western corn rootworm to *Bacillus thuringiensis* crystalline pesticidal proteins in transgenic maize roots. *BMC Genom.* **2021**, *22*, 639. [CrossRef] [PubMed]
365. Coates, B.S.; Walden, K.K.O.; Lata, D.; Vellichirammal, N.N.; Mitchell, R.F.; Andersson, M.N.; McKay, R.; Lorenzen, M.D.; Grubbs, N.; Wang, Y.-H.; et al. A draft *Diabrotica virgifera virgifera* genome: Insights into control and host plant adaption by a major maize pest insect. *BMC Genom.* **2023**, *24*, 19. [CrossRef] [PubMed]

 *insects*

Review

Models of *Diabrotica* Populations: Demography, Population Genetics, Geographic Spread, and Management

David W. Onstad [1,*], Michael A. Caprio [2] and Zaiqi Pan [3]

[1] Corteva Agriscience, Johnston, IA 50131, USA

[2] Department of Biochemistry, Molecular Biology, Entomology and Plant Pathology, Mississippi State University, Mississippi State, MS 39762, USA; mac24@msstate.edu

[3] Corteva Agriscience, Chestnut Run Plaza 735/4175-3, 974 Centre Rd, Wilmington, DE 19805, USA; zaiqi.pan@corteva.com

* Correspondence: david.onstad@corteva.com

Received: 9 September 2020; Accepted: 14 October 2020; Published: 17 October 2020

Simple Summary: Two beetles that are serious pests of maize, *Diabrotica virgifera virgifera* and *Diabrotica barberi*, have caused problems for farmers in the USA and Europe for many years. Because both species have developed resistance to several management tactics, including insecticides and crop rotation, mathematical modeling has been used to evaluate their life cycles for weaknesses and new tactics for value. This review highlights lessons learned from the past 35 years. Some models have focused on the probability of the beetles spreading across regions. Other models have been developed to estimate the risk of the evolution of resistance. These models are thoroughly reviewed with respect to the biological attributes incorporated in these models and the impact of those attributes on the evolution of resistance.

Abstract: Both *Diabrotica virgifera virgifera* LeConte and *D. barberi* Smith and Lawrence are among the most damaging insects impacting corn in North America. *D. virgifera virgifera* has also invaded Europe and has become an important pest in that region. Computer models have become an important tool for understanding the impact and spread of these important pests. Over the past 30 years, over 40 models have been published related to these pests. The focus of these models range from occupancy models (particularly for Europe), impact of climate change, range expansion, economics of pest management, phenology, to the evolution of resistance to toxins and crop rotation. All of these models share characteristics. We elaborate on the methods in which modelers have incorporated the biology of these pests, including density-dependence, movement, fecundity and overwintering mortality. We discuss the utility of both spatially-explicit, complex models and spatially-implicit, generational models and where each might be appropriate. We review resistance models that either explain past evolution to crop rotation, insecticides or insecticidal traits or attempt to predict the consequences of resistance management strategies.

Keywords: *Diabrotica barberi*; *Diabrotica virgifera virgifera*; insect resistance management; population dynamics

---

1. Introduction

Mathematical modeling allows one to explore scenarios and conditions that would be difficult to study with experiments. Time horizons can be much longer and spatial scales much larger. Models can be used to explain the past or predict the future. They can be used solely for biological purposes, or modeling can help us improve management and evaluate economics. In this review, we focus on the history of modeling populations of *Diabrotica* species with an emphasis on biology, but with consideration of management.

Diabrotica (Coleoptera: Chrysomelidae) are important pests of corn (*Zea mays*) in the Americas and, over the past few decades, also in Europe [1]. The larval stage feeds on plant roots, making data collection more difficult compared to species with larvae that feed above ground. An interesting aspect of *Diabrotica barberi* Smith and Lawrence (DB) and *Diabrotica virgifera virgifera* LeConte (DVV) in North America, with regard to both management and population biology, is their adaptation to crop rotation [2].

We present a narrative that describes the purpose of each model and highlights the unique contributions made by modelers as they create new models that often, as much as science does, build upon the experimental and modeling work of earlier scientists. We have chosen to ignore or de-emphasize conclusions drawn by authors concerning management, because these are likely specific to a given product and a *Diabrotica* phenotype that are not likely to be encountered by modelers in the future [3]. The lessons about how to create models are what we hope to share with the next generation of modelers.

One of the earliest models of *Diabrotica* addressed whether adulticide applications could reduce populations sufficiently so that subsequent larval damage would be ameliorated [4]. This early model illustrated the important interplay between models and field research. Although the paradigm is often assumed to be field research followed by model development, this model also illustrated the alternative paradigm of model development (first to identify data gaps and weaknesses) followed by prioritization of field research to address those gaps. In this scenario, the models identify, through processes such as sensitivity analysis or risk assessments, not just data gaps, but the data gaps that are most important to answer the particular questions being addressed.

This review is divided into several sections. First, we review models that emphasized the population biology of DVV. Then, we do the same for models of DB. Next, we review models simulating evolution of resistance to crop rotation. Fourth, we consider models simulating evolution of resistance to insecticides and insecticidal traits. Fifth, we review models of geographic spread of DVV. Finally, we draw conclusions in the discussion.

The models reviewed here consider environmental space in a variety of ways. Spatially implicit models are usually the default choice because they are easier to create and analyze when modeling populations. However, when the goal of the project requires, or knowledge of reality motivates, we often model populations moving and interacting in explicitly defined space, such as when we need to study geographic spread of pests or dispersal across heterogeneous landscapes. In these cases, spatially explicit models should be built, if supporting data are available.

2. Fundamental Models of *Diabrotica virgifera virgifera*

Many different types of models have been useful in addressing a variety of questions regarding *Diabrotica* biology. The parameters and functional forms of the parameters vary with the questions being addressed by the models. Models that address impacts of climate change on *Diabrotica* distribution will likely include temperature dependent functional forms for many parameters, while models focusing on evolution are less likely to do so. Before construction of any model it is key to understand how the model is to be used and the questions or hypotheses to be addressed.

To date, only one complete age-specific life table has been published for DVV [5]. There are several observations from this study that may be useful for modeling DVV. The first is that while the greatest single mortality factor in the life table is related to the establishment of first instars, there is little variance between years in this estimate. The greatest impacts on differences in growth rates between years were the overwintering egg mortality rate and mortality of second and third instars. Secondly, there were large differences between potential fecundity, measured as the fecundity of adults held under ideal laboratory conditions, compared to realized fecundity, the fecundity measured from similar adults held under field conditions in cages. The realized fecundity ranged from 13.1% to

16.1% of the potential fecundity (estimated at 353 eggs/female rather than the maximal observed egg production). Third, at least in southern Hungary, the natural enemy complex and pathogens had very little impact on growth rates. This may be related to the relatively recent introduction of DVV in Hungary. Ultimately, all of the life tables developed in Toepfer and Kuhlman [5] had growth rates less than one, suggesting the population sizes were decreasing. Use of the absolute parameter estimates from these tables may not be warranted when modeling areas where DVV commonly achieves pest status, but the overall pattern of life table parameters may still be useful.

2.1. Density Dependent Mortality

DVV has long been known to have decreased survival when densities of larvae are high [6–12]. Elliot and Hein [13] incorporated density dependence into an early population dynamics model of western corn rootworm, focusing on interactions with the corn plant growth stage. They suggested density-dependent mortality was negligible below 466 eggs/m^2. Genetics was not considered in this model. Onstad et al. [14] developed the first DVV resistance model that incorporated density dependence. Based on Branson and Sutter [6], Gray and Tollefson [15] and Elliot et al. [10], Onstad et al. [14] modeled density-dependent survivorship as $1/(1 + 2.42 \times 10^{0.7})$, where E is the egg density in millions/ha. The density-dependent survivorship in this model of resistance to transgenic plants occurred after mortality due to the toxins. This same equation was adopted in a model by Storer [16]. Note that because the denominator approaches 1 as the egg density decreases, this equation assumes that other density-independent survival factors are incorporated in other equations of the model. Crowder and Onstad [17], further developed a model incorporating rotation resistant DVV [18] using $1/(2.59 + 1.29 \times 10^{0.88})$ for density-dependent survival. Caprio et al. [19] split the survivorship curve into two sections (piece-wise regression), assuming that density dependent-survival was 1.0 if egg densities were below 288 million eggs/4.05 ha (ca. 960 eggs/plant). Above this cutoff, density-dependent survival was calculated as the maximum of 0 or $0.012 \times 10^{(\times 10^{-0.004})}$, where E is the egg density in millions/4.05 ha.

Ultimately, while all these approaches to density-dependent survival vary in specifics, it is perhaps most important that density-dependence is considered, particularly in multi-patch models where densities between patch types can vary [20–22]. Martinez et al. [23] did note that the actual form of density-dependence can impact rates of resistance evolution. Haridas et al. [24] modeled the interaction of the mean and variance of soil temperature with density dependent and independent mortality utilizing larval survivorship data from Hibbard et al. [12]. They suggested these interactions could result in populations that exhibit population size equilibrium, cycling or extinction. Onstad et al. [11] summarize many studies on density dependent survivorship in *Diabrotica*.

2.2. Oviposition and Fecundity

DVV females go through an extended, circa 13-day, preovipositional period after emergence. Oviposition rates are high after this period, declining as females age [25]. Elliott and Hein [13] used a maximal rate of 310 eggs/female [26], modified by the plant growth stage, insect age and temperature. Onstad and collaborators [11,14,17,18,27–29] used a fecundity rate of 440 eggs/female (or 220 female eggs/female). Storer [16] assumed females laid a maximal rate of 23 eggs/day for a total female fecundity of 118 eggs/female. This was further modified by the age and genotype of the female and her mate. Caprio et al. [19] used an oviposition rate of 29 eggs/day for a 28 day stage following a 13 day pre-ovipositional period. Older females produced 7.5 eggs/day. Based on the daily adult mortality rates in the life tables, this resulted in a net reproductive rate of 11.35 female adults produced on average per female (ignoring overwintering egg mortality which averaged 50%). Given the average survivorship to the adult stage of 3.0%, the mean number of female eggs oviposited by an average female was 378.3. Pan et al. [30] assumed a maximal rate of 356 eggs/female, though actual fecundity depended on genotype, natal habitat and dominance of resistance alleles.

To more accurately model oviposition over the season, Onstad et al. [14] accounted for adult feeding on corn and noncorn plants and aging of adults when calculating adult mortality. Based on data collected by Elliott et al. [31], Onstad et al. [14] created a function for nutrition-based mortality of adults. As the corn crop matures beyond the early reproductive stages, adult mortality increases.

2.3. Adult Dispersal

Dispersal can be measured in several different ways depending on the form of the model. Some models use a generational time-step, others use a daily time-step, e.g., Crowder and Onstad [17], Crowder et al. [28]. Some models are spatially explicit, that is there is a specific ordering and structure to patches (fields) so some patches are closer while others are more distant (concepts of distance, dispersal kernels and spread become important) [14,16,19,29,30,32–36]. Other models are spatially implicit and model populations with no spatial structure and movement is represented as an exchange of individuals between these populations without regards to distance [37–42]. Dispersal estimates for DVV are also complicated by the fact that several flight mill studies indicated both a trivial flight and sustained flight behavior [43–45]. Different models may focus on different aspects of these two behaviors. Recent studies using mark-recapture techniques in response to the introduction of DVV into Europe have provided additional data that might be useful in modeling movement of DVV [46–48]. Estimates from all the above studies have suggested that between 24–38% of the adult DVV population engages in density-independent, long-distance dispersal. Higher densities may increase the proportion of beetles engaging in long distance dispersal [48], although Naranjo [49] found a negative relationship between larval density and sustained flight behavior in a flight mill study.

Elliott & Hein [13] used a daily migration rate of 15%/day for females aged 5–10 days and a rate of 7.5% for similar aged males. This was a model of a single population, so emigration of adults was treated mathematically as mortality. Onstad et al. [14] used a daily between field dispersal rate of 2% for females and 0.5% for males. Storer [16] varied the daily dispersal rate depending on crop and crop phenology. The daily dispersal rates were 5%, 2.5% and 15% from preflowering, flowering and mature stage corn respectively. Due to the nature of the dispersal kernel not all of these dispersal events resulted in movement between fields. Crowder et al. [28] assumed a panmictic distribution over the fields they simulated, at least for individuals not resistant to crop rotation. Crowder and Onstad [17], using a generational time step model, simulated a similar panmictic movement pattern but with the addition of some male dispersal prior to mating (all females mated in their natal field). They noted "that the results of the daily time step and generational time step models were similar and that the complexity involving dispersal, mating, emergence, and oviposition needed to develop the daily model did not affect the results as significantly as other assumptions". Nonetheless, should one wish to approach a problem tied to a daily timestep, such as developmental delay of larvae, a daily timestep model might allow elucidation of that factor with fewer abstractions and more easily measured assumptions than a simpler generational model. Caprio et al. [19] only simulated long distance dispersal, assuming populations within patches/fields were panmictic. They used a dispersal rate of 1.13%/day for preovipositional adults (15% of all adults dispersed over the stage) and 0.2%/day for older adults (resulting in an additional 7.4% of adults engaging in long distance dispersal), for a total of 22.4% of all adults engaging in long distance dispersal. A different measure of dispersal would be the location of where adults placed their eggs, combining age-specific dispersal, mortality and fecundity rates. Caprio et al. [19] estimated that 83.7% of the eggs in their model were oviposited in the natal field. As with most multi-patch models, movement of mated females actually results in the movement of two genomes, that of the female and the male with which she had mated. Because DVV females only mate once or twice, this can result in greater gene flow than might be observed in species that mate frequently.

Other models have focused on the implications of fine-scale movement for local population dynamics. Pan et al. [30] focused on within field dispersal using estimates from mark-recapture studies. They simulated 80 ha corn fields that were further subdivided into a 10 × 8 grid of 1 ha cells.

Movement was simulated with an exponential distribution where the probability of moving a given distance is a function of the mean daily dispersal distance, assumed to be 15 m/d for both males and females) and the direction was drawn from a uniform circular distribution. Males dispersed upon emergence while females did not disperse until mated. The goal of this model was to more realistically describe the effects of local dispersal on insect resistance managenent (IRM). Caprio and Glaser [35] focused on local dispersal within fields using a simulation system of 0.09 ha (30 × 30 m) patches. They assumed that 55% of beetles moved per day and moved an average of 29 m/day. This value is similar to the value used in Pan et al. [30] but only measures the distance moved by the 55% of the beetles that engaged in movement. Because beetles were assigned a random position in the 30 × 30 m block and movement was in a random direction, not every movement led to a change in the block the insect was assigned to. The distance moved for each movement was based directly on random draws from binned movement data [50], and therefore was a direct translation of field measurements. These two models demonstrated that incorporating fine-scale measurements of movement with mating, fecundity, oviposition and mortality can alter the evolutionary trajectory of local populations compared to simulations that assume random mating and oviposition within patches or fields.

2.4. Overwintering Mortality

Overwintering mortality is an important factor in the yearly univoltine dynamics of DVV. Most authors have used a value in the range of 50% mortality during the overwintering phase [11,14,17,28,29]. Caprio et al. [19] also used 50% overwintering mortality but in an attempt to simulate yearly differences in population densities, that value was varied as a random draw from a normal distribution with a mean of 0.5 and a standard deviation of 0.25, with limits bounded by (0.05,1.0) Storer [16] does not explicitly state an overwintering mortality, but does use a density-independent egg survivorship of 0.05.

3. Fundamental Models of *Diabrotica barberi*

DB is well-known for variation in length of diapause in natural populations. Instead of diapause occurring only over one winter, mutants maintain diapause over two to three winters (and time in between). When crop rotation became a common practice, prolonged or extended diapause became more prevalent. Adults feed on many host plants besides corn, and may move long distances from corn fields [51].

In a series of papers, Naranjo described the creation and analysis of a relatively detailed model of DB adult dynamics over time in a single field of corn (*Zea mays*). In an excellent paper, Naranjo and Sawyer [52] described their core submodel for DB reproduction. Their model had temperature-dependent maturation through the adult stage and time and age-dependent oviposition. They used temperature to determine mean rate of maturation (physiological time) and included a temperature-independent distributed delay to determine the proportion of each cohort developing to the next stage. The stages modeled included the pre-reproductive, reproductive and post-reproductive stages of the female population. Mean rates of maturation were nonlinear functions of temperature. The model kept track of the number of individuals in each cohort and their physiological age. They calibrated the model and all of its equations by fitting curves to maturation and oviposition data that they collected. An hourly time scale was used to calculate physiological time and a daily time step was used to determine oviposition and cohort transition to the next life stage. The model was based on studies performed at constant temperature. Naranjo and Sawyer [52] validated the model against independent data collected at constant and fluctuating temperatures. Their model fit the observations well. A sensitivity analysis demonstrated that results were very sensitive to errors and variability in input temperatures.

Naranjo and Sawyer [53] expanded their core submodel of adulthood to quantify and validate the effects of crop phenology on adult populations and oviposition and to study the role of beetle dispersal in the dynamics within a single cornfield. They added a variable for density of adult males and modeled three growth stages of corn. The time steps remained the same as for the original core submodel [52]. Adult mortality was calculated from a function of nutrient availability in the corn crop. All functions were fit to data collected by Naranjo. The flowering stage of corn provides the necessary nutrients and influences not only mortality but also adult dispersal as described by Naranjo and Sawyer [52]. The model results were extensively tested against independent field data collected by Naranjo. Naranjo and Sawyer [53] urged entomologists to measure DB dispersal and feeding behavior (the weakest parts of their model) to improve the model and management.

Naranjo and Sawyer [54] performed extensive sensitivity analyses on their expanded model. One goal was to examine the relationship between adult abundance, oviposition and corn phenology. Another goal was to formalize hypothetical improvements in DB management based on adult sampling. Model results were very sensitive to changes in age-specific oviposition rates and to factors influencing the timing and duration of the flowering period of corn. Model results were moderately sensitive to changes in the timing of female emergence, rates of pre-reproductive female dispersal at peak flower, fecundity and maturation rates during reproduction.

Mitchell and Riedell [55] expanded on the work of Naranjo and Sawyer [53] in two major ways. First, they emphasized the variable and stochastic nature of temperature effects on DB maturation. Second, they included maturation and mortality of the immature stages (egg, larvae, and pupae). They incorporated stochastic density-dependent mortality of larvae based on field data. They believed that adult movement affects adult mortality because of feeding throughout the landscape not just in a single field. Daily rate of adult mortality depends on the availability of pollinating corn in all the fields in a region. Mitchell and Riedell [55] used the Naranjo and Sawyer [53] pollination model to calculate the proportion of plants flowering in the field of emergence. They expanded this with a submodel that calculates the availability of pollinating corn in all fields in the landscape. They fit a regression model to data on fields of corn and then converted to amount of plants pollinating.

Mitchell and Riedell [55] compared their results to a variety of field-data sets. Because of nonlinear equations, they concluded that realistic, variable temperatures produce better simulations of population dynamics compared to deterministic simulations using mean temperatures. A sensitivity analysis indicated that model results were extremely sensitive to some of the parameters defining crop phenology and its effects on adult mortality.

Mitchell and Onstad [56] created a generational time-step model of the entire life cycle that accounted for prolonged diapause in a population of DB in a landscape consisting of various patches of corn, soybean and wheat with and without rotation of corn. The purpose of the modeling was to study the evolution of resistance to transgenic insecticidal corn in this population. They calibrated a density-dependent larval survival function using field data [57]. Because Naranjo and Sawyer [53] concluded that their approach to calculating fecundity underestimated realistic values, Mitchell and Onstad [56] used the data of Boetel and Fuller [58] to choose a relatively high constant mean fecundity. Although previous models had emphasized the role of pollination and corn phenology in the movement of adults, Mitchell and Onstad [56] simplified the model and assumed eggs are uniformly distributed over all corn patches. They also assumed that some adults mate within their natal patch and a larger proportion mate randomly throughout the corn landscape.

Mitchell and Onstad [56] emphasized the influence of extended diapause on the evolution of resistance to transgenic insecticidal corn. Based on data, they assumed that the total hatch rate for eggs was 60%. The standard simulations either had no extended diapause (60% hatch after first winter) or distributed this total as 40% hatch for eggs after one winter, 15% after two winters, and 5% after three winters. In their extensive analysis, they varied and increased the proportion hatching after two or three winters but did not model evolution during a simulation. Mitchell and Onstad [56] concluded that extended diapause in DB has two offsetting effects on evolution of resistance to insecticidal corn. First, extended diapause injects older alleles with lower resistance allele frequencies into the breeding population, which slows resistance. Second, extended diapause speeds the population's recovery from perturbations (reduces the undercompensating, density-dependent larval survival) which accelerates resistance.

4. Modeling Evolution of Resistance to Crop Rotation

Alternating corn with another crop every other year was a very good design for managing *Diabrotica* in North America. However, DB and DVV adapted to corn-soybean rotation schedules in the United States. As noted above, prolonged or extended diapause in the egg stage became more prevalent in DB populations. DVV evolved oviposition behaviors that allowed mutants to lay eggs not only in cornfields but also in most other crops. Essentially, DVV lost its fidelity to corn as a location for oviposition because of the fitness advantages within a two-year crop rotation schedule [59].

The only model to consider rotation-resistance for DB is described above [56], but it does not explain or predict the evolution of resistance to crop rotation. A series of models produced by the Onstad lab considered the evolution of rotation-resistance by DVV and its management [59].

Onstad et al. [60] modeled the evolution of behavioral resistance by DVV to crop rotation that first occurred in Illinois in the USA. It was a simple, frequency-based (population genetics) model with a generational (1 year) time step. The model consisted of four plant patches: corn, rotated corn, soybean, and extra noncorn plants. The proportion of soybean in the landscape equaled the proportion of rotated corn. They assumed one gene for resistance with three alleles. Beetle phenotypes were corn-only wild-type individuals, soybean specialists, and those ovipositing in all vegetation. Onstad et al. [60] also defined the complex behaviors of genotypes with additive alleles. Mating was random in each patch, but nonrandom in the landscape. Movement and distribution of eggs was determined by the phenotype of the beetle and the proportional areas of the host plants in the landscape. All larvae hatching in noncorn patches including soybean die. Onstad et al. [60] determined selection by accounting for relative fitness (0–1) for each of the twelve genotypes. Relative fitness was calculated from logical formulas for relative fecundity and larval survival.

Onstad et al. [60] tested the model results against the first observation of resistance, which occurred about 16 years after the beetle invaded Illinois. Results showed that the greater the level of corn-soybean rotation in the landscape, the faster the evolution of the ovipositing generalists. Several parameter sets allowed the model results to match reality. The conclusion that this phenotype is the one that evolved was supported by the observation that beetles can be found in crops other than soybean. Model results demonstrated that the soybean specialist would have more difficulty evolving than the ovipositing generalist. Therefore, Onstad et al. [60] concluded that the soybean specialist did not evolve in Illinois.

Onstad et al. [18] updated and extended the model of Onstad et al. [60] based on data collected after the first publication. They modeled fecundity as total eggs per female beetle and incorporated density-dependent larval survival that is applied once per generation after overwintering and insecticidal mortality. Onstad et al. [18] used the improved, density-based model to evaluate seven management strategies including the typical two-year crop rotation. The six alternative system designs included host-plant resistance by repellency or insecticidal corn and unusual cropping schedules such as a three-year rotation or less rotation and more continuously planted corn.

Crowder et al. [28] and Crowder and Onstad [17] extended the models of Onstad et al. [14,18] to include more details about transgenic insecticidal corn or crop rotation. They analyzed multiple scenarios: both technologies/practices in areas with rotation-resistance or areas without rotation-resistance, and only one of the technologies/practices in each area. Crowder et al. [28] increased the proportional dispersal away from natal patches to allow results to match published data. Based on results of Onstad et al. [14], they simplified the mating behavior of the beetles. Crowder et al. [28] used the daily time-step model to study how results compared to observations of field emergence and apparent early mortality of adults in transgenic insecticidal cornfields. Crowder and Onstad [17] provided an interesting comparison of results from both the daily and generational models.

5. Modeling Evolution of Resistance to Insecticides and Insecticidal Traits

This section focuses on models that either explained past evolution of DVV to insecticides or insecticidal traits or attempted to predict the consequences of IRM strategies. Some models are hypothetical, while others are applied to particular insecticides or corn traits. The spatial aspects of the models and their complexity are determined by the goals of the projects (Table 1). We describe the processes that influenced model results, particularly expression of resistance, adult dispersal, and density-dependent survival. All the models described below assume that insect resistance is conferred by an autosomal, di-allelic major gene for each insecticide or insecticidal trait. Unless noted otherwise, resistance to each toxin is complete (100% survival by homozygous resistant beetles). In the models, initial resistance allele frequency is usually 0.0001–0.001.

Table 1. Purpose of model relative to integrated pest management (IPM) determines the complexity of model structure.

Reference	Time Step	Larval Movement	Adult Dispersal	Simulated Landscape	Purpose of Model
Onstad et al. [14]	Day	No	Cross patches	Spatially implicit, 100 ha with patches of corn and soybean	Estimate durability of rootworm traits with various structured refuge proportion
Storer [16]	Day	No	Intra-field and inter-field	Spatially explicit, 100 fields with each field of 25 ha planting corn or soybean	Estimate durability of rootworm traits with various structured refuge proportion
Onstad et al. [18]	Generation	No	Cross patches	Spatially implicit, 100 ha with patches of corn and soybean	Rotation resistance
Crowder and Onstad [17]	Generation	No	Cross patches	Spatially implicit, 100 ha with patches of corn and soybean	Transgenic resistance and rotation resistance
Crowder et al. [28]	Day	No	Cross patches	Spatially implicit, 100 ha with patches of corn and soybean	Transgenic resistance and rotation resistance
Onstad [61]	Generation	Yes	Cross patches	Spatially implicit, 100 ha with patches of corn and soybean	Comparison of blended refuge vs. block refuge
Caprio et al. [19]	Day	No	Cross patches	Spatially explicit, matrix of 25 fields	Retrospectively validated evolution of methyl-parathion resistance
Onstad and Meinke [29]	Generation	No	Cross patches	Spatially implicit, 100 ha with patches of corn and soybean	Durability of pyramid
Pan et al. [30]	Generation for larval stage and day for adult	Yes	Intra-field	Spatially explicit, 80 ha field gridded with 1 ha cells	Comparison of blended refuge vs. block refuge
Caprio and Glaser [35]	Day	No	Intra-field and long-range dispersal	Spatially explicit, multiple fields with four different field sizes	Evaluate block refuge effectiveness
Kang et al. [36]	Day	No	Ignored	Spatially implicit, seed blend in 100 ha field	Evaluate emergence delay and skewed sex ratio
Martinez and Caprio [32]	Day	Yes	Inter-field	Spatially explicit, 51 × 51 field matrix with 50 ha field	Evaluate durability benefit of IPM strategies to transgenic corn

When refuge of non-Bt corn is deployed as a seed blend (blended refuge) with Bt corn in the same field, larval behavior and related mortality factors are usually included in models. Although much attention is placed on movement from plant to plant, the most important question is Does survival, and therefore selection, differ between genotypes because of movement? Thus, movement per se may not be significant if heterozygotes do not have an advantage due to feeding and movement amongst refuge and insecticidal plants. Onstad [61] and Pan et al. [30] assumed that (1) only a fraction of the young larvae move and (2) movement occurs once per larval stage, and (3) most feeding by DVV larvae occurs after the one movement. Only minor mortality occurs due to tasting Bt corn before dispersal away from those plants in their models. Caprio and Glaser [35] modeled larval movement in their frequency-based deterministic model. They permitted daily chances for movement for all larvae. Thus, some larvae will move a few times and others will move often. Some will move later and others earlier in the larval stage.

Onstad et al. [14] created a deterministic simulation model of the population dynamics and genetics of DVV for a landscape of corn, soybean, and other crops. The model was published before any *Bt* transgenic rootworm-resistant corn was commercialized and was used to evaluate potential resistance management plans for transgenic corn. Results helped to identify research priorities and assisted those creating resistance management plans. The model simulated a crop landscape with continuous corn, rotated corn, soybean, and an extra noncorn crop. The model used a daily time step and was spatially implicit with most DVV biology of egg, larva, pupa and adult stages. Adult dispersal between different crops was also simulated in the model. The model assumed that no first instar DVV moved away from transgenic corn roots as only block and strip refuge was simulated. The density-dependent survival of larvae occurred after overwintering survival and survival due to conventional and transgenic plant toxins. This model was unique in its incorporation of mating complexity. Mating amongst DVV emerging in block refuge and fields of Bt corn is a concern as female beetles fly short distances before mating [62]. Based on lab data, the model assumed that female adults could mate in a second phase after the first oviposition period. Furthermore, Onstad et al. [14] calculated the probability of a female mating based on male density and search capability. They evaluated the sensitivity of results to male density and searching capability in blocks of corn. The standard search area was 100 m^2. Changing the area between 30 m^2 and 170 m^2 searched by a male each day for a mate had insignificant effect. Therefore, later models assumed mating would not be a concern and assumed that almost all realistic male densities could guarantee all females would be mated.

Onstad et al. [14] found the gene expression in DVV and toxin dose in the corn plant were the two most crucial factors affecting resistance development. The model results were not very sensitive to the refuge size between 5% and 30%. Because this might be due to density-dependent mortality in block refuges, the authors explored density-dependent survival in their future papers.

Storer [16] created a stochastic, spatially explicit computer model to simulate the adaptation by DVV to insecticidal traits in corn. The model included functions for crop development, egg and larval mortality, adult emergence, mating, egg laying, mortality and dispersal, and alternative methods of DVV control, to simulate the population dynamics of the DVV. The spatial unit in the Storer [16] model is a 25-hectare field and the whole landscape represents $10 \times 10 = 100$ fields. Adult dispersal is classified as inter-field dispersal and in-field flight. The time step is a day. Expression of resistance varies from incompletely recessive to incompletely dominant. The functional dominance value has a standard value of 0.1 with a range between 0.0001 and 0.88, depending on the efficacy of the Bt corn trait. The density-dependent mortality of larvae was calculated with a function from Onstad et al. [14]. The model was used to compare the rate at which the adaptation allele spread through the population under different nonresistant corn refuge deployment scenarios, and under various dose levels of the resistance trait. The model output is the relative rate of adaption of the specific scenario, which is the rate of adaption of each run divided by the rate of adaption for the baseline (20% refuge in locations rerandomized each year with default parameter sets). For a given refuge size, the model indicated that placing the nonresistant refuge in a block within a Bt corn field would be likely to delay DVV adaptation rather longer than planting the refuge in separate fields in varying locations. If a portion of the refuge were to be planted in the same fields or in-field blocks each year, DVV adaptation would be delayed substantially. DVV adaptation rate was also predicted to be greatly affected by the level of crop resistance, because of the expectation of dependence of functional dominance on dose. If the dose of the insecticidal protein in the corn is sufficiently high to kill 90% of heterozygotes and 100% of susceptible homozygotes, the trait is predicted to be much more durable than if the dose is lower. A sensitivity analysis showed that parameters relating to adult dispersal affected the rate of pest adaptation. Storer [16] also modeled soil insecticide and crop rotation with different block refuge strategies. Uniform distribution of block refuge and crop rotation is beneficial to delay resistance evolution.

Onstad et al. [18] expanded a simple model of adult behavior and population genetics to explain how rotation resistance may have developed and to study ways to manage the DVV in a landscape of corn, soybean, and winter wheat. The time step in this model is a generation. The resistance evolution to insecticide or insecticidal trait was not modeled. The model focused on a strategy involving a 2-year rotation of corn and soybean in 85% of the landscape, also evaluated six alternative management strategies over a 15-year time horizon to investigate their effectiveness. Results indicated the rate of evolution increases as the level of rotated landscape increases. The results were most sensitive to increases in the initial allele frequency, the function dominance of the resistance allele and modifications of the density-dependent survival function.

Crowder and Onstad [17] further expanded the simulation model described in Onstad et al. [18] to study the simultaneous development of resistance to both crop rotation and transgenic corn. The model supports two resistance genes. One gene is for crop rotation resistance and the other is for transgenic corn resistance. The dose of transgenic traits ranged from low dose with 20% survival to high dose with 0% survival. The dominance value of crop rotation resistance was simulated as dominant, additive and recessive. In simulations of areas with rotation-resistant populations, planting transgenic corn to only rotated cornfields (2 year crop rotation) was a robust strategy to prevent resistance to both traits. In areas with rotation resistance risk, planting transgenic corn to only continuous corn fields and planting conventional corn on rotated cornfields was not an effective strategy for preventing adaptation to crop rotation or transgenic corn. In areas without rotation-resistant phenotypes, the gene expression of the allele for resistance to transgenic corn was the most crucial factor affecting the development of resistance to transgenic corn. The model also concluded the initial allele frequency and density dependence were the two most crucial factors affecting the evolution of resistance.

Crowder et al. [28] expanded the deterministic model described in Onstad et al. [14], so the model could evaluate the risk of resistance to both transgenic crops and crop rotation in landscapes with and without rotation-resistant phenotypes as well. This complex model used a daily time-step and had a similar parameter set as Crowder and Onstad [17]. It included daily information of corn phenology, insect phenology, adult survival, emergence, dispersal and mating.

Results from Crowder et al. [28] indicated that planting transgenic corn to first-year cornfields is a robust strategy to prevent resistance to both crop rotation and transgenic corn in areas where rotation-resistant populations are currently a problem or may be a problem in the future. In areas without rotation-resistant populations, gene expression of the allele for resistance to transgenic corn is the most important factor affecting the evolution of resistance. If expression is recessive, resistance can be delayed longer than 15 year. If expression is dominant, resistance may be difficult to prevent. In a sensitivity analysis, results indicate that density dependence, rotational level in the landscape, and initial allele frequency are the three most important factors affecting the results.

Overall simulation results from a simpler model with a generational time-step [17] compared favorably with a more complex model [28] with a daily time-step. It may be assumed that the complexity involving dispersal, mating, emergence, and oviposition needed to develop the daily model did not affect the results as significantly as other assumptions about gene expression, refuge size, and toxin dose.

Onstad [61] expanded the population dynamics and genetics model published in 2005 by Crowder and Onstad [17] to include larval survival and movement to evaluate the role of mixtures of transgenic and nontransgenic corn seed for resistance management of DVV. The primary concern about seed blends in Onstad [61] was that blends may provide conditions for additional differential selection against susceptibles relative to heterozygotes leading to faster evolution of the pest. The survival of homozygous susceptible individuals (SS), or heterozygotes (RS) with R recessive, is 0, 0.001, 0.05, and 0.20. With R partially recessive, survival of the heterozygotes is 0, 0.01, 0.50, and 0.60 with a theoretical high, practical high, medium, and low toxin dose, respectively. The probability of leaving a nontransgenic plant and the proportion surviving dispersal as neonates are simulated as one of three sets (0.9, 0.06), (0.5, 0.5), and (0.35, 0.95). The probability of leaving a transgenic plant is either 0.9 or 0.5. The survival of first instars due to pre-dispersal tasting of transgenic roots, depends on genotype and allele expression. Onstad [61] studied pre-dispersal tasting survival values of 1 and 0.5 for RR and SS larvae. For heterozygotes, values of 1, 0.75, and 0.55 were simulated respectively. The results for the standard model were not very sensitive to refuge size. The comparison of blocks and seed blends with partially recessive expression was similar. Onstad [61] concluded, given the lack of random mating by DVV inhabiting blocks of transgenic and nontransgenic corn, the seed blends provide a positive effect to the trait durability.

Caprio et al. [19] developed and retrospectively validated a stochastic, individual-based, multifield, simulation model that predicts the evolution of methyl-parathion resistance in DVV populations of Nebraska. The life table parameters of DVV included neonate, larva, pupa, pre-ovipositional adult, young adult and old adult. Resistance to the methyl-parathion aerial adulticide was expressed as a dominant monogenic trait. A single application of methyl-parathion would kill 98.5% of susceptible individuals in the field, whereas 50% of resistant individuals with homozygous or heterozygous genotypes also were killed. Efficacy from a single spray decayed with a half-life of 10 d, and a total period of residual activity equivalent to 20 d. Resistance was determined by the failure of the spray program to reduce beetle populations in the field below a maximum acceptable density of 0.5 gravid females per plant. The landscape in the simulation was a matrix of 25 fields, all treated uniformly. The average plant population per field was 30,000 plants per acre. A consultant survey was conducted to estimate the field control failure of methyl-parathion aerial spray. The population dynamics properties of the model were qualitatively corroborated by comparing the simulated population dynamics with one of several sets of empirical population dynamics published data. Multiple, stepwise regressions were conducted for each of the two resistance measurements to determine model sensitivity to variation in parameters. Bayesian inference was used to estimate the candidate frequency most likely, given reported times to field control failures. Spray timing and multiple spray application strategy were also explored in the simulation.

The default parameters they incorporated into the model suggested that resistance could have evolved within 8.5 year, while the consultant survey reported control failures at some time with methyl-parathion, averaging 7.6 +/− 1.85 year, after it was initially used. Each decrease of an order of magnitude in the initial resistance allele frequency increased the time until resistance evolved by 2 year. Bayesian inference concluded the initial allele frequency of 10^{-4} was most likely (29%), 10^{-3} was less likely (28%). The model results indicated the time to control failure incorporates both the time it took for resistance to evolve, as well as some time for the population to rebuild to threshold levels. This validated model could also be used to help assess resistance risk for management strategies designed to sustain novel DVV control tactics that may be implemented in the future.

Onstad and Meinke [29] created a simulation model of the population dynamics and genetics of DVV to evaluate the use of refuges in the management of resistance to transgenic insecticidal corn, expressing one or two toxin traits. Hypothetical scenarios and a case study of a corn hybrid pyramided with two common insecticidal traits were simulated. This model extended the work of Crowder and Onstad [17]. The time step for calculation is a single generation (1 year). The main additions to Onstad and Meinke [29] model were (1) a second major gene for resistance, (2) possible reductions in fecundity for survivors of insecticidal corn, and (3) reduction in effective refuge because of limited compliance with refuge requirements. Landscape in Onstad and Meinke [29] is spatially implicit and it is a homogeneous region of cropland consisting only of continuous corn planted with transgenic and nontransgenic hybrids. Two field types were simulated, a field with a block refuge and a block of corn expressing two insecticidal traits, and a field with a block refuge and corn expressing only one insecticidal trait. Functional dominance values for each resistance gene is additive (0.5). The larval survival from density-dependent competition after mortality due to overwintering and toxin exposure was calculated by using a function in Crowder and Onstad [17] and fully explained in Onstad et al. [11]. Onstad and Meinke [29] evaluated required refuge levels of 5, 10, and 20%. The durability of transgenic corn products is defined as the number of years when all resistance allele frequencies reach 0.5.

In the hypothetical situations, results demonstrated that evolution is generally delayed by pyramids compared with deployment of a single-toxin corn hybrid. However, soil insecticide use in the refuge reduced this delay and quickened the evolution of resistance. Results were sensitive to the degree of male beetle dispersal before mating and to the effectiveness of both toxins in the pyramid. Resistance evolved faster as fecundity increased for survivors of insecticidal corn. Thus, effects on fecundity must be measured to predict which resistance management plans will work well. Evolution of resistance also occurred faster if the survival rate due to exposure to the two toxins was not calculated by multiplication of two independent survival rates (one for each insect gene) but was equivalent to the minimum of the two. Furthermore, when single-trait and pyramided corn hybrids were planted within rootworm-dispersal distance of each other, the toxin traits lost efficacy more quickly than they did in scenarios without single-trait corn. For the case study, the pyramid delayed evolution longer than a single trait corn hybrid and longer than a sequence of toxins based on at least one resistance-allele frequency remaining below 50%.

Pan et al. [30] created a simulation model of the temporal and spatial dynamics and population genetics of DVV to evaluate the use of block refuges and seed blends in the management of resistance to transgenic insecticidal corn. The model in Pan et al. [30] is based on the work of Onstad et al. [14], Crowder et al. [28], and Onstad [61]. This model combines an adult submodel calculated on a daily time step (adult emergence, dispersal, mating, and oviposition) with a larval submodel calculated once per generation (larval movement and survival). The resistance gene was modeled with the functional dominance value of 0.05. Pan et al. [30] modeled corn rootworm behavior in a set of 80 ha fields planted in corn every year. In the simulation, cornfields are adjacent to each other in a region simulated as a torus; hence, beetles leaving one side of a rectangular region are matched by beetles entering on the opposite side. The 80 ha field was divided into a grid with 10 columns by 8 rows, total 80 cells with each cell of 1 ha. For the fixed-location block refuge, the model chose to place the refuge in the middle of the field. For the scenario in which the refuge is relocated annually within the cornfield, the model alternated the refuge from middle to edge every other year. The daily adult emergence, 7 day protandry and developmental delay of susceptible genotypes were simulated in the adult submodel. In the adult dispersal submodel, the dispersal distance follows an exponential distribution and the dispersal direction is random. The male dispersal was modeled as a mean dispersal rate of 15 m per day and sensitivity analysis varied the mean dispersal rate from 5 to 45 m per day. The mated female adults disperse at a mean dispersal rate of 15 m per day, the same as male. Furthermore, the influence of more frequent random-walking with constant dispersal distance per step was also modeled. In the larval submodel, overwintering egg survival, larval movement and survival was modeled. The density-dependent survival occurs after mortality due to overwintering and Bt corn root exposure was adjusted by using a function in Onstad [61].

Pan et al. [30] concluded that the seed-blend scenarios in many cases produced equal or greater durability than block refuges that were relocated each year. The standard analysis presumed complete adoption of Bt corn by all farmers in the region, no crop rotation, and 100% compliance with IRM regulations. As compliance levels declined, resistance evolved faster when block refuges were deployed. Seed treatments that killed the pest when applied to all seeds in a seed blend or just to seeds in Bt corn blocks delayed evolution of resistance.

Caprio and Glaser [35] developed two simulation models to evaluate the DVV resistance evolution and management. The two models consisted of a modification of a spatially-explicit, stochastic model POPGEN-S2 [19] and a simpler, frequency-based deterministic model, POPGEN-D. The latter could be run in a single simulation mode with a graphical interface to enter parameters or in a risk-assessment mode capable of running thousands of simulations to estimate the effects of parameter uncertainty.

The biological parameters of the stochastic model were primarily derived from Caprio's previous DVV model [19] except for mortality rates, development delay, block size and dispersal rates. The stochastic model in Caprio and Glaser [35] was used only for evaluations of four different block refuge scenarios. The functional dominance values of 0.05 and 0.2 were simulated in the stochastic model. The simulation landscape consisted of a 40×40 matrix of 30 m $\times$ 30 m blocks, total 144 ha. Four different field sizes were simulated as 5×5 blocks, 10×10 blocks, 20×20 blocks and 40×40 blocks. Refuge was annually planted on one edge or opposite edge of each field randomly. Adult dispersal was modeled with the movement of an average of 41 m/day. Movement was assumed as a Brownian diffusion-based random walk. Additional rates for mortality and developmental delay of larvae on transgenic plants were adopted from a model, submitted to USEPA for product registration (DuPont Pioneer unpublished data). The stochastic model indicated the field size was highly significant to determine the resistance development. The results also indicated that rather than nonrandom mating being the threat to a block refuge system, it is nonrandom oviposition and refuge placement that are most important.

The deterministic model is a simple frequency-based model that incorporates noncompliance, nonrandom mating in compliant fields and larval movement [35]. It can simulate block refuge and blended refuge. The parameters in the deterministic model were drawn from uncertainty distributions, Pert distributions characterized by estimates of the most likely, minimum and maximum values. While each run of the simulation was deterministic, 1000 runs were conducted, each with randomly drawn parameters from the relevant uncertainty distributions. One of the key parameters in the deterministic model is the effective refuge size of block refuge. The worst-case effective refuge sizes for the four different field sizes were (from smallest to largest field size): 15%, 11%, 3% and 0.055% when the refuge was planted in the area of the field where the egg density was the lowest. For blended refuge, larval movement composed of base larval movement and asymmetrical movement, which accounts for larvae tending to taste and move from transgenic tissue at a greater rate than on non-Bt tissue. The movement rate is varied from 0% to 50%, with 30% as the most likely value. This model supports two loci and can compare the relative longevity of two genes stacked into the same plant versus using the same two genes sequentially in single gene plants.

Caprio and Glaser [35] concluded the durability of the block refuge is very sensitive to field size and the block refuge is better than the blended refuge at delaying resistance in the simulation of the deterministic model. The deterministic model is a frequency-based model and does not support population information, so it does not incorporate density dependent survival. This might overestimate the longevity of 20% refuge as the larval density is high enough to trigger density dependent mortality. Model results also indicated the expected lifetime of the dual gene pyramid product is much longer than if the two toxins had been used sequentially regardless of block or blend refuge deployment.

Kang et al. [36] created a daily time step population genetics model to study the effects of temporal separation between male and female beetle emergence from Bt and non-Bt corn and a female-skewed sex ratio for DVV emerging from Bt corn on the evolution of Bt resistance. The model simulated a 100 ha landscape with 20% blended refuge for single-trait Bt corn and 5% blended refuge for pyramided Bt corn. Two unlinked, loci in DVV were assumed to determine resistance to transgenic corn expressing Bt toxins. The toxin mortality was applied before density-dependent mortality occurred. The standard survival value for homozygous susceptible items was set at 0.01, and the standard dominance was set to 0.5. The density-dependent mortality occurred after toxin mortality. Variant sex-specific toxin mortality was assumed to determine the sex ratio of beetles emerging from Bt corn.

The standard values of male and female developmental delay were 0–3 and 0–2 days, the average of the differences between the mean emergence time for beetles from transgenic corn and the non-Bt corn. The sensitivity analyses ranged from 0 to 28 days. The effect of great emergence delays (>14 days) on resistance evolution was studied for theoretical purposes.

Kang et al. [36] concluded the effect of skewed toxin mortality in one sex on evolution of Bt resistance was insignificant. An emergence delay among resistant beetles from Bt corn slowed resistance evolution. A shift in the time of emergence for homozygous susceptible beetles from Bt corn did not have a significant effect on the evolution of Bt resistance in DVV.

Adult emergence delay has been a critical issue in DVV resistance management. Onstad et al. [14] simulated an emergence delay for homozygous susceptible beetles of 3–9 d in transgenic corn in fields with row strip or block refuges. They concluded that the emergence delay by susceptible beetles and the configuration of the refuge significantly accelerated the evolution of resistance by DVV. Storer [16] included developmental delays for susceptible and heterozygous beetles. For susceptible homozygotes and heterozygotes surviving on transgenic corn, the delay is 7 d and 3.5 d respectively. Caprio and Glaser [35] modeled 5% larvae feeding on transgenic corn were held back from each age group (though they did experience mortality) and did not age. This resulted in a mean developmental delay of 2.3 days. Pan et al. [30] included a 7-d delay in emergence for susceptible adults and heterozygotes emerging in transgenic corn compared with those insects emerging in refuge while there is no developmental delay for resistant adults. They showed that this type of developmental delay slightly delayed the evolution of Bt resistance by DVV. Kang et al. [36] modeled developmental delays in seed blends. The standard developmental delays for males and females were 3 d and 2 d and the sensitivity analysis evaluated 0 to 28 d. The results from Kang et al. [36] supported Pan et al. [30].

Martinez and Caprio [32] developed a two-gene, discrete generational and spatially explicit model with a probabilistic and deterministic mode for DVV. The model consisted of six separate, successive developmental and behavioral phases that occurred once per year: egg stage, larval stage, density-dependent mortality, pre-mating adult dispersal followed by mating and post-mating dispersal. This model is used to investigate the effects of IPM strategies along with planting of a structured refuge or seed blend on the lifetime of a hypothetical nonhigh dose Bt toxin pyramid.

The landscape consisted of a 51 × 51 field matrix (one field represented 50 ha (707 × 707 m) of corn with 4 million plants; matrix size = 36.1 km × 36.1 km) and was designed as a torus. A two-dimensional Gaussian redistribution kernel was used to simulate adult movement through space under the assumption of Brownian motion. The diffusivity parameter D (field^2/time step) describes the pest's propensity to disperse. Based on the daily dispersal rate 29 m/day and 775 m for lifetime dispersal, the diffusivity D was chosen as 2 field^2/generation for pre-reproductive adult dispersal and 4 field^2/generation for post-mating female dispersal. The model had only 10% adults perform pre-mating dispersal and 15% female performed long-distance post-mating dispersal.

Density dependent survival in the model occurred after natural and *Bt* mortality took place and was simulated as scramble competition using the Hassell equation to regulate population densities on refuge and Bt plants (separately) in each field. More details of the density dependent survival under scramble competition can be found in Martinez et al. [23]. The parameters were adjusted to result in an overall egg-to-larval survival of 4.5% in refuge under standard parameter set. The oviposition was assumed to be uniformly distributed across the 50-ha fields (blocks and fields with blended seed). Base larval movement in blended refuge was set to 30%.

Martinez and Caprio [32] concluded that crop rotation was the most effective strategy, followed by increasing the non-Bt refuge size from 5 to 20%. Soil applied insecticide use for Bt corn did not increase the durability compared with planting Bt with refuges alone, and both projected lower durabilities. The mitigation with random selection of strategies was ineffective at slowing resistance, unless crop rotation occurred immediately; regional mitigation was superior to random mitigation in the area with high levels of resistance and reduced the observed resistance allele frequencies in the neighborhood.

6. Models of *Diabrotica* Geographic Spread (>1 km/Year for Multiple Years)

D. virgifera virgifera (DVV) can fly several km using its own power and have flight distance lengthened by wind and storms. It is also possible that long-distance movement can be assisted by human activity. This section describes the wide variety of models that have been used to explain or predict geographic spread.

Onstad et al. [63] created and analyzed a set of simple meteorological and behavioral models that can be used to predict the spread of populations of DVV that had lost fidelity to corn, and therefore, commonly infested soybean (*Glycine max*) throughout the north central United States starting in the 1990s. These crop rotation-resistant beetles spread outwards from a focus in Illinois to infest multiple states over several years. This work extended the modeling performed by Onstad et al. [64] to determine how far and in what directions the population would spread in Illinois, Indiana, Michigan, and Ohio. Their primary hypothesis was that increased landscape diversity slows the rate of regional spread of the rotation-resistant DVV over several years. Thus, some of the models invoked a landscape-diversity function that included the proportion of noncorn, nonrotated-soybean vegetation on farmland in each county [63,64] called this extra vegetation).

Using logic and some flight mill data [43], Onstad et al. [63] determined that DVV adults can fly 3.9 km without wind support during the 1.30 h available on a typical summer day (based on temperature). Because the wave front of beetle dispersal will be determined by those flying downwind, the mean wind speeds for the 18 directions were calculated from the climate data. From these and the logic of the number of hours of flight per day, Onstad et al. [63] calculated the maximum possible wind-supported spread for each direction.

They used summer data for temperature, wind speed, and direction for several north central states to calculate wind-supported movement. In addition, to determine the role of storms, Onstad et al. [63] used published observations of heavy rainstorms and heavy rain cells (i.e., areas through which rain passes) to calculate probability distributions for storm tracks typical for the north central United States. Most movement is eastward or northeastward (coming from the west) with a maximum distance of 33 km [64]. This wave front rate was hypothesized to be valid for any threshold of measurement used to define the spread of the DVV. One sector (direction) was allowed maximum distance of spread; all others had reduced distances based on the proportion of storms observed for a given sector. Dispersal out of each sector was given the maximum of either wind-supported or storm-supported movement.

Finally, Onstad et al. [63] used data on the amount of farmland that was not corn or rotated soybean to calculate the proportional area of extra vegetation, which will act as a sink for the spreading population because offspring will die in fields without corn roots. Onstad et al. [63] used this information for each county to adjust the quality of each 1 km^2 cell. Their standard model used a linear effect to adjust the radii of dispersal used for each cell, depending on which of eight categories of extra vegetation that cell was in. For all cells within each category, the maximum radius for each sector was multiplied by (1-MEV), where MEV represents the mean value of extra vegetation for that category. They simulated the model for 16 years from 1986 to 2001 using geographic information system software to determine the presence of rotation-resistant DVV in grid cells in several states. The results were tested with observations from 1997–2001.

The best model for the period from 1997 to 2001 is based on heavy-storm data, with distance that beetles spread each year reduced by the proportion of extra vegetation in a county [63]. This model was superior to the previously published model and to two other models that do not consider landscape diversity. Most of the models predicted spread at too high a rate between 1997 and 2001, compared with observations, but a few new models with rates of spread reduced by a landscape-diversity function matched the observations relatively well.

Hemerik et al. [65] combined a temperature-based population dynamics model with a dispersal model [66] to predict how temperature differences would influence range expansion after the year 2000 in Europe, the initial year in the simulations. They created a mechanistic model of DVV maturation and population growth based on information in the literature. Adult flight behavior was estimated from flight mill studies of Coats et al. [43]. Hemerik et al. [65] also used literature information to calculate net reproduction rate based on temperature and then calculate velocity of range expansion, which they considered to be a random process.

Based on observations made from 1992–2000 in Europe, Hemerick et al. [65] determined that the mean velocity of range expansion was 33 km/year. Hemerik et al. [65] found that their simpler model underestimated the velocity of range expansion compared to observations, whereas their more-detailed model overestimated the same. They concluded that reliance on laboratory data to predict dynamics of field populations is a major problem.

Carrasco et al. [47] created and tested a model that incorporated natural and human-assisted long-distance dispersal of DVV in Austria. Their purpose was to determine the most important mechanisms explaining the observed pattern of spread of the DVV invasion in Austria. They also used the model to evaluate how effective the European regional measures for stopping the spread of DVV were likely to be. DVV was first detected in Austria in 2002–2003.

Human-assisted dispersal of DVV was believed to be relevant because isolated outbreaks of DVV occurred far beyond the main body of the invasion near important Danube River harbors and along highways. Thus, the model accounted for highways, railways and rivers. The Danube River basin is especially relevant because it is one of the most common means of international transport in Europe. Furthermore, farmers tend to grow continuous corn along the riverbanks.

The model by Carrasco et al. [47] calculated within-field population dynamics and various types of dispersal. Patches were cornfields and "macro-fields" which represented the area of corn grown in each municipality as a single field. Altitude or elevation of fields was considered in the model. For natural dispersal, they used two alternative probability distributions to calculate radial distance for a given direction. These probabilities were influenced by field size and growth stage of the corn crop.

Carrasco et al. [7] considered two types of human-assisted, long-distance dispersal. Cities were modelled as attractors drawing some DVV away from farmland. DVV adults within one day flight to the transportation lines connecting the 10 largest cities in Austria could move long-distances towards these cities. Human-assisted dispersal related to international transport through the Danube River basin was modeled by assuming that DVV adults within one day flight of the Danube River could also move long distances. Both of these types of dispersal included distances based on a negative power law dispersal kernel.

Carrasco et al. [47] calibrated the model by using maximum likelihood techniques combined with spatial stochastic simulation to fit model results to observed data from the historical invasion of Austria in 2002–2004. They then tested the model predictions against independent observations from 2005–2008.

Carrasco et al. [47] found that the model that included the human-assisted, long-distance dispersal through the Danube River basin matched the independent data set the best. In fact, results of this model had a very good fit between predictions and observations. Other models that only considered natural dispersal or only the influence of cities on dispersal were inadequate. Therefore, the authors concluded that the spatial pattern of the invasion could only be explained by a combination of natural dispersal and human-assisted dispersal along the Danube River.

To facilitate European Pest Risk Analyses, Robinet et al. [67] developed two generic spread models that consist of a set of submodels for population dynamics and dispersal, with linkages to niche maps that are based on climate and habitat suitability. They used the outputs of a CLIMEX model [68] run with climate data to define the area of potential establishment.

One model determines occupancy of cells, not density of DVV. It simulates radial range expansion at a constant rate, similar to reaction diffusion models. It has a single parameter, the constant expansion rate, which accounts for distance of dispersal and population growth. The rate is the same in all directions. All cells within the circular predicted range are considered invaded. For DVV, Robinet et al. [67] used a constant radial rate of expansion of 80 km/year.

The other model of Robinet et al. [67] calculates population density. The population growth process is logistic with two parameters: carrying capacity and growth rate. They modeled dispersal using a t-distribution with two parameters: a length scale and a shape parameter. They recommend the rotated t-distribution for its versatility in studies of geographic spread. The fatness of the tails, which reflects the likelihood of long-distance dispersal events, is determined by the shape parameter. A kernel with fat tails may be used to represent a situation in which human activity, for example, can be responsible for occasional spread events over much longer distances than are attained by biological spread mechanisms. Unfortunately, the shape parameter is very difficult to measure and is almost impossible to estimate from dispersal data according to Robinet et al. [67]. Robinet et al. [67] used a shape parameter that formed a fat-tailed dispersal kernel and caused a large proportion of DVV to disperse over long distances within Europe. They assumed the same radial expansion rate as in their first model (length scale = 80 km).

Robinet et al. [67] concluded that the key difficulty in using these simple models is estimating the parameters. For DVV, values of parameters are not directly published in the literature, and a good understanding of the parameters' meanings in the context of the model are needed to extract the required values from data. They also concluded that, for their density-dispersal model, the shape parameter governing the proportion of long-distance dispersers does not have the strongest effect on the predicted area invaded. This is the opposite of what Carrasco et al. [47] concluded.

Ecological site occupancy models used climate data combined with DVV distribution data to predict whether a given site is likely to be occupied (or capable of being occupied) by DVV. These have been particularly important in areas such as Europe where DVV is a relatively recent invasive species to determine the regions where DVV is likely to become established. These models can also be important when attempting to predict distribution shifts due to climate change. Aragon and Lobo [69] used a multidimensional envelope model customized for DVV while Bernardi et al. [70] used a geographic information system and agroclimate data to estimate possible European distributions. Eitsinger et al. [71] expanded this to include soil temperature and ground cover data to estimate lower developmental temperature thresholds. Dupin et al. [72] developed a training set of presence/absence data that they combined with 19 climatic features and estimated DVV distribution using several envelope models as well as support vector machine, a supervised machine learning technique. In this case the envelope models, with more domain-specific knowledge, outperformed the support vector machine, though all had significant misclassifications. Grozea et al. [73] used trap captures and climate data in Romania to predict presence and abundance. Feature or variable selection can be difficult when developing site occupancy models as significant collinearity can exist [74]. More conventional models can be combined with these climate-based models. For example, a degree-day driven model was combined with global warming estimates to estimate distribution shifts of both DB and DVV [75], warning that DVV could "become dominant in new areas such as northern Minnesota and the Dakotas".

7. Discussion

We believe that model simulation is essentially virtual experimentation [3,33,76]. Such "in silico" models are often used when experiments on the entire system, such as world climate or resistance to toxins, are either infeasible or ethically challenging.

The biological data and processes included in any model depend upon its intended use. The modeling goal helps modelers focus on the real ecological system and sharpens the focus on the major components of interest. Some models can be quite simple with only a few parameters. However, in some cases, these parameters are actually meta-parameters, summarizing many underlying parameters, such as the exponential growth rate for a population. A population's generational growth rate, for example, would encompass at the very least fecundity, natural mortality, overwintering survivorship and perhaps diffusion as well.

Onstad [77] noted that every model of the same system has the same total number of implicit and explicit assumptions. For example, a density-based, population-genetics model provides the ratio of male and female offspring and explicitly calculates both male and female adults. However, an allele-frequency based model would not provide explicit information about males and females and therefore would need to make implicit assumptions about them. In general, if one model has a mathematical function (explicit assumption) for a process or component, then a different model of the same ecological system, but without that function, can be considered to have one more implicit assumption regarding the system.

The eminent statistician GeorgeE. Box [78] noted that "all models are wrong, but some are useful". While the implications of this statement can be debated, it is clear that Box preferred the most parsimonious model possible, favoring simple models over more complex models if simple models can draw the same conclusion as complex models. In a tribute to RonaldA. Fisher [79], Box stated that the modeler "... should seek an economical description of natural phenomena. Just as the ability to devise simple but evocative models is the signature of the great scientist so overelaboration and over-parameterization is often the mark of mediocrity". Onstad [80] addressed the issues of simplicity, generality, realism and precision. He concluded that generality is not a property of a model that can be identified nor proclaimed at the time of a model's creation. A model is determined to be general after it has been tested against many systems. A model's accuracy and precision can be evaluated when it is tested against independent data. Complex and simple models can produce the same degree of accuracy and precision. Some complex models can also estimate, through stochastic variability, variance components beyond those identified through sensitivity analysis. Nor are these two approaches exclusive of each other. Occasionally, in mathematical models, we have found unexpected results in complex models and using those insights, developed simplified models to explore the same process in a less complicated environment [42].

Onstad [77] suggested that models should be made as realistic and as simple as possible to achieve the goal. A model that permits all individuals to reproduce without accounting for the differences between males and females or between immatures and adults cannot be considered as real as a model that has more realistic reproduction. For example, exponential growth of a population may seem to work fine at some gross scales of time, but the same function cannot be used when only immatures are alive over a given period.

The modeling purpose determines the complexity of the models, so perhaps it is not surprising that the reviewed models for DVV IRM tended to have similar model structures. Most were deterministic, but some involved stochastic processes. The genes resistant to transgenic traits and chemical compounds were assumed to have unlinked, diallelic autosomal loci. Onstad et al. [14] and Storer [16] provided guidelines for DVV models and had a complete list of DVV biological parameters. Although they used very different durability definitions, critical parameters to determine the durability were the same. Onstad et al. [18,63] Modeled rotation resistance and concluded high rotation adoption is a key factor to drive resistance. Crowder and Onstad [17] and Crowder et al. [28] extended this model approach and incorporated transgenic corn and crop rotation into the crop system. Onstad [61] modeled blended refuge and concluded larval movement might not decrease the durability of blended refuge significantly and blended refuge could overcome the less than complete compliance with block refuge requirements. Caprio et al. [19] used a model to retrospectively validate the evolution of resistance to adulticidal sprays of methyl-parathion. This model was expanded by Caprio and Grasser [35] to evaluate block refuge effectiveness with different field sizes. The model concluded that block refuge is less effective when the field size increases. Onstad and Meinke [29] expanded on the Crowder and Onstad [28] model and evaluated the resistance to single trait and pyramided transgenic corn. They concluded that novel traits are better than existing commercial traits when pyramids are commercialized. The benefit decreases further if survival from the pyramid is the minimum survival on either single trait instead of the product of survival of each trait. The modeling by Pan et al. [30] indicated that blended refuge produced greater durability than block refuges when the blocks either had compliance issues or were relocated each year due to nonuniform oviposition, which was consistent with findings in Caprio and Glasser [35]. Kang et al. [36] modeled temporal separation between male and female beetle emergence due to developmental delay on Bt corn and female-skewed sex ratio for adults emerging from Bt corn. They concluded the effects of these two factors on trait durability were insignificant. Martinez and Caprio [32] simulated different IPM strategies that could support IRM plans for DVV. They concluded that crop rotation was the most effective strategy and soil applied insecticide used for Bt corn would not help to boost the durability. Regional mitigation was a better strategy than random mitigation in an area with high levels of resistance.

It is noteworthy that, in spite of the difficulties of quantitatively studying insects that spend their immature stages underground and have only one generation per year with obligate diapause, scientists have found many reasons and many ways to model the *Diabrotica* described here. All resistance management models we reviewed are enriched with detailed demography and population genetics, which significantly impact model results. Other general models ignoring population biology might not be suitable for predicting evolution of resistance. The models of geographic spread demonstrate that modeling can involve many approaches and styles. We are inspired by the intellectual challenges these pests provide and the creative responses by the scientists who model them.

8. Conclusions

Many papers were reviewed and found to be valuable for guiding research in the future. Modeling is important because both of these species are difficult to manage.

Author Contributions: All authors contributed equally to all aspects of the review, except DWO was the one dealing directly with publisher. All authors have read and agreed to the published version of the manuscript.

Funding: This research received no external funding.

Conflicts of Interest: The authors declare no conflict of interest.

References

1. Bermond, G.; Blin, A.; Vercken, E.; Ravigné, V.; Rieux, A.; Mallez, S.; Guillemaud, T.; Morel-Journel, T. Estimation of the dispersal of a major pest of maize by cline analysis of a temporary contact zone between two invasive outbreaks. *Mol. Ecol.* **2013**, *22*, 5368–5381. [CrossRef]

2. Spencer, J.L.; Hughson, S.A.; Levine, E. Insect Resistance to Crop Rotation. Chapter 7. In *Insect Resistance Management: Biology, Economics and Prediction*; Onstad, D.W., Ed.; Academic Press: Cambridge, MA, USA, USA, 2014; pp. 233–278.

3. Onstad, D.; Pan, Z.; Tang, M.; Flexner, J.L. Economics of long-term IPM for western corn rootworm. *Crop. Prot.* **2014**, *64*, 60–66. [CrossRef]

4. Mooney, E.D.; Turpin, F.T. *ROWSIM, A GASP IV Based Rootworm Simulator*; Purdue University, Agricultural Experiment Station: West Lafayette, IN, USA, 1976; Volume 938.

5. Toepfer, S.; Kuhlmann, U. Constructing life-tables for the invasive maize pest *Diabrotica virgifera virgifera* (Col.; Chrysomelidae) in Europe. *J. Appl. Entomol.* **2006**, *130*, 193–205. [CrossRef]

6. Branson, T.F.; Sutter, G.R. Influence of Population Density of Immatures on Size, Longevity, and Fecundity of Adult *Diabrotica virgifera virgifera* (Coleoptera: Chrysomelidae). *Environ. Entomol.* **1985**, *14*, 687–690. [CrossRef]

7. Gray, M.E.; Tollefson, J.J. Influence of Tillage and Western and Northern Corn Rootworm (Coleoptera: Chrysomelidae) Egg Populations on Larval Populations and Root Damage. *J. Econ. Entomol.* **1987**, *80*, 911–915. [CrossRef]

8. Gray, M.E.; Tollefson, J.J. Influence of tillage systems on egg populations of western and northern corn rootworms (Coleoptera: Chrysomelidae). *J. Kansas Entomol. Soc.* **1988**, *61*, 186–194.

9. Gray, M.E.; Tollefson, J.J. Emergence of the Western and Northern Corn Rootworms (Coleoptera: Chrysomelidae) from Four Tillage Systems. *J. Econ. Entomol.* **1988**, *81*, 1398–1403. [CrossRef]

10. Elliott, N.C.; Sutter, G.R.; Branson, T.F.; Fisher, J.R. Effect of population density of immatures on survival and development of the western corn rootworm (coleoptera: Chrysomelidae). *J. Entomol. Sci.* **1989**, *24*, 209–213. [CrossRef]

11. Onstad, D.W.; Hibbard, B.E.; Clark, T.L.; Crowder, D.W.; Carter, K.G. Analysis of Density-Dependent Survival of *Diabrotica* (Coleoptera: Chrysomelidae) in Cornfields. *Environ. Entomol.* **2006**, *35*, 1272–1278. [CrossRef]

12. Hibbard, B.E.; Meihls, L.N.; Ellersieck, M.R.; Onstad, D.W. Density-dependent and density-independent mortality of the western corn rootworm: Impact on dose calculations of rootworm-resistant Bt corn. *J. Econ. Entomol.* **2010**, *103*, 77–84. [CrossRef] [PubMed]

13. Elliott, N.; Hein, G. Population dynamics of the western corn rootworm: Formulation, validation, and analysis of a simulation model. *Ecol. Model.* **1991**, *59*, 93–122. [CrossRef]

14. Onstad, D.W.; Guse, C.A.; Spencer, J.L.; Levine, E.; Gray, M.E. Modeling the dynamics of adaptation to transgenic corn by western corn rootworm (Coleoptera: Chrysomelidae). *J. Econ. Entomol.* **2001**, *94*, 529–540. [CrossRef]

15. Gray, M.E.; Tollefson, J.J. Survival of the Western and Northern Corn Rootworms (Coleoptera: Chrysomelidae) in Different Tillage Systems throughout the Growing Season of Corn. *J. Econ. Entomol.* **1988**, *81*, 178–183. [CrossRef]

16. Storer, N.P. A spatially explicit model simulating western corn rootworm (Coleoptera: Chrysomelidae) adaptation to insect-resistant maize. *J. Econ. Entomol.* **2003**, *96*, 1530–1547. [CrossRef] [PubMed]

17. Crowder, D.W.; Onstad, D.W. Using a generational time-step model to simulate dynamics of adaptation to transgenic corn and crop rotation by western corn rootworm (Coleoptera: Chrysomelidae). *J. Econ. Entomol.* **2005**, *98*, 518–533. [CrossRef]

18. Onstad, D.W.; Crowder, D.W.; Mitchell, P.D.; Guse, C.A.; Spencer, J.L.; Levine, E.; Gray, M.E. Economics versus alleles: Balancing IPM and IRM for rotation-resistant western corn rootworm (Coleoptera: Chrysomelidae). *J. Econ. Entomol.* **2003**, *96*, 1872–1885. [CrossRef]

19. Caprio, M.A.; Nowatzki, T.; Siegfried, B.; Meinke, L.J.; Wright, R.J.; Chandler, L.D. Assessing risk of resistance to aerial applications of methyl-parathion in western corn rootworm (Coleoptera: Chrysomelidae). *J. Econ. Entomol.* **2006**, *99*, 483–493. [CrossRef]

20. Comins, H.N. The development of insecticide resistance in the presence of migration. *J. Theor. Biol.* **1977**, *64*, 177–197. [CrossRef]

21. Comins, H.N. The management of pesticide resistance. *J. Theor. Biol.* **1977**, *65*, 399–420. [CrossRef]

22. Comins, H.N. Analytic methods for the management of pesticide resistance. *J. Theor. Biol.* **1979**, *77*, 171–188. [CrossRef]

23. Martinez, J.C.; Caprio, M.A.; Friedenberg, N.A. Density Dependence and Growth Rate: Evolutionary Effects on Resistance Development to Bt (Bacillus thuringiensis). *J. Econ. Entomol.* **2017**, *111*, 382–390. [CrossRef] [PubMed]

24. Haridas, C.V.; Meinke, L.J.; Hibbard, B.E.; Siegfried, B.D.; Tenhumberg, B. Effects of temporal variation in temperature and density dependence on insect population dynamics. *Ecosphere* **2016**, *7*. [CrossRef]

25. Hein, G.L.; Tollefson, J.J. Model of the Biotic Potential of Western Corn Rootworm (Coleoptera: Chrysomelidae) Adult Populations, and Its Use in Studying Population Dynamics. *Environ. Entomol.* **1987**, *16*, 446–452. [CrossRef]

26. Elliott, N.; Lance, D.; Hanson, S. Quantitative description of the influence of fluctuating temperatures on the reproductive biology and survival of the western corn rootworm, *Diabrotica virgifera virgifera* leconte (coleoptera: Chrysomelidae). *Can. Entomol.* **1990**, *122*, 59–68. [CrossRef]

27. Crowder, D.W.; Onstad, D.W.; Gray, M.E. Planting transgenic insecticidal corn based on economic thresholds: Consequences for integrated pest management and insect resistance management. *J. Econ. Entomol.* **2006**, *99*, 899–907. [CrossRef]

28. Crowder, D.W.; Onstad, D.W.; Gray, M.E.; Pierce, C.M.F.; Hager, A.G.; Ratcliffe, S.T.; Steffey, K.L. Analysis of the Dynamics of Adaptation to Transgenic Corn and Crop Rotation by Western Corn Rootworm (Coleoptera: Chrysomelidae) Using a Daily Time-Step Model. *J. Econ. Entomol.* **2005**, *98*, 534–551. [CrossRef] [PubMed]

29. Onstad, D.W.; Meinke, L.J. Modeling evolution of *Diabrotica virgifera virgifera* (Coleoptera: Chrysomelidae) to transgenic corn with two insecticidal traits. *J. Econ. Entomol.* **2010**, *103*, 849–860. [CrossRef]

30. Pan, Z.; Onstad, D.W.; Nowatzki, T.M.; Stanley, B.H.; Meinke, L.J.; Flexner, J.L. Western Corn Rootworm (Coleoptera: Chrysomelidae) Dispersal and Adaptation to Single-Toxin Transgenic Corn Deployed With Block or Blended Refuge. *Environ. Entomol.* **2011**, *40*, 964–978. [CrossRef]

31. Elliott, N.C.; Gustin, R.D.; Hanson, S.L. Influence of adult diet on the reproductive biology and survival of the western corn rootworm, *Diabrotica virgifera virgifera*. *Entomologia Experimentalis et Applicata* **1990**, *56*, 15–21. [CrossRef]

32. Martinez, J.C.; Caprio, M.A. IPM Use with the Deployment of a Non-High Dose Bt Pyearamid and Mitigation of Resistance for Western Corn Rootworm (*Diabrotica virgifera virgifera*). *Environ. Entomol.* **2016**, *45*, 747–761. [CrossRef]

33. Caprio, M.A.; Tabashnik, B.E. Gene Flow Accelerates Local Adaptation Among Finite Populations: Simulating the Evolution of Insecticide Resistance. *J. Econ. Entomol.* **1992**, *85*, 611–620. [CrossRef]

34. Caprio, M.A.; Storer, N.P.; Sisterson, M.S.; Peck, S.L.; Maia, A.H.N. Assessing the Risk of the Evolution of Resistance to Pesticides Using Spatially Complex Simulation Models. In *Global Pesticide Resistance in Arthropods*; CABI Publishing: Cambridge, MA, USA, 2008; pp. 90–117.

35. Caprio, M.A.; Glaser, J.A. Simulation Models Evaluation of Pest Resistance Development to Refuge in the Bag Concepts Related to Pioneer Submission. In *Letter Report from the US EPA ORD, Sustainable Technology Division, National Risk Management Research Laboratory, Cincinnati, OH, to the US EPA*; Office of Pesticide Programs, Biopesticides and Pollution Prevention Division: Washington, DC, USA, 2010. Available online: http://nepis.epa.gov/Adobe/PDF/P100EBX2.pdf (accessed on 14 June 2020).

36. Kang, J.; Krupke, C.H.; Murphy, A.F.; Spencer, J.L.; Gray, M.E.; Onstad, D.W. Modeling a western corn rootworm,*Diabrotica virgifera virgifera* (Coleoptera: Chrysomelidae), maturation delay and resistance evolution inBtcorn. *Pest Manag. Sci.* **2014**, *70*, 996–1007. [CrossRef] [PubMed]

37. Caprio, M.A.; Luttrell, R.G.; MacIntosh, S.; Rice, M.E.; Siegfried, B.; Witkowski, J.F.; van Duyn, J.W.; Moellenbeck, D.; Sachs, E.; Stein, J. *An Evaluation of Insect Resistance Management in Bt Field Corn: A Science-Based Framework for Risk Assessment and Risk Management*; International Life Sciences Institute/Health and Environmental Sciences Institute: Washington, DC, USA, 1999.

38. Caprio, M.A.; Martinez, J.C.; Porter, P.A.; Bynum, E.D. The Impact of Inter-Kernel Movement in the Evolution of Resistance to Dual-Toxin Bt-Corn Varieties in *Helicoverpa zea* (Lepidoptera: Noctuidae). *J. Econ. Entomol.* **2016**, *109*, 307–319. [CrossRef]

39. Ives, A.R.; Andow, D.A. Evolution of resistance to Bt crops: Directional selection in structured environments. *Ecol. Lett.* **2002**, *5*, 792–801. [CrossRef]

40. Ives, A.R.; Glaum, P.; Ziebarth, N.L.; Andow, D.A. The evolution of resistance to two-toxin pyearamid transgenic crops. *Ecol. Appl.* **2011**, *21*, 503–515. [CrossRef]

41. Caprio, M.A.; Martinez, J.C. Using seed mixes in conjunction with structured refuges. *Midsouth Entomol.* **2013**, *6*, 1–11.

42. Caprio, M.A. Source-sink dynamics between transgenic and non-transgenic habitats and their role in the evolution of resistance. *J. Econ. Entomol.* **2001**, *94*, 698–705. [CrossRef]

43. Coats, S.A.; Tollefson, J.J.; Mutchmor, J.A. Study of Migratory Flight in the Western Corn Rootworm (Coleoptera: Chrysomelidae). *Environ. Entomol.* **1986**, *15*, 620–625. [CrossRef]

44. Naranjo, S.E. Comparative flight behavior of *Diabrotica virgifera virgifera* and *Diabrotica barberi* in the laboratory. *Entomologia Experimentalis et Applicata* **1990**, *55*, 79–90. [CrossRef]

45. Stebbing, J.A.; Meinke, L.J.; Naranjo, S.E.; Siegfried, B.D.; Wright, R.J.; Chandler, L.D. Flight Behavior of Methyl-Parathion-Resistant and -Susceptible Western Corn Rootworm (Coleoptera: Chrysomelidae) Populations from Nebraska. *J. Econ. Entomol.* **2005**, *98*, 1294–1304. [CrossRef]

46. Toepfer, S.; Levay, N.; Kiss, J. Adult movements of newly introduced alien *Diabrotica virgifera virgifera* (Coleoptera: Chrysomelidae) from non-host habitats. *Bull. Entomol. Res.* **2006**, *96*, 327–335. [PubMed]

47. Carrasco, L.R.; Mumford, J.; MacLeod, A.; Harwood, T.; Grabenweger, G.; Leach, A.; Knight, J.; Baker, R. Unveiling human-assisted dispersal mechanisms in invasive alien insects: Integration of spatial stochastic simulation and phenology models. *Ecol. Model.* **2010**, *221*, 2068–2075. [CrossRef]

48. Levay, N.; Terpo, I.; Kiss, J.; Toepfer, S. Quantifying inter-field movements of the western corn rootworm (Diabrotica virgifera virgiferaLeConte)—A Central European field study. *Cereal Res. Commun.* **2014**, *43*, 155–165. [CrossRef]

49. Naranjo, S.E. Influence of temperature and larval density on flight performance of *Diabrotica virgifera virgifera* leconte (coleoptera: Chrysomelidae). *Can. Entomol.* **1991**, *123*, 187–196. [CrossRef]

50. Nowatzki, T.; Siegfried, B.D.; Meinke, L.J. Comparative Movement and Mating Behavior of Adult Western Corn Rootworm (Coleoptera: Chrysomelidae) in a YieldGard Rootworm Transgenic and A Conventional Cornfield. In Proceedings of the National Meeting of the Entomological Society of America, Cincinnati, OH, USA, 26–29 October 2003.

51. Haddock, R.C. Orientation and Movement of the Northern Corn Rootworm, *Diabrotica Barberi* (Coleoptera: Chrysomelidae) over Large and Small Distances. Ph.D. Thesis, Cornell University, Ithaca, NY, USA, 1984.

52. Naranjo, S.E.; Sawyer, A.J. A Temperature- and age-dependent simulation model of reproduction for the northern corn rootworm, diabrotica barberi smith and lawrence (Coleoptera: Chrysomelidae). *Can. Entomol.* **1988**, *120*, 1–17. [CrossRef]

53. Naranjo, S.E.; Sawyer, A.J. A simulation model of northern corn rootworm, *Diabrotica barberi* smith and lawrence (coleoptera: chrysomelidae), population dynamics and oviposition: Significance of host plant phenology. *Can. Entomol.* **1989**, *121*, 169–191. [CrossRef]

54. Naranjo, S.E.; Sawyer, A.J. Analysis of a simulation model of northern corn rootworm, *Diabrotica barberi* smith and lawrence (Coleoptera: chrysomelidae), dynamics in field corn, with implications for population management. *Can. Entomol.* **1989**, *121*, 193–208. [CrossRef]

55. Mitchell, P.D.; Riedell, W.E. Stochastic dynamic population model for northern corn rootworm (Coleoptera: Chrysomelidae). *J. Econ. Entomol.* **2001**, *94*, 599–608. [CrossRef]

56. Mitchell, P.D.; Onstad, D.W. Effect of extended diapause on evolution of resistance to transgenic Bacillus thuringiensis corn by northern corn rootworm (Coleoptera: Chrysomelidae). *J. Econ. Entomol.* **2005**, *98*, 2220–2234. [CrossRef]

57. Fisher, J.R. Comparison of Controlled Infestations of Diabrotica virgifera virgifera and Diabrotica barberi (Coleoptera: Chrysomelidae) on Corn. *J. Econ. Entomol.* **1985**, *78*, 1406–1408. [CrossRef]

58. Boetel, M.A.; Fuller, B.W. Seasonal Emergence-Time Effects on Adult Longevity, Fecundity, and Egg Viability of Northern and Western Corn Rootworms (Coleoptera: Chrysomelidae). *Environ. Entomol.* **1997**, *26*, 1208–1212. [CrossRef]

59. Onstad, D.W.; Guse, C.A.; Crowder, D.W. Heterogeneous Landscapes and Variable Behavior: Modeling Rootworm Evolution and Geographic Spread. Chapter 8. In *Western Corn Rootworm: Ecology and Management*; Vidal, S., Kuhlmann, U., Edwards, C.R., Eds.; CABI Publishing: Wallingford, UK, 2005; pp. 155–167.

60. Onstad, D.W.; Spencer, J.; Guse, C.; Levine, E.; Isard, S. Modeling evolution of behavioral resistance by an insect to crop rotation. *Entomologia Experimentalis et Applicata* **2001**, *100*, 195–201. [CrossRef]

61. Onstad, D.W. Modeling Larval Survival and Movement to Evaluate Seed Mixtures of Transgenic Corn for Control of Western Corn Rootworm (Coleoptera: Chrysomelidae). *J. Econ. Entomol.* **2006**, *99*, 1407–1414. [CrossRef] [PubMed]

62. Spencer, J.L.; Onstad, D.; Krupke, C.; Hughson, S.; Pan, Z.; Stanley, B.; Flexner, L. Isolated females and limited males: Evolution of insect resistance in structured landscapes. *Entomologia Experimentalis et Applicata* **2013**, *146*, 38–49. [CrossRef]

63. Onstad, D.W.; Crwoder, D.W.; Isard, S.A.; Levine, E.; Spencer, J.L.; O'Neal, M.E.; Ratcliffe, S.T.; Gray, M.E.; Bledsoe, L.W.; Di Fonzo, C.D.; et al. Does Landscape Diversity Slow the Spread of Rotation-Resistant Western Corn Rootworm (Coleoptera: Chrysomelidae)? *Environ. Entomol.* **2003**, *32*, 992–1001. [CrossRef]

64. Onstad, D.W.; Joselyn, M.G.; Isard, S.A.; Levine, E.; Spencer, J.L.; Bledsoe, L.W.; Edwards, C.R.; Di Fonzo, C.D.; Willson, H. Modeling the Spread of Western Corn Rootworm (Coleoptera: Chrysomelidae) Populations Adapting to Soybean-Corn Rotation. *Environ. Entomol.* **1999**, *28*, 188–194. [CrossRef]

65. Hemerik, L.; Busstra, C.; Mols, P. Predicting the temperature-dependent natural population expansion of the western corn rootworm, *Diabrotica virgifera*. *Entomologia Experimentalis et Applicata* **2004**, *111*, 59–69. [CrossRef]

66. Van den Bosch, F.; Hengeveld, R.; Metz, J.A.J. Analyzing the velocity of animal range expansion. *J. Biogeogr.* **1992**, *19*, 135–150. [CrossRef]

67. Robinet, C.; Kehlenbeck, H.; Kriticos, D.J.; Baker, R.H.A.; Battisti, A.; Brunel, S.; Dupin, M.; Eyeare, D.; Faccoli, M.; Ilieva, Z.; et al. A Suite of Models to Support the Quantitative Assessment of Spread in Pest Risk Analysis. *PLoS ONE* **2012**, *7*, e43366. [CrossRef]

68. Sutherst, R.W. Prediction of species geographical ranges. *J. Biogeogr.* **2003**, *30*, 805–816. [CrossRef]

69. Aragón, P.; Lobo, J.M. Predicted effect of climate change on the invasibility and distribution of the Western corn root-worm. *Agric. For. Entomol.* **2012**, *14*, 13–18. [CrossRef]

70. Bernardi, M. Linkages between FAO agroclimatic data resources and the development of GIS models for control of vector-borne diseases. *Acta Trop.* **2001**, *79*, 21–34. [CrossRef]

71. Eitzinger, J.; Schaumberger, A.; Grabenweger, P.; Schaumberger, J.; Thaler, S.; Murer, E.; Krammer, C.; Grabenweger, G.; Christina, P.; Kahrer, A. Spatial Simulation of Crop Conditions-Application for Crop Risk Monitoring, Crop Management and Water Footprint Estimation. In Proceedings of the Environmental Changes and Adaptation Strategies, Slovakia, Skalica, 9–11 September 2013; Conf., Šiška, B., Nejedlík, P., Hájková, L., Kožnárová, V., Eds.;

72. Dupin, M.; Reynaud, P.; Jarošík, V.; Baker, R.; Brunel, S.; Eyeare, D.; Pergl, J.; Makowski, D. Effects of the Training Dataset Characteristics on the Performance of Nine Species Distribution Models: Application to Diabrotica virgifera virgifera. *PLoS ONE* **2011**, *6*, e20957. [CrossRef] [PubMed]

73. Grozea, I.; Chis, C.; Carabet, A.; Virteiu, A.M.; Grozea, A.; Stef, R.; Corcionivoschi, N. Mathematical model to analyze the population changes of *Diabrotica virgifera* in terms of geographical coordinates and climatic factors. *Romanian Biotechnol. Lett.* **2017**, *22*, 12630–12642.

74. Júnior, P.D.M.; Nóbrega, C.C. Evaluating collinearity effects on species distribution models: An approach based on virtual species simulation. *PLoS ONE* **2018**, *13*, e0202403. [CrossRef]

75. Diffenbaugh, N.S.; Krupke, C.H.; White, M.A.; Alexander, C.E. Global warming presents new challenges for maize pest management. *Environ. Res. Lett.* **2008**, *3*, 044007. [CrossRef]

76. Peck, S.L. Simulation as experiment: A philosophical reassessment for biological modeling. *Trends Ecol. Evol.* **2004**, *19*, 530–534. [CrossRef] [PubMed]

77. Onstad, D.W. Modeling for Prediction and Management. Chapter 14. In *Insect Resistance Management: Biology, Economics and Prediction*; Onstad, D.W., Ed.; Academic Press: Cambridge, MA, USA, 2014; pp. 453–483.

78. Box, G.E.P. Robustness in the Strategy of Scientific Model Building. In *Robustness in Statistics*; Elsevier BV: Amsterdam, The Netherlands, 1979; pp. 201–236.

79. Box, G.E.P. Science and Statistics. *J. Am. Stat. Assoc.* **1976**, *71*, 791–799. [CrossRef]

80. Onstad, D.W. Population-dynamics theory: The roles of analytical, simulation, and supercomputer models. *Ecol. Model.* **1988**, *43*, 111–124. [CrossRef]

Publisher's Note: MDPI stays neutral with regard to jurisdictional claims in published maps and institutional affiliations.

Review

Western Corn Rootworm, Plant and Microbe Interactions: A Review and Prospects for New Management Tools

Kyle J. Paddock [1], Christelle A. M. Robert [2,3], Matthias Erb [2,3] and Bruce E. Hibbard [4,*]

[1] Division of Plant Sciences, University of Missouri, Columbia, MO 65211, USA; paddockk@mail.missouri.edu
[2] Institute of Plant Sciences, University of Bern, 3013 Bern, Switzerland; christelle.robert@ips.unibe.ch (C.A.M.R.); matthias.erb@ips.unibe.ch (M.E.)
[3] Oeschger Centre for Climate Change Research, University of Bern, 3013 Bern, Switzerland
[4] Plant Genetics Research Unit, United States Department of Agriculture, Agricultural Research Service, Columbia, MO 65211, USA
* Correspondence: bruce.hibbard@usda.gov

Simple Summary: Over 90 million acres of US cropland are planted with corn, *Zea mays*, annually. The western corn rootworm, *Diabrotica virgifera virgifera*, causes significant economic damage by feeding on corn roots and the insect has populations that have adapted to nearly all management techniques in some regions. Additional tools are needed. Significant research on the basic biology of this pest has added new possibilities. Here, we summarize research that we believe has potential for future management of this major pest.

Abstract: The western corn rootworm, *Diabrotica virgifera virgifera* LeConte, is resistant to four separate classes of traditional insecticides, all *Bacillius thuringiensis* (Bt) toxins currently registered for commercial use, crop rotation, innate plant resistance factors, and even double-stranded RNA (dsRNA) targeting essential genes via environmental RNA interference (RNAi), which has not been sold commercially to date. Clearly, additional tools are needed as management options. In this review, we discuss the state-of-the-art knowledge about biotic factors influencing herbivore success, including host location and recognition, plant defensive traits, plant-microbe interactions, and herbivore-pathogens/predator interactions. We then translate this knowledge into potential new management tools and improved biological control.

Keywords: Western corn rootworm; belowground herbivory; pest management strategies; push-pull; plant defenses; biological control; soil health

Citation: Paddock, K.J.; Robert, C.A.M.; Erb, M.; Hibbard, B.E. Western Corn Rootworm, Plant and Microbe Interactions: A Review and Prospects for New Management Tools. *Insects* **2021**, *12*, 171. https://doi.org/10.3390/insects12020171

Academic Editors: Lance J. Meinke and Joseph L. Spencer

Received: 22 January 2021
Accepted: 13 February 2021
Published: 17 February 2021

Publisher's Note: MDPI stays neutral with regard to jurisdictional claims in published maps and institutional affiliations.

1. Introduction

Management of the western corn rootworm (*Diabrotica virgifera virgifera* LeConte) in maize cropping systems has a long, complex history. After its discovery as a pest of corn in 1909 [1], western corn rootworm (WCR) populations rapidly expanded eastward as corn was planted, reaching New England and, more recently, through multiple establishments and subsequent spread, European regions [2–4]. Annual costs of damage due to yield loss and management practices are estimated to be over $2 billion in the United States [5]. A diversity of management options exists [6], but management has been complicated by the continual adaptation of WCR to control tactics. WCR have developed resistance to four separate classes of traditional insecticides, all Bt toxins currently registered for commercial use, crop rotation, and even to dsRNA and innate plant resistance traits [7–17]. In some regions of Europe, management options are even more limited as use of transgenic maize producing Bt targeting rootworms are restricted, and limitations have been placed on neonicotinoids [18].

New tools for sustainable and economical management of this elusive pest are crucially needed. In this review, we highlight the status and potential of several prospective tools based on recent advances in the understanding of the biology and chemical ecology of the pest. These tools include push-pull strategies, plant defenses and nutrition, beneficial plant-microbial partners, and microbial control agents.

2. Disrupting WCR Establishment

Considerable efforts investigating the chemical ecology of maize-rootworm interactions have illuminated complex mechanisms, including physical and chemical processes, involved in host plant attraction, recognition, and in feeding stimulation. WCR larvae hatch in spring from eggs laid in the previous year. The period between eclosion and host plant establishment is critical for WCR. It is estimated 95% of hatching larvae die before establishment, but the factors responsible for high mortality remain unknown [19–21]. Physical and chemical factors have been demonstrated to affect first-instar larval movement and host plant establishment. Larval movement is limited with increasing soil bulk density [22,23] with first instar larvae traveling farther in finer textured soils compared to more coarse textured soils [24]. Increased egg distance to maize roots can limit root damage and adult emergence [25]. Questions still remain as to why most viable eggs fail to produce established larvae. For example, it is unknown the extent to which first-instar larvae can burrow through soil and instead rely on pre-existing soil pores and air channels [26]. If this major mortality factor was better understood, it might be possible to manipulate it for management.

It may also be possible to utilize knowledge of the factors influencing host plant establishment for management in the future. WCR larvae orient towards maize roots following CO_2 gradients and can detect concentrations as low as 2 mmol/mol [27–29]. In choice tests, significantly more neonate WCR larvae were attracted to synthetic CO_2 with a concentration of 11.2 mmol/mol than to growing maize with a CO_2 concentration of 1.36 mmol/mol [29]. Encapsulated CO_2 sources were tested as a means to confuse western corn rootworm larvae. The treatment resulted in significantly less damage than untreated controls and resulted in damage similar to a soil insecticide control [30]. CO_2 has also been evaluated with insecticides in an attract-and-kill approach [31]. However, neither approach has been adopted by industry. Other volatiles, including *(E)*-β-caryophyllene and ethylene, can also be detected and used by the WCR to locate suitable hosts [32]. For instance, *(E)*-β-caryophyllene can be used as a cue to orient towards roots attacked by conspecific larvae and to aggregate in a density-dependent manner [33]. *(E)*-β-caryophyllene does not attract neonates [34], so its usefulness as a volatile to confuse WCR larvae in the field remains unclear.

Once maize roots are located, larval feeding is triggered by host recognition cues. Strnad and Dunn [35] were the first to document the existence of host recognition factors. One such recognition factor was isolated [36] and identified by Bernklau et al. [37] as monogalactosyldiacylglycerol (MGDG). Bernklau et al. [37] discovered that the proportion of larvae exhibiting the tight-turning behavior elicited by MGDG was higher for larvae exposed to MGDG-saturated discs previously fed upon by WCR larvae. The authors concluded that WCR larvae were responding to byproducts of MGDG breakdown in addition to MGDG itself, which generates questions in regard to salivary enzymatic functions and plant-insect interactions involving WCR. Previously, Bernklau and Bjostad [38] isolated and identified a blend of glucose (30 mg/mL), fructose (4 mg/mL) and sucrose (4 mg/mL) plus linoleic or oleic acid (0.3 mg/mL) as feeding stimulants for WCR larvae. Subsequent investigations revealed sucrose to be the preferred sugar of WCR larvae [39]. The addition of free fatty acids to feeding blends significantly increased staying behavior of larvae, but high concentrations were toxic [40]. In addition to primary metabolites, complexes between micronutrients and maize secondary metabolites shape the foraging behavior of the WCR within a root system [7,41]. Specifically, complexes between soil iron (Fe) and the exuded 7-O-methylated, N-hydroxylated benzoxazinoid (DIMBOA) elicited WCR feeding preferences [41]. Interestingly, the application of the $Fe(III)(DIMBOA)_3$ complex on rice or barley, two non-host plant species for the WCR, was sufficient to trigger WCR feeding [41].

Experiments with benzoxazinoid-deficient plants and WCR larvae with impaired capacities to detect sugars confirmed the importance of the individual and combined cues for WCR foraging, but also revealed considerable WCR robustness to disruption of individual cues [42]. This may complicate attempts to use single cues for foraging disruption. Plant roots also produce compounds that repel foraging WCR larvae. Bernklau et al. [43] identified that small amounts (1 ug) of methyl anthranilate could prevent WCR larvae from approaching CO_2 sources and maize roots. Although the identification of compounds involved in feeding stimulation and staying behavior are useful, a suite of compounds is likely at play, as responses to crude maize extracts are generally stronger.

This basic understanding of attraction, recognition, and feeding stimulation could be utilized to improve existing management strategies or aid the development of alternative control tactics. Strategies such as (i) fine-tuning the production of cues involved in WCR attraction, establishment, and feeding, (ii) combining attractants with pesticides, or (iii) using attractant and repellent chemicals in push-pull programs should be considered. Manipulating the production of attractants, recognition factors, or feeding stimulants remains a very delicate avenue for pest management. One should carefully consider the impact of any shifts in primary or secondary plant metabolites on the plant and their interactions with the environment. For instance, because maize plants use Fe(III)(DIMBOA)$_3$ complex for iron uptake and benzoxazinoids for protection against generalist herbivores, disrupting the benzoxazinoid pathway for WCR management could potentially have significant consequences on plant growth and yield, as well as on herbivore outbreaks [44]. The application of attractant volatiles can disrupt host location and, in turn, establishment and damage by the WCR. CO_2-generating materials are strong enough to disrupt the host-location ability of WCR larvae and significantly reduce damage under laboratory and field conditions [30]. Combining attractant cues with pesticides has been reported to be extremely effective in a laboratory setting. For example, Bernklau et al. [45] increased insecticidal activity of thiamethoxam by 10,000-fold when added to a feeding stimulant blend. The addition of 6-methoxy-2-benzoxazolinone (MBOA) to insecticides also improved field activity [46]. Adding host recognition cues, feeding stimulants, CO_2, or other attractants to insecticides could increase efficacy and perhaps even provide a "pull" factor for a push-pull management strategy [47]. Bernklau et al. [48] documented that methyl anthranilate acts as a repellent for foraging WCR larvae in soil. Methyl anthranilate-saturated carriers could be placed in-row with maize seedlings and function as a "push" factor away from roots. Similarly, susceptible corn with repellent seed treatments could be used in conjunction with high-dose transgenic corn treated with attractants and/or feeding stimulants. Push-pull strategies have been used successfully in pest management with other crop systems [47].

3. Selecting for Maize Lines with Effective Defenses against WCR

Selecting for plant natural defenses to insects has been a successful management strategy for many pests, with hundreds of insect-resistant crop cultivars grown around the world [49,50]. Plant defenses include resistance and tolerance traits. Resistance traits allow plants to reduce herbivore damage [51,52] and are further categorized into antibiosis and antixenosis. Antibiosis refers to a reduction in growth and/or reproduction of the insect due to feeding on a resistant plant, whereas antixenosis limits damage to the host plant by decreasing the attractiveness of the host as food or shelter. Resistance traits might involve structural changes (root architecture, lignin content, trichomes) or production of allelochemicals (tannins, alkaloids, glucosinolates) [49,53–55]. Tolerance traits allow the plant to maintain productivity in spite of sustaining damage [56]. Tolerance to root herbivores, for instance, includes changes in photosynthesis, resource reallocation, and delayed compensatory growth [57–60].

Public breeding efforts to develop or select plant defenses to WCR have been intermittent for the past 85 years (Table 1) [61–114]. Initial breeding programs began in response to observations that different maize strains varied in their response to WCR pressure [61,62]. In recent breeding programs, selection for resistant and tolerant strains of maize has been based on several criteria. These criteria include plant lodging, vertical pull resistance, and yield, which serve as indirect measures of WCR damage. Root damage ratings and WCR survival provide an estimate of antibiosis and/or antixenosis capacity. Root size and root regrowth, although influenced by environmental factors, provide an estimate of tolerance to WCR damage. Early resistant hybrids (1980s) had larger roots and experienced lower levels of lodging upon WCR feeding [80]. More recently, several germplasm lines with mechanisms of resistance beyond tolerance have been identified [75,85,92,94,99,100,104,113]. Hibbard et al. [94] released CRW3(S1)C6 that had damage ratings not significantly different than a Cry3Bb1 hybrid when crossed to an elite inbred line. El Khishen et al. [99] and Bernklau et al. [100] clearly documented that the commercial maize hybrids SUM2162 and SUM2068 had relatively strong antibiosis resistance. Unfortunately, these hybrids did not compete with elite hybrids for yield when rootworm pressure was lacking, and at this time, there are no commercially available hybrids providing natural and effective host plant resistance or tolerance to WCR.

Table 1. Breeding efforts to develop native plant defense to the western corn rootworm over the past 85 years.

Years	Location	References
1935–1945	Illinois Natural History Survey	[61,62]
1970–2007	Iowa State University	[63,64,73,84,95,106,108–112]
1963–2010	USDA-ARS, Brookings, South Dakota	[65–72,74–81,113]
1990–1997	University of Ottawa	[82,83,85–89]
1992–present	USDA-ARS, Columbia, MO	[90–94,96–100]
1995–present	University of J. J. Strossmayer	[101–104,114]
2002–present	University of Illinois-Champaign	[105,107]

Genomic work in maize has revealed a bounty of natural diversity across germplasm lines [115]. Screening of landraces, populations, and inbreds by insect-resistance breeding programs has revealed that genetic bins containing insect resistance quantitative trait loci (QTLs) are widespread, likely meaning there is great complexity and diversity in maize responses to herbivores [116]. However, because much of the work of insect-resistance breeding programs has focused on stem and leaf feeding traits, confirmed resistance against leaf herbivores is more prevalent than resistance to root feeding insects. This does not necessarily exclude these QTLs from conferring resistance to root-feeding insects. Recent work from Bohn et al. [107] revealed that QTLs associated with differences in root damage by WCR overlapped with QTLs involved with insect resistance previously identified by Meihls et al. [116]. Many of these QTLs (chromosome 1, 3, 6–10) contained gene/genes predicted to code for proteins involved in L-ascorbate and (*E*)-β-caryophyllene biosynthesis, in addition to the detoxification of reactive oxygen species. Investigations by Brkić et al. [114] found chromosome 1 and 6 contain several QTLs for maize resistance to WCR. Studies investigating the QTL regions previously described may provide a knowledge base for breeding programs aimed at increasing maize native resistance to WCR. Specifically, the QTLs correlated with resistance to WCR were located in the same genomic bins as two previously described insect resistance genes, *aoc1* (bin 1.04) and *mir1* (bin 6.02). *mir1* encodes an insecticidal protease, Maize Insect Resistance 1- Cysteine Protease (MIR1-CP), that can disrupt the peritrophic matrix of caterpillars and even act synergistically with Bt toxins [117,118]. Separate investigations have revealed *mir1* expression increases upon WCR feeding in the inbred Mp708 [119]. In addition, transcript levels of several defense genes (*tps23*, *fpps3*, *rip2*, *mpi*) significantly increased in conjunction with jasmonic acid (JA) levels, potentially contributing to reduced larval recovery and reduced root damage [119]. Unfortunately, Castano-Duque et al. [119] did not utilize resistant and susceptible maize controls, so it is unclear how well this resistance would translate into a field setting. Other data suggest Mp708 is highly susceptible to natural rootworm feeding in the field (BEH, unpublished data). Given the high degree of WCR host adaptation, we estimate that the efficacy of generalized defense traits present across most commercial maize lines have limited potential to serve as WCR resistance factors.

As mentioned above, there are no publicly available hybrids conferring natural resistance or tolerance to WCR damage. The lack of correlation between inbred performance and hybrid performance [98] likely has contributed to this. Genotype-by-environment interaction (GEI) is high for natural rootworm resistance, resulting in low heritability of traits [107]. Likely contributing to GEI variability are differences in methodology. Infestation levels (natural variability vs. artificial) can generate high amounts of variation between environments and drown out trait effects. The paucity of effective natural defenses against WCR in commercial hybrids could also be a function of private breeding programs largely controlling for the pest via crop rotation or soil insecticides in their yield trials. This effectively removed the selection pressure and potentially decoupled yield and WCR tolerance/resistance traits. In contrast, seed industry yield trials rarely control for herbivores such as the European corn borer, *Ostrinia nubilalis* (Hübner), and therefore indirectly increase tolerance to this pest over time (James Bing, Corteva, personal communication). Despite the lack of current resistant and/or tolerant hybrids available to growers, native resistance traits from exotic sources likely do have potential to improve WCR management. Full genome sequencing and improved breeding methods coupled with modern gene editing technologies examining direct effects of specific gene/genes involved with defensive traits could potentially increase the speed and efficiency of elite cultivar development with native plant defense [120]. Ultimately though, the success relies on investing time in screening these plants for their capacity to cope with WCR damage.

4. Altering Maize Nutritional Value for the WCR

The nutritional dimension of WCR biology is central to management, but research efforts to understand this aspect have been intermittent. Assessing chemical profiles of host and non-host plant species may allow determination of key compounds or compound blends involved in WCR nutrition. Branson and Ortman [121,122] observed larval survival for at least 10 days on grass species. WCR neonates developed to the second instar on 18 of 44 grass species screened, whereas no larvae developed on any of the 27 broadleaf species screened. Clark and Hibbard [123], Oyediran et al. [124], and Wilson and Hibbard [125] further refined the host range of WCR larvae. Moeser and Vidal [126] developed a food conversion index to evaluate alternate hosts and several maize varieties. Selecting for maize lines possessing some key characteristics of non-host plants may limit WCR damage in the field.

Evaluating WCR larval ability to pupate and to emerge as adults when feeding on maize plants of different ages showed promising results. WCR consume root resources near where they initially establish, before moving to larger, more nutritious nodal roots that form on the side of the stalk [7,22]. Not only do later instar larvae prefer younger, nodal roots, but larvae require these younger roots for proper development [127]. Hibbard et al. [127] conducted greenhouse and field trials to determine what root phenology was optimal for the establishment and development of WCR larvae. In the field, plants were infested weekly with WCR eggs starting on the initial planting date and continuing until plants matured to ~V13 [128]. As predicted, plant damage gradually decreased with later infestation dates, because larger root systems can better withstand attack. Interestingly, larval recovery did not differ between infestation dates, but adult emergence did. Significantly fewer adults emerged from later infestation dates, suggesting larvae can establish on late vegetative stage (V13) plants, but nutrition is insufficient to produce adults. Given that WCR larvae perform poorly in the absence of $Fe(III)(DIMBOA)_3$ complexes and that DIMBOA is mostly exuded by young node roots of young plants [7,41], it is tempting to speculate that iron, known to be an essential micronutrient for insects [129], and/or DIMBOA are key factors in limiting WCR development to adulthood. Results from alternative host plant species also point towards nutritional inadequacies of mature plants [130]. Further understanding of WCR nutritional requirements to successfully achieve pupation may allow selection for plants that do not support WCR larval development to adults.

Further efforts to understand the nutrition requirements of WCR larvae have resulted in the development of artificial diets [131]. Current efforts to improve artificial diet further are focused on understanding the metabolomic responses to maize, in addition to good and poor artificial diets (Huynh et al., unpublished). If successful in gaining this understanding, reverse engineering maize varieties with roots of poor nutritional quality may also be possible. Changing such traits is typically accompanied by large pleiotropic effects, as herbivore nutrients also serve essential roles in plants. Thus, such approaches should try to disrupt WCR nutrition without impairing plant vigor; strategies to reach this aim are currently not in place.

5. Plant-Mediated RNA Interference

RNA interference (RNAi) is a biological response to double-stranded RNA (dsRNA) that triggers sequence-specific gene silencing [132]. This conserved machinery is present in many eukaryotes, including insects [133]. Silencing essential genes in insects can reduce herbivore damage and survival [134–139].

Baum et al. [140] and Bolognesi et al. [141] characterized the mechanism of action of dsRNA in WCR larvae. Baum et al. [140] identified 125 genes whose silencing led to significant WCR mortality. From these 125 genes, 14 caused mortality in 50% (LC_{50}) of WCR at doses lower than 5.2 ng dsRNA/cm^2 of artificial diet. These genes included putative *V- ATPase* A and D subunits, *ESCRT I Vps28*, *III Vps2*, and *III Snf* orthologs, a β-subunit of a *COPI* coatomer, ribosomal proteins, a proteosome ortholog, α-*actin*, *tubulin*, and an *RNA polymerase II* ortholog [140]. Adult WCR exhibit similar responses to orally ingested dsRNA. Using artificial diet overlaid with dsRNA targeting *V-ATPase* subunit A, Rangasamy and Siegfried [142] successfully knocked down gene expression in adults and achieved significant mortality in 14 days. Knockdown of the gene target *Sec23* resulted in significant mortality in adults after only six days of feeding [143]. Additional gene targets have successfully altered adult gene expression, specifically ones targeting genes involved in reproduction such as the chromatin remodeling gene *brahma* (brm), and the gap gene *hunchback* (hb) [144,145]. Eggs from RNAi targeted adults also experience downregulation of targeted genes, which could provide transgenerational control [144]. RNAi is now routinely used to identify key genes regulating WCR survival and fitness [41,144,146,147].

Transgenic maize plants using RNAi exhibit enhanced protection against WCR larvae when targeting essential genes such as α-*tubulin* gene, *V-ATPase* subunits A and C genes, an intracellular protein trafficking pathway gene *snf7*, a subunit of the coat protein complex II *Sec23*, and a midgut expressed gene *ssj1* [140,141,143,147,148]. In addition, Niu et al. [149] demonstrated the potential for WCR management by silencing genes involved in female fecundity. Using transgenic plants to silence the *Boule (Dvbol)* gene in WCR larvae resulted in reduced egg production and egg hatchability in adults [149]. Yet, Khajuria et al. [16] demonstrated the ability of the WCR to adapt to RNAi. WCR selected on *DvSnf7* dsRNA displayed an impaired luminal uptake [16]. Intriguingly, *DvSnf7* dsRNA resistant WCR also displayed cross resistance against three other dsRNA sequences but not to the *Bacillus thuringiensis* Cry3Bb1 protein [16].

Transgenic crops using RNAi may therefore be a promising tool when combined with other strategies for pest management. Bayer Crop Science and Corteva developed new transgenic lines expressing the combination of microbial compounds with RNAi targets [150,151] (see the section below about pathogenic microbials). Large scale application and potential resistance development in the field have yet to be evaluated. Deployment of stacked dsRNA targeting immature and adult life stages might better capture WCR surviving single traits. However, efficacy of transgenic plants expressing dsRNA against adult WCR has not been thoroughly evaluated.

6. Enhancing Plant Health-Promoting Microbes

Plant-insect-microbe interactions occurring in the rhizosphere can have dramatic effects across trophic levels, above and below ground, and can shape plant, herbivore, and microbial communities [152]. Beneficial microbial communities can provide plants with increased pest resistance [153]. Plants release a suite of chemicals from roots upon insect damage to which specific microbes respond. Over multiple generations, plants can then refine microbial communities that provide beneficial functions [154]. In the case of maize, benzoxazinoids have been shown to alter the rhizosphere microbiome, providing potential benefits to maize in the form of pest suppression in following generations [155]. Organic practices that promote soil health can also alter plant resistance to aboveground pest pressure [153]. However, there has been little work to investigate how these rhizosphere interactions affect western corn rootworm. Several groups are currently evaluating the potential of root-associated microbes to manage WCR. As a soil-dwelling root herbivore, WCR larvae are likely well adapted to maize rhizosphere microbial communities. Therefore, introducing non-native microbes that are compatible with maize, but not WCR physiology, may be a promising path.

Arbuscular mycorrhizal fungi (AMF) are one of the most integral groups of microorganisms promoting plant health. It is estimated >80% of plant species form symbiotic relationships with AMF [156]. Plants associated with AMF exhibit increased nutrient absorption of P and other micronutrients [157]. AMF colonization can also increase induced jasmonic acid pathways involved in plant defense against herbivores [158]. However, predicting whether AMF colonization negatively or positively affects herbivore performance is complex [159]. A meta-analysis by Koricheva et al. [160] revealed chewing insects experience reduced performance on AMF associated plants while piercing-sucking insects experience an increase in performance. Many mechanisms are likely at play as AMF can reconfigure the plant primary and secondary metabolisms [161]. Jaffuel et al. [162] examined protection ability of a seven-species AMF seed treatment against WCR in the field. AMF treatment had no effect on root damage, WCR fitness, or yield. The extent of AMF association with roots was not measured, which makes predictions about AMF species effects difficult to interpret. Future investigations should consider promoting native AMF abundance and examine species-specific responses of WCR to AMF-colonized maize. Winter cover crops increase soil health by reducing erosion, limiting nutrient loss, and increasing microbial abundance and diversity [163,164]. Different species of cover crop can refine AMF and other microbial communities in distinct ways [165–168], and the legacy effects of cover crops can increase AMF colonization of cash crop roots [158,167]). Winter cover crops can also increase predator populations and correlate to reductions in root damage and WCR larvae [169]. This broader approach through ecological intensification could be employed by combining management techniques to sustainably manage populations and reduce damage. Work in this field is in its infancy but has potential to expand into new management applications.

7. Using Soil Microbials to Disrupt WCR Gut Microbiome

Douglas [170] theorized the exploitation of insect microbiomes could provide new pest management techniques. Studies have shown the western corn rootworm actively selects for microorganisms it harbors [171,172]. Larvae reared in two different soils harbor similar bacterial communities even though the soil samples vary widely in community composition [172]. The WCR bacterial community commonly consists of species of *Serratia*, *Pseudomonas*, *Klebsiella*, *Acinetobacter*, *Streptomyces*, and *Tsukamurella*, with other species appearing in high abundance but sporadically [171–174]. Studies have focused on surveying the bacterial community of WCR but have done little to try to characterize functionality of that community. Robert et al. [175] evaluated fitness of multigenerational antibiotic-treated WCR and found no significant difference in weight gain or survival on conventional corn. Antibiotics were given to adults, and only the presence of *Wolbachia* was analyzed using PCR, so it is unclear what other bacteria remained after antibiotic treatment, or what bacteria were acquired from the soil during larval feeding. A majority of WCR populations also carry a high proportion of the maternally transmitted endosymbiont, *Wolbachia* [171,172,176]. *Wolbachia* can play an influential role in insect reproduction by inducing cytoplasmic incompatibility, parthenogenesis, feminization, and male killing [177]. Reproductive isolation caused by cytoplasmic incompatibility can result in speciation events at a much greater speed than traditional genetic elements [178,179]. Two subspecies of *Diabrotica virgifera*, *D. v. virgifera* (WCR) and *D. v. zeae* Krysan and Smith (Mexican corn rootworm), are a result of *Wolbachia*-induced cytoplasmic incompatibility that occurred after the ancestral population reached the area of modern-day Arizona less than 1100 years ago [180,181]. *Wolbachia* has also been shown to influence the composition of the host microbiome [182] and even protect the host from viral infection in populations of *Drosophila melanogaster* [183]. A role outside of reproductive incompatibility has not been found for *Wolbachia* in WCR. *Wolbachia* does appear to modulate plant gene expression, but this does not seem to impact major defenses or WCR resistance [175,184]. Nonetheless, it appears that some of the bacteria that inhabit WCR display functional capacity in overcoming plant defenses. Chu et al. [173] identified alterations in the gut microbial community of rotation-resistant populations of WCR. These shifts in the bacterial community were accompanied by increased cysteine protease activity in the gut that facilitated adult survival on soybean foliage [173]. In another study, bacterial isolates from abdomens of females influenced oviposition preference in choice tests [185]. The number of examples illustrating the role of the herbivore microbiome in overcoming plant defenses in other systems is increasing [186–189].

Mechanistic studies investigating the role of WCR gut microbiome in WCR ecological success are crucial to develop effective pest management strategies. A possible avenue would be to inoculate the soil with specific microbes that would shift WCR gut microbiome communities and hinder their ability to overcome plant defenses.

8. Using Pathogenic Microbials to Reduce WCR Populations

Management of arthropods through the use of microbes has a long history [190,191]. More recently, greater emphasis has been placed on limiting damage to non-target organisms and, in turn, has significantly increased the attractiveness of microbes as biocontrol agents. Classically, there are five main categories recognized under the term microbial control agents (MCAs): bacteria, viruses, fungi, protozoa, and nematodes [192]. Each has a unique mode of action and requires careful application to maximize benefit within the integrated pest management (IPM) framework. As advancements have been made in the microbial control of several pest species, their applications for WCR have mostly focused on transgenic approaches and entomopathogenic nematodes (EPNs).

The entomopathogenic bacteria from the genus *Bacillus* are some of the most widely used microbial biocontrol agents. These bacteria produce crystal toxins (Cry) that cause mortality in the insect by inducing cell lysis in the midgut [193]. Corn rootworm management has largely depended on *Bacillus thuringiensis* (Bt) for several decades through the planting of transgenic corn that express Cry toxins. There are currently four different Bt toxins commercially available as in-plant corn traits. In addition, Bayer Crop Science has developed SmartStax PRO expressing Cry3Bb1, Cry34Ab1/Cry35Ab1 and DvSnf7 (*Diabrotica virgifera* (Dv) + sucrose-non-fermenting (SNF) locus), a novel RNAi-based trait which targets a specific RNA sequence of WCR [150]. This product is approved by the EPA and recently received import approval from China. These germplasm will be widely planted for the first time in 2022. Bayer also recently discovered an additional Bt protein [194] and a protein from *Brevibacillus laterosporus* [195], each with strong activity against western corn rootworm larvae and no cross resistance to current rootworm toxins. Corteva has a new transgenic maize line producing a toxin derived from *Pseudomonas chlororaphis* pyramided with dsRNA available [151], but the feasibility for large scale application has yet to be evaluated. Other toxins displaying activity against WCR have also been documented. These toxins were originally isolated from the entomopathogenic bacteria, *Chromobacterium piscinae*, *Pseudomonas mosselii*, *Alcaligenes faecalis*, *Photorhabdus luminescens* [196–199], and entomopathogenic fungi from the genus *Pleurotus* [200]. Examination of bacterial species that display toxicity in other Coleopteran species (*Yersinia entomophaga*, *Paenobacillus spp.*, *Serratia entomophila*) are lacking for the western corn rootworm [191,201–203].

Entomopathogenic virus research has largely focused on the family Baculoviridae. This diverse viral family has shown promise as an MCA for Lepidopterans but seems to lack efficacy in Coleopterans. As such, Coleopteran viral research has focused on non-Baculoviridae species. Liu et al. [204–206] have identified two single-strand RNA viruses and an iflavirus present in WCR, but functional characterization has yet to be elucidated. Some success using virus as an MCA in beetles was demonstrated with the *Oryctes rhinoceros nudivirus* control coconut palm rhinoceros beetle (*Oryctes rhinoceros*) [207]. Fungi have also been under-evaluated as an MCA in corn rootworm. Strains of *Metarhizium anisopliae*, *Beauveria bassiana*, and *Beauveria brongniartii* display toxicity in larvae and adults [208], and *M. anisopliae* can significantly reduce adult emergence in field settings [209]. *B. bassiana* is commonly available as an MCA from several biopesticide companies, but it remains to be seen if treatments can be an economically viable control measure for WCR. Alternatively, both *M. anisopliae* and *B. bassiana* are found in agricultural soils and with proper soil management, could serve as a type of conservation biocontrol [210].

Entomopathogenic nematodes from the families Steinernematidae and Heterorhabditidae display high virulence against WCR [211] and, consequently, have been used to limit damage to maize infested with WCR [162,212,213]. Infective juveniles have a relatively short shelf life, making formulations difficult to use [214]. However, inducing a state of quiescence can prolong the infectiveness of EPN juveniles [215]. Root cap extracts of maize and pea contain potent amounts of quiescence factors that could be utilized to increase shelf life [216]. In addition, releases of nematodes for control rely on annual releases. No published studies have examined the long-term persistence of EPNs in WCR-maize systems, but such evidence exists in other systems [217]. Furthermore, field formulations and WCR specific strains show promising effects in the field and are available commercially [218–220].

Understanding the ecology of EPNs and identifying the infochemicals involved in their success will facilitate development of improved IPM strategies [221]. EPNs locate their host using universal, plant- and insect-derived chemical cues [222]. For instance, *H. megidis* can use the herbivore-induced root volatile *(E)*-β-caryophyllene [223] but see [224]. Hiltpold et al. [225] showed that EPNs can be selected for increased responsiveness towards the terpenoid volatile within six generations. The enhanced EPN responsiveness to *(E)*-β-caryophyllene increased EPN success in controlling WCR populations, but only to maize hybrids that emit the volatile. *H. bacteriophora* is strongly attracted to WCR cadaver cues, such as butylated hydroxytoluene [226]. Interestingly, EPN-infected cadavers are not only attractive to EPNs but also to the herbivores themselves [226,227] and trigger a plant defensive response [228,229].

WCR larvae can redirect plant defenses against nematodes [230]. Specifically, WCR larvae accumulate two benzoxazinoid glucosides: 6-methoxy-2-benzoxazolinone N-glucoside (MBOA-Glc) and 2-hydroxy-4,7-dimethoxy-1,4-benzoxazin-3-one O-glucoside (HDMBOA-Glc). MBOA-Glc is exuded by WCR onto its cuticle and is present in large amount in its frass [230]. This insect-specific detoxification benzoxazinone repels the EPN, *H. bacteriophora*. HDMBOA-Glc accumulates in the insect hemolymph and can be reactivated upon attack by nematodes to form MBOA [230], a toxic compound for both EPNs and their endosymbiotic bacteria [230]. By comparing EPN populations from the primary (US and Mexico) and invasive (Europe, Asia, Africa) ranges of the herbivore and conducting real-time selection assays, Zhang et al. [231] demonstrated that the herbivore adaptation to hijack plant defenses can shape the evolution of resistance in nematodes. Although *H. bacteriophora* EPNs from the invasive WCR range were again repelled by MBOA-Glc and susceptible to HDMBOA-Glc, EPNs from the original range of WCR were neither repelled nor susceptible to the sequestered compounds. Rearing a susceptible EPN strain in benzoxazinoid-sequestering hosts was sufficient for the nematode to evolve a complete resistance to benzoxazinoid-dependent defenses. The ability of EPNs to overcome these defenses was further associated with higher infectivity rates of WCR. Similarly, a screening of *H. bacteriophora* Mexican isolates showed that most were resistant to the benzoxazinoid defenses of WCR larvae [232]. Interestingly, the variability in infectiveness of these isolates in the benzoxazinoid-sequestering WCR larvae suggests other WCR defensive mechanisms may exist and require further investigation. The relative contribution of genetic variation and epigenetic effects in the nematodes and its endosymbionts is currently under investigation. Early results demonstrate that engineering bacterial symbionts that are resistant to the WCR benzoxazinoid-defenses can improve EPN infectivity [233]. Resistance to benzoxazinoids of five *Photorhabdus* strains was successfully enhanced through experimental evolution. Strikingly, the evolution of resistance was acquired through multiple mechanisms in the different bacterial strains, because the observed insertions and nonsynonymous point mutations did not overlap. The insertions and mutations were located in genes encoding for a DNA-directed RNA polymerase, a transcriptional regulator of porins, a regulator of unsaturated fatty acid biosynthesis, a ligase involved in the biosynthesis of the outer membrane, and in an aquaporin gene, *aqpZ*, involved in membrane permeability. Further characterization of *aqpZ* confirmed its impact in benzoxazinoid resistance, as complementation of the mutated strain with the wild-type gene restored the bacterial susceptibility to MBOA. Reestablishing symbiosis between EPNs and the enhanced *Photorhabdus* strains increased EPN infectivity towards WCR by over 50%. Efforts are now underway to test this strategy in the field and to assess whether EPNs compatible with commercial application can be enhanced in this manner.

The growing body of literature documenting factors shaping EPN success in killing WCR larvae will surely enhance the efficacy of EPN-mediated strategies. Using attractant signals for EPNs may allow for the maintenance of elevated numbers of EPNs in the field. This solution was tested with transgenic maize plants that constitutively release *(E)*-β-caryophyllene, which resulted in effective WCR suppression in the presence of EPNs [234]. However, the overexpression of the associated terpene synthase also had a number of physiological and ecological costs [235]. Releasing the pure compound synthetically may be an alternative option, but the long-term impact of the sometimes-deceptive strategy requires careful investigation. Adding EPN-infected cadavers that attract both WCR larvae and EPNs [226,227] is a promising avenue, at least for high-value smaller fields. Placing EPN-infected cadavers in the field confers multiple advantages as the insect cadavers will provide the EPNs with shelter until favorable soil conditions are reached [236], thereby optimizing EPN survival, dispersal, and virulence after application [237,238]. In a greenhouse assay, Shapiro-Ilan et al. [238] demonstrated that the release of EPN-infected insect cadavers reduced the survival of the root weevil, *D. abbreviates*, and the black vine weevil, *Otiorhynchus sulcatus*, two times better than EPN suspensions within 7 days. The application of EPN-infected cadavers eliminated the herbivore population within 28 days, whereas EPN suspensions only reduced the herbivore populations by about 50%. Because EPN-infected cadavers induce plant resistance against leaf pathogens through volatile chemical cues [228,229], this strategy may be valuable for IPM. Alternatively, it is possible to apply nematodes in encapsulated hydrocapsules containing EPN quiescence factors covered with herbivore attractants and feeding stimulants. Such a strategy would attract WCR larvae to feed on the capsules, thereby directly releasing EPNs [239,240]. When applied in the field, these hydrocapsules effectively controlled WCR populations and reduced damage [240]. Deployment of EPN-containing hydrocapsules for control also appears feasible in other herbivore species [220]. Finally, priming, selecting, or engineering EPNs for increased responsiveness to WCR-indicating chemical cues or for increased resistance to WCR defenses appears promising. Priming of EPNs can be achieved through exposure to insect cues, such as insect macerates or pheromones, prior to application in the field [241–243]. Selection of EPNs or their endosymbiotic bacteria for enhanced responsiveness or resistance to insect chemicals can be obtained within a few generations in laboratory conditions [231,233]. The growing understanding of EPN biology and of their interactions with prey will only enhance the efficacy of integrated pest management strategies in general.

9. Conclusions

Reliance on one dimensional management techniques has resulted in failures, both in terms of technologies and population suppression. In dealing with this established and persistent pest in the US and Europe, additional tools and a multifaceted approach to management are needed (Figure 1). The intense research involving WCR biology and chemical ecology has yielded knowledge that could translate into effective management strategies in the near- and long-term. Here, we reviewed a number of potential WCR management possibilities that, if implemented, have promise for new, effective, and sustainable WCR management. Results from lab-based studies sometimes fail to translate to field-based studies. Many of the strategies discussed here need additional testing in field settings under varying environmental conditions to properly assess their commercial viability but hold promise nonetheless.

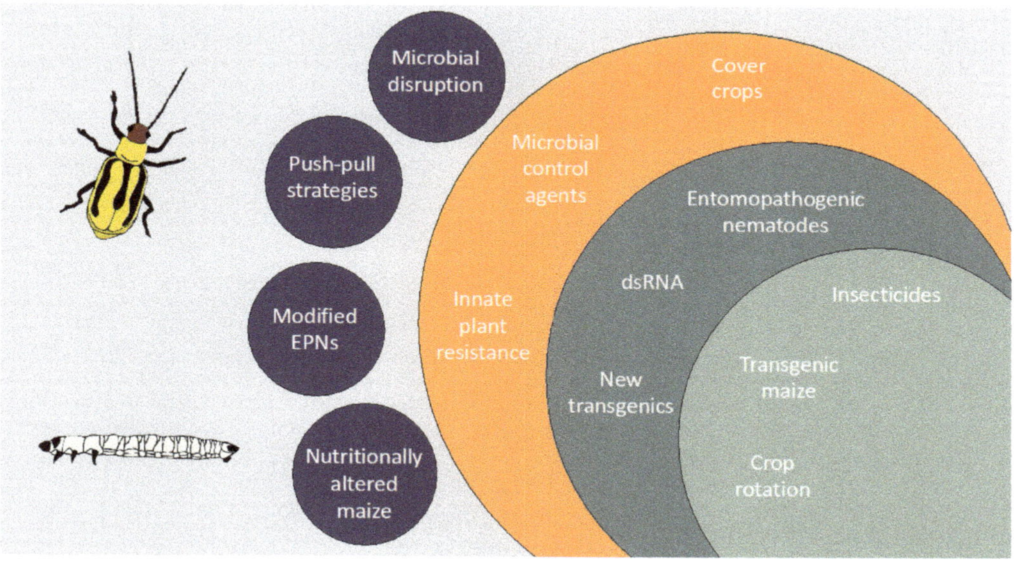

Figure 1. Broadening of management tactics aimed at controlling western corn rootworm (*Diabrotica virgifera virgifera* LeConte) and reducing damage to maize. The inner most circle (light green) represents the most common and widely adopted management techniques, all of which have seen failures in the field due to evolved resistance by western corn rootworm. The middle circle (dark green) represents management techniques less frequently adopted but have demonstrated effectiveness in laboratory or small field trials. The outermost circle (orange) represents management tactics that show promise and could be adopted using existing technology. Orbiting circles (blue) represent future management tactics that could be used if developed further.

Author Contributions: K.J.P., C.A.M.R., M.E., B.E.H. contributed to all drafts of the manuscript. All authors have read and agreed to the published version of the manuscript.

Funding: This research received no external funding.

Institutional Review Board Statement: Not applicable.

Data Availability Statement: Not applicable.

Acknowledgments: The work of C.A.M. Robert was supported by the European Research Council (ERC) under the European Union's Horizon 2020 research and innovation program (grant agreement no. ERC-2020-STG 949595). The work of K.J. Paddock was supported by the University of Missouri Division of Plant Sciences and USDA-ARS. We thank the editor and three anonymous reviewers for their help in improving the manuscript.

Conflicts of Interest: Authors declare no conflict of interest.

References

1. Gillette, C.P. *Diabrotica virgifera* Lec. as a corn rootworm. *J. Econ. Entomol.* **1912**, *5*, 364–366. [CrossRef]
2. Metcalf, R.L. Foreword. In *Methods for the Study of the Pest Diabrotica*; Krysan, J.L., Miller, T.A., Eds.; Springer: New York, NY, USA, 1986; pp. 7–15.
3. Chiang, H.C. Bionomics of the northern and western corn rootworms. *Annu. Rev. Entomol.* **1973**, *18*, 47–72. [CrossRef]
4. Miller, N.; Estoup, A.; Toepfer, S.; Bourguet, D.; Lapchin, L.; Derridj, S.; Kim, K.S.; Reynaud, P.; Furlan, L.; Guillemaud, T. Multiple transatlantic introductions of the western corn rootworm. *Science* **2005**, *310*, 992. [CrossRef]
5. Wechsler, S.; Smith, D. Has resistance taken root in US corn fields? Demand for insect control. *Am. J. Agric. Econ.* **2018**, *100*, 1136–1150. [CrossRef]
6. Veres, A.; Wyckhuys, K.A.G.; Kiss, J.; Tóth, F.; Burgio, G.; Pons, X.; Avilla, C.; Vidal, S.; Razinger, J.; Bazok, R.; et al. An update of the Worldwide Integrated Assessment (WIA) on systemic pesticides. Part 4: Alternatives in major cropping systems. *Environ. Sci. Pollut. Res.* **2020**, *27*, 29867–29899. [CrossRef]
7. Robert, C.A.M.; Veyrat, N.; Glauser, G.; Marti, G.; Doyen, G.R.; Villard, N.; Gaillard, M.D.P.; Köllner, T.G.; Giron, D.; Body, M.; et al. A specialist root herbivore exploits defensive metabolites to locate nutritious tissues. *Ecol. Lett.* **2012**, *15*, 55–64. [CrossRef] [PubMed]
8. Ball, H.J.; Weekman, G.T. Insecticide resistance in the adult western corn rootworm in Nebraska. *J. Econ. Entomol.* **1962**, *55*, 439–441. [CrossRef]
9. Meinke, L.J.; Siegfried, B.D.; Wright, R.J.; Chandler, L.D. Adult susceptibility of Nebraska western corn rootworm (Coleoptera: Chrysomelidae) populations to selected insecticides. *J. Econ. Entomol.* **1998**, *91*, 594–600. [CrossRef]
10. Pereira, A.E.; Wang, H.; Zukoff, S.N.; Meinke, L.J.; French, B.W.; Siegfried, B.D. Evidence of field-evolved resistance to bifenthrin in western corn rootworm (*Diabrotica virgifera virgifera* LeConte) populations in western Nebraska and Kansas. *PLoS ONE* **2015**, *10*, e0142299. [CrossRef]
11. Meihls, L.N.; Higdon, M.L.; Siegfried, B.D.; Miller, N.J.; Sappington, T.W.; Ellersieck, M.R.; Spencer, T.A.; Hibbard, B.E. Increased survival of western corn rootworm on transgenic corn within three generations of on-plant greenhouse selection. *Proc. Natl. Acad. Sci. USA* **2008**, *105*, 19177. [CrossRef]
12. Meihls, L.N.; Higdon, M.L.; Ellersieck, M.; Hibbard, B.E. Selection for resistance to mCry3A-expressing transgenic corn in western corn rootworm. *J. Econ. Entomol.* **2011**, *104*, 1045–1054. [CrossRef] [PubMed]
13. Deitloff, J.; Dunbar, M.W.; Ingber, D.A.; Hibbard, B.E.; Gassmann, A.J. Effects of refuges on the evolution of resistance to transgenic corn by the western corn rootworm, *Diabrotica virgifera virgifera* LeConte. *Pest Manag. Sci.* **2016**, *72*, 190–198. [CrossRef] [PubMed]
14. Frank, D.L.; Zukoff, A.; Barry, J.; Higdon, M.L.; Hibbard, B.E. Development of resistance to eCry3.1Ab-expressing transgenic maize in a laboratory-selected population of western corn rootworm (Coleoptera: Chrysomelidae). *J. Econ. Entomol.* **2013**, *106*, 2506–2513. [CrossRef]
15. Levine, E.; Spencer, J.L.; Isard, S.A.; Onstad, D.W.; Gray, M.E. Adaptation of the western corn rootworm to crop rotation: Evolution of a new strain in response to a management practice. *Am. Entomol.* **2002**, *48*, 94–107. [CrossRef]
16. Khajuria, C.; Ivashuta, S.; Wiggins, E.; Flagel, L.; Moar, W.; Pleau, M.; Miller, K.; Zhang, Y.; Ramaseshadri, P.; Jiang, C.; et al. Development and characterization of the first dsRNA-resistant insect population from western corn rootworm, *Diabrotica virgifera virgifera* LeConte. *PLoS ONE* **2018**, *13*, 7059. [CrossRef]
17. Meinke, L.J.; Souza, D.; Siegfried, B.D. The Use of Insecticides to Manage the Western Corn Rootworm, *Diabrotica virgifera virgifera*, LeConte: History, Field-Evolved Resistance, and Associated Mechanisms. *Insects* **2021**, *12*, 112. [CrossRef]
18. EU No. 485. Available online: https://eur-lex.europa.eu/eli/reg_impl/2013/485/oj (accessed on 15 January 2021).
19. Storer, N.P. A spatially explicit model simulating western corn rootworm (Coleoptera: Chrysomelidae) adaptation to insect-resistant maize. *J. Econ. Entomol.* **2003**, *5*, 1530–1547. [CrossRef]
20. Hibbard, B.E.; Higdon, M.L.; Duran, D.P.; Schweikert, Y.M.; Ellersieck, M.R. Role of egg density on establishment and plant-to-plant movement by western corn rootworm larvae (Coleoptera: Chrysomelidae). *J. Econ. Entomol.* **2004**, *97*, 871–882. [CrossRef]
21. Hibbard, B.E.; Meihls, L.N.; Ellersieck, M.R.; Onstad, D.W. Density-dependent and density-independent mortality of the western corn rootworm: Impact on dose calculations of rootworm-resistant Bt corn. *J. Econ. Entomol.* **2010**, *103*, 77–84. [CrossRef]
22. Strnad, S.P.; Bergman, M.K. Movement of first-instar western corn rootworms (Coleoptera: Chrysomelidae) in soil. *Environ. Entomol.* **1987**, *4*, 975–978. [CrossRef]
23. Ellsbury, M.M.; Schumacher, T.E.; Gustin, R.D.; Woodson, W.D. Soil compaction effect on corn rootworm populations in maize artificially infested with eggs of western corn rootworm (Coleoptera: Chrysomelidae). *Environ. Entomol.* **1994**, *4*, 943–948. [CrossRef]
24. Macdonald, P.J.; Ellis, C.R. Survival time of unfed, first-instar western corn rootworm (Coleoptera: Chrysomelidae) and the effects of soil type, moisture, and compaction on their mobility in soil. *Environ. Entomol.* **1990**, *3*, 666–671. [CrossRef]

25. Chaddha, S. Influence of Placement of Western Corn Rootworm Eggs on Survivorship, Root Injury and Yield. Master's Thesis, University of Minnesota, Minneapolis, MN, USA, 1990.

26. Gustin, R.D.; Schumacher, T.E. Relationship of some soil pore parameters to movement of first-instar western corn rootworm (Coleoptera: Chrysomelidae). *Environ. Entomol.* **1989**, *18*, 343–346. [CrossRef]

27. Strnad, S.P.; Bergman, M.K. Distribution and orientation of western corn rootworm (Coleoptera: Chrysomelidae) larvae in corn roots. *Environ. Entomol.* **1987**, *16*, 1193–1198. [CrossRef]

28. Arce, C.C.M.; Theepan, V.; Schimmel, B.C.J.; Jaffuel, G.; Erb, M.; Machado, R.A.R. Plant-derived CO_2 mediates long-distance host location and quality assessment by a root herbivore. *BioRxiv* **2020**. [CrossRef]

29. Bernklau, E.J.; Bjostad, L.B. Behavioral responses of first-instar western corn rootworm (Coleoptera: Chrysomelidae) to carbon dioxide in a glass bead bioassay. *J. Econ. Entomol.* **1998**, *91*, 444–456. [CrossRef]

30. Bernklau, E.J.; Fromm, E.A.; Bjostad, L.B. Disruption of host location of western corn rootworm larvae (Coleoptera: Chrysomelidae) with carbon dioxide. *J. Econ. Entomol.* **2004**, *97*, 330–339. [CrossRef] [PubMed]

31. Schumann, M.; Patel, A.; Vidal, S. Soil application of an encapsulated CO_2 source and its potential for management of western corn rootworm larvae. *J. Econ. Entomol.* **2014**, *107*, 230–239. [CrossRef] [PubMed]

32. Robert, C.A.M.; Erb, M.; Duployer, M.; Zwahlen, C.; Doyen, G.R.; Turlings, T.C.J. Herbivore-induced plant volatiles mediate host selection by a root herbivore. *N. Phytol.* **2012**, *194*, 1061–1069. [CrossRef]

33. Robert, C.A.M.; Erb, M.; Hibbard, B.E.; Wade French, B.; Zwahlen, C.; Turlings, T.C.J. A specialist root herbivore reduces plant resistance and uses an induced plant volatile to aggregate in a density-dependent manner. *Funct. Ecol.* **2012**, *26*, 1429–1440. [CrossRef]

34. Hiltpold, I.; Hibbard, B.E. Neonate larvae of the specialist herbivore *Diabrotica virgifera virgifera* do not exploit the defensive volatile *(E)*-β-caryophyllene in locating maize roots. *J. Pest Sci.* **2016**, *89*, 853–858. [CrossRef]

35. Strnad, S.P.; Dunn, P.E. Host search behaviour of neonate western corn rootworm (*Diabrotica virgifera virgifera*). *J. Insect Physiol.* **1990**, *36*, 201–205. [CrossRef]

36. Bernklau, E.J.; Hibbard, B.E.; Bjostad, L.B. Isolation and characterization of host recognition cues in corn roots for larvae of the western corn rootworm (Coleoptera: Chrysomelidae). *J. Econ. Entomol.* **2013**, *106*, 2354–2363. [CrossRef]

37. Bernklau, E.J.; Hibbard, B.E.; Dick, D.L.; Rithner, C.D.; Bjostad, L.B. Monogalactosyldiacylglycerols as host recognition cues for western corn rootworm larvae (Coleoptera: Chrysomelidae). *J. Econ. Entomol.* **2015**, *108*, 539–548. [CrossRef] [PubMed]

38. Bernklau, E.J.; Bjostad, L.B. Identification of feeding stimulants in corn roots for western corn rootworm (Coleoptera: Chrysomelidae) larvae. *J. Econ. Entomol.* **2008**, *101*, 341–351. [CrossRef]

39. Bernklau, E.J.; Hibbard, B.E.; Bjostad, L.B. Sugar preferences of western corn rootworm larvae in a feeding stimulant blend. *J. Appl. Entomol.* **2018**, *142*, 947–958. [CrossRef]

40. Bernklau, E.J.; Hibbard, B.E.; Bjostad, L.B. Toxic and behavioural effects of free fatty acids on western corn rootworm (Coleoptera: Chrysomelidae) larvae. *J. Appl. Entomol.* **2016**, *140*, 725–735. [CrossRef]

41. Hu, L.; Mateo, P.; Ye, M.; Zhang, X.; Berset, J.D.; Handrick, V.; Radisch, D.; Grabe, V.; Köllner, T.G.; Gershenzon, J.; et al. Plant iron acquisition strategy exploited by an insect herbivore. *Science* **2018**, *361*, 694–697. [CrossRef] [PubMed]

42. Machado, R.A.R.; Theepan, V.; Robert, C.A.M.; Züst, T.; Hu, L.; Su, Q.; Schimmel, B.C.J.; Erb, M. Complex plant metabolomes guide fitness-relevant foraging decisions of a specialist herbivore. *BioRxiv* **2020**. [CrossRef]

43. Bernklau, E.J.; Hibbard, B.E.; Norton, A.P.; Bjostad, L.B. Methyl anthranilate as a repellent for western corn rootworm larvae (Coleoptera: Chrysomelidae). *J. Econ. Entomol.* **2016**, *109*, 1683–1690. [CrossRef]

44. Wouters, F.C.; Blanchette, B.; Gershenzon, J.; Vassão, D.G. Plant defense and herbivore counter-defense: Benzoxazinoids and insect herbivores. *Phytochem. Rev.* **2016**, *15*, 1127–1151. [CrossRef] [PubMed]

45. Bernklau, E.J.; Bjostad, L.B.; Hibbard, B.E. Synthetic feeding stimulants enhance insecticide activity against western corn rootworm larvae, *Diabrotica virgifera virgifera* (Coleoptera: Chrysomelidae). *J. Appl. Entomol.* **2011**, *135*, 47–54. [CrossRef]

46. Hibbard, B.E.; Peairs, F.B.; Pilcher, S.D.; Schroeder, M.E.; Jewett, D.K.; Bjostad, L.B. Germinating corn extracts and 6-methoxy-2-benzoxazolinone: Western corn rootworm (Coleoptera: Chrysomelidae) larval attractants evaluated with soil insecticides. *J. Econ. Entomol.* **1995**, *88*, 716–724. [CrossRef]

47. Khan, Z.R.; Pickett, J.A. The "push-pull" strategy for stemborer management: A case study in exploiting biodiversity and chemical ecology. In *Ecological Engineering for Pest Management: Advances in Habitat Manipulation for Arthropods*; Gurr, G.M., Wratten, S.D., Altieri, M.A., Eds.; CABI: Oxon, UK, 2004; pp. 155–164.

48. Bernklau, E.J.; Hibbard, B.E.; Bjostad, L.B. Repellent effects of methyl anthranilate on western corn rootworm larvae (Coleoptera: Chrysomelidae) in soil bioassays. *J. Econ. Entomol.* **2019**, *112*, 683–690. [CrossRef] [PubMed]

49. Puebla, F.A.A.; Bernal, J.S. Resistance and tolerance to root herbivory in maize were mediated by domestication, spread, and breeding. *Front. Plant Sci.* **2020**, *27*, 223. [CrossRef]

50. Smith, C.M. *Plant Resistance to Insects: A Fundamental Approach*; Wiley: New York, NY, USA, 1989.

51. Painter, R.H. *Insect Resistance in Crop Plants*; Macmillian: New York, NY, USA, 1951. [CrossRef]

52. Howe, G.A.; Jander, G. Plant immunity to insect herbivores. *Annu. Rev. Plant Biol.* **2008**, *59*, 41–66. [CrossRef]

53. Loranger, J.; Meyer, S.T.; Shipley, B.; Kattge, J.; Loranger, H.; Roscher, C.; Weisser, W.W. Predicting invertebrate herbivory from plant traits: Evidence from 51 grassland species in experimental monocultures. *Ecology* **2012**, *93*, 2674–2682. [CrossRef] [PubMed]

54. Iason, G.R.; Dicke, M.; Hartley, S.E. *The Ecology of Plant Secondary Metabolites: From Genes to Global Processes*; Cambridge University Press: Cambridge, UK, 2012.

55. Hanley, M.E.; Lamont, B.B.; Fairbanks, M.M.; Rafferty, C.M. Plant structural traits and their role in anti-herbivore defence. *Perspect. Plant Ecol. Evol. Syst.* **2007**, *8*, 157–178. [CrossRef]

56. Strauss, S.Y.; Agrawal, A.A. The ecology and evolution of plant tolerance to herbivory. *Trends Ecol. Evol.* **1999**, *14*, 179–185. [CrossRef]

57. Newingham, B.A.; Callaway, R.M.; Bassirirad, H. Allocating nitrogen away from a herbivore: A novel compensatory response to root herbivory. *Oecologia* **2007**, *153*, 913–920. [CrossRef]

58. Xue, K.; Serohijos, R.C.; Devare, M.; Duxbury, J.; Lauren, J.; Thies, J.E. Short-term carbon allocation and root lignin of Cry3Bb Bt and nonBt corn in the presence of corn rootworm. *Appl. Soil Ecol.* **2012**, *57*, 16–22. [CrossRef]

59. Robert, C.A.M.; Ferrieri, R.A.; Schirmer, S.; Babst, B.A.; Schueller, M.J.; Machado, R.A.R.; Arce, C.C.M.; Hibbard, B.E.; Gershenzon, J.; Turlings, T.C.J.; et al. Induced carbon reallocation and compensatory growth as root herbivore tolerance mechanisms. *Plant Cell Environ.* **2014**, *37*, 2613–2622. [CrossRef]

60. Robert, C.A.M.; Schirmer, S.; Barry, J.; French, W.B.; Hibbard, B.E.; Gershenzon, J. Belowground herbivore tolerance involves delayed overcompensatory root regrowth in maize. *Entomol. Exp. Appl.* **2015**, *157*, 113–120. [CrossRef]

61. Bigger, J.H.; Holbert, J.R.; Flint, W.P.; Lang, A.L. Resistance of certain corn hybrids to attack of southern corn rootworm. *J. Econ. Entomol.* **1938**, *21*, 103–107. [CrossRef]

62. Bigger, J.H.; Snelling, R.O.; Blanchard, R.A. Resistance of corn strains to the southern corn rootworm, *Diabrotica duodecimpunctata* F. *J. Econ. Entomol.* **1941**, *34*, 605–613. [CrossRef]

63. Wilson, R.L.; Peters, D.C. Plant Introductions of *Zea mays* as sources of corn rootworm tolerance. *J. Econ. Entomol.* **1973**, *66*, 101–104. [CrossRef]

64. Tollefson, J.J. Evaluating maize for resistance to *Diabrotica virgifera virgifera* Leconte (Coleoptera: Chrysomelidae). *Maydica* **2007**, *52*, 311–318.

65. Branson, T.F. Resistance in the grass tribe Maydeae to larvae of the western corn rootworm. *Ann. Entomol. Soc. Am.* **1971**, *64*, 861–863. [CrossRef]

66. Branson, T.F.; Guss, P.L. Potential for utilizing resistance from relatives of cultivated crops. *Proc. North Cent. Branch Entomol. Soc. Am.* **1972**, *27*, 91–95.

67. Branson, T.F.; Reyes, R.J. The Association of *Diabrotica* spp. with *Zea diploperennis*. *J. Kans. Entomol. Soc.* **1983**, *56*, 97–99.

68. Branson, T.F.; Fisher, J.R.; Kahler, A.L.; Sutter, G.R. Host plant resistance to corn rootworms. In *Proceedings of the 17th Annual Illinois Corn Breeder School*; Allerton House: Champaign, IL, USA, 1981.

69. Branson, T.F.; Sutter, G.R.; Fisher, J.R. Comparison of a tolerant and a susceptible maize inbred under artificial infestations of *Diabrotica virgifera virgifera*: Yield and adult emergence. *Environ. Entomol.* **1982**, *11*, 371–372. [CrossRef]

70. Branson, T.F. Larval feeding behavior and host-plant resistance in maize. In *Methods for the Study of Pest Diabortica*; Krysan, J.L., Miller, T.A., Eds.; Springer: New York, NY, USA, 1986; pp. 159–182.

71. Fitzgerald, P.J.; Ortman, E.E. Breeding for resistance to the western corn rootworm. In *Proceedings of the Annual Hybrid Corn Industry Research Conference*; Heckendorn, W., Sutherland, J.I., Eds.; American Seed Trade Association: Washington, DC, USA, 1964; pp. 46–60.

72. Fitzgerald, P.J.; Ortman, E.E. Two-year performance of inbreds and their single crosses grown under corn rootworm infestation. *Proc. North Cent. Branch Entomol. Soc. Am.* **1965**, *20*, 46–47.

73. Hills, T.M.; Peters, D.C. A method of evaluating postplanting insecticide treatments for control of western corn rootworm larvae. *J. Econ. Entomol.* **1971**, *64*, 764–765. [CrossRef]

74. Kahler, A.L.; Olness, A.E.; Sutter, G.R.; Dybing, C.D.; Devine, O.J. Root damage by western corn rootworm and nutrient content in maize. *Agron. J.* **1985**, *77*, 769–774. [CrossRef]

75. Kahler, A.L.; Telkamp, R.E.; Penny, L.H.; Branson, T.F.; Fitzgerald, P.J. Registration of NGSDCRW1(S2)C4 Maize Germplasm. *Crop Sci.* **1985**, *25*, 202. [CrossRef]

76. Ortman, E.E.; Gerloff, E.D. Rootworm resistance: Problems in measuring and its relationship to performance. In Proceedings of the 25th Annual Corn and Sorghum Research Conference, Washington, DC, USA, 8–10 December 1970; Sutherland, J.I., Falasca, R.J., Eds.; pp. 161–174.

77. Ortman, E.E.; Branson, T.F. Growth pouches for studies of host plant resistance to larvae of corn rootworms. *J. Econ. Entomol.* **1976**, *69*, 380–382. [CrossRef]

78. Ortman, E.E.; Branson, T.F.; Gerloff, E.D. Techniques, accomplishments, and future potential of host plant resistance to Diabrotica. In *Proceedings of the Summer Institute on Biological Control of Plant Insects and Diseases*; Maxwell, F.G., Harris, F.A., Eds.; University Press: Jackson, MS, USA, 1974; pp. 344–358.

79. Riedell, W.E. Western corn rootworm damage in maize: Greenhouse technique and plant response. *Crop Sci.* **1989**, *29*, 412–415. [CrossRef]

80. Riedell, W.E.; Evenson, P.D. Rootworm feeding tolerance in single-cross maize hybrids from different eras. *Crop Sci.* **1993**, *33*, 951–955. [CrossRef]

81. Prischmann, D.A.; Dashiell, K.E.; Schneider, D.J.; Hibbard, B.E. Field screening maize germplasm for resistance and tolerance to western corn rootworms (Col.: Chrysomelidae). *J. Appl. Entomol.* **2007**, *131*, 406–415. [CrossRef]

82. Assabgui, R.A.; Arnason, J.T.; Hamilton, R.I. Hydroxamic acid content in maize (*Zea mays*) roots of 18 Ontario recommended hybrids and prediction of antibiosis to the western corn rootworm, *Diabrotica virgifera virgifera* LeConte (Coleoptera: Chrysomelidae). *Can. J. Plant Sci.* **1993**, *73*, 359–363. [CrossRef]

83. Assabgui, R.A.; Arnason, J.T.; Hamilton, R.I. Hydroxamic acid content and plant development of maize (*Zea mays* L.) in relation to damage by the western corn rootworm, *Diabrotica virgifera virgifera* LeConte. *Can. J. Plant Sci.* **1995**, *75*, 51–856. [CrossRef]

84. Rogers, R.R.; Owens, J.C.; Tollefson, J.J.; Witkowski, J.F. Evaluation of commercial corn hybrids for tolerance to corn rootworms. *Environ. Entomol.* **1975**, *4*. [CrossRef]

85. Assabgui, R.A.; Arnason, J.T.; Hamilton, R.I. Field evaluations of hydroxamic acids as antibiosis factors in elite maize inbreds to the western corn rootworm (Coleoptera: Chrysomelidae). *J. Econ. Entomol.* **1995**, *88*, 1482–1493. [CrossRef]

86. Xie, Y.S.; Arnason, J.T.; Philogene, B.J.R.; Lambert, J.D.H.; Atkinson, J.; Morand, P. Role of 2, 4-dihydroxy-7-methoxy-1, 4-benzoxazin-3-one (DIMBOA) in the resistance of maize to western corn rootworm, *Diabrotica virgifera virgifera* (Leconte) (Coleoptera: Chrysomelidae). *Can. Entomol.* **1990**, *122*, 1177–1186. [CrossRef]

87. Xie, Y.; Arnason, T.J.; Philogène, B.J.R.; Olechowski, H.T.; Hamilton, R. Variation of Hydroxamic acid content in maize roots in relation to geographic origin of maize germplasm and resistance to western corn rootworm (Coleoptera: Chrysomelidae). *J. Econ. Entomol.* **1992**, *85*, 2478–2485. [CrossRef]

88. Xie, Y.; Arnason, J.T.; Philogéne, B.J.R.; Atkinson, J.; Morand, P. Behavioral responses of western corn rootworm larvae to naturally occurring and synthetic hydroxamic acids. *J. Chem. Ecol.* **1992**, *18*, 945–957. [CrossRef] [PubMed]

89. Arnason, J.T.; Larsen, J.; Assabgui, R.; Xie, Y.; Atkinson, J.; Philogene, B.J.R.; Hamilton, R.I. Mechanisms of resistance in maize to western corn rootworm. In Insect resistant maize: Recent advances and utilization. In Proceedings of the International Symposium Held at the International Maize and Wheat Improvement Center (CIMMYT), El Bátan, Tunisia, 27 November–3 December 1994; pp. 96–100.

90. Moellenbeck, D.J.; Bergvinson, B.D.; Darrah, L.L. Advances in rating and phytochemical screening for corn rootworm resistance. In Insect resistant maize: Recent advances and utilization. In Proceedings of the International Symposium Held at the International Maize and Wheat Improvement Center (CIMMYT), El Bátan, Tunisia, 27 November–3 December 1994; pp. 203–210.

91. Praiswater, T.W.; Hibbard, B.E.; Barry, B.D.; Darrah, L.L.; Smith, V.A. An implement for dislodging maize roots from the soil for corn rootworm (Coleoptera: Chrysomelidae) damage evaluations. *J. Kans. Entomol. Soc.* **1997**, *70*, 335–338. [CrossRef]

92. Hibbard, B.E.; Darrah, L.L.; Barry, B.D. Combining ability of resistance leads and identification of a new resistance source for western corn rootworm (Coleoptera: Chrysomelidae) larvae in corn. *Maydica* **1999**, *44*, 133–139.

93. Hibbard, B.E.; Barry, B.D.; Darrah, L.L.; Jackson, J.J.; Chandler, L.D.; French, L.K.; Mihm, J.A. Controlled field infestations with western corn rootworm (Coleoptera: Chrysomelidae) eggs in Missouri: Effects of egg strains, infestation dates, and infestation levels on corn root damage. *J. Kans. Entomol. Soc.* **1999**, *72*, 214–221. [CrossRef]

94. Hibbard, B.E.; Willmot, D.B.; Garcia, F.S.A.; Darrah, L.L. Registration of the maize germplasm CRW3(S1)C6 with resistance to western corn rootworm. *J. Plant Regist.* **2007**, *1*, 151–152. [CrossRef]

95. Rogers, R.R.; Russell, W.A.; Owens, J.C. Evaluation of a vertical-pull technique in population improvement of maize for corn rootworm tolerance. *Crop Sci.* **1976**, *16*, 591–594. [CrossRef]

96. Knutson, R.J.; Hibbard, B.E.; Barry, B.D.; Smith, V.A.; Darrah, L.L. Comparison of screening techniques for western corn rootworm (Coleoptera: Chrysomelidae) host-plant resistance. *J. Econ. Entomol.* **1999**, *92*, 714–722. [CrossRef]

97. Abel, C.A.; Berhow, M.A.; Wilson, R.L.; Binder, B.F.; Hibbard, B.E. Evaluation of conventional resistance to European corn borer (Lepidoptera: Crambidae) and western corn rootworm (Coleoptera: Chrysomelidae) in experimental maize lines developed from a backcross breeding program. *J. Econ. Entomol.* **2000**, *93*, 1814–1821. [CrossRef]

98. Garcia, F.S.A.; Dashiell, K.E.; Prischmann, D.A.; Bohn, M.O.; Hibbard, B.E. Conventional screening overlooks resistance sources: Rootworm damage of diverse inbred lines and their B73 hybrids is unrelated. *J. Econ. Entomol.* **2009**, *102*, 1317–1324. [CrossRef]

99. El Khishen, A.A.; Bohn, M.O.; Voldseth, P.D.A.; Dashiell, K.E.; French, B.W.; Hibbard, B.E. Native resistance to western corn rootworm (Coleoptera: Chrysomelidae) larval feeding: Characterization and mechanisms. *J. Econ. Entomol.* **2009**, *102*, 2350–2359. [CrossRef]

100. Bernklau, E.J.; Hibbard, B.E.; Bjostad, L.B. Antixenosis in maize reduces feeding by western corn rootworm larvae (Coleoptera: Chrysomelidae). *J. Econ. Entomol.* **2010**, *103*, 2052–2060. [CrossRef]

101. Šimić, D.; Ivezić, M.; Brkić, I.; Raspudić, E.; Brmež, M.; Majić, I.; Brkić, A.; Ledenčan, T.; Tollefson, J.J.; Hibbard, B.E. Environmental and genotypic effects for western corn rootworm tolerance traits in American and European maize trials. *Maydica* **2007**, *52*, 425–430.

102. Ivezic, M.; Tollefson, J.J.; Raspudic, E.; Brkic, I.; Brmez, M.; Hibbard, B.E. Evaluation of corn hybrids for tolerance to corn rootworm (*Diabrotica virgifera virgifera* LeConte) larval feeding. *Cereal Res. Commun.* **2006**, *34*, 1101–1107. [CrossRef]

103. Ivezić, M.; Tollefson, J.J.; Raspudić, E.; Hibbard, B.E.; Brkić, I. Evaluation of Croation corn hybrids for tolerance to corn rootworm (*Diabrotica virgifera virgifera* LeConte) larval feeding. In Proceedings of the 21st IWGO Conference and VIII Diabrotica subgroup meeting, Venice, Italy, 27 October–3 November 2001; pp. 205–212.

104. Ivezić, M.; Raspudić, E.; Brmež, M.; Majić, I.; Brkić, I.; Tollefson, J.J.; Bohn, M.; Hibbard, B.E.; Šimić, D. A review of resistance breeding options targeting western corn rootworm (*Diabrotica virgifera virgifera* LeConte). *Agric. For. Entomol.* **2009**, *11*, 307–311. [CrossRef]

105. Bohn, M. How to improve rootworm resistance in corn. In *Proceedings of the 41st Illinois Corn Breeders School*; Illinois Corn Breeders School: Champaign-Urbana, IL, USA, 2005; pp. 183–194.

106. Rogers, R.R.; Russell, W.A.; Owens, J.C. Relationship of corn rootworm (*Diabrotica*) tolerance to yield in the Isss (Iowa stiff stalk synthetic) maize population. *Iowa State J. Res.* **1976**, *51*, 125–129.

107. Bohn, M.O.; Marroquin, J.J.; Garcia, F.S.; Dashiell, K.; Willmot, D.B.; Hibbard, B.E. Quantitative trait loci mapping of western corn rootworm (Coleoptera: Chrysomelidae) host plant resistance in two populations of doubled haploid lines in maize (*Zea mays* L.). *J. Econ. Entomol.* **2018**, *111*, 435–444. [CrossRef]

108. Russell, W.A.; Owens, J.C.; Peters, D.C.; Rogers, R.R. Registration of maize germplasm 1 (Reg. Nos. GP 72 and GP 73). *Crop Sci.* **1976**, *16*, 886–887. [CrossRef]

109. Rogers, R.R.; Russell, W.A.; Owens, J.C. Expected gains from selection in maize for resistance to corn rootworms. *Maydica* **1977**, *22*, 27–36.

110. Russell, W.A.; Penny, L.H.; Guthrie, W.D.; Dicke, F.F. Registration of maize germplasm inbreds 1 (Reg. Nos. GP 1 to 5). *Crop Sci.* **1971**, *11*, 140. [CrossRef]

111. Russell, W.A.; Penny, L.H.; Sprague, G.F.; Guthrie, W.D.; Dicke, F.F. Registration of corn parental lines (Reg. nos. PL 1 to 13). *Crop Sci.* **1971**, *11*, 143. [CrossRef]

112. Owens, J.C.; Peters, D.C.; Hallauer, A.R. Corn rootworm tolerance in maize. *Environ. Entomol.* **1974**, *3*, 767–772. [CrossRef]

113. Branson, T.F.; Welch, V.A.; Sutter, G.R.; Fisher, J.R. Resistance to larvae of *Diabrotica virgifera virgifera* in three experimental maize hybrids. *Environ. Entomol.* **1983**, *12*, 1509–1512. [CrossRef]

114. Brkić, A.; Šimić, D.; Jambrović, A.; Zdunić, Z.; Ledenčan, T.; Raspudić, E.; Brmeţ, M.; Brkić, J.; Mazur, M.; Galić, V. QTL analysis of western corn rootworm resistance traits in maize ibm population grown in continuous maize. *Genetika* **2020**, *52*, 137–148. [CrossRef]

115. Prasanna, B.M. Diversity in global maize germplasm: Characterization and utilization. *J. Biosci.* **2012**, *37*, 843–855. [CrossRef]

116. Meihls, L.N.; Kaur, H.; Jander, G. Natural variation in maize defense against insect herbivores. *Cold Spring Harb. Symp. Quant. Biol.* **2012**, *77*, 269–283. [CrossRef] [PubMed]

117. Mohan, S.; Ma, P.W.K.; Pechan, T.; Bassford, E.R.; Williams, W.P.; Luthe, D.S. Degradation of the *S. frugiperda* peritrophic matrix by an inducible maize cysteine protease. *J. Insect Physiol.* **2006**, *52*, 21–28. [CrossRef] [PubMed]

118. Mohan, S.; Ma, P.W.K.; Williams, W.P.; Luthe, D.S. A naturally occurring plant cysteine protease possesses remarkable toxicity against insect pests and synergizes *Bacillus thuringiensis* toxin. *PLoS ONE* **2008**, *3*, e1786. [CrossRef]

119. Duque, C.L.; Loades, K.W.; Tooker, J.F.; Brown, K.M.; Williams, P.W.; Luthe, D.S. A maize inbred exhibits resistance against western corn rootwoorm, *Diabrotica virgifera virgifera*. *J. Chem. Ecol.* **2017**, *43*, 1108–1123. [CrossRef]

120. Ramstein, G.P.; Jensen, S.E.; Buckler, E.S. Breaking the curse of dimensionality to identify causal variants in Breeding 4. *Theor. Appl. Genet.* **2019**, *132*, 559–567. [CrossRef]

121. Branson, T.F.; Ortman, E.E. Host range of larvae of the western corn rootworm. *J. Econ. Entomol.* **1967**, *60*, 201–203. [CrossRef]

122. Branson, T.F.; Ortman, E.E. The host range of larvae of the western corn rootworm: Further studies. *J. Econ. Entomol.* **1970**, *63*, 800–803. [CrossRef]

123. Clark, T.L.; Hibbard, B.E. Comparison of nonmaize hosts to support western corn rootworm (Coleoptera: Chrysomelidae) larval biology. *Environ. Entomol.* **2004**, *33*, 681–689. [CrossRef]

124. Oyediran, I.O.; Hibbard, B.E.; Clark, T.L. Prairie grasses as hosts of the western corn rootworm (Coleoptera: Chrysomelidae). *Environ. Entomol.* **2004**, *33*, 740–747. [CrossRef]

125. Wilson, T.A.; Hibbard, B.E. Host suitability of nonmaize agroecosystem grasses for the western corn rootworm (Coleoptera: Chrysomelidae). *Environ. Entomol.* **2004**, *33*, 1102–1108. [CrossRef]

126. Moeser, J.; Vidal, S. How to measure the food utilization of subterranean insects: A case study with the western corn rootworm (*Diabrotica virgifera virgifera*). *J. Appl. Entomol.* **2005**, *129*, 60–63. [CrossRef]

127. Hibbard, B.E.; Schweikert, Y.M.; Higdon, M.L.; Ellersieck, M.R. Maize phenology affects establishment, damage, and development of the western corn rootworm (Coleoptera: Chrysomelidae). *Environ. Entomol.* **2008**, *37*, 1558–1564. [CrossRef] [PubMed]

128. Ritchie, W.S.; Hanway, J.J.; Benson, G.O. *How a Corn Plant Develops*; Iowa State University: Ames, IA, USA, 1992.

129. Law, J.H. Insects, oxygen, and iron. *Biochem. Biophys. Res. Commun.* **2002**, *292*, 1191–1195. [CrossRef]

130. Chege, P.G.; Clark, T.L.; Hibbard, B.E. Alternate host phenology affects survivorship, growth, and development of western corn rootworm (Coleoptera: Chrysomelidae) larvae. *Environ. Entomol.* **2005**, *34*, 1441–1447. [CrossRef]

131. Huynh, M.P.; Hibbard, B.E.; Vella, M.; Lapointe, S.L.; Niedz, R.P.; Shelby, K.S.; Coudron, T.A. Development of an improved and accessible diet for western corn rootworm larvae using response surface modeling. *Sci. Rep.* **2019**, *9*, 6009. [CrossRef] [PubMed]

132. Hannon, G.J. RNA interference. *Nature* **2002**, *418*, 244–251. [CrossRef]

133. Katoch, R.; Sethi, A.; Thakur, N.; Murdock, L.L. RNAi for insect control: Current perspective and future challenges. *Appl. Biochem. Biotechnol.* **2013**, *171*, 847–873. [CrossRef]

134. Ivashuta, S.; Zhang, Y.; Wiggins, B.E.; Ramaseshadri, P.; Segers, G.C.; Johnson, S.; Meyer, S.E.; Kerstetter, R.A.; McNulty, B.C.; Bolognesi, R.; et al. Environmental RNAi in herbivorous insects. *RNA* **2015**, *21*, 840–850. [CrossRef]

135. Turner, C.T.; Davy, M.W.; Diarmid, M.R.M.; Plummer, K.M.; Birch, N.P.; Newcomb, R.D. RNA interference in the light brown apple moth, *Epiphyas postvittana* (Walker) induced by double-stranded RNA feeding. *Insect Mol. Biol.* **2006**, *15*, 383–391. [CrossRef] [PubMed]

136. Rajagopal, R.; Sivakumar, S.; Agrawal, N.; Malhotra, P.; Bhatnagar, R.K. Silencing of midgut aminopeptidase N of *Spodoptera litura* by double-stranded RNA establishes its role as *Bacillus thuringiensis* toxin receptor. *J. Biol. Chem.* **2002**, *6*, 46849–46851. [CrossRef]

137. Bucher, G.; Scholten, J.; Klingler, M. Parental RNAi in tribolium (Coleoptera). *Curr. Biol.* **2002**, *5*, R85–R86. [CrossRef]

138. Tomoyasu, Y.; Denell, R.E. Larval RNAi in *Tribolium* (Coleoptera) for analyzing adult development. *Dev. Genes Evol.* **2004**, *214*, 575–578. [CrossRef]

139. Soares, C.A.G.; Lima, C.M.R.; Dolan, M.C.; Piesman, J.; Beard, C.B.; Zeidner, N.S. Capillary feeding of specific dsRNA induces silencing of the isac gene in nymphal *Ixodes scapularis* ticks. *Insect Mol. Biol.* **2005**, *14*, 443–452. [CrossRef]

140. Baum, J.A.; Bogaert, T.; Clinton, W.; Heck, G.R.; Feldmann, P.; Ilagan, O.; Johnson, S.; Plaetinck, G.; Munyikwa, T.; Pleau, M.; et al. Control of coleopteran insect pests through RNA interference. *Nat. Biotechnol.* **2007**, *25*, 1322–1326. [CrossRef]

141. Bolognesi, R.; Ramaseshadri, P.; Anderson, J.; Bachman, P.; Clinton, W.; Flannagan, R.; Ilagan, O.; Lawrence, C.; Levine, S.; Moar, W.; et al. Characterizing the mechanism of action of double-stranded RNA activity against western corn rootworm (*Diabrotica virgifera virgifera* LeConte). *PLoS ONE* **2012**, *7*, e47534. [CrossRef]

142. Rangasamy, M.; Siegfried, B.D. Validation of RNA interference in western corn rootworm *Diabrotica virgifera virgifera* LeConte (Coleoptera: Chrysomelidae) adults. *Pest Manag. Sci.* **2012**, *68*, 587–591. [CrossRef] [PubMed]

143. Vélez, A.M.; Fishilevich, E.; Rangasamy, M.; Khajuria, C.; McCaskill, D.G.; Pereira, A.E.; Gandra, P.; Frey, M.L.F.; Worden, S.E.; Whitlock, S.L.; et al. Control of western corn rootworm via RNAi traits in maize: Lethal and sublethal effects of Sec23 dsRNA. *Pest Manag. Sci.* **2020**, *76*, 1500–1512. [CrossRef] [PubMed]

144. Khajuria, C.; Vélez, A.M.; Rangasamy, M.; Wang, H.; Fishilevich, E.; Frey, M.L.F.; Carneiro, N.P.; Gandra, P.; Narva, K.E.; Siegfried, B.D. Parental RNA interference of genes involved in embryonic development of the western corn rootworm, *Diabrotica virgifera virgifera* LeConte. *Insect Biochem. Mol. Biol.* **2015**, *62*, 54–62. [CrossRef]

145. Vélez, A.M.; Fishilevich, E.; Matz, N.; Storer, N.P.; Narva, K.E.; Siegfried, B.D. Parameters for successful parental RNAi as an insect pest management tool in western corn rootworm, *Diabrotica virgifera virgifera*. *Genes* **2017**, *8*, 7. [CrossRef]

146. Ramaseshadri, P.; Segers, G.; Flannagan, R.; Wiggins, E.; Clinton, W.; Ilagan, O.; McNulty, B.; Clark, T.; Bolognesi, R. Physiological and cellular responses caused by RNAi-mediated suppression of Snf7 orthologue in western corn rootworm (*Diabrotica virgifera virgifera*) larvae. *PLoS ONE* **2013**, *8*, e54270. [CrossRef]

147. Hu, X.; Richtman, N.M.; Zhao, J.Z.; Duncan, K.E.; Niu, X.; Procyk, L.A.; Oneal, M.A.; Kernodle, B.M.; Steimel, J.P.; Crane, V.C.; et al. Discovery of midgut genes for the RNA interference control of corn rootworm. *Sci. Rep.* **2016**, *6*, 542. [CrossRef]

148. Li, H.; Khajuria, C.; Rangasamy, M.; Gandra, P.; Fitter, M.; Geng, C.; Woosely, A.; Hasler, J.; Schulenberg, G.; Worden, S.; et al. Long dsRNA but not siRNA initiates RNAi in western corn rootworm larvae and adults. *J. Appl. Entomol.* **2015**, *139*, 432–445. [CrossRef]

149. Niu, X.; Kassa, A.; Hu, X.; Robeson, J.; McMahon, M.; Richtman, N.M.; Steimel, J.P.; Kernodle, B.M.; Crane, V.C.; Sandahl, G.; et al. Control of western corn rootworm (*Diabrotica virgifera virgifera*) reproduction through plant-mediated RNA interference. *Sci. Rep.* **2017**, *7*, 12591. [CrossRef] [PubMed]

150. Head, G.P.; Carroll, M.W.; Evans, S.P.; Rule, D.M.; Willse, A.R.; Clark, T.L.; Storer, N.P.; Flannagan, R.D.; Samuel, L.W.; Meinke, L.J. Evaluation of SmartStax and SmartStax PRO maize against western corn rootworm and northern corn rootworm: Efficacy and resistance management. *Pest Manag. Sci.* **2017**, *73*, 1883–1899. [CrossRef]

151. Anderson, J.A.; Mickelson, J.; Challender, M.; Moellring, E.; Sult, T.; TeRonde, S.; Walker, C.; Wang, Y.; Maxwell, C.A. Agronomic and compositional assessment of genetically modified DP23211 maize for corn rootworm control. *GM Crop. Food* **2020**, *11*, 206–214. [CrossRef]

152. Wardle, D.A.; Bardgett, R.D.; Klironomos, J.N.; Setälä, H.; Van Der Putten, W.H.; Wall, D.H. Ecological linkages between aboveground and belowground biota. *Science* **2004**, *304*, 1629–1633. [CrossRef] [PubMed]

153. Blundell, R.; Schmidt, J.E.; Igwe, A.; Cheung, A.L.; Vannette, R.L.; Gaudin, A.C.M.; Casteel, C.L. Organic management promotes natural pest control through altered plant resistance to insects. *Nat. Plants* **2020**, *6*, 483–491. [CrossRef]

154. Mueller, U.G.; Sachs, J.L. Engineering microbiomes to improve plant and animal health. *Trends Microbiol.* **2015**, *23*, 606–617. [CrossRef] [PubMed]

155. Hu, L.; Robert, C.A.M.; Cadot, S.; Zhang, X.; Ye, M.; Li, B.; Manzo, D.; Chervet, N.; Steinger, T.; Van Der Heijden, M.G.A.; et al. Root exudate metabolites drive plant-soil feedbacks on growth and defense by shaping the rhizosphere microbiota. *Nat. Commun.* **2018**, *9*, 2738. [CrossRef] [PubMed]

156. Willis, A.; Rodrigues, B.F.; Harris, P.J.C. The ecology of arbuscular mycorrhizal fungi. *CRC Crit. Rev. Plant Sci.* **2013**, *32*, 1–20. [CrossRef]

157. Javaid, A. Arbuscular mycorrhizal mediated nutrition in plants. *J. Plant Nutr.* **2009**. [CrossRef]

158. Murrell, E.G.; Ray, S.; Lemmon, M.E.; Luthe, D.S.; Kaye, J.P. Cover crop species affect mycorrhizae-mediated nutrient uptake and pest resistance in maize. *Renew. Agric. Food Syst.* **2020**, *35*, 467–474. [CrossRef]

159. Gehring, C.; Bennett, A. Mycorrhizal fungal-plant-insect interactions: The importance of a community approach. *Environ. Entomol.* **2009**, *38*, 93–102. [CrossRef] [PubMed]

160. Koricheva, J.; Gange, A.C.; Jones, T. Effects of mycorrhizal fungi on insect herbivores: A meta-analysis. *Ecology* **2009**, *90*, 2088–2097. [CrossRef] [PubMed]

161. Bennett, A.E.; Garcia, A.J.; Bever, J.D. Three-way interactions among mutualistic mycorrhizal fungi, plants, and plant enemies: Hypotheses and synthesis. *Am. Nat.* **2006**, *167*, 141–152. [CrossRef]

162. Jaffuel, G.; Imperiali, N.; Shelby, K.; Herrera, C.R.; Geisert, R.; Maurhofer, M.; Loper, J.; Keel, C.; Turlings, T.C.J.; Hibbard, B.E. Protecting maize from rootworm damage with the combined application of arbuscular mycorrhizal fungi, *Pseudomonas* bacteria and entomopathogenic nematodes. *Sci. Rep.* **2019**, *9*, 3127. [CrossRef]

163. Vukicevich, E.; Lowery, T.; Bowen, P.; Torres, Ú.J.R.; Hart, M. Cover crops to increase soil microbial diversity and mitigate decline in perennial agriculture. A review. *Agron. Sustain. Dev.* **2016**, *36*, 48. [CrossRef]

164. Hartwig, N.L.; Ammon, H.U. Cover crops and living mulches. *Weed Sci.* **2002**, *50*, 688–699. [CrossRef]

165. Bainard, L.D.; Bainard, J.D.; Hamel, C.; Gan, Y. Spatial and temporal structuring of arbuscular mycorrhizal communities is differentially influenced by abiotic factors and host crop in a semi-arid prairie agroecosystem. *FEMS Microbiol. Ecol.* **2014**, *88*, 333–344. [CrossRef] [PubMed]

166. Benitez, M.S.; Taheri, W.I.; Lehman, R.M. Selection of fungi by candidate cover crops. *Appl. Soil Ecol.* **2016**, *103*, 72–82. [CrossRef]

167. Hontoria, C.; González, G.I.; Quemada, M.; Roldán, A.; Alguacil, M.M. The cover crop determines the AMF community composition in soil and in roots of maize after a ten-year continuous crop rotation. *Sci. Total Environ.* **2019**, *660*, 913–922. [CrossRef]

168. Cloutier, M.L.; Murrell, E.; Barbercheck, M.; Kaye, J.; Finney, D.; González, G.I.; Bruns, M.A. Fungal community shifts in soils with varied cover crop treatments and edaphic properties. *Sci. Rep.* **2020**, *10*, 6198. [CrossRef] [PubMed]

169. Lundgren, J.G.; Fergen, J.K. The effects of a winter cover crop on *Diabrotica virgifera* (Coleoptera: Chrysomelidae) populations and beneficial arthropod communities in no-till maize. *Environ. Entomol.* **2010**, *39*, 1816–1828. [CrossRef] [PubMed]

170. Douglas, A.E. Symbiotic microorganisms: Untapped resources for insect pest control. *Trends Biotechnol.* **2007**, *25*, 338–342. [CrossRef]

171. Dematheis, F.; Kurtz, B.; Vidal, S.; Smalla, K. Microbial communities associated with the larval gut and eggs of the western corn rootworm. *PLoS ONE* **2012**, *7*, 4685. [CrossRef]

172. Ludwick, D.C.; Ericsson, A.C.; Meihls, L.N.; Gregory, M.L.J.; Finke, D.L.; Coudron, T.A.; Hibbard, B.E.; Shelby, K.S. Survey of bacteria associated with western corn rootworm life stages reveals no difference between insects reared in different soils. *Sci. Rep.* **2019**, *9*, 1–11. [CrossRef] [PubMed]

173. Chu, C.C.; Spencer, J.L.; Curzi, M.J.; Zavala, J.A.; Seufferheld, M.J. Gut bacteria facilitate adaptation to crop rotation in the western corn rootworm. *Proc. Natl. Acad. Sci. USA* **2013**, *110*, 11917–11922. [CrossRef] [PubMed]

174. Perlatti, B.; Luiz, A.L.; Prieto, E.L.; Fernandes, J.B.; da Silva, M.F.d.G.F.; Ferreira, D.; Costa, E.N.; Júnior, B.A.L.; Forim, M.R. MALDI-TOF MS identification of microbiota associated with pest insect *Diabrotica speciosa*. *Agric. For. Entomol.* **2017**, *19*, 408–417. [CrossRef]

175. Robert, C.A.M.; Frank, D.L.; Leach, K.A.; Turlings, T.C.J.; Hibbard, B.E.; Erb, M. Direct and indirect plant defenses are not suppressed by endosymbionts of a specialist root herbivore. *J. Chem. Ecol.* **2013**, *39*, 507–515. [CrossRef]

176. Clark, T.L.; Meinke, L.J.; Skoda, S.R.; Foster, J.E. Occurrence of Wolbachia in selected Diabroticite (Coleoptera: Chrysomelidae) beetles. *Ann. Entomol. Soc. Am.* **2001**, *94*, 877–885. [CrossRef]

177. Werren, J.H.; Baldo, L.; Clark, M.E. Wolbachia: Master manipulators of invertebrate biology. *Nat. Rev. Microbiol.* **2008**, *6*, 741–751. [CrossRef]

178. Werren, J.H. Wolbachia and Speciation. In *Endless Forms: Species and Speciation*; Berlocher, S.H., Ed.; Oxford University Press: New York, NY, USA, 1998; pp. 245–260.

179. Bordenstein, S.R. Symbiosis and the origin of species. In *Insect Symbiosis*; CRC Press: Boca Raton, FL, USA, 2003; pp. 283–304.

180. Giordano, R.; Jackson, J.J.; Robertson, H.M. The role of Wolbachia bacteria in reproductive incompatibilities and hybrid zones of *Diabrotica* beetles and *Gryllus* crickets. *Proc. Natl. Acad. Sci. USA* **1997**, *94*, 11439–11444. [CrossRef]

181. Lombaert, E.; Ciosi, M.; Miller, N.J.; Sappington, T.W.; Blin, A.; Guillemaud, T. Colonization history of the western corn rootworm (*Diabrotica virgifera virgifera*) in North America: Insights from random forest ABC using microsatellite data. *Biol. Invasions* **2018**, *20*, 665–677. [CrossRef]

182. Ye, Y.H.; Seleznev, A.; Flores, H.A.; Woolfit, M.; McGraw, E.A. Gut microbiota in Drosophila melanogaster interacts with Wolbachia but does not contribute to Wolbachia-mediated antiviral protection. *J. Invertebr. Pathol.* **2017**, *143*, 18–25. [CrossRef] [PubMed]

183. Hedges, L.M.; Brownlie, J.C.; O'Neill, S.L.; Johnson, K.N. Wolbachia and virus protection in insects. *Science* **2008**, *322*, 702. [CrossRef]

184. Barr, K.L.; Hearne, L.B.; Briesacher, S.; Clark, T.L.; Davis, G.E. Microbial symbionts in insects influence down-regulation of defense genes in maize. *PLoS ONE* **2010**, *5*, e11339. [CrossRef] [PubMed]

185. Lance, D.R. Odors influence choice of oviposition sites by *Diabrotica virgifera virgifera* (Coleoptera: Chrysomelidae). *J. Chem. Ecol.* **1992**, *18*, 1227–1237. [CrossRef] [PubMed]

186. Shukla, S.P.; Beran, F. Gut microbiota degrades toxic isothiocyanates in a flea beetle pest. *Mol. Ecol.* **2020**, *29*, 4692–4705. [CrossRef]

187. Itoh, H.; Tago, K.; Hayatsu, M.; Kikuchi, Y. Detoxifying symbiosis: Microbe-mediated detoxification of phytotoxins and pesticides in insects. *Nat. Prod. Rep.* **2018**, *35*, 434–454. [CrossRef]

188. van den Bosch, T.J.M.; Welte, C.U. Detoxifying symbionts in agriculturally important pest insects. *Microb. Biotechnol.* **2017**, *10*, 531–540. [CrossRef]

189. Hammer, T.J.; Bowers, M.D. Gut microbes may facilitate insect herbivory of chemically defended plants. *Oecologia* **2015**, *179*, 1–14. [CrossRef]

190. Steinhaus, E. Microbial Control-The emergence of an idea. *Hilgardia* **1956**, *26*, 107–160. [CrossRef]

191. Lacey, L.A.; Grzywacz, D.; Ilan, S.D.I.; Frutos, R.; Brownbridge, M.; Goettel, M.S. Insect pathogens as biological control agents: Back to the future. *J. Invertebr. Pathol.* **2015**, *132*, 1–41. [CrossRef] [PubMed]

192. Vega, F.; Kaya, H. (Eds.) *Insect Pathology*; Academic Press: San Diego, CA, USA, 2012; ISBN 9780123849847.

193. Knowles, B.H.; Ellar, D.J. Colloid-osmotic lysis is a general feature of the mechanism of action of *Bacillus thuringiensis* δ-endotoxins with different insect specificity. *BBA Gen. Subj.* **1987**, *924*, 509–518. [CrossRef]

194. Yin, Y.; Flasinski, S.; Moar, W.; Bowen, D.; Chay, C.; Milligan, J.; Kouadio, J.L.; Pan, A.; Werner, B.; Buckman, K.; et al. A new *Bacillus thuringiensis* protein for western corn rootworm control. *PLoS ONE* **2020**, *15*, e0242791. [CrossRef]
195. Bowen, D.; Yin, Y.; Flasinski, S.; Chay, C.; Bean, G.; Milligan, J.; Moar, W.; Pan, A.; Werner, B.; Buckman, K.; et al. Cry75Aa (Mpp75Aa) insecticidal proteins for controlling the western corn rootworm, *Diabrotica virgifera virgifera*, (Coleoptera: Chrysomelidae), isolated from the insect pathogenic bacteria *Brevibacillus laterosporus*. *Appl. Environ. Microbiol.* **2020**. [CrossRef]
196. Sampson, K.; Zaitseva, J.; Stauffer, M.; Berg, V.B.; Guo, R.; Tomso, D.; McNulty, B.; Desai, N.; Balasubramanian, D. Discovery of a novel insecticidal protein from *Chromobacterium piscinae*, with activity against western corn rootworm, *Diabrotica virgifera virgifera*. *J. Invertebr. Pathol.* **2017**, *142*, 34–43. [CrossRef]
197. Wei, J.Z.; Rear, O.J.; Schellenberger, U.; Rosen, B.A.; Park, Y.J.; McDonald, M.J.; Zhu, G.; Xie, W.; Kassa, A.; Procyk, L.; et al. A selective insecticidal protein from *Pseudomonas mosselii* for corn rootworm control. *Plant Biotechnol. J.* **2018**, *16*, 649–659. [CrossRef]
198. Yalpani, N.; Altier, D.; Barry, J.; Kassa, A.; Nowatzki, T.M.; Sethi, A.; Zhao, J.Z.; Diehn, S.; Crane, V.; Sandahl, G.; et al. An *Alcaligenes* strain emulates *Bacillus thuringiensis* producing a binary protein that kills corn rootworm through a mechanism similar to Cry34Ab1/Cry35Ab1. *Sci. Rep.* **2017**, *7*, 3063. [CrossRef]
199. Bowling, A.J.; Pence, H.E.; Li, H.; Tan, S.Y.; Evans, S.L.; Narva, K.E. Histopathological effects of Bt and TcdA insecticidal proteins on the midgut epithelium of western corn rootworm larvae (*Diabrotica virgifera virgifera*). *Toxins (Basel)* **2017**, *9*, 156. [CrossRef] [PubMed]
200. Panevska, A.; Hodnik, V.; Skočaj, M.; Novak, M.; Modic, Š.; Pavlic, I.; Podržaj, S.; Zarić, M.; Resnik, N.; Maček, P.; et al. Pore-forming protein complexes from *Pleurotus* mushrooms kill western corn rootworm and Colorado potato beetle through targeting membrane ceramide phosphoethanolamine. *Sci. Rep.* **2019**, *9*, 5073. [CrossRef]
201. Hurst, M.R.H.; van Koten, C.; Jackson, T.A. Pathology of *Yersinia entomophaga* MH96 towards *Costelytra zealandica* (Coleoptera; Scarabaeidae) larvae. *J. Invertebr. Pathol.* **2014**, *115*, 102–107. [CrossRef]
202. Ruiu, L.; Satta, A.; Floris, I. Emerging entomopathogenic bacteria for insect pest management. *Bull. Insectology* **2013**, *66*, 181–186.
203. Jackson, T.A.; Pearson, J.F.; Callaghan, O.M.; Mahanty, H.K.; Willocks, M.J. Pathogen to product - development of *Serratia entomophila* (Enterobacteriaceae) as a commercial biological control agent for the New Zealand grass grub (*Costelytra zealandica*). In *Use of Pathogens in Scarab Management*; Jackson, T.A., Glare, T.R., Eds.; Intercept Ltd.: Andover, MA, USA, 1992; pp. 191–198.
204. Liu, S.; Chen, Y.; Sappington, T.W.; Bonning, B.C. Genome sequence of a novel positive-sense, single-stranded RNA virus isolated from western corn rootworm, *Diabrotica virgifera virgifera* LeConte. *Genome Announc.* **2017**, *5*, 17. [CrossRef] [PubMed]
205. Liu, S.; Chen, Y.; Sappington, T.W.; Bonning, B.C. Genome sequence of the first coleopteran iflavirus isolated from western corn rootworm, *Diabrotica virgifera virgifera* LeConte. *Genome Announc.* **2017**, *5*, 16. [CrossRef]
206. Liu, S.; Chen, Y.; Sappington, T.W.; Bonning, B.C. Genome sequence of *Diabrotica virgifera virgifera* virus 2, a novel small RNA virus of the western corn rootworm, *Diabrotica virgifera virgifera* LeConte. *Genome Announc.* **2017**, *5*, 17. [CrossRef] [PubMed]
207. Huger, A.M. The Oryctes virus: Its detection, identification, and implementation in biological control of the coconut palm rhinoceros beetle, *Oryctes rhinoceros* (Coleoptera: Scarabaeidae). *J Invertebr. Pathol.* **2005**, *89*, 78–84. [CrossRef]
208. Pilz, C.; Wegensteiner, R.; Keller, S. Selection of entomopathogenic fungi for the control of the western corn rootworm *Diabrotica virgifera virgifera*. *J. Appl. Entomol.* **2007**, *131*, 426–431. [CrossRef]
209. Pilz, C.; Keller, S.; Kuhlmann, U.; Toepfer, S. Comparative efficacy assessment of fungi, nematodes and insecticides to control western corn rootworm larvae in maize. *BioControl* **2009**, *54*, 671–684. [CrossRef]
210. Meyling, N.V.; Eilenberg, J. Ecology of the entomopathogenic fungi *Beauveria bassiana* and *Metarhizium anisopliae* in temperate agroecosystems: Potential for conservation biological control. *Biol. Control* **2007**, *43*, 145–155. [CrossRef]
211. Toepfer, S.; Gueldenzoph, C.; Ehlers, R.U.; Kuhlmann, U. Screening of entomopathogenic nematodes for virulence against the invasive western corn rootworm, *Diabrotica virgifera virgifera* (Coleoptera: Chrysomelidae) in Europe. *Bull. Entomol. Res.* **2005**, *95*, 473–482. [CrossRef]
212. Kurtz, B.; Toepfer, S.; Ehlers, R.U.; Kuhlmann, U. Assessment of establishment and persistence of entomopathogenic nematodes for biological control of western corn rootworm. *J. Appl. Entomol.* **2007**, *131*, 420–425. [CrossRef]
213. Kurtz, B.; Hiltpold, I.; Turlings, T.C.J.; Kuhlmann, U.; Toepfer, S. Comparative susceptibility of larval instars and pupae of the western corn rootworm to infection by three entomopathogenic nematodes. *BioControl* **2009**, *54*, 255. [CrossRef]
214. Ilan, S.D.I.; Gouge, D.H.; Piggott, S.J.; Fife, J.P. Application technology and environmental considerations for use of entomopathogenic nematodes in biological control. *Biol. Control* **2006**, *38*, 124–133. [CrossRef]
215. Hiltpold, I.; Jaffuel, G.; Turlings, T.C.J. The dual effects of root-cap exudates on nematodes: From quiescence in plant-parasitic nematodes to frenzy in entomopathogenic nematodes. *J. Exp. Bot.* **2015**, *66*, 603–611. [CrossRef]
216. Jaffuel, G.; Hiltpold, I.; Turlings, T.C.J. Highly potent extracts from pea (*Pisum sativum*) and maize (*Zea mays*) roots can be used to induce quiescence in entomopathogenic nematodes. *J. Chem. Ecol.* **2015**, *41*, 793–800. [CrossRef]
217. Shields, E.J.; Testa, A.M.; Neil, O.W.J. Long-term Persistence of Native New York Entomopathogenic Nematode Isolates Across Crop Rotation. *J. Econ. Entomol.* **2018**, *111*, 2592–2598. [CrossRef]
218. Ehlers Ralf, U.; Hiltpold, I.; Kulhmann, U.; Toepfer, S. Field results on the use of Heterorhabditis bacteriophage against the invasive maize pest *Diabrotica virgifera virgeifera*. *Insect Pathog. Insect Parasit Nematodes* **2008**, *31*, 332–335.
219. Toepfer, S.; Peters, A.; Ehlers, R.U.; Kuhlmann, U. Comparative assessment of the efficacy of entomopathogenic nematode species at reducing western corn rootworm larvae and root damage in maize. *J. Appl. Entomol.* **2008**, *132*, 337–348. [CrossRef]

220. Jaffuel, G.; Sbaiti, I.; Turlings, T.C.J. Encapsulated entomopathogenic nematodes can protect maize plants from *Diabrotica balteata* larvae. *Insects* **2020**, *11*, 27. [CrossRef]

221. Griffin, C.T. Perspectives on the behavior of entomopathogenic nematodes from dispersal to reproduction: Traits contributing to nematode fitness and biocontrol efficacy. *J Nematol.* **2012**, *44*, 177–184.

222. Dillman, A.R.; Guillermin, M.L.; Lee, J.H.; Kim, B.; Sternberg, P.W.; Hallem, E.A. Olfaction shapes host-parasite interactions in parasitic nematodes. *Proc. Natl. Acad. Sci. USA* **2012**, *109*, E2324–E2333. [CrossRef] [PubMed]

223. Rasmann, S.; Köllner, T.G.; Degenhardt, J.; Hiltpold, I.; Toepfer, S.; Kuhlmann, U.; Gershenzon, J.; Turlings, T.C.J. Recruitment of entomopathogenic nematodes by insect-damaged maize roots. *Nature* **2005**, *434*, 732–737. [CrossRef] [PubMed]

224. Anbesse, S.; Ehlers, R.U. Attraction of *Heterorhabditis* sp. toward synthetic (E)-beta-cariophyllene, a plant SOS signal emitted by maize on feeding by larvae of *Diabrotica virgifera virgifera. Commun. Agric. Appl. Biol. Sci.* **2010**, *75*, 455–458.

225. Hiltpold, I.; Baroni, M.; Toepfer, S.; Kuhlmann, U.; Turlings, T.C.J. Selection of entomopathogenic nematodes for enhanced responsiveness to a volatile root signal helps to control a major root pest. *J. Exp. Biol.* **2010**, *213*, 2417–2423. [CrossRef]

226. Zhang, X.; Machado, R.A.R.; Van Doan, C.; Arce, C.C.M.; Hu, L.; Robert, C.A.M. Entomopathogenic nematodes increase predation success by inducing cadaver volatiles that attract healthy herbivores. *eLife* **2019**, *8*, 6668. [CrossRef] [PubMed]

227. Baiocchi, T.; Lee, G.; Choe, D.H.; Dillman, A.R. Host seeking parasitic nematodes use specific odors to assess host resources. *Sci. Rep.* **2017**, *7*, 6270. [CrossRef]

228. Jagdale, G.B.; Kamoun, S.; Grewal, P.S. Entomopathogenic nematodes induce components of systemic resistance in plants: Biochemical and molecular evidence. *Biol. Control* **2009**, *51*, 102–109. [CrossRef]

229. Helms, A.M.; Ray, S.; Matulis, N.L.; Kuzemchak, M.C.; Grisales, W.; Tooker, J.F.; Ali, J.G. Chemical cues linked to risk: Cues from below-ground natural enemies enhance plant defences and influence herbivore behaviour and performance. *Funct. Ecol.* **2019**, *33*, 798–808. [CrossRef]

230. Robert, C.A.M.; Zhang, X.; Machado, R.A.R.; Schirmer, S.; Lori, M.; Mateo, P.; Erb, M.; Gershenzon, J. Sequestration and activation of plant toxins protect the western corn rootworm from enemies at multiple trophic levels. *eLife* **2017**, *6*, 9307. [CrossRef] [PubMed]

231. Zhang, X.; Van Doan, C.; Arce, C.C.M.; Hu, L.; Gruenig, S.; Parisod, C.; Hibbard, B.E.; Hervé, M.R.; Nielson, C.; Robert, C.A.M.; et al. Plant defense resistance in natural enemies of a specialist insect herbivore. *Proc. Natl. Acad. Sci. USA* **2019**, *116*, 23174–23181. [CrossRef] [PubMed]

232. Bruno, P.; Machado, R.A.R.; Glauser, G.; Köhler, A.; Herrera, C.R.; Bernal, J.; Toepfer, S.; Erb, M.; Robert, C.A.M.; Arce, C.C.M.; et al. Entomopathogenic nematodes from Mexico that can overcome the resistance mechanisms of the western corn rootworm. *Sci. Rep.* **2020**, *10*, 8257. [CrossRef] [PubMed]

233. Machado, R.A.R.; Thönen, L.; Arce, C.C.M.; Theepan, V.; Prada, F.; Wüthrich, D.; Robert, C.A.M.; Vogiatzaki, E.; Shi, Y.M.; Schaeren, O.P.; et al. Engineering bacterial symbionts of nematodes improves their biocontrol potential to counter the western corn rootworm. *Nat. Biotechnol.* **2020**, *38*, 600–608. [CrossRef]

234. Degenhardt, J.; Hiltpold, I.; Köllner, T.G.; Frey, M.; Gierl, A.; Gershenzon, J.; Hibbard, B.E.; Ellersieck, M.R.; Turlings, T.C.J. Restoring a maize root signal that attracts insect-killing nematodes to control a major pest. *Proc. Natl. Acad. Sci. USA* **2009**, *106*, 13213–13218. [CrossRef]

235. Robert, C.A.M.; Erb, M.; Hiltpold, I.; Hibbard, B.E.; Gaillard, M.D.P.; Bilat, J.; Degenhardt, J.; Cambet, P.J.X.; Turlings, T.C.J.; Zwahlen, C. Genetically engineered maize plants reveal distinct costs and benefits of constitutive volatile emissions in the field. *Plant Biotechnol. J.* **2013**, *11*, 628–639. [CrossRef] [PubMed]

236. Lewis, E.E.; Ilan, S.D.I. Host cadavers protect entomopathogenic nematodes during freezing. *J. Invertebr. Pathol.* **2002**, *81*, 25–32. [CrossRef]

237. Ilan, S.D.I.; Brown, I. Earthworms as phoretic hosts for *Steinernema carpocapsae* and *Beauveria bassiana*: Implications for enhanced biological control. *Biol. Control* **2013**, *66*, 41–48. [CrossRef]

238. Ilan, S.D.I.; Lewis, E.E.; Son, Y.; Tedders, W.L. Superior efficacy observed in entomopathogenic nematodes applied in infected-host cadavers compared with application in aqueous suspension. *J. Invertebr. Pathol.* **2003**, *83*, 270–272. [CrossRef]

239. Kaya, H.K.; Nelsen, C.E. Encapsulation of Steinernematid and Heterorhabditid nematodes with calcium alginate: A new approach for insect control and other applications. *Environ. Entomol.* **1985**, *14*, 572–574. [CrossRef]

240. Hiltpold, I.; Hibbard, B.E.; French, B.W.; Turlings, T.C.J. Capsules containing entomopathogenic nematodes as a Trojan horse approach to control the western corn rootworm. *Plant Soil* **2012**, *358*, 11–25. [CrossRef]

241. Wu, S.; Kaplan, F.; Lewis, E.; Alborn, H.T.; Ilan, S.D.I. Infected host macerate enhances entomopathogenic nematode movement towards hosts and infectivity in a soil profile. *J. Invertebr. Pathol.* **2018**, *159*, 141–144. [CrossRef]

242. Ilan, S.D.I.; Kaplan, F.; Hofman, O.C.; Schliekelman, P.; Alborn, H.T.; Lewis, E.E. Conspecific pheromone extracts enhance entomopathogenic infectivity. *J. Nematol.* **2019**, *51*, 1–5. [CrossRef]

243. Hofman, O.C.; Kaplan, F.; Stevens, G.; Lewis, E.; Wu, S.; Alborn, H.T.; Gentil, P.A.; Ilan, S.D.I. Pheromone extracts act as boosters for entomopathogenic nematodes efficacy. *J. Invertebr. Pathol.* **2019**, *164*, 38–42. [CrossRef] [PubMed]

Review

The Use of Insecticides to Manage the Western Corn Rootworm, *Diabrotica virgifera virgifera*, LeConte: History, Field-Evolved Resistance, and Associated Mechanisms

Lance J. Meinke [1,*], Dariane Souza [2] and Blair D. Siegfried [2]

[1] Department of Entomology, University of Nebraska, Lincoln, NE 68583, USA
[2] Entomology and Nematology Department, University of Florida, Gainesville, FL 32611, USA; dariane.souza@ufl.edu (D.S.); bsiegfried1@ufl.edu (B.D.S.)
* Correspondence: lmeinke1@unl.edu

Simple Summary: The structure of agricultural enterprises in the western United States Corn Belt (large irrigated monocultures, continuous planting of maize, strong aerial pesticide application and livestock industries) has led to a tradition of extensive insecticide use over time to manage the western corn rootworm, *Diabrotica virgifera virgifera* LeConte (Dvv) a key insect pest of maize. Dvv damages maize roots, which can cause maize plant instability, reduced plant growth, and significant yield loss. Long-term insecticide use has contributed to Dvv becoming resistant to cyclodiene, organophosphate, carbamate, and pyrethroid insecticides since the 1950s. This paper reviews the historical and current use of insecticides in Dvv management programs and Dvv adaptation to insecticide use. Currently, insecticides have a reduced role in Dvv management programs but are still used as complementary tactics with other management approaches. Past history suggests that the probability of selecting for resistance to any future Dvv control technology will be high if it is not used within an integrated pest management framework with other tactics.

Abstract: The western corn rootworm, *Diabrotica virgifera virgifera* LeConte (Dvv) is a significant insect pest of maize in the United States (U.S.). This paper reviews the history of insecticide use in Dvv management programs, Dvv adaptation to insecticides, i.e., field-evolved resistance and associated mechanisms of resistance, plus the current role of insecticides in the transgenic era. In the western U.S. Corn Belt where continuous maize is commonly grown in large irrigated monocultures, broadcast-applied soil or foliar insecticides have been extensively used over time to manage annual densities of Dvv and other secondary insect pests. This has contributed to the sequential occurrence of Dvv resistance evolution to cyclodiene, organophosphate, carbamate, and pyrethroid insecticides since the 1950s. Mechanisms of resistance are complex, but both oxidative and hydrolytic metabolism contribute to organophosphate, carbamate, and pyrethroid resistance facilitating cross-resistance between insecticide classes. History shows that Dvv insecticide resistance can evolve quickly and may persist in field populations even in the absence of selection. This suggests minimal fitness costs associated with Dvv resistance. In the transgenic era, insecticides function primarily as complementary tools with other Dvv management tactics to manage annual Dvv densities/crop injury and resistance over time.

Keywords: chemical control; pest management; insecticide metabolism; *Diabrotica virgifera virgifera*; insecticide resistance

Citation: Meinke, L.J.; Souza, D.; Siegfried, B.D. The Use of Insecticides to Manage the Western Corn Rootworm, *Diabrotica virgifera virgifera*, LeConte: History, Field-Evolved Resistance, and Associated Mechanisms. *Insects* **2021**, *12*, 112. https://doi.org/10.3390/insects12020112

Academic Editor: Katarina M. Mikac
Received: 8 January 2021
Accepted: 25 January 2021
Published: 28 January 2021

Publisher's Note: MDPI stays neutral with regard to jurisdictional claims in published maps and institutional affiliations.

1. Introduction

The western corn rootworm, *Diabrotica virgifera virgifera* LeConte (Dvv) is a galerucerine Chrysomelid beetle (Figure 1A) that is one of the most significant insect pests of maize (*Zea mays* L.) in the United States (U.S.). Annually, this species is responsible for over

$1 billion in control costs and yield losses [1,2]. Similar to other *Diabrotica* species, the larvae are root feeders and adults feed on above-ground plant tissues [3,4]. Dvv larvae survive only on a small number of grass species [3–8] while adults feed primarily on pollen and reproductive tissues of a variety of plants [9,10]. Maize is the primary Dvv host in modern agroecosystems [10,11] and is strongly attractive to wild-type adult Dvv [12–14]. The initial Dvv species description was made in 1868 from collections made in what today is Wallace Co., KS [15] but phylogenetic research points toward a species origin in Mexico or Central America [16,17]. It has been hypothesized that Dvv may have survived at low densities on native grasses such as western wheatgrass, *Pascopyrum smithii* (Rydb.) at the time of initial discovery [7]. The pheromone of Dvv is very efficient at ultra-low levels, which supports the low Dvv density hypothesis before adaptation to maize monocultures [11]. It is unclear when Dvv arrived in what today is the southwestern U.S. but records document that Dvv did not occur east of western Kansas, Colorado, and southwestern Nebraska, U.S. prior to the 1920s [18].

Figure 1. (**A**) Adult *Diabrotica virgifera virgifera* LeConte; (**B**) example of severe root injury from *Diabrotica virgifera virgifera* LeConte (Dvv) larval feeding that can occur when Dvv larval density is high (right) versus uninjured root (left); photos by L. J. Meinke.

Dvv was first recorded feeding on maize roots at Ft. Collins, Co. in 1909 [19]. Annual rotation from maize to a crop that would not support Dvv larval survival was the recommended Dvv management tactic as early as 1930 in southwestern Nebraska [20,21] but the profitability of maize led some growers to start planting continuous maize (maize planted for ≥2 years in one location). As agriculture developed in western areas of Nebraska during the 1930s sporadic reports of larval Dvv injury were reported in continuous maize [20–22]. By the 1940s, Dvv injury to continuous maize (Figure 1B) in central Nebraska was common and was facilitated in part by the introduction of irrigation systems and synthetic fertilizer [23–26]. This agricultural system became very profitable and helped meet the demand for maize from a growing confined livestock industry. Continuous maize provided optimal conditions for the build-up of Dvv densities, increasing larval injury and greater adult dispersal from infested fields [18]. The large monocultures of continuous maize may have been the bridge needed to jump-start the fairly rapid range expansion that progressed through the U.S. Corn Belt reaching NJ, USA by the 1980's [18,27].

The emergence of Dvv as a major insect pest coincided with the post-World War II emergence of agrochemical companies that synthesized and manufactured synthetic organic insecticides. The need for Dvv control in continuous maize created a niche that developed into a large insecticide market in the U.S. Corn Belt [28]. Evaluation of insecticide efficacy targeting larvae (soil insecticides) or adults (foliar applications) was a major focus of both industry and academia in the 1940s–1990s [25,29–38]. Seed treatments and transgenic plants that express proteins that are toxic to Dvv (plant incorporated protectants) were introduced and widely adopted in the 2000s, which reduced the role of insecticides as management tactics in continuous maize [26,39–41]. This paper reviews the history of insecticide use in Dvv management programs, Dvv adaption to insecticides (Figure 2), i.e., field-evolved resistance and associated mechanisms of resistance, plus the current role of insecticides in the transgenic era.

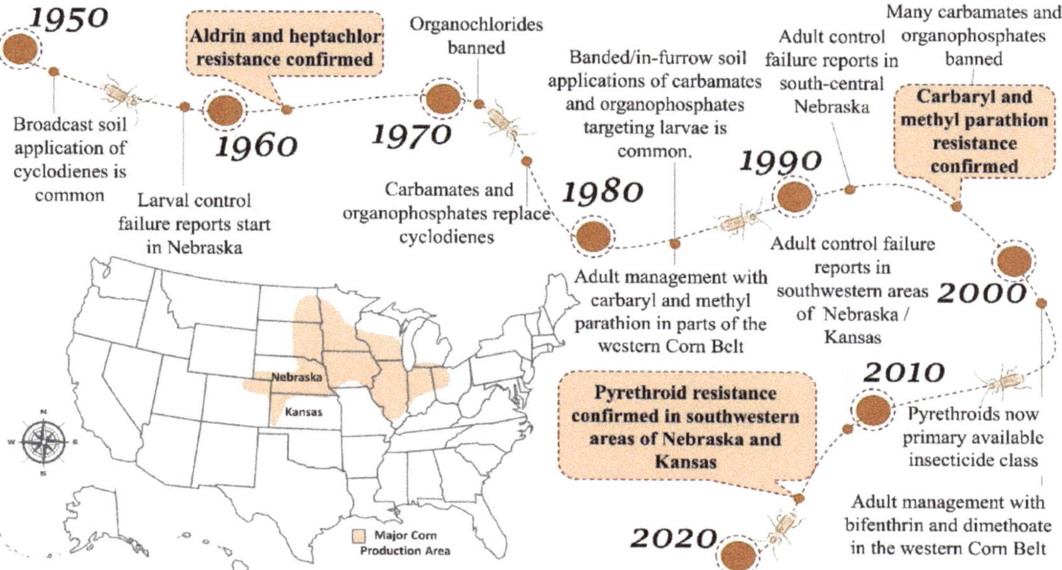

Figure 2. The agricultural region in the midwestern U.S. that is a major producer of maize is called the Corn Belt. This figure shows the major maize grain production area within this region, a Dvv insecticide use/field-evolved resistance timeline associated with continuous maize, and the geographic location of states Nebraska and Kansas where Dvv resistance to multiple insecticides has occurred. Maize grain production area is based on USDA-NASS 2015–2019 data [42,43].

2. Western Corn Rootworm—Insecticide History

2.1. Soil Insecticides

The use of soil insecticides to manage Dvv in field maize was first demonstrated by Hill et al. (1948) [23] and Muma et al. (1949) [29] in Nebraska. Broadcast planting-time applications of the chlorinated hydrocarbons DDT and the gamma isomer of benzene hexachloride (BHC) were initially evaluated. BHC (lindane) significantly reduced Dvv larval densities, plant lodging caused by larval Dvv feeding, and in some trials, increased yield when compared to untreated maize. Soil application of DDT was found to be comparatively ineffective. Muma et al. (1949) [29] also reported that BHC persisted in the soil providing Dvv control the following season after initial application. In subsequent years, cyclodienes were shown to have activity against Dvv larvae [44] and along with BHC were recommended to growers for larval control (aldrin, chlordane: [45]; heptachlor: [46]). Large-scale broadcast soil applications of BHC and cyclodiene insecticides were commonly made to field maize during the 1950s. By 1959, almost 1000 metric tons of aldrin had been applied as a soil insecticide in Nebraska alone [47].

Ineffective control of Dvv larvae after application of a soil insecticide was initially observed in south-central Nebraska in 1959 [48]. The rapid development and continuance of the problem became more apparent by 1960–1961 [49–51]. In lab bioassays, LD50s of a Dvv population from a control failure area of south-central Nebraska were 78.8 and 43.2-fold greater for heptachlor and aldrin, respectively, than a population in eastern Nebraska where no larval control issues had been observed [52]. This was the first direct evidence of field-evolved resistance in Dvv to an insecticide. A follow-up study documented that susceptibility of Dvv to aldrin was highly correlated with previous insecticide use patterns in Nebraska. The LD50s of populations from high insecticide use areas around the Platte River Valley were up to 1000-fold greater than some populations from far western NE where little soil insecticide had been used [53].

Cyclodiene resistance in Dvv populations was limited to the western U.S. Corn Belt at this time because the eastern Dvv expansion across the U.S had only reached the Nebraska—Iowa border by 1954 [54] and western Wisconsin by 1964 [18]. Cyclodiene resistance spread throughout the existing Dvv range and was maintained in Dvv populations as the range expansion across the U.S. Corn Belt progressed. Populations along the expanding species boundary exhibited similar resistance levels to cyclodienes by 1964 [55] and high levels of resistance were present in the eastern U.S. during the 1980s even in areas where cyclodienes were not widely used before removal from the market by the U.S. Environmental Protection Agency (USEPA) in the 1970s [56]. Metcalf (1986) [57,58] noted that the rate of geographic spread was slower before than right after resistance evolved to cyclodiene insecticides. Metcalf inferred that a behavioral change associated with resistance may have led to increased movement of the range expansion front [57,58]. This hypothesis turned out to be unsupported as Dvv exhibited stratified dispersal with wide variation in rate of expansion depending on location and year [18].

Following widespread Dvv resistance to the environmentally persistent cyclodiene insecticides, there was a concentrated effort by state, federal, and industry scientists to discover and commercialize new insecticides with activity against Dvv [33]. Chlorinated hydrocarbon and cyclodiene insecticides were gradually removed from commercial use by regulation in the U.S. and replaced by organophosphate and carbamate insecticides [59]. The replacement insecticide classes were highly toxic, more expensive, and had shorter residual activity than the cyclodienes [33,59]. A shift from broadcast soil applications to applications banded over the row or placed in the seed furrow were adopted to reduce rates applied and subsequent cost [33]. Most soil-applied insecticides targeting larval Dvv are applied at planting-time, which often is 4–8 weeks before Dvv eggs begin to hatch. Therefore, historically, many products were granular formulations that would extend residual activity. Eventually, prophylactic use of soil-applied organophosphate or carbamate insecticides at planting became the primary Dvv management approach in continuous maize throughout the U.S. Corn Belt [28]. Prophylactic applications were often made without knowledge of the pest density present, which led to some unnecessary insecticide applications [60–62]. It was not until the 1980s that several insecticides in the pyrethroid insecticide class (bifenthrin, tefluthrin) were registered as Dvv soil insecticides in the U.S. [63,64].

During the 1970s to early 1990s, there were numerous studies conducted to understand environmental and agronomic factors that could impact soil insecticide efficacy and increase understanding of Dvv population dynamics when soil insecticides were used in continuous maize. Performance of insecticide applications of all classes can be inconsistent from year to year in the soil environment [59,65]. The response of an insecticide in soil is a complex interaction of the physicochemical properties of an insecticide, soil and environmental conditions, and the susceptibility of the target insect pest [65]. Very dry or excessively wet soils, high soil pH, or soils with high organic matter content have all contributed to variable soil insecticide efficacy when targeting Dvv larvae [28,66–69].

Organophosphate and carbamate insecticides are biodegraded by microbiological metabolism [59,70,71]. This is normally viewed as a positive characteristic since it prevents persistence of insecticides in the soil over time but can negatively impact soil insect control if biodegradation occurs rapidly and reduces insecticide residual in the soil [72]. Declining Dvv control with some carbamate or organophosphate soil insecticides was reported during the 1970s–1980s after consecutive years of application [59]. A key example was the carbamate carbofuran where studies documented poor Dvv control in soil plots treated with carbofuran at planting for successive years versus plots with no history of carbofuran use [73]. Subsequent research determined that rapid degradation of carbofuran by soil microbes in conditioned soils led to poor control and that field-evolved resistance to carbofuran in Dvv probably was not the causative factor [74,75]. A similar scenario occurred with the organophosphate isophenphos. This soil insecticide was initially marketed as Amaze® in 1981 but by 1983, Dvv control failures were reported from various locations in the U.S. Corn Belt in fields where the product had been used the previous year [59]. Enhanced microbial biodegradation was found to be the cause of rapid loss of isophenphos in the soil [76,77] and the formulation was removed from the market soon thereafter [59]. In both of these examples, microbial degradation of the soil insecticide occurred between planting time and the Dvv egg hatch period reducing active ingredient in the soil.

Planting date and associated timing of insecticide application relative to Dvv population dynamics can also impact efficacy of soil-applied insecticides. Soil insecticides often perform better when planting date is later and insecticide application is closer to the occurrence of Dvv larvae in the field [78–80]. Before minimum or no-till practices were widely adopted, application of insecticides at cultivation during the Dvv egg hatch period was demonstrated as an alternative to planting-time applications especially when maize was planted very early in the season [81]. Reduced tillage practices do not appear to significantly impact Dvv control with soil insecticides [28,82,83].

The placement of soil-applied insecticides in a band over the row or placed in the seed furrow protects the main maize root system (reduces injury, lodging) but does not manage the Dvv population [28,83,84]. Little to no reduction in adult Dvv emergence from soil insecticide treated plots versus untreated plots is commonly reported [83–86]. This has been attributed to survival of Dvv larvae feeding on maize roots between rows that have grown outside the treated zone and in some cases differences in larval density-dependent mortality between treatments [83,86]. Therefore, there is often an inconsistent relationship between soil insecticide efficacy as measured by level of root injury and subsequent adult emergence. This built-in refuge is probably a major reason why field-evolved Dvv resistance attributed to direct selection of Dvv larvae has only been documented with broadcast-applied cyclodiene insecticides but not insecticides applied in-furrow or banded over the row [83,86,87].

The fitness of Dvv adults that emerge from soil insecticide treated fields is highly variable. Later mean adult emergence time, increased longevity, increased egg production, and variable sex ratios have been reported after Dvv larval exposure to organophosphate and carbamate soil insecticides [85,88–91], but results are inconsistent within products over years. Soil insecticide exposure does not appear to significantly affect egg viability [89,90]. Extended mean development time can also be caused by high Dvv larval densities [89,92,93], therefore, the variability in Dvv life history parameters associated with soil insecticide environments may be caused by the complex interaction of Dvv density, sublethal exposure/insecticide dose, and environmental conditions.

Liquid formulations of some carbamate, organophosphate, pyrethroid, and phenylpyrazole insecticides applied at planting or during the Dvv larval period have provided an alternative to granular formulations. Many liquid formulations can be applied in-furrow in starter fertilizer at planting [31,94] or chemigated through center pivot irrigation systems to control Dvv larvae [95]. In 1996, the Food Quality Protection Act altered the regulation of pesticides in the U.S. and the USEPA canceled uses of a number of organophosphate and carbamate insecticides, which has greatly reduced the insecticide options available for larval Dvv control.

2.2. Seed Treatments

The neonicotinoid insecticides thiamethoxam and clothianidin were initially registered as maize seed treatments in the early 2000s in the U.S. [96–98]. These seed treatments were primarily marketed at low rates (0.25–0.5 mg/seed) to control seedling maize insect pests and were quickly adopted by the seed industry. Today, most field-maize seed planted in the U.S. is treated with a low-rate neonicotinoid insecticide [39,99]. A higher rate (1.25 mg/seed) of thiamethoxam or clothianidin has been marketed for Dvv control. Clothianidin is very toxic to Dvv neonate larvae and some natural variability in susceptibility occurs among Dvv populations [100]. Field trial data consistently show that the 1.25 mg/seed neonicotinoid rate provides effective maize root protection when Dvv densities are low to moderate but protection is reduced under high Dvv larval densities [101,102]. To date there have not been any reports of field-evolved Dvv resistance to neonicotinoids deployed as seed treatments [102]. The future of neonicotinoids as seed treatments is unclear as environmental and nontarget concerns [103–106] have led to neonicotinoid bans in the European Union, restrictions in Canada, and a formal review of use in the U.S. [107–109]. Questions have also been raised as to whether an insecticidal seed treatment is necessary in all field situations because of the sporadic nature of many seedling pests [110]. Therefore, there is a need for greater understanding of actual pest pressure applied by the seedling pest complex in various crops/regions, methodology to assess the risk posed by seedling pests, and where appropriate, alternative technologies or strategies that can effectively protect seeds and seedlings from arthropods [110,111].

2.3. Foliar Insecticides

The application of foliar insecticides to suppress adult Dvv densities is used primarily for two purposes: (1) Protect maize during the pollination period from excessive adult Dvv silk feeding, which can interfere with pollination and result in poorly filled ears, and/or (2) reduce Dvv female density and associated oviposition to reduce potential larval injury the following season in continuous maize [112,113]. Initial experiments conducted by Hill et al. (1948) [23] demonstrated that DDT provided excellent control of adult Dvv. DDT was a key compound used to reduce Dvv silk feeding during the 1950s in Nebraska [25]. During the 1960s–1970s after field-evolved Dvv cyclodiene resistance limited larval control options, there was increasing interest in using adult foliar insecticide applications to reduce adult densities and oviposition in continuous maize fields to prevent the need for a soil insecticide the following year. Organophosphate and carbamate insecticides replaced DDT and provided control of adult Dvv [32,34,114]. A constraint of many products was low residual activity in the field because adequate residual activity coupled with proper application timing was needed to effectively manage rootworm densities [32]. Eventually, formulations of carbaryl (Sevin 4-Oil®, used mainly during the 1970s) and encapsulated methyl parathion (Penncap-M®, used primarily during 1980s–1990s) were developed that provided up to three weeks residual activity and became products of choice in many Dvv adult management programs [115]. The viable aerial application industry and the growing number of professional crop consultants led producers in some areas of south-central Nebraska to exclusively use adult control to manage Dvv in continuous maize [32,112,114].

The adult management strategy worked well in the western U.S. Corn Belt from the 1960s to early 1990s but reports of adult control failures increased during the 1990s in south-central Nebraska [115–118]. In areas where intensive use of Penncap-M® was common, both rates of application and the number of applications per season increased as a result of reduced product efficacy [115]. Topical bioassays were used to document initial 16- and 9-fold field-evolved Dvv resistance to methyl parathion and carbaryl, respectively in 1995 [115], and vial bioassays were developed to monitor changes in geographical distribution of Dvv resistance in Nebraska over time [117,119,120]. Initially, Dvv resistance centered around two distinct focal areas separated by about 145 km in south central NE; Phelps/Kearney Counties and York County [117]. Dvv populations collected in counties between the focal areas were highly susceptible to methyl parathion suggesting that resistance may have evolved independently in each area. From 1996 to 2001, resistance intensity increased and areas between the initial focal areas previously identified as susceptible became highly resistant expanding the distribution of resistant Dvv populations throughout the Platte River Valley [121]. The use of adult control as a stand-alone Dvv management strategy gradually was abandoned during the late 1990s–2000s in Nebraska.

Research on the chemical ecology of *Diabrotica* species during the 1980s identified a variety of *Diabrotica* semiochemicals that provided new opportunities to manage adult Dvv populations with greatly reduced rates of insecticide. Semiochemicals included a variety of plant-derived attractants [122–124] and cucurbitacins (tetracyclic triterpenoids found in many cucurbits), which are arrestants and feeding stimulants for adult *Diabrotica* species [125–127]. A bait-type concept was developed in which semiochemicals were used to facilitate adult Dvv location and feeding on formulations that contain minute amounts of insecticide [128–131]. Initial formulations were granular and carbaryl was often used as the insecticide because of its efficacy as a Dvv oral poison. Efficacy of granular formulations was shown to be greatly affected by bait location and vertical distribution of adult Dvv in the plant canopy. In field maize, semiochemical-based baits were very effective up in the canopy but ineffective when placed on the ground [132,133]. To circumvent this problem, industry developed sprayable formulations that would adhere and dry on the plant versus granules that would roll off of plant leaves. MicroFlo Company (Lakeland FL) was issued a registration in 1992 for a sprayable microsphere formulation (SLAM) that contained cucurbitacin (*Cucurbita foetidissima* HBK root powder) and carbaryl [134]. Broadcast aerial applications of SLAM often provided >90% reduction of adult Dvv 24 h post application using only 10–13% of the active ingredient normally applied with conventional carbaryl applications [135,136]. Other companies subsequently developed adjuvant products that contained either *C. foetidissima* root powder (COMPEL: Scentry Inc. Billings, MT, USA; Cidetrak: Trece, Salinas, CA, USA) or bitter Hawksbury watermelon (*Citrullus vulgaris* Schrad.) juice (Invite: Florida Food Products, Eustis, FL, USA) that were marketed for tank mixing with very low rates of insecticide [134,137].

Pruess et al. (1974) [34] demonstrated in the late 1960s that application of ULV malathion over a 41.4 sq km area could reduce adult Dvv densities so larval densities the following year would be greatly reduced and not cause economic loss. This concept was revisited in a USDA-sponsored pilot areawide management program that was conducted at four 41.4 sq km locations in the Corn Belt from 1996 to 2002 to evaluate the efficacy of the semiochemical bait SLAM to manage Dvv population densities and crop injury [138,139]. The program clearly demonstrated that annual use of SLAM could effectively reduce adult Dvv and subsequent larval feeding injury the following season comparable to the level of control obtained with soil insecticides. Population management of Dvv was variable across sites as annual immigration of adults into the managed area prevented long-term suppression of Dvv over time. However, at some sites, the number of fields requiring treatment was significantly reduced over years in the managed area versus control companion area. The semiochemical bait approach significantly reduced the amount of ai applied by up to 20-fold and did not significantly affect densities of nontarget insects [139,140]. Previous research had shown that cucurbitacin was actually a repellent/antifeedant to some nontarget insects so nontargets could avoid the discrete bait droplets and not consume the toxin [141].

Part of the Dvv areawide management program was a project to monitor potential shifts in susceptibility of Dvv populations during the life of the pilot program. Annual collections of adult Dvv from managed and control companion areas of each program location were bioassayed with a diagnostic concentration of carbaryl (LC99). Significant reduction in Dvv susceptibility to carbaryl was detected in three of the four managed areas and a significant reduction in the responsiveness of Dvv to cucurbitacin was also observed at the same three sites [142]. Zhu et al. (2001) [143] conducted a separate monitoring study at the Kansas location and also reported rapid changes in susceptibility to carbaryl in the managed area after selection with SLAM over years. A similar shift in susceptibility was not observed in the companion control area. These results collectively suggest that areawide programs have the potential to select for resistance and that a strategy for managing and reducing selection pressure (e.g., rotation of insecticide ai) should be implemented. When the toxicity of various insecticide semiochemical bait mixtures was evaluated comparing Dvv organophosphate resistant and susceptible populations, results indicated that bait efficacy may be compromised by previously identified resistance and by insecticides that antagonize the feeding stimulation of the cucurbitacin in the bait [144]. Therefore, careful selection of insecticide active ingredient should be made if rotation of insecticides in baits over time is a resistance management goal.

The semiochemical bait concept or Dvv areawide management were not widely adopted in the U.S. Corn Belt. A number of factors may have contributed to this. Only small companies with limited research budgets and sales/marketing personnel developed the commercialized bait products. Larger agrochemical companies were committed to development of transgenic plants and companion seed treatment technologies as future Dvv management tools during the 1990's–2000's. Adult management is more knowledge and labor intensive than using a soil insecticide or planting a transgenic hybrid. In Nebraska, professional consultants are often hired to scout fields and properly time insecticide applications. Areawide programs involve a lot of organizational complexity and cooperation to uniformly apply specific management tactics over defined areas [145]. Therefore, farmers are required to relinquish some freedom to make independent insect management decisions on their farms. The end of the Dvv areawide pilot program coincided with the introduction of Dvv-active transgenic plant technologies and complementary neonicotinoid seed treatments that were quickly adopted [39,41]. The ability of farmers to incorporate a new technology with current practices and the perceived complexity of the innovation often impacts adoption [146]. The commercialization of transgenic plants simplified insect control by eliminating the need for soil-applied insecticides and provided excellent larval control [147]. While the semiochemical-bait concept still has merit, only a small niche market developed, and most bait or adjuvant products have been removed by industry from the market.

Since 2003, aerial or chemigated application of foliar-applied insecticides to manage arthropod pests is still commonly practiced in the western U.S. Corn Belt. However, planting of hybrids expressing Dvv-active Bacillus thuringiensis Berliner (Bt) proteins are often the centerpiece of Dvv management programs with fewer growers relying exclusively on insecticides to manage Dvv. Stand-alone adult Dvv management programs used in the 1960s–1990s have been replaced by periodic use of foliar applications to manage Dvv densities in continuous maize as a complement to other management tactics [148]. Regulatory reviews by the USEPA of older registered chemistries gradually removed many insecticides that were labelled for adult Dvv control leaving only pyrethroids and a few organophosphates as primary options. In the continuous maize system, the pyrethroid bifenthrin and organophosphate dimethoate are routinely used to manage other arthropod pests (i.e., western bean cutworm *Striacosta albicosta* (Smith), two-spotted spider mite *Tetranychus urticae* Koch, and banks grass mite *Oligonychus pratensis* (Banks) as well as Dvv, therefore, Dvv can receive selection pressure as a target or nontarget insect over time [148,149].

Results of active ingredient lab bioassays revealed that adult Dvv were very susceptible to bifenthrin during the 1990s in south-central Nebraska and southwestern Kansas [38,115]. However, from 2002–2014 there was a 40% increase in the use of bifenthrin in Nebraska [150]. From 2010–2012, anecdotal reports of reduced Dvv control with bifenthrin in parts of Nebraska and Kansas became more common. Follow-up active ingredient lab adult bioassays of Dvv field collections made in 2013–2014 from the affected areas of Kansas and Nebraska detected emerging Dvv field-evolved resistance to bifenthrin [149,151]. Initial resistance ratios were relatively low for adults (<10-fold), but subsequent active ingredient lab bioassays conducted with 2016 collections from the same Nebraska county as 2013–2014 collections revealed bifenthrin resistance ratios in the 4 to 55-fold range depending on the susceptible reference population used [148]. Dimethoate active ingredient lab bioassays with the same 2016-collected populations revealed a low level of Dvv resistance as well (RR: 3 to 16-fold) [148]. Indoxacarb, a recently registered insecticide labeled for Dvv adult control, was toxic to all adult Dvv populations bioassayed [148]. The bifenthrin resistance detected in the active ingredient lab bioassays was confirmed with commercial formulated bifenthrin using a simulated aerial application assay while the low level of dimethoate resistance detected in active ingredient bioassays was not apparent in formulated product assays [148]. This Dvv resistance problem is restricted primarily to southwestern areas of Nebraska and Kansas (and possibly northeastern Colorado).

Bifenthrin diagnostic concentration (LC99) bioassays of adult Dvv indicated that all populations tested from Corn Belt states east of Nebraska were very susceptible with increased survival in northeast NE and highest survival in southwestern areas of Kansas and Nebraska [149]. Dvv cross-resistance with the pyrethroids tefluthrin and cyfluthrin was also documented in lab bioassays [86,149]. Tefluthrin and bifenthrin are commonly used Dvv soil insecticides often applied in-furrow at planting throughout the Corn Belt while extensive use of aerial application of bifenthrin is limited to the western Corn Belt. Because resistance evolution attributed to direct selection of larvae has not been documented with soil insecticides placed in-furrow or over the row, the significant difference in Dvv susceptibility to bifenthrin between southwestern areas of Kansas and Nebraska versus geographic areas east of Nebraska strongly supports the working hypothesis that selection of adults is the main driver of observed Dvv pyrethroid resistance.

The Dvv organophosphate/carbamate resistance during the 1990s and Dvv pyrethroid resistance during the 2010s are the only known cases where the efficacy of Dvv soil insecticide active ingredients in bioassays or formulated product applied in-furrow or banded over the row was reduced by Dvv resistance [86,152]. In each case, resistance evolution was tied to selection of Dvv adults with foliar applied formulations leading to reduced efficacy of one or more soil-applied insecticides targeting larvae. Adult selection with the organophosphate methyl parathion reduced larval control efficacy with methyl parathion (not used as a soil insecticide), tefluthrin, and carbofuran [152]. The more recent pyrethroid selection of adults led to significant reductions in larval control efficacy with bifenthrin, tefluthrin, and cyfluthrin [86].

3. Mechanisms of Dvv Resistance to Insecticides

3.1. Organochlorides

The first investigations on mechanisms of Dvv cyclodiene resistance were mostly focused on comparative susceptibility levels and metabolic responses in resistant individuals. Experiments performed with field collected Dvv populations in the mid-1970s showed high levels of resistance to aldrin and heptachlor but not to other organochlorides such as DDT or methoxychlor indicating that the major selection for resistance in this species had resulted from cyclodiene soil treatments and suggested different metabolic pathways and/or modes of action between some organochloride insecticides [35]. An early investigation of in vivo insecticide detoxification mechanisms in Dvv showed that relative to organophosphates and carbamates, aldrin was very resistant to total detoxification but that it was readily converted to the more stable epoxide dieldrin [153]. Further examinations revealed a highly active cytochrome P450-dependent aldrin epoxidase system in cyclodiene resistant Dvv [154], which could not clearly explain the resistance phenotype since the epoxide dieldrin is equal to or more toxic to Dvv than the parental compound [155]. Another assessment showed that the susceptibility of Dvv to aldrin was increased by 9-fold when adults were maintained on diets other than corn suggesting that host-dependent alterations in aldrin metabolizing enzymes could contribute to the resistance mechanism [56]. Then a laboratory study demonstrated that aldrin resistant Dvv adults displayed a much higher conversion of epoxide dieldrin into aldrin trans-diol in the central nervous system relative to the whole body and also displayed cross-resistance to picrotoxinin [156], a compound thought to share a similar target site with cyclodiene insecticides [157,158].

Cyclodienes mode of action involves antagonism of the γ-aminobutyric acid (GABA) receptor in the nervous system thereby suppressing inhibitory post synaptic graded potentials [159,160]. The GABA receptors are ligand-gated chloride channels that regulate inhibitory potentials by controlling the flux of chloride ions through nerve cell membranes, and are important targets for several insecticides [161–164]. Target site mutations in the GABA-receptor (*Rdl*) were first detected in a dieldrin resistant strain of the fruit fly

Drosophila melanogaster Meigen [165] and then in several other insect species [163,166]. In most cases studied, cyclodiene resistance was caused by a conserved point mutation which results in an amino acid substitution of an alanine either to serine or glycine within the second transmembrane domain (M2) of the GABA-receptor [163,166,167]. Investigations on cyclodiene-resistant Dvv populations collected throughout the U.S. Corn Belt from 2006 to 2011 confirmed a non-synonymous single nucleotide polymorphism (SNP) G/T at the GABA receptor cDNA position 838, which resulted in the alanine to serine change [168]. Results collected from Wang et al. (2013) [168] also suggested that a phenotypic gradient of increasing resistance levels from west to east in the U.S. Corn Belt historically documented [53,87] was correlated with higher frequencies of the resistance-conferring allele in the eastern-most populations.

For a period of nine years (1952–1961), the extensive use of aldrin in Nebraska provided high selection pressure over local Dvv populations and left considerable residue of this persistent organochloride and its epoxide, dieldrin, in the soil [47]. Data collected in Nebraska from 1962 through 1981 suggested a possible correlation between LD50 values for Dvv to aldrin and amounts of aldrin (<0.01 ppm) and dieldrin (0.02 ppm) remaining in the soils [47]. However, similarly high levels of aldrin resistance were still detected in the early 2000s in field-collected Dvv adults [87] in areas where soil samples contained residues as low as 0.002 ppm for aldrin and 0.006 ppm for dieldrin [169], which is likely below an effective concentration that would cause mortality even in susceptible populations. Additionally, Parimi et al. (2006) [87] found high levels of aldrin resistance in field collected Dvv populations that had been laboratory reared without insecticide exposure for 7–8 generations. Models of insecticide resistance evolution often share the central assumption that resistance is associated with a fitness cost because it involves a significant modification of a common phenotype, resistance genes generally have a low initial frequency, and resistance is rarely fixed in natural populations in the absence of selection [170–172]. Despite evidences for fitness cost associated with cyclodiene resistance in some insect species [173,174], cyclodiene resistant populations of *Drosophila* spp., *Blattella germanica* (L), and *Anopheles gambiae* Giles illustrate that *Rdl* mutations in the GABA-receptor can persist in the absence of selection [175–177]. The relatively high frequencies of *Rdl* mutations persisting in Dvv populations [168] is remarkable and may indicate lack of fitness cost associated with the resistance trait in this species [87].

3.2. Carbamates and Organophosphates

Carbamates and organophosphates are neurotoxins that bind to and inhibit acetylcholinesterases and prevent these enzymes from hydrolyzing the excitatory neurotransmitter, acetylcholine, thereby prolonging neuroexcitation of postsynaptic receptors in the nervous system [178,179]. Although both insecticide classes are acetylcholinesterase inhibitors, the pharmacokinetics of these compounds are not the same. The majority of organophosphate insecticides used for Dvv control contain the P = S moiety (e.g., parathion) and are metabolically activated by cytochrome P450 monooxygenases into more toxic true phosphates (e.g., paraoxon), which forms very stable bonds with serine residues of acetylcholinesterases [180]. The process of acetylcholinesterase recovery from organophosphate binding i.e., dephosphorylation, is very slow and mortality occurs before significant recovery [180,181]. Carbamates are direct inhibitors of acetylcholinesterases and metabolic activation is not required [180]. Moreover, the rate of acetylcholinesterase decarbamylation is significantly faster than dephosphorylation such that carbamates are generally less effective inhibitors of acetylcholinesterase [181]. Carbamates and organophosphates are both generally apolar compounds with ester bonds and can be converted in vivo into more water soluble and less toxic metabolites by hydrolytic esterases [182,183]. Additionally, several P450-dependent oxidative processes are involved in detoxification of these insecticides [184].

The most common mechanism of carbamate and organophosphate resistance in insects is through enhanced insecticide detoxification [184], but mutations in acetylcholinesterases that minimize insecticide binding have also been documented [185,186]. Mechanisms involved in carbamate and organophosphate resistance in Nebraska Dvv populations exhibiting high levels of resistance to methyl parathion and carbaryl have been extensively studied. Cross-resistance has been identified not only to other organophosphates and carbamates, but also to pyrethroids, which were not used extensively in affected areas, suggesting that common metabolic pathways shared between these different insecticide classes were involved in the resistance mechanism [115,152,187]. Additional in vivo and in vitro experiments confirmed significant differences in oxidative (i.e., P450-based) and hydrolytic (i.e., esterase-based) insecticide metabolism in the resistant Dvv populations [119,152,187,188] and provided collective evidence for a number of different attributes of the resistant populations.

Dvv populations with similar levels of resistance to methyl parathion and carbaryl in insecticide bioassays [115] displayed variable isoforms and activity levels of both P450s and esterases suggesting that resistance mechanisms were different among the populations examined [187]. Enhanced esterase activity detected in resistant Dvv populations was involved in more efficient detoxification of both methyl parathion and carbaryl [187,188]. P450-dependent bioactivation of methyl parathion into the more toxic metabolite methyl paraoxon was consistently reduced among resistant populations and in combination with enhanced esterase activity conferred high resistance levels [187]. Cytochrome P450-dependent N-demethylation was enhanced in at least one resistant Dvv population and was a primary metabolic event involved in carbaryl detoxification [188]. Both methyl parathion and carbaryl resistance that was selected in Dvv adults conferred resistance to larvae but did not result in equal cross-resistance to soil-applied O-ethyl-substituted organophosphates (terbufos and chlorpyrifos) and carbamates (carbofuran) [152]. Enhanced esterase activity was the most consistent resistance mechanism identified among both adult and larval stages, and cross-resistance with pyrethroids was higher in Dvv larvae (tefluthrin) than in adults (bifenthrin) [115,152]. The complexity of these results indicates that it is probably inappropriate to generalize a common Dvv resistance mechanism although common attributes of enhanced metabolic detoxification were detected.

Further investigations of the organophosphate- and carbamate-resistant Dvv identified through Northern blotting experiments revealed higher expression of CYP4 P450 genes in Dvv populations that were significantly resistant to methyl parathion and carbaryl [189]. In addition, three distinct esterase isozyme groups were identified in Dvv and the activity of a specific group of esterase isozymes (group II) was consistently higher in all resistant Dvv populations [190]. The activities of these isozymes based on native gel electrophoresis were identified as a reliable marker for detection of resistance [120]. Purification and characterization of group II esterases demonstrated that they were present in all Dvv populations, but were overproduced in resistant phenotypes indicating quantitative genetic differences rather than qualitative changes in physical chemical properties of these enzymes [191]. Reciprocal crosses of resistant Dvv populations then revealed that inheritance of methyl parathion resistance was not exclusively correlated with inheritance of elevated esterase activity [192] confirming earlier findings that additional factors other than just enhanced hydrolytic metabolism were involved in the resistance, such as differential P450-mediated oxidation [187–189]. Similar to cyclodiene resistance, Dvv resistance levels to methyl parathion remained relatively stable over many years in the absence of selection in both laboratory and field populations suggesting lack of fitness cost associated with the resistance trait and persistence of correspondent resistance alleles [87]. However, it is unclear whether there is a lack of fitness cost or if the univoltine life cycle prevented/delayed the fitness cost to manifest.

3.3. Pyrethroids

The pyrethroid insecticide class includes neurotoxins that bind and disrupt voltage-gated sodium channels associated with axonal transmission [193–195], although voltage-gated calcium and chloride channels have also been identified as secondary targets [196–201].

Once pyrethroids bind to sodium channels, axons experience a higher influx of sodium ions, prolonged sodium inactivation and repetitive electrical discharges that disrupt the normal flow of information through the central nervous system [202]. Pyrethroids are generally divided into two classes based on the symptomology, and by the presence or absence of an α-cyano group in the molecular structure. Type I compounds (e.g., permethrin, tefluthrin, bifenthrin) are characterized by causing restlessness, incoordination, and prostration, whereas Type II (e.g., deltamethrin, λ-cyhalothrin, β-cyfluthrin) contain an α-cyano group in the molecule structure and cause incoordination, convulsions, and intense hyperactivity [200,203,204]. Although this distinction among the two subclasses exists, it is possible to find some pyrethroids exhibiting intermediate properties [203–206].

Multiple studies in different insect species have shown that enhanced activity of some cytochrome P450s, glutathione *S*-transferases and esterases can provide more efficient pyrethroid detoxification in resistant insects, while sodium channel mutations (*kdr*) can affect the binding of these insecticides to the target site [184,207–209]. Lab bioassays have demonstrated that Dvv resistance to bifenthrin is partially suppressed by inhibitors of esterases and P450s, and that bifenthrin resistant Dvv populations are cross-resistant to a soil applied pyrethroid (tefluthrin) and to the organochloride DDT [151]. DDT was banned and no longer used in the U.S. since the early 1970s, but it targets the voltage gated sodium channels similarly to pyrethroids making it useful to diagnose potential target site mutations in cross-resistance lab bioassays [210–212]. Interestingly, Dvv resistance to DDT has not been previously reported. The initial screening for mechanisms of Dvv pyrethroid resistance suggested that both enhanced metabolism and sodium channel mutations could be involved in the resistance trait [151].

Further investigations revealed that bifenthrin- and tefluthrin-resistant Dvv populations also showed resistance to a Type II pyrethroid (cyfluthrin), reduced susceptibility to the organophosphate dimethoate, and increased susceptibility to the oxadiazine indoxacarb [86,148]. Pyrethroids, carbamates, and organophosphates all possess ester bonds and can be detoxified by the same hydrolytic enzymes [182,183]. Conversely, indoxacarb becomes more toxic through esterase/amidase hydrolytic bioactivation [213–215]. Therefore, reduced efficacy of the organophosphate dimethoate and negative cross-resistance with indoxacarb observed in pyrethroid-resistant Dvv populations [148] suggested enhanced hydrolytic enzyme activity as has been reported in other insect species [216–220]. A biochemical evaluation of total protein collected from Dvv adults confirmed significantly higher activity of esterases and P450s in pyrethroid-resistant Dvv populations [221] supporting previous work that suggested the involvement of these enzyme groups in the resistance trait [148,151].

Laboratory selection of insecticide resistant populations is often used to examine the genetic basis of resistance, such as mechanisms, heritability, and time for resistance development [222–224]. Laboratory selection of a pyrethroid-resistant Dvv population demonstrated a major genetic contribution to the resistant phenotype and suggested that pyrethroid resistance evolution may occur rapidly in this species (~7 generations) under continuous adult selection [221]. Toxicological and biochemical comparisons between lab-selected and field-derived pyrethroid resistant individuals suggested that variable esterase and P450 isoforms/activity may be selected across Dvv populations showing similar levels of resistance [148,221]. RNA-seq gene expression analysis confirmed major overexpression of several CYP6 P450s in pyrethroid resistant Dvv populations, possible contribution of proteins with nervous system functions, and that enhanced activity of esterases previously detected could be a qualitative difference rather than quantitative [225]. However, *kdr* sodium channel mutations were not detected in pyrethroid resistant Dvv populations suggesting that pyrethroid resistance and DDT cross-resistance observed in Dvv populations is not associated with target site insensitivity [221,225].

4. Conclusions

The history of insecticide use to manage Dvv clearly reveals a common thread among all documented cases of Dvv resistance evolution, namely population management. The broadcast cyclodiene soil applications and the organophosphate, carbamate, and pyrethroid foliar applications all exposed a large part of the target Dvv population to some level of toxicant without any structured refuge. This combined with repeated use of the same insecticides over Dvv generations led to field-evolved resistance. The structure of agricultural enterprises in the western Corn Belt (large irrigated monocultures, continuous maize, strong aerial application and livestock industries) has led to a tradition of extensive foliar insecticide use over time to manage relatively high annual densities of Dvv and other secondary insect pests. This has contributed to the sequential occurrence of Dvv resistance evolution to different active ingredients not observed in other parts of the U.S. Corn Belt. The history of Dvv resistance to insecticides illustrates the contrasting selection intensity that insecticide application strategies may provide to different life stages of insect pests and shows that insecticide resistance may evolve quickly in field populations if adequate insecticide resistance management practices are not adopted. Dvv cross-resistance to different insecticide classes and long persistence of insecticide resistance mechanisms even in the absence of selection exemplifies how levels of resistance estimated in the laboratory can have variable practical implications on the field efficacy of formulated insecticides.

In the transgenic era, foliar insecticides will still be used as a tool to prevent excessive Dvv silk-clipping when needed, especially in seed production fields. The stand-alone adult management programs are no longer recommended partly because many longer residual insecticide products have been removed from the market and the adult emergence and oviposition periods associated with transgenic seed blends ("refuge in the bag") are extended after sublethal exposure to Cry toxins in transgenic plants [226–229]. This makes it difficult to obtain the required level of oviposition suppression to prevent economic injury the following season without applying multiple adult insecticide applications. A key message from the Dvv insecticide use story presented in this review is that more holistic Dvv management strategies are needed, which combine and rotate multiple tactics at the farm level. Foliar and soil insecticides still have a place as complementary tactics with crop rotation, conventional and transgenic hybrids, and potentially biological control (e.g., nematodes [230,231]) if used within an integrated pest management framework to manage annual densities/crop injury and resistance over time [41,232]. History has clearly documented that Dvv is highly adaptable to selection pressure from management tactics [52,115,149,233–235] so the probability of selecting for field-evolved resistance to any future Dvv control technology will be high if it is repeatedly used as a stand-alone management tactic.

Author Contributions: Conceptualization and writing project administration, L.J.M.; writing, reviewing, and editing, L.J.M., D.S., and B.D.S. All authors have read and agreed to the published version of the manuscript.

Funding: This was partially supported by The Nebraska Agricultural Experiment Station with funding from the Hatch Act (Accession Number 1007272) through the USDA National Institute of Food and Agriculture.

Institutional Review Board Statement: Not applicable.

Informed Consent Statement: Not applicable.

Data Availability Statement: Data sharing is not applicable to this article as new data were not created or analyzed.

Acknowledgments: The authors thank Emily E. Reinders for assistance with manuscript formatting.

Conflicts of Interest: The authors declare no conflict of interest.

References

1. Sappington, T.W.; Siegfried, B.D.; Guillemaud, T. Coordinated *Diabrotica* genetics research: Accelerating progress on an urgent insect pest problem. *Am. Entomol.* **2006**, *52*, 90–97. [CrossRef]
2. Wechsler, S.; Smith, D. Has resistance taken root in U.S. corn fields? Demand for insect control. *Am. J. Agric. Econ.* **2018**, *100*, 1136–1150. [CrossRef]
3. Branson, T.F.; Ortman, E.E. Host range of larvae of the western corn rootworm. *J. Econ. Entomol.* **1967**, *60*, 201–203. [CrossRef]
4. Branson, T.F.; Ortman, E.E. The host range of larvae of the western corn rootworm: Further studies. *J. Econ. Entomol.* **1970**, *63*, 800–803. [CrossRef]
5. Branson, T.F. Resistance in the grass tribe Maydeae to larvae of the western corn rootworm. *Ann. Entomol. Soc. Am.* **1971**, *64*, 861–863. [CrossRef]
6. Clark, T.L.; Hibbard, B.E. Comparison of nonmaize hosts to support western corn rootworm (Coleoptera: Chrysomelidae) larval biology. *Environ. Entomol.* **2004**, *33*, 681–689. [CrossRef]
7. Oyediran, I.O.; Hibbard, B.E.; Clark, T.L. Prairie grasses as hosts of the western corn rootworm (Coleoptera: Chrysomelidae). *Environ. Entomol.* **2004**, *33*, 740–747. [CrossRef]
8. Moeser, J.; Vidal, S. Do Alternative host plants enhance the invasion of the maize pest *Diabrotica virgifera virgifera* (Coleoptera: Chrysomelidae, Galerucinae) in Europe? *Environ. Entomol.* **2004**, *33*, 1169–1177. [CrossRef]
9. Campbell, L.A.; Meinke, L.J. Seasonality and adult habitat use by four *Diabrotica* species at prairie-corn interfaces. *Environ. Entomol.* **2006**, *35*, 922–936. [CrossRef]
10. Moeser, J.; Hibbard, B.E. A synopsis of the nutritional ecology of larvae and adults of *Diabrotica virgifera virgifera* (LeConte) in the new and old world—Nouvelle cuisine for the invasive maize pest *Diabrotica virgifera virgifera* in Europe? In *Western Corn Rootworm: Ecology and Management*; Vidal, S., Kuhlmann, U., Edwards, C.R., Eds.; CABI Publishing: Wallingford, UK, 2005; pp. 41–65.
11. Branson, T.F.; Krysan, J.L. Feeding and oviposition behavior and life cycle strategies of *Diabrotica*: An evolutionary view with implications for pest management. *Environ. Entomol.* **1981**, *10*, 826–831. [CrossRef]
12. Prystupa, B.; Ellis, C.R.; Teal, P.E.A. Attraction of adult *Diabrotica* (Coleoptera: Chrysomelidae) to corn silks and analysis of the host-finding response. *J. Chem. Ecol.* **1988**, *14*, 635–651. [CrossRef] [PubMed]
13. Naranjo, S.E. Flight orientation of *Diabrotica virgifera virgifera* and *D. barberi* (Coleoptera: Chrysomelidae) at habitat interfaces. *Ann. Entomol. Soc. Am.* **1994**, *87*, 383–394. [CrossRef]
14. Darnell, S.J.; Meinke, L.J.; Young, L.J. Influence of corn phenology on adult western corn rootworm (Coleoptera: Chrysomelidae) distribution. *Environ. Entomol.* **2000**, *29*, 587–595. [CrossRef]
15. LeConte, J.L. New Coleoptera collected on the survey for the extension of the Union Pacific Railway, E. D. from Kansas to Fort Craig, New Mexico. In *Transactions of the American Entomological Society*; The Society: Philadelphia, PA, USA, 1868; Volume 2, pp. 58–59.
16. Segura-Leon, O.L. Phylogeography of *Diabrotica virgifera virgifera* LeConte and *Diabrotica virgifera zeae* Krysan and Smith (Coleoptera: Chrysomelidae). Ph.D Thesis, The University of Nebraska-Lincoln, Lincoln, NE, USA, 2004; pp. 1–143.
17. Lombaert, E.; Ciosi, M.; Miller, N.J.; Sappington, T.W.; Blin, A.; Guillemaud, T. Colonization history of the western corn rootworm (*Diabrotica virgifera virgifera*) in North America: Insights from random forest ABC using microsatellite data. *Biol. Invasions* **2018**, *20*, 665–677. [CrossRef]
18. Meinke, L.J.; Sappington, T.W.; Onstad, D.W.; Guillemaud, T.; Miller, N.J.; Komáromi, J.; Levay, N.; Furlan, L.; Kiss, J.; Toth, F. Western corn rootworm (*Diabrotica virgifera virgifera* LeConte) population dynamics. *Agric. For. Entomol.* **2009**, *11*, 29–46. [CrossRef]
19. Gillette, C.P. *Diabrotica virgifera* Lec. as a corn root-worm. *J. Econ. Entomol.* **1912**, *5*, 364–366. [CrossRef]
20. Bare, O.S. *Corn Rootworm Does Damage in Southwestern Nebraska*; Annual Report of Cooperative Extension Work in Agriculture and Home Economics, Entomology, State of Nebraska; The University of Nebraska-Lincoln: Lincoln, NE, USA, 1930; p. 21.
21. Bare, O.S. *Corn Rootworms*; University Nebraska Cooperative Extension Circular 1506; The University of Nebraska-Lincoln: Lincoln, NE, USA, 1931.
22. Tate, H.D.; Bare, O.S. *Corn Rootworms*; Bulletin of the Agricultural Experiment Station of Nebraska; The University of Nebraska-Lincoln: Lincoln, NE, USA, 1946; pp. 1–12.
23. Hill, R.E.; Hixson, E.; Muma, M.H. Corn rootworm control tests with benzene hexachloride DDT, nitrogen fertilizers and crop rotations. *J. Econ. Entomol.* **1948**, *41*, 392–401. [CrossRef]
24. Lilly, J.H. Soil insects and their control. *Annu. Rev. Entomol.* **1956**, *1*, 203–222. [CrossRef]
25. Ball, H.J. *Larval and Adult Control Recommendations and Insecticide Resistance Data for Corn Rootworms in Nebraska (1948–1981)*; Department of Entomology Report; The University of Nebraska-Lincoln: Lincoln, NE, USA, 1981; pp. 1–16.
26. Wangila, D.S.; Gassmann, A.J.; Petzold-Maxwell, J.L.; French, B.W.; Meinke, L.J. Susceptibility of Nebraska western corn rootworm (Coleoptera: Chrysomelidae) populations to Bt corn events. *J. Econ. Entomol.* **2015**, *108*, 742–751. [CrossRef]
27. Gray, M.E.; Sappington, T.W.; Miller, N.J.; Moeser, J.; Bohn, M.O. Adaptation and invasiveness of western corn rootworm: Intensifying research on a worsening pest. *Annu. Rev. Entomol.* **2009**, *54*, 303–321. [CrossRef]
28. Levine, E.; Oloumi-Sadeghi, H. Management of diabroticite rootworms in corn. *Annu. Rev. Entomol.* **1991**, *36*, 229–255. [CrossRef]
29. Muma, M.H.; Hill, R.E.; Hixson, E. Soil treatments for corn rootworm control. *J. Econ. Entomol.* **1949**, *42*, 822–824. [CrossRef]

30. Munson, J.D.; Hill, R.E. Insecticides applied at planting time for control of corn rootworm larvae in Nebraska. *J. Econ. Entomol.* **1970**, *63*, 1614–1617. [CrossRef]

31. Mayo, Z.B. *A Five-Year Comparison of Insecticides Applied to Control Larvae of Western and Northern Corn Rootworms*; Agricultural Experiment Station, University of Nebraska-Lincoln: Lincoln, NE, USA, 1975; pp. 1–15.

32. Mayo, Z.B. *Aerial Suppression of Rootworm Adults for Larval Control*; Agricultural Experiment Station, University of Nebraska-Lincoln: Lincoln, NE, USA, 1976; pp. 1–12.

33. Mayo, Z.B. Field evaluation of insecticides for control of larvae of corn rootworms. In *Methods for the Study of Pest Diabrotica*; Springer Series in Experimental Entomology; Springer: New York, NY, USA, 1986; pp. 183–203. ISBN 978-1-4612-9338-5.

34. Pruess, K.P.; Witkowski, J.F.; Raun, E.S. Population suppression of western corn rootworm by adult control with ULV malathion. *J. Econ. Entomol.* **1974**, *67*, 651–655. [CrossRef]

35. Chio, H.; Chang, C.-S.; Metcalf, R.L.; Shaw, J. Susceptibility of four species of *Diabrotica* to insecticides. *J. Econ. Entomol.* **1978**, *71*, 389–393. [CrossRef]

36. Sutter, G.R. Comparative toxicity of insecticides for corn rootworm (Coleoptera: Chrysomelidae) larvae in a soil bioassay. *J. Econ. Entomol.* **1982**, *75*, 489–491. [CrossRef]

37. Fuller, B.W.; Boetel, M.A.; Walgenbach, D.D.; Grundler, J.A.; Hein, G.L.; Jarvi, K.J.; Keaster, A.J.; Landis, D.A.; Meinke, L.J.; Oleson, J.D.; et al. Optimization of soil insecticide rates for managing corn rootworm (Coleoptera: Chrysomelidae) larvae in the North Central United States. *J. Econ. Entomol.* **1997**, *90*, 1332–1340. [CrossRef]

38. Zhu, K.Y.; Wilde, G.E.; Sloderbeck, P.E.; Buschman, L.L.; Higgins, R.A.; Whitworth, R.J.; Bowling, R.A.; Starkey, S.R.; He, F. Comparative susceptibility of western corn rootworm (Coleoptera: Chrysomelidae) adults to selected insecticides in Kansas. *J. Econ. Entomol.* **2005**, *98*, 2181–2187. [CrossRef]

39. Douglas, M.R.; Tooker, J.F. Large-scale deployment of seed treatments has driven rapid increase in use of neonicotinoid insecticides and preemptive pest management in U.S. field crops. *Environ. Sci. Technol.* **2015**, *49*, 5088–5097. [CrossRef]

40. USEPA. Current & Previously Registered Section 3 PIP Registrations. Available online: https://www3.epa.gov/pesticides/chem_search/reg_actions/pip/pip_list.htm (accessed on 16 December 2020).

41. Andow, D.A.; Pueppke, S.G.; Schaafsma, A.W.; Gassmann, A.J.; Sappington, T.W.; Meinke, L.J.; Mitchell, P.D.; Hurley, T.M.; Hellmich, R.L.; Porter, R.P. Early detection and mitigation of resistance to Bt maize by western corn rootworm (Coleoptera: Chrysomelidae). *J. Econ. Entomol.* **2016**, *109*, 1–12. [CrossRef]

42. USDA-NASS. United States: Corn Production. Available online: https://ipad.fas.usda.gov/rssiws/al/crop_production_maps/us/USA_Corn_Total_Lev2_Prod.png (accessed on 30 December 2020).

43. USDA-NASS. Corn for Grain 2019 Production by County for Selected States. Available online: https://www.nass.usda.gov/Charts_and_Maps/Crops_County/cr-pr.php (accessed on 30 December 2020).

44. Cox, H.C.; Lilly, J.H. Chemical control of the corn rootworm. *J. Econ. Entomol.* **1953**, *46*, 217–224. [CrossRef]

45. Ball, H.J.; Hill, R.E. You can control corn rootworm. *Neb. Exp. Stn. Quart.* **1953**, *1*, 3–4.

46. Ball, H.J.; Roselle, R.E. *You Can Control Corn Rootworms*; University Nebraska College of Agricultural Extension Circular; University of Nebraska-Lincoln: Lincoln, NE, USA, 1954.

47. Ball, H.J. Aldrin and dieldrin residues in Nebraska soils as related to decreasing LD50 values for the adult western corn rootworm—1962–1981. *J. Environ. Sci. Health B* **1983**, *18*, 735–744. [CrossRef]

48. Roselle, R.E.; Anderson, L.W.; Simpson, R.C.; Webb, M.C. *Annual Report for 1959*; Cooperative Extension Work in Entomology; University of Nebraska-Lincoln: Lincoln, NE, USA, 1959.

49. Roselle, R.E.; Anderson, L.W.; Simpson, R.C.; Bergman, P.W. *Annual Report for 1960*; Cooperative Extension Work in Entomology; University of Nebraska-Lincoln: Lincoln, NE, USA, 1960.

50. Roselle, R.E.; Anderson, L.W.; Bergman, P.W. *Annual Report for 1961*; Cooperative Extension Work in Entomology; University of Nebraska-Lincoln: Lincoln, NE, USA, 1961.

51. Weekman, G.T. Problem in rootworm control. *Proc. North Cent. Branch Entomol. Soc. Am.* **1961**, *16*, 32–34.

52. Ball, H.J.; Weekman, G.T. Insecticide resistance in the adult western corn rootworm in Nebraska. *J. Econ. Entomol.* **1962**, *55*, 439–441. [CrossRef]

53. Ball, H.J.; Weekman, G.T. Differential resistance of corn rootworms to insecticides in Nebraska and adjoining states. *J. Econ. Entomol.* **1963**, *56*, 553–555. [CrossRef]

54. Ball, H.J. On the biology and egg-laying habits of the western corn rootworm. *J. Econ. Entomol.* **1957**, *50*, 126–128. [CrossRef]

55. Hamilton, E.W. Aldrin resistance in corn rootworm beetles. *J. Econ. Entomol.* **1965**, *58*, 296–300. [CrossRef]

56. Siegfried, B.D.; Mullin, C.A. Influence of alternative host plant feeding on aldrin susceptibility and detoxification enzymes in western and northern corn rootworms. *Pestic. Biochem. Physiol.* **1989**, *35*, 155–164. [CrossRef]

57. Metcalf, R.L. Foreword. In *Methods for the Study of Pest Diabrotica*; Krysan, J.L., Miller, T.A., Eds.; Springer: New York, NY, USA, 1986; pp. vii–xv.

58. Metcalf, R.L. The ecology of insecticides and the chemical control of insects. In *Ecological Theory and Integrated Pest Management Practice*; Kogan, M., Ed.; Wiley: New York, NY, USA, 1986; pp. 251–297.

59. Felsot, A.S. Enhanced biodegradation of insecticides in soil: Implications for agroecosystems. *Ann. Rev. Entomol.* **1989**, *34*, 453–476. [CrossRef]

60. Turpin, F.T.; Thieme, J.M. Impact of soil insecticide usage on corn production in Indiana: 1972–19741. *J. Econ. Entomol.* **1978**, *71*, 83–86. [CrossRef]

61. Stamm, D.E.; Mayo, Z.B.; Campbell, J.B.; Witkowski, J.F.; Andersen, L.W.; Kozub, R. Western corn rootworm (Coleoptera: Chrysomelidae) beetle counts as a means of making larval control recommendations in Nebraska. *J. Econ. Entomol.* **1985**, *78*, 794–798. [CrossRef]

62. Gray, M.E.; Steffey, K.L.; Oloumi-Sadeghi, H. Participatory on-farm research in Illinois cornfields: An evaluation of established soil insecticide rates and prevalence of com rootworm (Coleoptera: Chrysomelidae) injury. *J. Econ. Entomol.* **1993**, *86*, 1473–1482. [CrossRef]

63. USEPA. EPA Pesticide Fact Sheet: Tefluthrin. Available online: https://nepis.epa.gov/Exe/ZyPDF.cgi/P100XZAX.PDF?Dockey= P100XZAX.PDF (accessed on 16 December 2020).

64. Johnson, M.; Luukinen, B.; Gervais, J.; Buhl, K.; Stone, D. *Bifenthrin General Fact Sheet*; National Pesticide Information Center, Oregon State University Extension Services: Corvallis, OR, USA, 2010.

65. Harris, C.R. Factors influencing the effectiveness of soil insecticides. *Annu. Rev. Entomol.* **1972**, *17*, 177–198. [CrossRef]

66. Mayo, Z.B. Influences of rainfall and sprinkler irrigation on the residual activity of insecticides applied to corn for control of adult western corn rootworm (Coleoptera: Chrysomelidae). *J. Econ. Entomol.* **1984**, *77*, 190–193. [CrossRef]

67. Sutter, G.R.; Gustin, R.D. Environmental factors influencing corn rootworm biology and control. In Proceedings of the Illinois Agricultural Pesticides Conference; University of Illinois Cooperative Extension Service: Urbana-Champaign, IL, USA, 1989; pp. 43–48.

68. Sutter, G.R.; Branson, T.F.; Fisher, J.R.; Elliott, N.C.; Jackson, J.J. Effect of insecticide treatments on root damage ratings of maize in controlled infestations of western corn rootworms (Coleoptera: Chrysomelidae). *J. Econ. Entomol.* **1989**, *82*, 1792–1798. [CrossRef]

69. Riedell, W.E.; Sutter, G.R. Soil moisture and survival of western corn rootworm larvae in field plots. *J. Kans. Entomol. Soc.* **1995**, *68*, 80–84.

70. Gorder, G.W.; Dahm, P.A.; Tollefson, J.J. Carbofuran persistence in cornfield soils. *J. Econ. Entomol.* **1982**, *75*, 637–642. [CrossRef]

71. Harris, C.R.; Chapman, R.A.; Harris, C.; Tu, C.M. Biodegradation of pesticides in soil: Rapid induction of carbamate degrading factors after carbofuran treatment. *J. Environ. Sci. Health B* **1984**, *19*, 1–11. [CrossRef]

72. Walker, A.; Suett, D.L. Enhanced degradation of pesticides in soil: A potential problem for continued pest, disease and weed control. *Asp. Appl. Biol.* **1986**, *12*, 95–103.

73. Tollefson, J.J. Accelerated degradation of soil insecticides used for corn rootworm control. In Proceedings of the British Crop Protection Conference Pests and Diseases, Brighton Metropole, UK, 17–20 November 1986; pp. 1131–1136.

74. Williams, I.H.; Pepin, H.S.; Brown, M.J. Degradation of carbofuran by soil microorganisms. *Bull. Environ. Contam. Toxicol.* **1976**, *15*, 244–249. [CrossRef]

75. Felsot, A.S.; Wilson, J.G.; Kuhlman, D.E.; Steffey, K.L. Rapid dissipation of carbofuran as a limiting factor in corn rootworm (Coleoptera: Chrysomelidae) control in fields with histories of continous carbofuran use. *J. Econ. Entomol.* **1982**, *75*, 1098–1103. [CrossRef]

76. Abou-Assaf, N.; Coats, J.R.; Gray, M.E.; Tollefson, J.J. Degradation of isofenphos in cornfields with conservation tillage practices. *J. Environ. Sci. Health B* **1986**, *21*, 425–446. [CrossRef]

77. Racke, K.D.; Coats, J.R. Enhanced degradation of isofenphos by soil microorganisms. *J. Agric. Food Chem.* **1987**, *35*, 94–99. [CrossRef]

78. Hills, T.M.; Peters, D.C. Methods of applying insecticides for controlling western corn rootworm larvae. *J. Econ. Entomol.* **1972**, *65*, 1714–1718. [CrossRef]

79. Mayo, Z.B. Influence of planting dates on the efficacy of soil insecticides applied to control larvae of the western and northern corn rootworm. *J. Econ. Entomol.* **1980**, *73*, 211–212. [CrossRef]

80. Hoffmann, M.P.; Kirkwyland, J.J.; Gardner, J. Impact of western corn rootworm (Coleoptera: Chrysomelidae) on sweet corn and evaluation of insecticidal and cultural control options. *J. Econ. Entomol.* **2000**, *93*, 805–812. [CrossRef] [PubMed]

81. Mayo, Z.B.; Peters, L.L. Planting vs. cultivation time applications of granular soil insecticides to control larvae of corn rootworms in Nebraska. *J. Econ. Entomol.* **1978**, *71*, 801–803. [CrossRef]

82. Felsot, A.S.; Bruce, W.N.; Steffey, K.L. Degradation of terbufos (Counter) soil insecticide in cornfields under conservation tillage practices. *Bull. Environ. Contam. Toxicol.* **1987**, *38*, 369–376. [CrossRef]

83. Gray, M.E.; Felsot, A.S.; Steffey, K.L.; Levine, E. Planting time application of soil insecticides and western corn rootworm (Coleoptera: Chrysomelidae) emergence: Implications for long-term management programs. *J. Econ. Entomol.* **1992**, *85*, 544–553. [CrossRef]

84. Furlan, L.; Canzi, S.; Di Bernardo, A.; Edwards, C.R. The ineffectiveness of insecticide seed coatings and planting-time soil insecticides as *Diabrotica virgifera virgifera* LeConte population suppressors. *J. Appl. Entomol.* **2006**, *130*, 485–490. [CrossRef]

85. Boetel, M.A.; Fuller, B.W.; Evenson, P.D. Emergence of adult northern and western corn rootworms (Coleoptera: Chrysomelidae) following reduced soil insecticide applications. *J. Econ. Entomol.* **2003**, *96*, 714–729. [CrossRef] [PubMed]

86. Souza, D.; Peterson, J.A.; Wright, R.J.; Meinke, L.J. Field efficacy of soil insecticides on pyrethroid-resistant western corn rootworms (*Diabrotica virgifera virgifera* LeConte). *Pest Manag. Sci.* **2020**, *76*, 827–833. [CrossRef] [PubMed]

87. Parimi, S.; Meinke, L.J.; Wade French, B.; Chandler, L.D.; Siegfried, B.D. Stability and persistence of aldrin and methyl-parathion resistance in western corn rootworm populations (Coleoptera: Chrysomelidae). *Crop Prot.* **2006**, *25*, 269–274. [CrossRef]

88. Ball, H.J.; Su, P.P. Effect of sublethal dosages of carbofuran and carbaryl on fecundity and longevity of the female western corn rootworm. *J. Econ. Entomol.* **1979**, *72*, 873–876. [CrossRef]
89. Sutter, G.R.; Branson, T.F.; Fisher, J.R.; Elliott, N.C. Effect of insecticides on survival, development, fecundity, and sex ratio in controlled infestations of western corn rootworm (Coleoptera: Chrysomelidae). *J. Econ. Entomol.* **1991**, *84*, 1905–1912. [CrossRef]
90. Boetel, M.A.; Fuller, B.W.; Chandler, L.D.; Hovland, D.G.; Evenson, P.D. Fecundity and egg viability of northern and western corn rootworm (Coleoptera: Chrysomelidae) beetles surviving labeled and reduced soil insecticide applications. *J. Econ. Entomol.* **1998**, *91*, 274–279. [CrossRef]
91. Petzold-Maxwell, J.L.; Meinke, L.J.; Gray, M.E.; Estes, R.E.; Gassmann, A.J. Effect of Bt maize and soil insecticides on yield, injury, and rootworm survival: Implications for resistance management. *J. Econ. Entomol.* **2013**, *106*, 1941–1951. [CrossRef]
92. Weiss, M.J.; Seevers, K.P.; Mayo, Z.B. Influence of western corn rootworm larval densities and damage on corn rootworm survival, developmental time, size, and sex ratio (Coleoptera: Chrysomelidae). *J. Kans. Entomol. Soc.* **1985**, *58*, 397–402.
93. Elliott, N.C.; Sutter, G.R.; Branson, T.F.; Fisher, J.R. Effect of population density of immatures on survival and development of the western corn rootworm (Coleoptera: Chrysomelidae). *J. Entomol. Sci.* **1989**, *24*, 209–213. [CrossRef]
94. Mayo, Z.B. Control of larvae of the western and northern corn rootworm with liquid starter fertilizer-insecticide combinations and the influence of depth of placement. *J. Econ. Entomol.* **1977**, *70*, 234–236. [CrossRef]
95. Peters, L.L.; Lowry, S.R. Western corn rootworm (Coleoptera: Chrysomelidae) larval control with chlorpyrifos applied at planting time versus a post-planting chemigation application to corn growth under two different tillage systems. *J. Kans. Entomol. Soc.* **1991**, *64*, 451–454.
96. USEPA. Pesticide Fact Sheet: Clothianidin. Available online: https://www3.epa.gov/pesticides/chem_search/reg_actions/registration/fs_PC-044309_30-May-03.pdf (accessed on 16 December 2020).
97. Beyond Pesticides Appendix B. Thiamethoxam Pesticide Registrations. 2013. Available online: https://www.beyondpesticides.org/assets/media/documents/pollinators/documents/2013-03-21AppendixBThiamethoxamII.pdf (accessed on 16 December 2020).
98. Syngenta. Syngenta: Pioneering Seed Treatment Innovation. Available online: https://www.syngenta-us.com/prodrender/imagehandler.ashx?imid=a0209de7-27cf-4e31-8d4a-10c19f3b1194&fty=0&et=8 (accessed on 16 December 2020).
99. Mullin, C.A.; Saunders, M.C., II; Leslie, T.W.; Biddinger, D.J.; Fleischer, S.J. Toxic and behavioral effects to Carabidae of seed treatments used on Cry3Bb1- and Cry1Ab/c-protected corn. *Environ. Entomol.* **2005**, *34*, 1626–1636. [CrossRef]
100. Magalhaes, L.C.; French, B.W.; Hunt, T.E.; Siegfried, B.D. Baseline susceptibility of western corn rootworm (Coleoptera: Chrysomelidae) to clothianidin. *J. Appl. Entomol.* **2007**, *131*, 251–255. [CrossRef]
101. Steffey, K.; Gray, M.; Estes, R. Insecticidal seed treatments and soil insecticides for corn rootworm control. In Proceedings of the Illinois Crop Protection Technology Conference; University of Illinois Extension: Urbana-Champaign, IL, USA, 2005; pp. 35–40.
102. Van Rozen, K.; Ester, A. Chemical control of *Diabrotica virgifera virgifera* LeConte. *J. Appl. Entomol.* **2010**, *134*, 376–384. [CrossRef]
103. Krupke, C.H.; Long, E.Y. Intersections between neonicotinoid seed treatments and honey bees. *Curr. Opin. Insect Sci.* **2015**, *10*, 8–13. [CrossRef]
104. Schaafsma, A.; Limay-Rios, V.; Baute, T.; Smith, J.; Xue, Y. Neonicotinoid insecticide residues in surface water and soil associated with commercial maize (Corn) fields in southwestern ontario. *PLoS ONE* **2015**, *10*, e0118139. [CrossRef]
105. Douglas, M.R.; Rohr, J.R.; Tooker, J.F. Neonicotinoid insecticide travels through a soil food chain, disrupting biological control of non-target pests and decreasing soya bean yield. *J. Appl. Ecol.* **2015**, *52*, 250–260. [CrossRef]
106. Wood, T.J.; Goulson, D. The environmental risks of neonicotinoid pesticides: A review of the evidence post 2013. *Environ. Sci. Pollut. Res.* **2017**, *24*, 17285–17325. [CrossRef]
107. European Commission. Neonicotinoids. 2020. Available online: https://ec.europa.eu/food/plant/pesticides/approval_active_substances/approval_renewal/neonicotinoids_en (accessed on 14 December 2020).
108. Health Canada Update on the Neonicotinoid Pesticides. 2020. Available online: https://www.canada.ca/content/dam/hc-sc/documents/services/consumer-product-safety/reports-publications/pesticides-pest-management/fact-sheets-other-resources/update-neonicotinoid-pesticides-january-2020/neonic-update-sept2020-eng.pdf (accessed on 14 December 2020).
109. USEPA. Schedule for Review of Neonicotinoid Pesticides. Available online: https://www.epa.gov/pollinator-protection/schedule-review-neonicotinoid-pesticides (accessed on 14 December 2020).
110. Sappington, T.W.; Hesler, L.S.; Allen, K.C.; Luttrell, R.G.; Papiernik, S.K. Prevalence of sporadic insect pests of seedling corn and factors affecting risk of infestation. *J. Integr. Pest Manag.* **2018**, *9*. [CrossRef]
111. Veres, A.; Wyckhuys, K.A.G.; Kiss, J.; Tóth, F.; Burgio, G.; Pons, X.; Avilla, C.; Vidal, S.; Razinger, J.; Bazok, R.; et al. An update of the worldwide integrated assessment (WIA) on systemic pesticides. Part 4: Alternatives in major cropping systems. *Environ. Sci. Pollut. Res.* **2020**, *27*, 29867–29899. [CrossRef] [PubMed]
112. Meinke, L.J. Adult corn rootworm management. In *University of Nebraska Agricultural Research Division Miscellaneous Publication 63-C*; The University of Nebraska-Lincoln: Lincoln, NE, USA, 1995.
113. Meinke, L.J. Adult Corn Rootworm Suppression: Corn Rootworm Management in the Transgenic Era. Available online: http://www.plantmanagementnetwork.org/edcenter/seminars/corn/AdultRootworm/ (accessed on 20 December 2020).
114. Beardmore, B.W. Evaluation of western corn rootworm, *Diabrotica virgifera* LeConte, adult control on an individual field basis as an alternative to the use of soil insecticides in eastern Nebraska. Master's Thesis, University of Nebraska-Lincoln, Lincoln, NE, USA, 1975.

115. Meinke, L.J.; Siegfried, B.; Wright, R.J.; Chandler, L. Adult susceptibility of Nebraska western corn rootworm (Coleoptera: Chrysomelidae) populations to selected insecticides. *J. Econ. Entomol.* **1998**, *91*, 594–600. [CrossRef]
116. Wright, R.J.; Meinke, L.J.; Siegfried, B.D. Corn rootworm management. In *Proceedings Crop Protection Clinics*; The University of Nebraska-Lincoln: Lincoln, NE, USA, 1996; pp. 45–53.
117. Meinke, L.J.; Siegfried, B.D.; Wright, R.J.; Chandler, L.D. Western corn rootworm resistance to insecticides: Current situation in Nebraska. In Proceedings of the Illinois Agricultural Pesticides Conference, Champaign, IL, USA, 8–9 January 1997; pp. 88–92.
118. Siegfried, B.D.; Meinke, L.J.; Scharf, M.E. Resistance management concerns for areawide management programs. *J. Agric. Entomol.* **1998**, *15*, 359–369.
119. Scharf, M.E.; Meinke, L.J.; Siegfried, B.D.; Wright, R.J.; Chandler, L.D. Carbaryl susceptibility, diagnostic concentration determination, and synergism for U.S. populations of western corn rootworm (Coleoptera: Chrysomelidae). *J. Econ. Entomol.* **1999**, *92*, 33–39. [CrossRef]
120. Zhou, X.; Scharf, M.E.; Parimi, S.; Meinke, L.J.; Wright, R.J.; Chandler, L.D.; Siegfried, B.D. Diagnostic assays based on esterase-mediated resistance mechanisms in western corn rootworms (Coleoptera: Chrysomelidae). *J. Econ. Entomol.* **2002**, *95*, 1261–1266. [CrossRef]
121. Miller, N.J.; Guillemaud, T.; Giordano, R.; Siegfried, B.D.; Gray, M.E.; Meinke, L.J.; Sappington, T.W. Genes, Gene flow and adaptation of *Diabrotica virgifera virgifera*. *Agric. For. Entomol.* **2009**, *11*, 47–60. [CrossRef]
122. Anderson, J.F.; Metcalf, R.L. Identification of a volatile attractant for *Diabrotica* and *Acalymma* spp. from blossoms of *Cucurbita maxima* Duchesne. *J. Chem. Ecol.* **1986**, *12*, 87–699.
123. Lampman, R.L.; Metcalf, R.L.; Andersen, J.F. Semiochemical attractants of *Diabrotica undecimpunctata howardi* Barber, southern corn rootworm, and *Diabrotica virgifera virgifera* LeConte, the western corn rootworm (Coleoptera: Chrysomelidae). *J. Chem. Ecol.* **1987**, *13*, 959–975. [CrossRef]
124. Lampman, R.L.; Metcalf, R.L. Multicomponent kairomonal lures for southern and western corn rootworms (Coleoptera: Chrysomelidae: *Diabrotica* spp.). *J. Econ. Entomol.* **1987**, *80*, 1137–1142. [CrossRef]
125. Chambliss, O.L.; Jones, C.M. Cucurbitacins: Specific insect attractants in Cucurbitaceae. *Science* **1966**, *153*, 1392–1393. [CrossRef]
126. Metcalf, R.L. Changing role of insecticides in crop protection. *Ann. Rev. Entomol.* **1980**, *25*, 219–256. [CrossRef]
127. Metcalf, R.L.; Rhodes, A.M.; Metcalf, R.A.; Ferguson, J.; Metcalf, E.R.; Lu, P.-Y. Cucurbitacin contents and diabroticite (Coleoptera: Chrysomelidae) feeding upon *Cucurbita* spp. *Environ. Entomol.* **1982**, *11*, 931–937. [CrossRef]
128. Metcalf, R.L.; Ferguson, J.E.; Lampman, R.; Andersen, J.F. Dry cucurbitacin-containing baits for controlling diabroticite beetles (Coleoptera: Chrysomelidae). *J. Econ. Entomol.* **1987**, *80*, 870–875. [CrossRef]
129. Lance, D.R.; Sutter, G.R. Field-cage and laboratory evaluations of semiochemical-based baits for managing western corn rootworm (Coleoptera: Chrysomelidae). *J. Econ. Entomol.* **1990**, *83*, 1085–1090. [CrossRef]
130. Weissling, T.J.; Meinke, L.J. Potential of starch encapsulated semiochemical-insecticide formulations for adult corn rootworm (Coleoptera: Chrysomelidae) control. *J. Econ. Entomol.* **1991**, *84*, 601–609. [CrossRef]
131. Lance, D.R.; Sutter, G.R. Field tests of a semiochemical-based toxic bait for suppression of corn rootworm beetles (Coleoptera: Chrysomelidae). *J. Econ. Entomol.* **1992**, *85*, 967–973. [CrossRef]
132. Weissling, T.J.; Meinke, L.J. Semiochemical-insecticide bait placement and vertical distribution of corn rootworm (Coleoptera: Chrysomelidae) adults: Implications for management. *Environ. Entomol.* **1991**, *20*, 945–952. [CrossRef]
133. Lance, D.R.; Sutter, G.R. Semiochemical-based toxic baits for *Diabrotica virgifera virgifera* (Coleoptera: Chrysomelidae): Effects of particle size, location, and attractant content. *J. Econ. Entomol.* **1991**, *84*, 1861–1868. [CrossRef]
134. Meinke, L.J. Behavior-based adult corn rootworm management. In Proceedings of the 1992 Crop Pest Management Update; University of Illinois Extension: Urbana-Champaign, IL, USA, 1992; pp. 61–65.
135. Chandler, L.D. Comparison of insecticide-bait aerial application methods for management of corn rootworm (Coleoptera: Chrysomelidae). *Southwest. Entomol.* **1998**, *23*, 147–159.
136. Wilde, G.E.; Whitworth, R.J.; Shufran, R.A.; Zhu, K.Y.; Sloderbeck, P.E.; Higgins, R.A.; Buschman, L.L. Rootworm areawide management project in Kansas. *J. Agric. Entomol.* **1998**, *15*, 335–349.
137. Behle, R.W. Consumption of residue containing cucurbitacin feeding stimulant and reduced rates of carbaryl insecticide by western corn rootworm (Coleoptera: Chrysomelidae). *J. Econ. Entomol.* **2001**, *94*, 1428–1433. [CrossRef] [PubMed]
138. Chandler, L.D. Corn rootworm areawide management program: United States Department of Agriculture-Agricultural Research Service. *Pest Manag. Sci.* **2003**, *59*, 605–608. [CrossRef] [PubMed]
139. Chandler, L.D.; Coppedge, J.R.; Edwards, C.R.; Tollefson, J.J.; Wilde, G.R.; Faust, R.M. Corn rootworm areawide pest management in the midwestern USA. In *Areawide Pest Management: Theory and Implementation*; Koul, O., Cuperous, G., Elliott, N., Eds.; CAB International: Oxfordshire, UK, 2008; pp. 191–207.
140. Boetel, M.A.; Fuller, B.W.; Chandler, L.D.; Tollefson, J.J.; Mcmanus, B.L.; Kadakia, N.D.; Evenson, P.D.; Mishra, T.P. Nontarget arthropod abundance in areawide-managed corn habitats treated with semiochemical-based bait insecticide for corn rootworm (Coleoptera: Chrysomelidae) control. *J. Econ. Entomol.* **2005**, *98*, 1957–1968. [CrossRef]
141. Weissling, T.J.; Meinke, L.J.; Lytle, K.A. Effect of starch-based corn rootworm (Coleoptera: Chrysomelidae) baits on selected nontarget insect species: Influence of semiochemical composition. *J. Econ. Entomol.* **1991**, *84*, 1235–1241. [CrossRef]

142. Siegfried, B.D.; Meinke, L.J.; Parimi, S.; Scharf, M.E.; Nowatzki, T.J.; Zhou, X.; Chandler, L.D. Monitoring western corn rootworm (Coleoptera: Chrysomelidae) susceptibility to carbaryl and cucurbitacin baits in the areawide management pilot program. *J. Econ. Entomol.* **2004**, *97*, 1726–1733. [CrossRef] [PubMed]
143. Zhu, K.Y.; Wilde, G.E.; Higgins, R.A.; Sloderbeck, P.E.; Buschman, L.L.; Shufran, R.A.; Whitworth, R.J.; Starkey, S.R.; He, F. Evidence of evolving carbaryl resistance in western corn rootworm (Coleoptera: Chrysomelidae) in areawide-managed cornfields in north central Kansas. *J. Econ. Entomol.* **2001**, *94*, 929–934. [CrossRef] [PubMed]
144. Parimi, S.; Meinke, L.J.; Nowatzki, T.M.; Chandler, L.D.; Wade French, B.; Siegfried, B.D. Toxicity of insecticide-bait mixtures to insecticide resistant and susceptible western corn rootworms (Coleoptera: Chrysomelidae). *Crop Prot.* **2003**, *22*, 781–786. [CrossRef]
145. Chandler, L.D.; Faust, R.M. Overview of areawide management of insects. *J. Agric. Entomol.* **1998**, *15*, 319–325.
146. Rogers, E.M. *Diffusion of Innovations*, 3rd ed.; Free Press: New York, NY, USA, 1983.
147. Rice, M.E. Transgenic rootworm corn: Assessing potential agronomic, economic, and environmental benefits. *Plant Health Prog.* **2004**. [CrossRef]
148. Souza, D.; Vieira, B.C.; Fritz, B.K.; Hoffmann, W.C.; Peterson, J.A.; Kruger, G.R.; Meinke, L.J. Western corn rootworm pyrethroid resistance confirmed by aerial application simulations of commercial insecticides. *Sci. Rep.* **2019**, *9*, 6713. [CrossRef]
149. Pereira, A.E.; Wang, H.; Zukoff, S.N.; Meinke, L.J.; French, B.W.; Siegfried, B.D. Evidence of field-evolved resistance to bifenthrin in western corn rootworm (*Diabrotica virgifera virgifera* LeConte) populations in western Nebraska and Kansas. *PLoS ONE* **2015**, *10*, e0142299. [CrossRef] [PubMed]
150. NASS-USDA. QuickStats Ad-Hoc Query Tool. Available online: https://quickstats.nass.usda.gov/ (accessed on 20 December 2020).
151. Pereira, A.E.; Souza, D.; Zukoff, S.N.; Meinke, L.J.; Siegfried, B.D. Cross-resistance and synergism bioassays suggest multiple mechanisms of pyrethroid resistance in western corn rootworm populations. *PLoS ONE* **2017**, *12*, e0179311. [CrossRef] [PubMed]
152. Wright, R.J.; Scharf, M.E.; Meinke, L.J.; Zhou, X.; Siegfried, B.D.; Chandler, L.D. Larval susceptibility of an insecticide-resistant western corn rootworm (Coleoptera: Chrysomelidae) population to soil insecticides: Laboratory bioassays, assays of detoxification enzymes, and field performance. *J. Econ. Entomol.* **2000**, *93*, 7–13. [CrossRef] [PubMed]
153. Chio, H.; Metcalf, R.L. Detoxication mechanisms for aldrin, carbofuran, fonofos, phorate, and terbufos in four species of diabroticites. *J. Econ. Entomol.* **1979**, *72*, 732–738. [CrossRef]
154. Siegfried, B.D.; Mullin, C.A. Properties of a cytochrome P-450-dependent epoxidase in aldrin-resistant western corn rootworms, *Diabrotica virgifera virgifera* LeConte. *Pestic. Biochem. Physiol.* **1988**, *31*, 261–268. [CrossRef]
155. Hamilton, E.W. Relative toxicity of aldrin and dieldrin. *J. Agric. Food Chem.* **1971**, *19*, 863–864. [CrossRef]
156. Siegfried, B.D.; Mullin, C.A. Metabolism, penetration, and partitioning of [14C]aldrin in aldrin-resistant and susceptible corn rootworms. *Pestic. Biochem. Physiol.* **1990**, *36*, 135–146. [CrossRef]
157. Kadous, A.A.; Ghiasuddin, S.M.; Matsumura, F.; Scott, J.G.; Tanaka, K. Difference in the picrotoxinin receptor between the cyclodiene-resistant and susceptible strains of the german cockroach. *Pestic. Biochem. Physiol.* **1983**, *19*, 157–166. [CrossRef]
158. Tanaka, K.; Scott, J.G.; Matsumura, F. Picrotoxinin receptor in the central nervous system of the american cockroach: Its role in the action of cyclodiene-type insecticides. *Pestic. Biochem. Physiol.* **1984**, *22*, 117–127. [CrossRef]
159. Bloomquist, J.R.; Soderlund, D.M. Neurotoxic insecticides inhibit GABA-dependent chloride uptake by mouse brain vesicles. *Biochem. Biophys. Res. Commun.* **1985**, *133*, 37–43. [CrossRef]
160. Narahashi, T.; Frey, J.M.; Ginsburg, K.S.; Roy, M.L. Sodium and GABA-activated channels as the targets of pyrethroids and cyclodienes. *Toxicol. Lett.* **1992**, *64–65*, 429–436. [CrossRef]
161. Buckingham, S.D.; Biggin, P.C.; Sattelle, B.M.; Brown, L.A.; Sattelle, D.B. Insect GABA receptors: Splicing, editing, and targeting by antiparasitics and insecticides. *Mol. Pharmacol.* **2005**, *68*, 942–951. [CrossRef]
162. Ffrench-Constant, R.H.; Williamson, M.S.; Davies, T.G.E.; Bass, C. Ion channels as insecticide targets. *J. Neurosci.* **2016**, *30*, 163–177. [CrossRef] [PubMed]
163. Ffrench-Constant, R.H.; Anthony, N.; Aronstein, K.; Rocheleau, T.; Stilwell, G. Cyclodiene insecticide resistance: From molecular to population genetics. *Ann. Rev. Entomol.* **2000**, *45*, 449–466. [CrossRef] [PubMed]
164. Raymond-Delpech, V.; Matsuda, K.; Sattelle, B.M.; Rauh, J.J.; Sattelle, D.B. Ion channels: Molecular targets of neuroactive insecticides. *Invertebr. Neurosci.* **2005**, *3–4*, 119–133. [CrossRef]
165. Ffrench-Constant, R.H.; Roush, R.T. Cloning of a locus associated with cyclodiene resistance in *Drosophila*. In *Molecular Mechanisms of Insecticide Resistance*; ACS Symposium Series; American Chemical Society: Washington, DC, USA, 1992; Volume 505, pp. 90–98. ISBN 978-0-8412-2474-2.
166. Taylor-Wells, J.; Jones, A.K. Variations in the insect GABA receptor, RDL, and their impact on receptor pharmacology. In *Advances in Agrochemicals: Ion Channels and G Protein-Coupled Receptors (GPCRs) as Targets for Pest Control*; ACS Symposium Series; American Chemical Society: Washington, DC, USA, 2017; Volume 1265, pp. 1–21. ISBN 978-0-8412-3260-0.
167. Hosie, A.; Sattelle, D.; Aronstein, K.; Ffrench-Constant, R. Molecular biology of insect neuronal GABA receptors. *Trends Neurosci.* **1997**, *20*, 578–583. [CrossRef]
168. Wang, H.; Coates, B.S.; Chen, H.; Sappington, T.W.; Guillemaud, T.; Siegfried, B.D. Role of a gamma-aminobutryic acid (GABA) receptor mutation in the evolution and spread of *Diabrotica virgifera virgifera* resistance to cyclodiene insecticides. *Insect. Mol. Biol.* **2013**, *22*, 473–484. [CrossRef]

169. Chen, H.; Wang, H.; Siegfried, B.D. Genetic differentiation of western corn rootworm populations (Coleoptera: Chrysomelidae) relative to insecticide resistance. *Ann. Entomol. Soc. Am.* **2012**, *105*, 232–240. [CrossRef]

170. Coustau, C.; Chevillon, C.; Ffrench-Constant, R. Resistance to xenobiotics and parasites: Can we count the cost? *Trends Ecol. Evol.* **2000**, *15*, 378–383. [CrossRef]

171. Ffrench-Constant, R.H.; Bass, C. Does resistance really carry a fitness cost? *Curr. Opin. Insect Sci.* **2017**, *21*, 39–46. [CrossRef]

172. Taylor, M.; Feyereisen, R. Molecular biology and evolution of resistance of toxicants. *Mol. Biol. Evol.* **1996**, *13*, 719–734. [CrossRef] [PubMed]

173. Mckenzie, J.A. Selection at the dieldrin resistance locus in overwintering populations of *Lucilia cuprina* (Wiedemann). *Aust. J. Zool.* **1990**, *38*, 493–501. [CrossRef]

174. Rowland, M. Activity and mating competitiveness of ΓHCH/dieldrin resistant and susceptible male and virgin female *Anopheles gambiae* and *An. stephensi* mosquitoes, with assessment of an insecticide-rotation strategy. *Med. Vet. Entomol.* **1991**, *5*, 207–222. [CrossRef] [PubMed]

175. Aronstein, K.; Ode, P.; Ffrench-Constant, R.H. Direct comparison of PCR-based monitoring for cyclodiene resistance in *Drosophila* populations with insecticide bioassay. *Pestic. Biochem. Physiol.* **1994**, *48*, 229–233. [CrossRef]

176. Hansen, K.K.; Kristensen, M.; Jensen, K.-M.V. Correlation of a resistance-associated *Rdl* mutation in the german cockroach, *Blattella germanica* (L), with persistent dieldrin resistance in two Danish field populations. *Pest Manag. Sci.* **2005**, *61*, 749–753. [CrossRef] [PubMed]

177. Kwiatkowska, R.M.; Platt, N.; Poupardin, R.; Irving, H.; Dabire, R.K.; Mitchell, S.; Jones, C.M.; Diabaté, A.; Ranson, H.; Wondji, C.S. Dissecting the mechanisms responsible for the multiple insecticide resistance phenotype in *Anopheles gambiae* s.s., M Form, from Vallée Du Kou, Burkina Faso. *Gene* **2013**, *519*, 98–106. [CrossRef]

178. Aldridge, W.N. Some properties of specific cholinesterase with particular reference to the mechanism of inhibition by diethyl p-nitrophenyl thiophosphate (E 605) and analogues. *Biochem. J.* **1950**, *46*, 451–460. [CrossRef]

179. Aldridge, W.N.; Reiner, E. *Enzyme Inhibitors as Substrates: Interactions of Esterases with Esters of Organophosphorus and Carbamic Acids*; North-Holland Publishing Co.: Amsterdam, The Netherland, 1972.

180. Fukuto, T.R. Mechanism of action of organophosphorus and carbamate insecticides. *Environ. Health Perspect.* **1990**, *87*, 245–254. [CrossRef]

181. Čolović, M.B.; Krstić, D.Z.; Lazarević-Pašti, T.D.; Bondžić, A.M.; Vasić, V.M. Acetylcholinesterase inhibitors: Pharmacology and toxicology. *Curr. Neuropharmacol.* **2013**, *11*, 315–335. [CrossRef]

182. Montella, I.R.; Schama, R.; Valle, D. The classification of esterases: An important gene family involved in insecticide resistance—A Review. *Mem. Inst. Oswaldo Cruz* **2012**, *107*, 437–449. [CrossRef]

183. Sogorb, M.A.; Vilanova, E. Enzymes involved in the detoxification of organophosphorus, carbamate and pyrethroid insecticides through hydrolysis. *Toxicol. Lett.* **2002**, *128*, 215–228. [CrossRef]

184. Li, X.; Schuler, M.A.; Berenbaum, M.R. Molecular mechanisms of metabolic resistance to synthetic and natural xenobiotics. *Ann. Rev. Entomol.* **2007**, *52*, 231–253. [CrossRef] [PubMed]

185. Fournier, D. Mutations of acetylcholinesterase which confer insecticide resistance in insect populations. *Chem. Biol. Interact.* **2005**, *157–158*, 257–261. [CrossRef] [PubMed]

186. Lee, S.H.; Kim, Y.H.; Kwon, D.H.; Cha, D.J.; Kim, J.H. Mutation and duplication of arthropod acetylcholinesterase: Implications for pesticide resistance and tolerance. *Pestic. Biochem. Physiol.* **2015**, *120*, 118–124. [CrossRef]

187. Miota, F.; Scharf, M.E.; Ono, M.; Marçon, P.; Meinke, L.J.; Wright, R.J.; Chandler, L.D.; Siegfried, B.D. Mechanisms of methyl and ethyl parathion resistance in the western corn rootworm (Coleoptera: Chrysomelidae). *Pestic. Biochem. Physiol.* **1998**, *61*, 39–52. [CrossRef]

188. Scharf, M.E.; Meinke, L.J.; Wright, R.J.; Chandler, L.D.; Siegfried, B.D. Metabolism of carbaryl by insecticide-resistant and -susceptible western corn rootworm populations (Coleoptera: Chrysomelidae). *Pesti. Biochem. Physiol.* **1999**, *63*, 85–96. [CrossRef]

189. Scharf, M.E.; Parimi, S.; Meinke, L.J.; Chandler, L.D.; Siegfried, B.D. Expression and induction of three family 4 cytochrome P450 (CYP4)* genes identified from insecticide-resistant and susceptible western corn rootworms, *Diabrotica virgifera virgifera. Insect Mol. Biol.* **2001**, *10*, 139–146. [CrossRef]

190. Zhou, X.; Scharf, M.E.; Meinke, L.J.; Chandler, L.D.; Siegfried, B.D. Characterization of general esterases from methyl parathion-resistant and -susceptible populations of western corn rootworm (Coleoptera: Chrysomelidae). *J. Econ. Entomol.* **2003**, *96*, 1855–1863. [CrossRef]

191. Zhou, X.; Scharf, M.E.; Sarath, G.; Meinke, L.J.; Chandler, L.D.; Siegfried, B.D. Partial purification and characterization of a methyl-parathion resistance-associated general esterase in *Diabrotica virgifera virgifera* (Coleoptera: Chrysomelidae). *Pestic. Biochem. Physiol.* **2004**, *78*, 114–125. [CrossRef]

192. Parimi, S.; Scharf, M.E.; Meinke, L.J.; Chandler, L.D.; Siegfried, B.D. Inheritance of methyl-parathion resistance in Nebraska western corn rootworm populations (Coleoptera: Chrysomelidae). *J. Econ. Entomol.* **2003**, *96*, 131–136. [CrossRef]

193. Narahashi, T. Nerve membrane as a target of pyrethroids. *Pestic. Sci.* **1976**, *7*, 267–272. [CrossRef]

194. Soderlund, D.M.; Bloomquist, J.R. Neurotoxic actions of pyrethroid insecticides. *Ann. Rev. Entomol.* **1989**, *34*, 77–96. [CrossRef] [PubMed]

195. Vijverberg, H.P.M.; van der Zalm, J.M.; van den Bercken, J. Similar mode of action of pyrethroids and DDT on sodium channel gating in myelinated nerves. *Nature* **1982**, *295*, 601–603. [CrossRef] [PubMed]

196. Breckenridge, C.B.; Holden, L.; Sturgess, N.; Weiner, M.; Sheets, L.; Sargent, D.; Soderlund, D.M.; Choi, J.-S.; Symington, S.; Clark, J.M.; et al. Evidence for a separate mechanism of toxicity for the Type I and the Type II pyrethroid insecticides. *Neurotoxicology* **2009**, *30*, S17–S31. [CrossRef] [PubMed]

197. Hildebrand, M.E.; McRory, J.E.; Snutch, T.P.; Stea, A. Mammalian voltage-gated calcium channels are potently blocked by the pyrethroid insecticide allethrin. *J. Pharmacol. Exp. Ther.* **2004**, *308*, 805–813. [CrossRef] [PubMed]

198. Lawrence, L.J.; Casida, J.E. Stereospecific action of pyrethroid insecticides on the gamma-aminobutyric acid receptor-ionophore complex. *Science* **1983**, *221*, 1399–1401. [CrossRef] [PubMed]

199. Ray, D.E.; Sutharsan, S.; Forshaw, P.J. Actions of pyrethroid insecticides on voltage-gated chloride channels in neuroblastoma cells. *Neurotoxicology* **1997**, *18*, 755–760.

200. Soderlund, D.M. Molecular mechanisms of pyrethroid insecticide neurotoxicity: Recent advances. *Arch. Toxicol.* **2011**, *86*, 165–181. [CrossRef]

201. Symington, S.B.; Clark, J.M. Action of deltamethrin on N-type (Cav2.2) voltage-sensitive calcium channels in rat brain. *Pestic. Biochem. Physiol.* **2005**, *82*, 1–15. [CrossRef]

202. Vijverberg, H.P.M.; van den Bercken, J. Neurotoxicological effects and the mode of action of pyrethroid insecticides. *Crit. Rev. Toxicol.* **1990**, *21*, 105–126. [CrossRef]

203. Gammon, D.W.; Brown, M.A.; Casida, J.E. Two classes of pyrethroid action in the cockroach. *Pestic. Biochem. Physiol.* **1981**, *15*, 181–191. [CrossRef]

204. Lawrence, L.J.; Casida, J.E. Pyrethroid toxicology: Mouse intracerebral structure-toxicity relationships. *Pestic. Biochem. Physiol.* **1982**, *18*, 9–14. [CrossRef]

205. Scott, J.G.; Matsumura, F. Evidence for two types of toxic actions of pyrethroids on susceptible and DDT-resistant german cockroaches. *Pestic. Biochem. Physiol.* **1983**, *19*, 141–150. [CrossRef]

206. Soderlund, D.M.; Clark, J.M.; Sheets, L.P.; Mullin, L.S.; Piccirillo, V.J.; Sargent, D.; Stevens, J.T.; Weiner, M.L. Mechanisms of pyrethroid neurotoxicity: Implications for cumulative risk assessment. *Toxicology* **2002**, *171*, 3–59. [CrossRef]

207. Dong, K.; Du, Y.; Rinkevich, F.; Nomura, Y.; Xu, P.; Wang, L.; Silver, K.; Zhorov, B.S. Molecular biology of insect sodium channels and pyrethroid resistance. *Insect Biochem. Mol. Biol.* **2014**, *50*, 1–17. [CrossRef] [PubMed]

208. Liu, N.; Xu, Q.; Zhu, F.; Zhang, L. Pyrethroid resistance in mosquitoes. *Insect Sci.* **2006**, *13*, 159–166. [CrossRef]

209. Soderlund, D.M.; Knipple, D.C. The molecular biology of knockdown resistance to pyrethroid insecticides. *Insect Biochem. Mol. Biol.* **2003**, *33*, 563–577. [CrossRef]

210. Brengues, C.; Hawkes, N.J.; Chandre, F.; Mccarroll, L.; Duchon, S.; Guillet, P.; Manguin, S.; Morgan, J.C.; Hemingway, J. Pyrethroid and DDT cross-resistance in *Aedes aegypti* is correlated with novel mutations in the voltage-gated sodium channel gene. *Med. Vet. Entomol.* **2003**, *17*, 87–94. [CrossRef]

211. Schuler, T.H.; Martinez-Torres, D.; Thompson, A.J.; Denholm, I.; Devonshire, A.L.; Duce, I.R.; Williamson, M.S. Toxicological, electrophysiological, and molecular characterisation of knockdown resistance to pyrethroid insecticides in the diamondback moth, *Plutella Xylostella* (L.). *Pestic. Biochem. Physiol.* **1998**, *59*, 169–182. [CrossRef]

212. Williamson, M.S.; Denholm, I.; Bell, C.A.; Devonshire, A.L. Knockdown resistance (*kdr*) to DDT and pyrethroid insecticides maps to a sodium channel gene locus in the housefly (*Musca domestica*). *Mol. Gen. Genet.* **1993**, *240*, 17–22. [CrossRef]

213. Li, F.-G.; Ai, G.-M.; Zou, D.-Y.; Ji, Y.; Bao-Gen, G.; Gao, X.-W. Characterization of activation metabolism activity of indoxacarb in insects by liquid chromatography-triple quadrupole mass spectrometry. *Chin. J. Anal. Chem.* **2014**, *42*, 463–468. [CrossRef]

214. Wing, K.D.; Sacher, M.; Kagaya, Y.; Tsurubuchi, Y.; Mulderig, L.; Connair, M.; Schnee, M. Bioactivation and mode of action of the oxadiazine indoxacarb in insects. *Crop Prot.* **2000**, *19*, 537–545. [CrossRef]

215. Wing, K.D.; Schnee, M.E.; Sacher, M.; Connair, M. A novel oxadiazine insecticide is bioactivated in lepidopteran larvae. *Arch. Insect Biochem. Physiol.* **1998**, *37*, 91–103. [CrossRef]

216. Bisset, J.; Rodriguez, M.; Soca, A.; Pasteur, N.; Raymond, M. Cross-resistance to pyrethroid and organophosphorus insecticides in the southern house mosquito (Diptera: Culicidae) from Cuba. *J. Med. Entomol.* **1997**, *34*, 244–246. [CrossRef]

217. Cahill, M.; Byrne, F.J.; Gorman, K.; Denholm, I.; Devonshire, A.L. Pyrethroid and organophosphate resistance in the tobacco whitefly *Bemisia tabaci* (Homoptera: Aleyrodidae). *Bull. Entomol. Res.* **1995**, *85*, 181–187. [CrossRef]

218. Devonshire, A.L.; Moores, G.D. A carboxylesterase with broad substrate specificity causes organophosphorus, carbamate and pyrethroid resistance in peach-potato aphids (*Myzus persicae*). *Pestic. Biochem. Physiol.* **1982**, *18*, 235–246. [CrossRef]

219. Gunning, R.V.; Devonshire, A.L. Negative cross resistance between indoxacarb and pyrethroids in the cotton bollworm, *Helicoverpa armigera*, in Australia: A tool for resistance management. In Proceedings of the BCPC International Congress, Glasgow, Scotland, UK, 10–12 November 2003; Volume 2, pp. 789–794.

220. Ramasubramanian, T.; Regupathy, A. Pattern of cross-resistance in pyrethroid-selected populations of *Helicoverpa armigera* Hübner (Lep., Noctuidae) from India. *J. Appl. Entomol.* **2004**, *128*, 583–587. [CrossRef]

221. Souza, D.; Jiménez, A.V.; Sarath, G.; Meinke, L.J.; Miller, N.J.; Siegfried, B.D. Enhanced metabolism and selection of pyrethroid-resistant western corn rootworms (*Diabrotica virgifera virgifera* LeConte). *Pestic. Biochem. Physiol.* **2020**, *164*, 165–172. [CrossRef]

222. Georghiou, G.P.; Taylor, C.E. Genetic and biological influences in the evolution of insecticide resistance. *J. Econ. Entomol.* **1977**, *70*, 319–323. [CrossRef]

223. Roush, R.T.; McKenzie, J.A. Ecological genetics of insecticide and acaricide resistance. *Ann. Rev. Entomol.* **1987**, *32*, 361–380. [CrossRef]

224. Tabashnik, B.E. Resistance risk assessment: Realized heritability of resistance to *Bacillus thuringiensis* in diamondback moth (Lepidoptera: Plutellidae), tobacco budworm (Lepidoptera: Noctuidae), and Colorado potato beetle (Coleoptera: Chrysomelidae). *J. Econ. Entomol.* **1992**, *85*, 1551–1559. [CrossRef]
225. Souza, D.; Siegfried, B.D.; Meinke, L.J.; Miller, N.J. Molecular characterization of western corn rootworm pyrethroid resistance. *Pest Manag. Sci.* **2021**, *77*, 860–868. [CrossRef] [PubMed]
226. Crowder, D.W.; Onstad, D.W.; Gray, M.E.; Pierce, C.M.F.; Hager, A.G.; Ratcliffe, S.T.; Steffey, K.L. Analysis of the dynamics of adaptation to transgenic corn and crop rotation by western corn rootworm (Coleoptera: Chrysomelidae) using a daily time-step model. *J. Econ. Entomol.* **2005**, *98*, 534–551. [CrossRef] [PubMed]
227. Clark, T.L.; Frank, D.L.; French, B.W.; Meinke, L.J.; Moellenbeck, D.; Vaughn, T.T.; Hibbard, B.E. Mortality impact of MON863 transgenic maize roots on western corn rootworm larvae in the field. *J. Appl. Entomol.* **2012**, *136*, 721–729. [CrossRef]
228. Hitchon, A.J.; Smith, J.L.; French, B.W.; Schaafsma, A.W. Impact of the Bt corn proteins Cry34/35Ab1 and Cry3Bb1, alone or pyramided, on western corn rootworm (Coleoptera: Chrysomelidae) beetle emergence in the field. *J. Econ. Entomol.* **2015**, *108*, 1986–1993. [CrossRef]
229. Keweshan, R.S.; Head, G.P.; Gassmann, A.J. Effects of pyramided Bt corn and blended refuges on western corn rootworm and northern corn rootworm (Coleoptera: Chrysomelidae). *J. Econ. Entomol.* **2015**, *108*, 720–729. [CrossRef]
230. Shields, E.J.; Testa, A.M.; O'Neil, W.J. Long-term persistence of native New York entomopathogenic nematode isolates across crop rotation. *J. Econ. Entomol.* **2018**, *111*, 2592–2598. [CrossRef]
231. Shields, E.J. Biological Control of Corn Rootworm with Persistent Entomopathogenic Nematodes: An Opportunity to Try Them on Your Farm. Available online: https://blogs.cornell.edu/whatscroppingup/2019/03/22/biological-control-of-corn-rootworm-with-persistent-entomopathogenic-nematodes-an-opportunity-to-try-them-on-your-farm/ (accessed on 31 December 2020).
232. USEPA. Framework to Delay Corn Rootworm Resistance. Available online: https://www.epa.gov/regulation-biotechnology-under-tsca-and-fifra/framework-delay-corn-rootworm-resistance (accessed on 16 December 2020).
233. Levine, E.; Spencer, J.L.; Isard, S.A.; Onstad, D.W.; Gray, M.E. Adaptation of the western corn rootworm, *Diabrotica virgifera virgifera* LeConte (Coleoptera: Chrysomelidae), to crop rotation: Evolution of a new strain in response to a cultural management practice. *Am. Entomol.* **2002**, *48*, 94–107. [CrossRef]
234. Gassmann, A.J.; Petzold-Maxwell, J.L.; Clifton, E.H.; Dunbar, M.W.; Hoffmann, A.M.; Ingber, D.A.; Keweshan, R.S. Field-evolved resistance by western corn rootworm to multiple *Bacillus thuringiensis* toxins in transgenic maize. *Proc. Natl. Acad. Sci. USA* **2014**, *111*, 5141–5146. [CrossRef]
235. Reinders, J.D.; Hitt, B.D.; Stroup, W.W.; French, B.W.; Meinke, L.J. Spatial variation in western corn rootworm (Coleoptera: Chrysomelidae) susceptibility to Cry3 toxins in Nebraska. *PLoS ONE* **2018**, *13*, e0208266. [CrossRef]

Review

Resistance to Bt Maize by Western Corn Rootworm: Effects of Pest Biology, the Pest–Crop Interaction and the Agricultural Landscape on Resistance

Aaron J. Gassmann

Department of Entomology, Iowa State University, Ames, IA 50011, USA; aaronjg@iastate.edu

Simple Summary: Since the 1990s, an important innovation in the management of agricultural pest insects has been the commercial cultivation of genetically engineered crops that produce insecticidal toxins, which in turn act to protect plants from feeding injury by insects. To date, these transgenic crops, which include cotton, maize and soybean, have produced insecticidal proteins derived from the bacterium *Bacillus thuringiensis* (Bt). Benefits associated with planting of Bt crops include reduced feeding injury from pest insects, decreased yield losses from pests and less harm to the environment. However, the evolution of Bt resistance by insect pests can diminish these benefits. One serious insect pest currently managed with Bt maize is the western corn rootworm. The larval stage of this insect feeds on maize roots and can substantially reduce yield. In some parts of the US Corn Belt, western corn rootworm rapidly adapted to Bt maize, and currently, some populations show resistance to all commercially available Bt traits. This review summarizes the time course of resistance development in the field, key factors contributing to resistance evolution, and steps that biotechnology companies, farmers and regulatory agencies can take to delay additional cases of pest resistance to current and future transgenic technologies.

Citation: Gassmann, A.J. Resistance to Bt Maize by Western Corn Rootworm: Effects of Pest Biology, the Pest–Crop Interaction and the Agricultural Landscape on Resistance. *Insects* 2021, 12, 136. https://doi.org/10.3390/insects12020136

Academic Editors: Katarina M. Mikac
Received: 28 December 2020
Accepted: 1 February 2021
Published: 5 February 2021

Publisher's Note: MDPI stays neutral with regard to jurisdictional claims in published maps and institutional affiliations.

Abstract: The western corn rootworm, *Diabrotica virgifera virgifera* LeConte, is among the most serious pests of maize in the United States. Since 2003, transgenic maize that produces insecticidal toxins from the bacterium *Bacillus thuringiensis* (Bt) has been used to manage western corn rootworm by killing rootworm larvae, which feed on maize roots. In 2009, the first cases of field-evolved resistance to Bt maize were documented. These cases occurred in Iowa and involved maize that produced Bt toxin Cry3Bb1. Since then, resistance has expanded to include other geographies and additional Bt toxins, with some rootworm populations displaying resistance to all commercially available Bt traits. Factors that contributed to field-evolved resistance likely included non-recessive inheritance of resistance, minimal fitness costs of resistance and limited adult dispersal. Additionally, because maize is the primary agricultural crop on which rootworm larvae can survive, continuous maize cultivation, in particular continuous cultivation of Bt maize, appears to be another key factor facilitating resistance evolution. More diversified management of rootworm larvae, including rotating fields out of maize production and using soil-applied insecticide with non-Bt maize, in addition to planting refuges of non-Bt maize, should help to delay the evolution of resistance to current and future transgenic traits.

Keywords: dispersal; field-evolved resistance; fitness costs; inheritance; integrated pest management; pyramid strategy; refuge strategy; resistance management

1. Introduction

Planting of transgenic maize that produces insecticidal toxins derived from the bacterium *Bacillus thuringiensis* (Bt) has played a prominent role in the management of western corn rootworm, *Diabrotica virgifera virgifera* LeConte, for nearly two decades. The release of the first transgenic events for management of corn rootworm followed several years of successful management of some other key insect pests of maize and cotton [1,2]. However, within six years of the initial release of Bt maize targeting western corn rootworm, the first cases of field-evolved Bt resistance were observed [3]. Since then, field-evolved resistance to all available Bt traits has been documented, and resistance to some Bt traits appears to be widespread within certain regions of the US Corn Belt [4–7]. The goals of this paper are to review the time course of field-evolved resistance to Bt maize by western corn rootworm, discuss the factors associated with the evolution of resistance, and consider how the management of western corn rootworm could be improved for current and future transgenic technologies.

The western corn rootworm is one of the most serious pests of maize in the United States [8]. Most yield losses associated with this pest are from larval feeding on maize roots [9–11]. In the United States, annual economic losses associated with corn rootworm, including both management costs and yield losses, range between $1 to $2 billion [12]. Western corn rootworm is a univoltine pest and its primary larval host is maize [13]. As such, fields that are planted to maize for several consecutive years provide the ideal habitat for this pest, and can be associated with large populations of western corn rootworm and high levels of larval feeding injury [9].

2. Bt Maize and Resistance Management

Transgenic Bt maize has been used to manage corn rootworm since 2003. The first Bt maize produced a single Bt trait, Cry3Bb1, and subsequently three addition Bt traits were registered by the US Environmental Protection Agency (EPA): Cry34Ab1/Cry35Ab1 (now called Gpp34Ab1/Tpp35Ab1 [14]) in 2005, mCry3A in 2006 and eCry3.1Ab in 2012 [15–18]. Similar to Cry3Bb1, both Cry34/35Ab1 and mCry3A were used initially as single Bt traits targeting corn rootworm, while eCry3.1Ab was released as a pyramid with mCry3A [19]. Additionally, pyramids of Cry3Bb1 with Cry34/35Ab1 and mCry3A with Cry34/35Ab1 were registered by the US EPA in 2009 and 2012, respectively [20]. Replacement of single Bt traits targeting corn rootworm by pyramids of two Bt traits was a gradual process, and for multiple growing seasons, both single traits and pyramids occurred together in the agricultural landscape [21,22]. Moreover, resistance to some of these Bt traits, in particular Cry3Bb1 and mCry3A, was already present in the agricultural landscape prior to the release of pyramided events [3,23].

In the US, an insect resistance management (IRM) strategy is mandated by the US EPA for the commercial cultivation of any Bt crop, including Bt maize that targets corn rootworm [19]. Currently, EPA-mandated IRM approaches for Bt crops focus on the refuge strategy. Under the refuge strategy, a non-Bt host is provided for an insect pest, with the goal of producing Bt-susceptible individuals that can mate with any Bt-resistant individuals surviving on a Bt crop [24]. The use of refuges to delay resistance can be especially effective when combined with either high-dose Bt events or Bt crops that are pyramided with multiple Bt toxins targeting the same pest [25–28].

A high-dose Bt crop is defined as one that either produces 25 times more Bt toxin than necessary to kill a Bt-susceptible pest or kills 99.99% of susceptible individuals [29]. When a high dose is achieved, Bt resistance is rendered a functionally recessive trait because the dose of toxin produced is sufficient to kill not only homozygous susceptible individuals, but also heterozygous resistant individuals [28]. Thus, in the high-dose/refuge scenario, susceptible individuals from a refuge mate with resistant individuals surviving on a Bt crop, producing heterozygous progeny that are unable to survive on a high-dose Bt crop, which in turn delays resistance.

By contrast, a pyramid delays resistance through redundant killing, with insects that harbor alleles for resistance to one Bt toxin in a pyramid killed by the second toxin and vice versa [30]. In the pyramid/refuge scenario, susceptible individuals from a refuge mate with resistant individuals from a Bt crop, thereby reducing the proportion of individuals that harbor resistance alleles for both Bt toxins in a pyramid, and consequently delaying the evolution of resistance. However, for a pyramid to work effectively, there must be an absence of cross-resistance between the toxins, and alleles for resistance to either toxin must be at a low frequency within the population [31,32]. If a pest population has evolved resistance to one toxin in a pyramid, the delay in resistance achieved by combining two toxins in a pyramid will be greatly diminished or lost altogether [30,32].

A critical factor affecting how quickly a pest will evolve Bt resistance is the inheritance of resistance, and this is especially important when Bt traits are deployed singly or if one of two traits in a pyramid is compromised by resistance [32,33]. When resistance to a Bt trait is inherited in a non-recessive manner, some proportion of the heterozygous resistant individuals will not be killed by that Bt trait, and resistance will evolve faster than when resistance is recessive [33]. Furthermore, the rate of resistance evolution is expected to show a positive relationship with the genetic dominance of resistance, and occur at a faster rate as the fitness of heterozygous resistant individuals on a Bt crop increases [27,33]. Thus, understanding the inheritance of resistance is essential for predicting the success of the refuge strategy to delay resistance [27].

Additionally, whether or not Bt resistance has accompanying fitness costs also will affect the rate of resistance evolution [34]. Fitness costs arise, in the absence of Bt, when individuals with resistance alleles have lower fitness than Bt-susceptible individuals [34]. Fitness costs of Bt resistance impose a counter-acting selective force that removes resistance alleles from refuge populations and delays the rate of pest adaptation to a Bt crop [35–37]. However, if fitness costs of resistance are minimal or absent, resistance is expected to evolve more rapidly than when fitness costs are present [21]. As such, quantifying the extent to which fitness costs accompany Bt resistance is another key factor in determining whether or not a population will evolve resistance, and how rapidly resistance will evolve and spread.

3. Time Course and Current Status of Field-Evolved Resistance

In 2009, farmers in Iowa observed high levels of feeding injury by western corn rootworm to maize producing Cry3Bb1 [3]. Subsequent plant-based bioassays found that this feeding injury was associated with Cry3Bb1 resistance by western corn rootworm [3]. Additional cases of Cry3Bb1 resistance were identified in Iowa in 2010 [38]. In 2011, field populations were sampled from several fields that had high levels of feeding injury from western corn rootworm to either Cry3Bb1 maize or mCry3A maize, and results of plant-based bioassays revealed resistance to both mCry3A and Cry3Bb1, and cross-resistance between these Bt toxins [23]. In 2012, field populations were sampled across the northern half of Iowa from fields, where high levels of feeding injury to Cry3Bb1 maize were observed, and plant-based bioassays demonstrated that cross-resistance between Cry3Bb1 and mCry3A also extended to eCry3.1Ab [39].

Fields that were sampled in these studies were typified by a node or more of feeding injury to roots of Bt maize, and in some cases, more than two nodes, with each node of roots lost translating to a 15% to 17% reduction in yield [10,11]. This means that Bt resistance by western corn rootworm has practical significance for farmers by substantially reducing yield. It is noteworthy that field populations evaluated from 2009 to 2012 were not resistant to Cry34/35Ab1 [3,23,38,39]. However, resistance to the Cry3 traits meant that the IRM advantage of pyramiding had been greatly reduced when Cry34/35Ab1 was placed in a pyramid with a Cry3 trait.

Resistance to Cry3 maize by western corn rootworm is not limited to Iowa and has been documented in Illinois, Minnesota, Nebraska and North Dakota [40–43]. All four states have areas of intensive maize production, with fields commonly planted to maize for several consecutive years [44]. Additionally, some of these populations were tested for cross-resistance to other Bt traits, and similar to research from Iowa, cross-resistance was found among Cry3Bb1, mCry3A and eCry3.1Ab [40,41]. Field-evolved resistance in some of these states appears to have occurred at a similar time as the observation of Bt resistance in Iowa, suggesting that Bt resistance evolved independently in several locations throughout the Corn Belt.

The presence of cross-resistance among Cry3Bb1, mCry3A and eCry3.1Ab may be due to the structural similarities among these three-domain Bt toxins, such that genetic changes conferring resistance to one of the Cry proteins likely confer resistance to the others [23,31,39]. By contrast, Cry34/35Ab1 is a binary toxin and differs structurally from three-domain toxins, and thus also likely has a mechanism of toxicity that is independent of the Cry3 toxins [31,39]. To date, little is known about the mechanistic basis of Bt resistance in western corn rootworm [14]. However, the potential for western corn rootworm to develop Cry34/35Ab1 resistance appears similar to that of Cry3Bb1. For example, laboratory selection experiments have shown that western corn rootworm developed Cry34/35Ab1 resistance after three to seven generations of selection [45,46], which is similar to past selection studies with Cry3Bb1 maize [47,48].

As suggested by these laboratory selection experiments, following field-evolved resistance to Cry3 maize, studies provided evidence that resistance to Cry34/35Ab1 had emerged in field populations of western corn rootworm. In 2013, fields in Iowa were sampled where high levels of feeding injury from western corn rootworm were observed for Cry34/35Ab1 maize (>2 nodes of injury on average) and for maize pyramided with Cry34/35Ab1 and a Cry3 toxin (>1 node of injury on average). Plant-based bioassays with progeny of western corn rootworm from these fields showed elevated survival on Cry34/35Ab1 maize compared to Bt-susceptible controls, which indicated that the high levels of feeding injury observed in the field were associated with western corn rootworm resistance to Cry34/35Ab1 [49]. However, for these populations, larval survival was lower on Cry34/35Ab1 maize than on non-Bt maize, indicating that resistance was incomplete [49]. Similarly, a field population of western corn rootworm from Minnesota, which was sampled in 2013, was found to have incomplete resistance to Cry34/35Ab1 [50].

More recently, Gassmann et al. [51] examined field populations sampled in Iowa during 2017, which were collected from two fields with a high level of western corn rootworm feeding injury to maize pyramided with Cry3 and Cry34/35Ab1. Both populations were found to have resistance to Cry34/35Ab1 in addition to resistance to all three Cry3 toxins. For one population, no difference in larval survival or development was detected between non-Bt maize and maize with Cry34/35Ab1, either alone or in a pyramid with Cry3Bb1, suggesting complete resistance to Cry34/35Ab1. In addition to these field populations associated with injury to Bt maize, Gassmann et al. [51] also included three field populations that were not associated with injury to Bt maize, and one of these populations also displayed resistance to Cry34/35Ab1. Taken together, these data suggest that resistance to Cry34/35Ab1 has persisted in the agricultural landscape, and appears to be increasing in magnitude. A key factor affecting the future utility of Bt maize for management of western corn rootworm will be how quickly additional cases of resistance to Cry34/35Ab1 evolve.

4. Factors Affecting Resistance Evolution

Several factors likely contributed to the rapid development of field-evolved resistance to Bt maize by western corn rootworm. Limited movement of adult rootworm prior to mating and after mating likely reduced the effectiveness of refuges to delay resistance and enabled resistance to build within populations. The lack of a high dose for Bt toxins that target corn rootworm, and the resulting non-recessive inheritance of resistance, coupled with standing genetic variation for resistance within populations, facilitated rapid resistance evolution when rootworm populations were exposed to Bt maize. Additionally, the presence of minimal fitness costs of resistance also favored rapid resistance evolution.

It appears that adult western corn rootworm engages in limited dispersal within the agricultural landscape, and this likely contributed to field-evolved resistance in multiple ways. Available data suggest that the majority of adults only move about 40 m per day [52,53]. Furthermore, newly emerged, teneral females will often mate near the plant where they emerge [53,54]. As a result, when refuges are spatially structured, with blocks of Bt maize and non-Bt maize, there will be limited mating between Bt-selected individuals and refuge individuals, which in turn will reduce the ability of refuges to delay resistance. When Bt maize was initially released in 2003, only structured refuges were used, with integrated, or blended refuges first planted in 2011. Additionally, available data suggest limited compliance by farmers in the planting of structured refuges, an effect that is expected to further increase the rate of resistance evolution [33,55].

Another important consequence of limited adult dispersal is the role of local, within-field selection in driving resistance evolution. Studies of other species of pest insects indicate that limited dispersal can increase the rate of resistance evolution [56]. The western corn rootworm is a univoltine pest, with females mostly laying eggs in maize fields, and eggs then diapausing through the winter and hatching the following spring [9]. Because the primary host for western corn rootworm larvae is maize, continuous maize cultivation is necessary for populations to persist within a field. Limited adult dispersal means that most adult females emerging from a maize field will also oviposit in the same field. Consequently, continuous planting of maize containing the same rootworm trait leads to continuous selection for resistance. The first cases of Cry3Bb1 resistance were associated with continuous cultivation of Cry3Bb1 maize, and there was a positive correlation between the years that a field was planted to Cry3Bb1 maize and the level of Cry3Bb1 resistance [3]. As such, continuous maize cultivation and continuous use of the same Bt trait within a field appears to be an important factor affecting the rate of resistance evolution. This finding is concordant with laboratory selection experiments, which found that resistance to Bt maize can develop rapidly (e.g., within three generations) under continuous laboratory selection in the absence of refuges [45–48,57,58]. Importantly, the rapid evolution of resistance to Bt maize by western corn rootworm under laboratory selection was found for all currently available Bt traits (i.e., Cry3Bb1, eCry3.1Ab, mCry3A and Cry34/35Ab1).

One factor that likely contributed to the rapid evolution of resistance in these laboratory selection experiments, and in the field, is the initial frequency for resistance traits within populations. Onstad and Meinke [59] conducted a retrospective analysis of laboratory selection experiments described in Meihls et al. [47] and Lefko et al. [46] and concluded that the initial resistance allele frequency was in the range of 0.05 to 0.20. This is a higher frequency than was found for Bt-resistance alleles in several lepidopteran pests [33], or typically used in simulation models of pest resistance [60,61]. As a result of this higher resistance allele frequency, the rate of evolution is expected to be faster than would occur at lower resistance allele frequencies [28].

A second factor facilitating the development of resistance, in both the laboratory and field, is the inheritance of resistance. Because none of the Bt traits available for management of western corn rootworm produce a high dose of toxin, theory predicts that the inheritance of resistance traits will be non-recessive [21,28,62]. Studies on the inheritance of Bt resistance by western corn rootworm have used both laboratory-selected strains and strains with field-evolved Bt resistance (Table 1). Studies of strains with laboratory-selected Cry3Bb1 resistance found evidence of non-recessive inheritance [47,63]. Similarly, resistance to mCry3A and eCry3.1Ab in laboratory-selected strains was found to be non-recessive, and in some cases dominant [58,64].

Research on strains with field-evolved resistance to Cry3Bb1 maize also found evidence of non-recessive inheritance resistance [65–67] (Table 1). In three of four strains, where field-evolved resistance had been introgressed into a non-diapausing background, resistance to Cry3Bb1 maize was found to be non-recessive [65,66]. Additionally, in a study using diapausing western corn rootworm strains with field-evolved resistance to Cry3Bb1 maize, the general pattern was for resistance to be non-recessive, with three of four strains displaying non-recessive inheritance in plant-based bioassays [67].

Table 1. Studies testing the inheritance of resistance to *Bacillus thuringiensis* (Bt) maize by western corn rootworm.

Type of Resistance [1]	Strain	Resistant to Toxin [2]	Type of Assay [3]	Metric Used	Inheritance of Resistance	Heritability [4]	Reference
Laboratory Selected	Constant Exposure	CryBb1	Single Plant	Larval Survival	Non-Recessive	0.29	[47]
Laboratory Selected	Constant Exposure	Cry3Bb1	Single Plant	Survival to Adult	Non-Recessive	0.30	[47]
Laboratory Selected	Brookings Moderately Selected	Cry3Bb1	Seedling Mat	Larval Survival	Non-Recessive to Dominant	0.19 to 1.22	[63]
Laboratory Selected	Brookings Moderately Selected	Cry3Bb1	Seedling Mat	Larval Growth	Non-Recessive	0.51	[63]
Laboratory Selected	mCry3A selected	mCry3A	Single Plant	Larval Survival	Non-Recessive	0.66	[58]
Laboratory Selected	mCry3A selected	mCry3A	Single Plant	Survival to Adult	Dominant	1.03	[58]
Laboratory Selected	eCry3.1Ab selected	eCry3.1Ab	Seedling Mat	Larval Survival	Dominant	0.94 to 1.38	[64]
Field Evolved	Hopkinton	Cry3Bb1	Seedling Mat	Survival to Adult	Non-Recessive	0.37	[65]
Field Evolved	Cresco	Cry3Bb1	Seedling Mat	Survival to Adult	Recessive	0.27	[65]
Field Evolved	Elma	Cry3Bb1	Seedling Mat	Survival to Adult	Non-Recessive	0.14 to 0.29	[66]
Field Evolved	Monona	Cry3Bb1	Seedling Mat	Survival to Adult	Non-Recessive	0.45	[66]
Field Evolved	Monona	Cry3Bb1	Single Plant	Larval Survival	Non-Recessive	0.73	[66]
Field Evolved	Monona	Cry3Bb1	Diet Based	Larval Survival	Non-Recessive	—— [5]	[66]
Field Evolved	Central Iowa	Cry3Bb1	Single Plant	Larval Survival	Non-Recessive	0.23	[67]
Field Evolved	Eastern Iowa	Cry3Bb1	Single Plant	Larval Survival	Non-Recessive	0.50	[67]
Field Evolved	Northern Iowa	Cry3Bb1	Single Plant	Larval Survival	Non-Recessive	0.54	[67]
Field Evolved	Western Iowa	Cry3Bb1	Single Plant	Larval Survival	Recessive	0.08	[67]

[1] Describes whether a strain was generated by selecting a susceptible stain on Bt maize in the laboratory (Laboratory Selected) or by collecting Bt-resistant insects from the field (Field Evolved). [2] Type of Bt maize on which a rootworm strain was selected and to which it was resistant. [3] Describes the bioassay approach that was used to measure resistance. Details are provided within the references, but in general, these approaches involved measure survival on single plants in containers (single-plant assay), on a mat of maize roots generated by germinating a small number of maize seeds in a container (seedling-mat assay) or in an assay where Bt toxin was placed on top of an artificial diet (diet-based assay). [4] Heritability is a metric that describes the extent to which heterozygous individuals resemble the parental strains (i.e., Bt resistant and Bt susceptible) for survival on Bt maize. Specifics on each calculation are given within individual references, but in general, a score of 0 indicates equal survival on Bt maize between the parental Bt-susceptible strain and heterozygotes, 1 indicates equal survival on Bt maize between the parental Bt-resistant strain and heterozygotes, and 0.5 indicates that survival of the heterozygotes on Bt maize is halfway between that observed for Bt-resistant and Bt-susceptible parental strains. Scores greater than 1 occur when the heterozygotes have higher survival on Bt maize than their parental Bt-resistant strain. [5] Heritability was not calculated because an LC_{50} for the resistant strain could not be determined.

In the presence of non-Bt refuges, fitness costs of Bt resistance can act to delay the evolution of resistance [34]. Several studies have tested for fitness costs of Bt resistance in strains of western corn rootworm with laboratory-selected resistance and in strains with field-evolved resistance (Table 2). Strains with laboratory-selected resistance to Cry3Bb1 maize have displayed fitness costs in some cases [68,69] but not in others [63,70,71]. Additionally, fitness costs appeared to be absent in strains with laboratory-selected resistance to eCry3.1Ab and mCry3A [58,72].

Table 2. Studies testing for fitness costs of resistance to Bt maize by western corn rootworm.

Type of Resistance [1]	Strain	Resistant to Toxin [2]	Cost Present? [3]	Traits Affected [4]	Reference
Laboratory Selected	Brookings Moderately Selected	Cry3Bb1	No	—	[63]
Laboratory Selected	Brookings Moderately Selected (Strain 1)	Cry3Bb1	No	—	[71]
Laboratory Selected	Brookings Moderately Selected (Strain 2)	Cry3Bb1	No	—	[71]
Laboratory Selected	Brookings Moderately Selected (Strain 3)	Cry3Bb1	No	—	[71]
Laboratory Selected	Brookings Intensely Selected (Strain 1)	Cry3Bb1	No	—	[71]
Laboratory Selected	Brookings Intensely Selected (Strain 2)	Cry3Bb1	No	—	[71]
Laboratory Selected	Data Presented as Composite of Three Resistant Strains	CryBb1	Yes	Fecundity; Adult (male) Longevity	[68]
Laboratory Selected	Brookings Moderately Selected	CryBb1	No	—	[70]
Laboratory Selected	Brookings Moderately Selected	CryBb1	Yes	Larval Development; Egg Viability	[69]
Laboratory Selected	mCry3A Selected	mCry3A	No	—	[58]
Laboratory Selected	eCry3.1Ab Selected	eCry3.1Ab	No	—	[72]
Field Evolved	Hopkinton	Cry3Bb1	No	—	[65]
Field Evolved	Cresco	Cry3Bb1	Yes	Larval Development; Survival to Adulthood; Fecundity	[65]
Field Evolved	Elma	Cry3Bb1	Yes	Larval Development	[66]
Field Evolved	Monona	Cry3Bb1	No	—	[66]
Field Evolved	Cresco	Cry3Bb1	Yes	Decline in Resistance over Time	[73]
Field Evolved	Hopkinton	Cry3Bb1	Yes	Decline in Resistance over Time	[73]
Field Evolved	Data Presented as Composite of Eight Resistant Strains	Cry3B1	Yes	Adult Size	[67]

[1] Describes whether a strain was generated by selecting a susceptible stain on Bt maize in the laboratory (Laboratory Selected) or was generated from Bt-resistant insects collected from the field (Field Evolved). [2] Type of Bt maize on which the rootworm strain was selected and to which it was resistant. [3] States whether fitness costs of Bt resistant were detected for a specific strain in a study. [4] Life-history traits for which a fitness cost was detected or cases where resistance declined over time when a stain was not exposed to Bt maize.

Patterns of fitness costs associated with field-evolved Cry3Bb1 resistance were similar to results for strains with laboratory-selected resistance, with costs present in some cases but not in others. However, costs appear to be more common in strains with field-evolved resistance compared to laboratory-selected resistance (Table 2). Ingber and Gassmann [65] identified costs affecting larval development, survival to adulthood and fecundity in one strain (Cresco) but costs were absent in another strain (Hopkinton). In a study of two additional strains, Paolino and Gassmann [66] found a fitness cost in one strain (Elma) but not in a second strain (Monona). Shrestha and Gassmann [67] studied several field populations and detected a negative relationship between adult size and the level of Cry3Bb1 resistance, indicating a fitness cost of resistance affecting adult size. St. Clair et al. [73] assessed fitness costs of Cry3Bb1 resistance in Hopkinton and Cresco through a selection experiment, which tested for a loss of resistance over time in the absence of exposure to Cry3Bb1 maize, and this study found evidence of fitness costs in both strains. The contrasting results between Ingber and Gassmann [65], which measured individual life-history characteristics, and St. Clair et al. [73], which used a selection experiment, likely arose because selection experiments provide a more comprehensive metric for assessing fitness costs and are therefore more sensitive [34].

However, it is important to note that St. Clair et al. [73] also found that Cry3Bb1 resistance persisted for at least six generations in the absence of exposure to Cry3Bb1 maize, which translates to 6 years in the field, because western corn rootworm has one generation per year. Consequently, to the extent that fitness costs do accompany Cry3Bb1 resistance, it is likely that Bt resistance currently present in the agricultural landscape may remain for several years, even if farmers were to discontinue planting of Bt maize [4–7].

Taken together, the available data suggest that minimal fitness costs may often be associated with Bt resistance in western corn rootworm, and consequently, fitness costs may do little to delay the evolution of Bt resistance. In general, fitness costs tend to increase with the magnitude of resistance, with strains that have higher resistance ratios incurring more fitness costs than strains with lower resistance ratios [34]. Data from strains with field-evolved resistance to Cry3Bb1 maize indicate that resistance ratios tend to range from 2.5 to 19, which is substantially lower than resistance ratios found for pests targeted by high-dose Bt crops [65–67]. For example, in cases where a high dose was achieved, such as with Bt cotton that targets pink bollworm *Pectinophora gossypiella* and Bt maize that targets European corn borer *Ostrinia nubilalis*, resistance ratios were greater than 500 [74,75]. As such, for western corn rootworm, fitness costs of resistance to Bt maize may often be less than costs associated with Bt crops that are high dose and require pests to have a much higher resistance ratio to survive [23]. Consequently, fitness costs, and the corresponding delay in resistance evolution, are expected to be less for western corn rootworm than for other pests targeted by Bt crops where a high dose is achieved. Furthermore, fitness costs of Bt resistance in western corn rootworm appear insufficient to delay resistance development in the field, at least with the current refuge requirements.

Data are currently lacking on the inheritance and fitness costs of resistance to Cry34/35Ab1, in either laboratory-selected strains or in strains with field-evolved resistance. With the emergence of field-evolved resistance to Cry34/35Ab1 [49–51], such data would enable valuable insights into how quickly resistance will develop in the broader agricultural landscape. In the case of resistance to Cry3Bb1 maize, which in turn confers cross-resistance to mCry3A and eCry3.1Ab, it appears that several factors contributed to the evolution of resistance. In particular, continuous planting of maize containing the same Bt trait, coupled with limited adult dispersal, likely favored the evolution of resistance. Additionally, it appears that substantial standing genetic variation for resistance, non-recessive inheritance of resistance, and minimal fitness costs of resistance also contributed to resistance development in the field.

5. Resistance to Bt maize in the Agricultural Landscape

Initial characterization of Bt resistance focused on fields with high levels of rootworm feeding injury to Cry3 maize (i.e., Cry3Bb1 maize and mCry3A maize) [3,23,38–43]. However, this raised the question of how common Cry3 resistance was within the broader agricultural landscape, and how cropping practices might in turn influence patterns of pest abundance and pest injury.

Landscape-level patterns of Cry3 resistance were examined for western corn rootworm populations in Nebraska by Reinders et al. [5]. This study considered populations at a spatial scale of 2 to 10 Km, and examined two areas of intensive maize production. The authors found significant spatial variation among populations in the level of resistance to Cry3Bb1 maize and mCry3A maize, with a few populations showing no difference in survival from the susceptible controls. This work also looked at the association of various field-history variables with the level of Bt resistance, and found that the use of a Bt trait within a field showed a positive relationship with the level of resistance, which in turn highlights the importance of diversified management in delaying resistance [5].

Studies of Cry3 resistance in the agricultural landscape in Iowa include work by St. Clair et al. [6,7]. St. Clair et al. [6] tested for resistance in fields with a history of high levels of feeding injury to Cry3 maize (i.e., past problem fields) and fields that were in close proximity (within < 2.2 km) to past problem fields. This study found that both field types harbored Cry3Bb1-resistant populations of western corn rootworm [6]. Similarly, St. Clair et al. [7] compared counties in Iowa with and without a known history of past Cry3 problem fields. Bioassay data revealed the presence of Cry3Bb1 resistance in both types of counties and found similar levels of resistance. In both studies, there was some variability in the level of resistance, with some populations displaying complete resistance while others had incomplete resistance (i.e., survival or larval development was lower on Cry3Bb1 maize compared to non-Bt maize). These studies suggested widespread resistance to Cry3Bb1 maize in Iowa, although there was variation among populations in the level of resistance.

In a study examining landscape-level patterns of resistance and the effects of cropping practices on resistance, Shrestha et al. [4] measured Cry3Bb1 resistance in western corn rootworm from fields in Iowa with a variety of management histories, including (1) fields in continuous maize production, (2) rotated fields, (3) past Cry3 problem fields, and (4) current Cry3 problem fields. Data from plant-based and diet-based bioassays illustrated that all field types harbored Cry3Bb1-resistant western corn rootworm [4]. However, larval development on Cry3Bb1 maize was significantly reduced compared to non-Bt maize for rotated fields and past problem fields, but not for continuous maize fields or current problem fields, suggesting that crop rotation may help delay the development of Cry3Bb1 resistance [4].

St. Clair and Gassmann [76] analyzed landscape-level patterns of maize cultivation in past problem fields and in areas surrounding past problem fields in Iowa during the timeframe when these fields failures occurred. These patterns were compared with randomly selected agricultural fields in Iowa during the same time period [76]. This study found that, not only were past problem fields characterized by higher levels of continuous maize cultivation compared to randomly selected fields, but also that the local landscape around these past problem fields had more continuous maize than randomly selected points [76]. The local landscapes around past problem fields with elevated percentages of continuous maize cultivation included an area within 3.2 km of past problem fields. Furthermore, available data indicate that 57% of fields in continuous maize production during the timeframe examined contained Cry3Bb1 maize [22]. These studies point to the role of the local landscape in facilitating the evolution of resistance, and suggest that continuous Bt maize cultivation in the broader agricultural landscape contributed to Bt resistance and high levels of feeding injury in past problem fields.

With the emergence of field-evolved Cry34/35Ab1 resistance, a key question now becomes how widespread Cry34/35Ab1 resistance is within the agricultural landscape [49–51]. For Cry3Bb1 resistance, the initial occurrence of Cry3Bb1 resistance in 2009 was followed rapidly by widespread resistance within the agricultural landscape by 2015 [3,4]. As such, characterizing the distribution of Cry34/35Ab1 in the agricultural landscape and taking steps to delay the evolution of Cry34/35Ab1 resistance are of critical importance.

6. Approaches for Improving Resistance Management and for Managing Resistant Populations

The presence of Bt resistant populations within the agricultural landscape raises questions about how best to manage these populations, and how to delay additional cases of Bt resistance. When resistance to Cry3 maize developed, farmers responded by planting maize that contained a pyramid of Cry3 and Cry34/35Ab1 [22]. This approach was effective at mitigating Cry3 resistance, with western corn rootworm population size and root injury scores in these past problem fields returning to levels that were similar to other maize fields in the agricultural landscape [22]. However, resistance to Cry3Bb1 continued to persist in these fields [4]. With the more recent development of resistance to Cry34/35Ab1, there are now western corn rootworm populations that possess resistance to all commercially available Bt traits, and consequently, the challenge of managing western corn rootworm has become more difficult [51].

One approach used by farmers to mitigate the effects of Bt resistance has been to combine Bt maize with soil-applied insecticide [6,7,22]. Available data indicate that this approach has short comings both in terms of integrated pest management and insect resistant management. Specifically, if a rootworm population is not resistant to a Bt trait, the reduction in root injury achieved by applying soil insecticide to Bt maize is minimal, and the yield preserved does not appear to justify the cost of the insecticide application [77]. Furthermore, the reduction in survival achieved by adding insecticide does not appear sufficient to provide an effective pyramid with a Bt trait, and therefore, is not expected to delay the evolution of Bt resistance [77,78].

Additionally, lessons learned from studies where Cry3Bb1 maize and soil insecticides were combined to manage Cry3-resesitant populations cast light on the general short comings of combining Bt maize and soil insecticide to manage Bt-resistant populations. Shrestha et al. [78] studied Cry3Bb1-resistant populations and tested how the use of Cry3Bb1 maize with soil-applied insecticide affected root injury and survival of western corn rootworm. Applying soil-applied insecticide to Cry3Bb1 maize did not significantly reduce adult emergence compared to the use of Cry3Bb1 maize alone, suggesting that the number of Bt-resistant individuals produced within a field would not be reduced by adding insecticide. Furthermore, the reduction in root injury achieved by combining Cry3Bb1 maize with soil-applied insecticide did not differ from non-Bt maize with insecticide, indicating that farmers did not achieve a benefit in terms of root protection by adding insecticide to Bt maize compared to using non-Bt maize with soil insecticide. These data illustrate that, once a population develops Bt resistance, using soil-applied insecticide with a Bt trait that has been compromised by resistance is not a worthwhile strategy because it will continue to select for resistance, while providing little addition benefit to farmers in terms of preserving yield.

It is important to note that, when resistance to Cry3Bb1 maize arose, farmers with past problem fields responded by continuing to grow maize, but began using a pyramid of Cry34/35Ab1 with Cry3Bb1 [22]. Because these fields harbored Cry3Bb1-resistant western corn rootworm, the ability of this Bt pyramid to delay resistance was compromised [3,4,23,38,39]. Furthermore, the use of Bt pyramids to manage Cry3-resistant populations likely hastened the evolution of Cry34/35Ab1 resistance [49,51]. The application of a more integrated approach to management of corn rootworm in these fields, including the use of crop rotation and non-Bt maize with soil-applied insecticide, could have helped to delay the development of Cry34/35Ab1 resistance [3,62,79,80].

The use of soil-applied insecticide with non-Bt maize reduces selection for Bt resistance, reduces root injury, and permits the survival of corn rootworm to adulthood [77,78,81]. Specifically, the use of non-Bt maize with soil-applied insecticide produces a temporal refuge (i.e., a year in which selection for Bt resistance is absent) thereby enabling the survival of Bt-susceptible individuals in addition to relaxing selection for Bt resistance. In situations where maize is grown for several consecutive years, rotating Bt maize with non-Bt maize that has soil-applied insecticide should both preserve yield and delay the evolution of Bt resistance. This concept is supported by the results of a computer simulation study conducted by Martinez and Caprio [82], which found that use of non-Bt maize with soil-applied insecticide delayed Bt resistance by western corn rootworm. However, it is important to note that the resistance-management benefit of refuges is in delaying resistance [24,28]. Once resistance evolves and is prevalent within a population, refuges of any type (e.g., structured, integrated, naturally occurring or temporal) will be of minimal value in managing resistance.

Rotating fields out of maize production (i.e., crop rotation) may also aid in reducing Bt resistance and the high levels of feeding injury that can be associated with

Bt resistance [4,76,83]. Because maize is the primary larval host for western corn rootworm, crop rotation will eliminate a western corn rootworm population from a maize field, which should increase the ability of farmers to maintain rootworm population below the economic injury level [9]. Work by St. Clair and Gassmann [76] illustrated that high levels of rootworm feeding injury to Bt maize by western corn rootworm, and Bt resistance, were associated with continuous maize cultivation in the local landscape. This study also points to the potential of crop rotation to reduce the occurrence of high levels of feeding injury to Bt maize by Bt-resistant western corn rootworm [76]. Carrière et al. [83] found that increased crop rotation was associated with a reduction in high levels of feeding injury to Cry3Bb1 maize by western corn rootworm. In addition to reducing high levels of feeding injury by Bt-resistant rootworm, Shrestha and Gassmann [4] revealed that crop rotation also can delay the evolution of Bt resistance within a field, an effect that likely arises because of the recolonization of a field by rootworm in neighboring fields. However, the benefit crop rotation in delaying resistance will be contingent on the level of Bt resistance in rootworm populations from neighboring fields, because these populations will serve as a source of recolonizing individuals after a field is rotated back to maize. Work by Reinders et al. [5] highlights the potential for fields to display a high level of Bt resistance following crop rotation if resistance is prevalent in the surrounding landscape. As such, the use of integrated pest management within the broader agricultural landscape is likely to be an important factor influencing the level of Bt resistance in a field following crop rotation.

The success of future transgenic traits may be affected, in part, by resistance to current Bt traits, particularly in cases where current traits will be pyramided with future traits. The use of RNA interference (RNAi), induced by double-stranded RNA (dsRNA), represents a novel approach for managing rootworm, and DvSnf7 will likely be the first RNAi trait used to manage rootworm [84,85]. However, this RNAi trait will be pyramided with current Bt traits, specifically Cry3Bb1 and Cry34/35Ab1 [86]. Consequently, to the extent that resistance to Cry3Bb1 and Cry34/35Ab1 is present in the landscape, the IRM benefit of pyramiding will be compromised [30]. Widespread resistance to Cry3Bb1 in some regions of the Corn Belt, coupled with emerging resistance to Cry34/35Ab1, may enable the rapid evolution of resistance to RNAi. In a laboratory selection experiment, western corn rootworm was evaluated for resistance to RNAi after seven generations of selection and found to be resistant to multiple insecticidal dsRNA molecules [87]. An alternative approach, where an RNAi trait is coupled with a novel insecticidal protein derived *Pseudomonas chlororaphis*, may provide a more effective IRM approach [88,89]. However, because of the limited dispersal displayed by adult western corn rootworm, diversified management within a field will still be essential to delay resistance to this novel pyramid. Additionally, pyramiding of traits to delay resistance is dependent on the presence of refuges [30]. Because of the limited adult dispersal displayed by adult western corn rootworm, the use of integrated refuges (i.e., blended refuges) with these novel transgenic technologies, should improve mating between resistant individuals and refuge insects, thereby delaying resistance [53,54].

Past work has shown that refuges can delay Bt resistance in western corn rootworm [45]. However, widespread field-evolved resistance by this pest also illustrates that refuges alone are not sufficient to delay resistance [4], and fields in continuous maize cultivation with continuous use of the same Bt traits can serve as foci for resistance within the agricultural landscape [3,76]. The use of more integrated pest management, including crop rotation and the use of non-Bt maize with soil-applied insecticide, will be important for delaying additional cases of resistance to current and future transgenic technologies by this serious agricultural pest [90].

7. Conclusions

The evolution of resistance to Bt maize by western corn rootworm illustrates the potential for insect pests to develop resistance to Bt crops when a high dose is not present. Key factors that facilitated field-evolved resistance by the western corn rootworm included non-recessive inheritance of resistance and minimal fitness costs of resistance (Tables 1 and 2). Furthermore, the use of Bt events singly before pyramiding and cross-resistance among Bt traits, likely hastened resistance development [31,32]. The use of novel, pyramided transgenic traits, for which resistance allele frequency is low, should provide a more durable approach for managing this pest [25,30,32].

Another important factor facilitating resistance evolution is the limited dispersal displayed by western corn rootworm adults [52,53]. This factor is intrinsic to the biology of the pest and cannot be manipulated. However, resistance management approaches can be refined to take into account this important aspect of pest biology. Specifically, integrated refuges should be used to increase mating between Bt-selected individuals and those emerging from refuge plants [91]. Additionally, continuous cultivation of Bt maize, coupled with limited adult dispersal, appears to be an important driver of resistance development within the agricultural landscape [3,5,76]. Consequently, more diversified management, including crop rotation and use of non-Bt maize with soil-applied insecticide, should help to delay the development of resistance to current and future transgenic traits for management of western corn rootworm [82,90].

Funding: This work was supported by the National Institute of Food and Agricultural, Hatch Project IOW05617.

Acknowledgments: I thank Abigail Kropf, John McCulloch, Devin Radosevich and Eliott Smith for helpful comments on this manuscript.

Conflicts of Interest: A.J.G. has received research funding, not related to this work, from AMVAC, Bayer, Dow AgroSciences, DuPont, FMC, Monsanto, Syngenta, and Valent.

References

1. Siegfried, B.D.; Hellmich, R.L. Understanding successful resistance management: The European corn borer and Bt corn in the United States. *GM Crops Food* **2012**, *3*, 184–193. [CrossRef] [PubMed]
2. Tabashnik, B.E.; Patin, A.L.; Dennehy, T.J.; Liu, Y.-B.; Carrière, Y.; Sims, M.A.; Antilla, L. Frequency of resistance to *Bacillus thuringiensis* in field populations of pink bollworm. *Proc. Natl. Acad. Sci. USA* **2000**, *97*, 12980–12984. [CrossRef]
3. Gassmann, A.J.; Petzold-Maxwell, J.L.; Keweshan, R.S.; Dunbar, M.W. Field-evolved resistance to Bt maize by western corn rootworm. *PLoS ONE* **2011**, *6*, e22629. [CrossRef]
4. Shrestha, R.B.; Dunbar, M.W.; French, B.W.; Gassmann, A.J. Effects of field history on resistance to Bt maize by western corn rootworm, *Diabrotica virgifera virgifera* LeConte (Coleoptera: Chrysomelidae). *PLoS ONE* **2018**, *13*, e0200156. [CrossRef] [PubMed]
5. Reinders, J.D.; Hitt, B.D.; Stroup, W.W.; French, B.W.; Meinke, L.J. Spatial variation in western corn rootworm (Coleoptera: Chrysomelidae) susceptibility to Cry3 toxins in Nebraska. *PLoS ONE* **2018**, *3*, e0208266. [CrossRef]
6. St. Clair, C.R.; Head, G.P.; Gassmann, A.J. Western corn rootworm abundance, injury to corn, and resistance to Cry3Bb1 in the local landscape of previous problem fields. *PLoS ONE* **2020**, *15*, e0237094. [CrossRef] [PubMed]
7. St. Clair, C.R.; Head, G.P.; Gassmann, A.J. Comparing populations of western corn rootworm (Coleoptera: Chrysomelidae) in regions with and without a history of injury to Cry3 corn. *J. Econ. Entomol.* **2020**, *113*, 1839–1849. [CrossRef]
8. Gray, M.E.; Sappington, T.W.; Miller, N.J.; Moeser, J.; Bohn, M.O. Adaptation and invasiveness of western corn rootworm: Intensifying research on a worsening pest. *Annu. Rev. Entomol.* **2009**, *54*, 303–321. [CrossRef] [PubMed]
9. Meinke, L.J.; Sappington, T.W.; Onstad, D.W.; Guillemaud, T.; Miller, N.J.; Komáromi, J.; Levay, N.; Furlan, L.; Kiss, J.; Toth, F. Western corn rootworm (*Diabrotica virgifera virgifera* LeConte) population dynamics. *Agric. For. Entomol.* **2009**, *11*, 29–46. [CrossRef]
10. Tinsley, N.A.; Estes, R.E.; Gray, M.E. Validation of a nested error component model to estimate damage caused by corn rootworm larvae. *J. Appl. Entomol.* **2013**, *137*, 161–169. [CrossRef]
11. Dun, Z.; Mitchell, P.D.; Agosti, M. Estimating *Diabrotica virgifera virgifera* damage functions with field trial data: Applying an unbalanced nested error component model. *J. Appl. Entomol.* **2010**, *134*, 409–419. [CrossRef]
12. Wechsler, S.; Smith, D. Has resistance taken root in U.S. Corn Fields? Demand for insect control. *Am. J. Agric. Econ.* **2018**, *100*, 1136–1150. [CrossRef]
13. Moeser, J.; Hibbard, B.E. A synopsis of the nutritional ecology of larvae and adults of *Diabrotica virgifera virgifera* (LeConte) in the New and Old World—Nouvelle cuisine for the invasive maize pest *Diabrotica virgifera virgifera* in Europe? In *Western Corn Rootworm: Ecology and Management*; Vidal, S., Kuhlmann, U., Edwards, C.R., Eds.; CABI Publishing: Wallingford, CT, USA, 2005.
14. Jurat-Fuentes, J.L.; Heckel, D.G.; Ferré, J. Mechanisms of resistance to insecticidal proteins from *Bacillus thuringiensis*. *Annu. Rev. Entomol.* **2021**, *66*, 121–140. [CrossRef] [PubMed]
15. EPA [Environmental Protection Agency]. Biopesticides Registration Action Document: Bacillus thuringiensis Cry3Bb1 Protein and the Genetic Material Necessary for Its Production (Vector PV-ZMIR13L) in MON 863 Corn (OECD Unique Identifier: MON-

ØØ863-5). 2010. Available online: http://www3.epa.gov/pesticides/chem_search/reg_actions/pip/cry3bb1-brad.pdf (accessed on 30 November 2020).

16. EPA [Environmental Protection Agency]. Biopesticides Registration Action Document: Bacillus thuringiensis Cry34Ab1 and Cry35Ab1 Proteins and the Genetic Material Necessary for Their Production (PHP17662 T-DNA) in Event DAS-59122-7 Corn (OECD Unique Identifier: DAS-59122-7). 2005. Available online: http://www3.epa.gov/pesticides/chem_search/reg_actions/pip/cry3435ab1-brad.pdf (accessed on 30 November 2020).

17. EPA [Environmental Protection Agency]. Biopesticides Registration Action Document: Modified Cry3A Protein and the Genetic Material Necessary for Its Production (Via Elements of pZM26) in Event MIR604 Corn SYN-IR604-8. 2006. Available online: https://www3.epa.gov/pesticides/chem_search/reg_actions/pip/mcry3a-brad.pdf (accessed on 30 November 2020).

18. EPA [Environmental Protection Agency]. Draft Biopesticides Registration Action Document: Bacillus thuringiensis eCry3.1Ab Protein and the Genetic Material Necessary for Its Production (via elements of vector PSYN12274) in 5307 Corn (SYN-05307-1). 2012. Available online: https://www.regulations.gov/document?D=EPA-HQ-OPP-2012-0108-0010 (accessed on 30 November 2020).

19. EPA [Environmental Protection Agency]. Current and Previously Registered Section 3 Plant-Incorporated Protectant (PIP) Registrations. 2020. Available online: http://www.epa.gov/ingredients-used-pesticide-products/current-previously-registered-section-3-plant-incorporated (accessed on 20 November 2020).

20. EPA [Environmental Protection Agency]. Current & Previously Registered Section 3 Plant-Incorporated Protectant (PIP) Registrations. 2015. Available online: https://archive.epa.gov/pesticides/reregistration/web/html/current-previously-registered-section-3-plant-incorporated.html (accessed on 30 November 2020).

21. Gassmann, A.J. Resistance to Bt maize by western corn rootworm: Insights from the laboratory and the field. *Curr. Opin. Insect Sci.* **2016**, *15*, 111–115. [CrossRef]

22. Dunbar, M.W.; O'Neal, M.E.; Gassmann, A.J. Effects of field history on corn root injury and adult abundance of northern and western corn rootworm (Coleoptera: Chrysomelidae). *J. Econ. Entomol.* **2016**, *109*, 2096–2104. [CrossRef] [PubMed]

23. Gassmann, A.J.; Petzold-Maxwell, J.L.; Clifton, E.H.; Dunbar, M.W.; Hoffmann, A.M.; Ingber, D.A.; Keweshan, R.S. Field-evolved resistance by western corn rootworm to multiple *Bacillus thuringiensis* toxins in transgenic maize. *Proc. Natl. Acad. Sci. USA* **2014**, *111*, 5141–5146. [CrossRef] [PubMed]

24. Carrière, Y.; Crowder, D.W.; Tabashnik, B.E. Evolutionary ecology of insect adaptation to Bt crops. *Evol. Appl.* **2010**, *3*, 561–573. [CrossRef]

25. Roush, R.T. Bt-transgenic crops: Just another pretty insecticide or a chance for a new start in resistance management? *Pestic. Sci.* **1997**, *51*, 328–334. [CrossRef]

26. Taylor, C.E.; Georghiou, G.P. Suppression of insecticide resistance by alteration of gene dominance and migration. *J. Econ. Entomol.* **1979**, *72*, 105–109. [CrossRef]

27. Tabashnik, B.E.; Gould, F.; Carrière, Y. Delaying evolution of insect resistance to transgenic crops by decreasing dominance and heritability. *J. Evol. Biol.* **2004**, *17*, 904–912. [CrossRef]

28. Gould, F. Sustainability of transgenic insecticidal cultivars: Integrating pest genetics and ecology. *Annu. Rev. Entomol.* **1998**, *43*, 701–726. [CrossRef]

29. EPA [Environmental Protection Agency]. Final Report of the FIFRA Scientific Advisory Panel Subpanel on Bacillus thuringiensis (Bt) Plant-Pesticides and Resistance Management. 1998. Available online: http://archive.epa.gov/scipoly/sap/meetings/web/pdf/finalfeb.pdf (accessed on 30 November 2020).

30. Roush, R.T. Two-toxin strategies for management of insecticidal transgenic crops: Can pyramiding succeed where pesticide mixtures have not? *Philos. Trans. R. Soc. Lond. B Biol. Sci.* **1998**, *353*, 1777–1786. [CrossRef]

31. Carrière, Y.; Crickmore, N.; Tabashnik, B.E. Optimizing pyramided transgenic Bt crops for sustainable pest management. *Nat. Biotechnol.* **2015**, *33*, 161–168. [CrossRef] [PubMed]

32. Gressel, J.; Gassmann, A.J.; Owen, M.D.K. How well will stacked transgenic pest/herbicide resistances delay pests from evolving resistance? *Pest Manag. Sci.* **2017**, *73*, 22–34. [CrossRef] [PubMed]

33. Tabashnik, B.E.; Gassmann, A.J.; Crowder, D.W.; Carrière, Y. Insect resistance to Bt crops: Evidence versus theory. *Nat. Biotechnol.* **2008**, *26*, 199–202. [CrossRef] [PubMed]

34. Gassmann, A.J.; Carrière, Y.; Tabashnik, B.E. Fitness costs of insect resistance to *Bacillus thuringiensis*. *Annu. Rev. Entomol.* **2009**, *54*, 147–163. [CrossRef] [PubMed]

35. Carrière, Y.; Tabashnik, B.E. Reversing insect adaptation to transgenic insecticidal plants. *Proc. R. Soc. Lond. B Biol. Sci.* **2001**, *268*, 1475–1480. [CrossRef]

36. Pittendrigh, B.R.; Gaffney, P.J.; Huesing, J.E.; Onstad, D.W.; Roush, R.T.; Murdock, L.L. "Active" refuges can inhibit the evolution of resistance in insects towards transgenic insect-resistant plants. *J. Theor. Biol.* **2004**, *231*, 461–474. [CrossRef]

37. Pittendrigh, B.R.; Huesing, J.; Walters, K.; Olds, B.; Steele, L.D.; Sun, L.; Gaffney, P.; Gassmann, A.J. Negative cross-resistance: History, present status, and emerging opportunities. In *Insect Resistance Management: Biology, Economics and Predictions*, 2nd ed.; Onstad, D.W., Ed.; Elsevier: London, UK, 2014.

38. Gassmann, A.J.; Petzold-Maxwell, J.L.; Keweshan, R.S.; Dunbar, M.W. Western corn rootworm and Bt maize: Challenges of pest resistance in the field. *GM Crops Food* **2012**, *3*, 235–244. [CrossRef]

39. Jakka, S.R.K.; Shrestha, R.B.; Gassmann, A.J. Broad-spectrum resistance to Bacillus thuringiensis toxins by western corn rootworm (*Diabrotica virgifera virgifera*). *Sci. Rep.* **2016**, *6*, 1–9. [CrossRef]

40. Wangila, D.S.; Gassmann, A.J.; Petzold-Maxwell, J.L.; French, B.W.; Meinke, L.J. Susceptibility of Nebraska western corn rootworm populations (Coleoptera: Chrysomelidae) populations to Bt corn events. *J. Econ. Entomol.* **2015**, *108*, 742–751. [CrossRef]

41. Zukoff, S.N.; Ostlie, K.R.; Potter, B.; Meihls, L.N.; Zukoff, A.L.; French, L.; Ellersieck, M.R.; French, B.W.; Hibbard, B.E. Multiple Assays indicate varying levels of cross resistance in Cry3Bb1-selected field populations of the western corn rootworm to mCry3A, eCry3.1Ab, and Cry34/35Ab1. *J. Econ. Entomol.* **2016**, *109*, 1387–1398. [CrossRef]

42. Schrader, P.M.; Estes, R.E.; Tinsley, N.A.; Gassmann, A.J.; Gray, M.E. Evaluation of adult emergence and larval root injury for Cry3Bb1-resistant populations of the western corn rootworm. *J. Appl. Entomol.* **2016**, *141*, 41–52. [CrossRef]

43. Calles-Torrez, V.; Knodel, J.J.; Boetel, M.A.; French, B.W.; Fuller, B.W.; Ransom, J.K. Field-evolved resistance of northern and western corn rootworm (Coleoptera: Chrysomelidae) populations to corn hybrids expressing single and pyramided Cry3Bb1 and Cry34/35Ab1 Bt proteins in North Dakota. *J. Econ. Entomol.* **2019**, *112*, 1875–1886. [CrossRef]

44. USDA [United States Department of Agriculture, National Agricultural Statistics Service]. Corn for All Purposes 2019 Planted Acres by County for Selected States. 2020. Available online: https://www.nass.usda.gov/Charts_and_Maps/graphics/CR-PL-RGBChor.pdf (accessed on 1 December 2020).

45. Deitloff, J.; Dunbar, M.W.; Ingber, D.A.; Hibbard, B.E.; Gassmann, A.J. Effects of refuges on the evolution of resistance to transgenic corn by the western corn rootworm, *Diabrotica virgifera virgifera* LeConte. *Pest Manag. Sci.* **2016**, *72*, 190–198. [CrossRef]

46. Lefko, S.A.; Nowatzki, T.M.; Thompson, S.D.; Binning, R.R.; Pascual, M.A.; Peters, M.L.; Simbro, E.J.; Stanley, B.H. Characterizing laboratory colonies of western corn rootworm (Coleoptera: Chrysomelidae) selected for survival on maize containing event DAS-59122-7. *J. Appl. Entomol.* **2008**, *132*, 189–204. [CrossRef]

47. Meihls, L.N.; Higdon, M.L.; Siegfried, B.D.; Miller, N.J.; Sappington, T.W.; Ellersieck, M.R.; Spencer, T.A.; Hibbard, B.E. Increased survival of western corn rootworm on transgenic corn within three generations of on-plant greenhouse selection. *Proc. Natl. Acad. Sci. USA* **2008**, *105*, 19177–19182. [CrossRef] [PubMed]

48. Oswald, K.J.; French, B.W.; Nielson, C.; Bagley, M. Selection for Cry3Bb1 resistance in a genetically diverse population of nondiapausing western corn rootworm (Coleoptera: Chrysomelidae). *J. Econ. Entomol.* **2011**, *104*, 1038–1044. [CrossRef] [PubMed]

49. Gassmann, A.J.; Shrestha, R.B.; Jakka, S.R.K.; Dunbar, M.W.; Clifton, E.H.; Paolino, A.R.; Ingber, D.A.; French, B.W.; Masloski, K.E.; Doudna, J.W.; et al. Evidence of resistance to Cry34/35Ab1 corn by western corn rootworm (Coleoptera: Chrysomelidae): Root injury in the field and larval survival in plant-based bioassays. *J. Econ. Entomol.* **2016**, *109*, 1872–1880. [CrossRef] [PubMed]

50. Ludwick, D.C.; Meihls, L.N.; Ostlie, K.R.; Potter, B.D.; French, L.; Hibbard, B.E. Minnesota field population of western corn rootworm (Coleoptera: Chrysomelidae) shows incomplete resistance to Cry34Ab1/Cry35Ab1 and Cry3Bb1. *J. Appl. Entomol.* **2017**, *141*, 28–40. [CrossRef]

51. Gassmann, A.J.; Shrestha, R.B.; Kropf, A.L.; St. Clair, C.R.; Brenizer, B.D. Field-evolved resistance by western corn rootworm to Cry34/35Ab1 and other *Bacillus thuringiensis* traits in transgenic maize. *Pest Manag. Sci.* **2020**, *76*, 268–276. [CrossRef]

52. Spencer, J.L.; Mabry, T.R.; Vaughn, T.T. Use of transgenic plants to measure insect herbivore movement. *J. Econ. Entomol.* **2003**, *96*, 1738–1749. [CrossRef]

53. Hughson, S.A.; Spencer, J.L. Emergence and abundance of western corn rootworm (Coleoptera: Chrysomelidae) in Bt cornfields with structured and seed blend refuges. *J. Econ. Entomol.* **2015**, *108*, 114–125. [CrossRef] [PubMed]

54. Spencer, J.; Onstad, D.; Krupke, C.; Hughson, S.; Pan, Z.; Stanley, B.; Flexner, L. Isolated females and limited males: Evolution of insect resistance in structured landscapes. *Entomol. Exp. Appl.* **2013**, *146*, 38–49. [CrossRef]

55. Jaffe, G. *Complacency on the Farm: Significant Noncompliance with EPA's Refuge Requirements Threatens the Future Effectiveness of Genetically Engineered Pest-Protected Corn*; Center for Science in the Public Interest: Washington, DC, USA, 2009.

56. Denholm, I.; Rowland, M.W. Tactics for managing pesticide resistance in arthropods: Theory and practice. *Annu. Rev. Entomol.* **1992**, *37*, 91–112. [CrossRef] [PubMed]

57. Frank, D.L.; Zukoff, A.L.; Barry, J.; Higdon, M.L.; Hibbard, B.E. Development of resistance to eCry3.1Ab-expressing transgenic maize in a laboratory-selected population of western corn rootworm (Coleoptera: Chrysomelidae). *J. Econ. Entomol.* **2013**, *106*, 2506–2513. [CrossRef] [PubMed]

58. Meihls, L.N.; Frank, D.L.; Ellersieck, M.R.; Hibbard, B.E. Development and characterization of MIR604 resistance in a western corn rootworm population (Coleoptera: Chrysomelidae). *Environ. Entomol.* **2016**, *45*, 526–536. [CrossRef]

59. Onstad, D.W.; Meinke, L.J. Modeling evolution of *Diabrotica virgifera virgifera* (Coleoptera: Chrysomelidae) to transgenic corn with two insecticidal traits. *J. Econ. Entomol.* **2010**, *103*, 849–860. [CrossRef]

60. Storer, N.P. A spatially explicit model simulating western corn rootworm (Coleoptera: Chrysomelidae) adaptation to insect-resistant maize. *J. Econ. Entomol.* **2003**, *96*, 1530–1547. [CrossRef]

61. Gassmann, A.J.; Stock, S.P.; Sisterson, M.S.; Carrière, Y.; Tabashnik, B.E. Synergism between entomopathogenic nematodes and *Bacillus thuringiensis* crops: Integrating biological control and resistance management. *J. Appl. Ecol.* **2008**, *45*, 957–966. [CrossRef]

62. Andow, D.A.; Pueppke, S.G.; Schaafsma, A.W.; Gassmann, A.J.; Sappington, T.W.; Meinke, L.J.; Mitchell, P.D.; Hurley, T.M.; Hellmich, R.L.; Porter, R.P. Early detection and mitigation of resistance to Bt maize by western corn rootworm (Coleoptera: Chrysomelidae) *J. Econ. Entomol.* **2016**, *109*, 1–12. [CrossRef] [PubMed]

63. Petzold-Maxwell, J.L.; Cibils-Stewart, X.; French, B.W.; Gassmann, A.J. Adaptation by western corn rootworm (Coleoptera: Chrysomelidae) to Bt maize: Inheritance, fitness costs, and feeding preference. *J. Econ. Entomol.* **2012**, *105*, 1407–1418. [CrossRef]
64. Geisert, R.W.; Ellersieck, M.R.; Hibbard, B.E. Tolerance of eCry3.1Ab in reciprocal cross offspring of eCry3.1Ab-selected and control colonies of *Diabrotica virgifera virgifera* (Coleoptera: Chrysomelidae). *J. Econ. Entomol.* **2016**, *109*, 815–820. [CrossRef]
65. Ingber, D.A.; Gassmann, A.J. Inheritance and fitness costs of resistance to Cry3Bb1 corn by western corn rootworm (Coleoptera: Chrysomelidae). *J. Econ. Entomol.* **2015**, *108*, 2421–2432. [CrossRef] [PubMed]
66. Paolino, A.R.; Gassmann, A.J. Assessment of inheritance and fitness costs associated with field-evolved resistance to Cry3Bb1 maize by western corn rootworm. *Toxins* **2017**, *9*, 159. [CrossRef] [PubMed]
67. Shrestha, R.B.; Gassmann, A.J. Inheritance and fitness costs of Cry3Bb1 resistance in diapausing field strains of western corn rootworm (Coleoptera: Chrysomelidae). *J. Econ. Entomol.* **2020**, *113*, 2873–2882. [CrossRef]
68. Meihls, L.N.; Higdon, M.L.; Ellersieck, M.R.; Tabashnik, B.E.; Hibbard, B.E. Greenhouse-selected resistance to Cry3Bb1-producing corn in three western corn rootworm populations. *PLoS ONE* **2012**, *7*, e51055. [CrossRef]
69. Hoffmann, A.M.; French, B.W.; Hellmich, R.L.; Lauter, N.; Gassmann, A.J. Fitness costs of resistance to Cry3Bb1 maize by western corn rootworm. *J. Appl. Entomol.* **2015**, *139*, 403–415. [CrossRef]
70. Hoffmann, A.M.; French, B.W.; Jaronski, S.T.; Gassmann, A.J. Effects of entomopathogens on mortality of western corn rootworm and fitness costs of resistance to Cry3Bb1 maize. *J. Econ. Entomol.* **2014**, *107*, 352–360. [CrossRef]
71. Oswald, K.J.; French, B.W.; Nielson, C.; Bagley, M. Assessment of fitness costs in Cry3Bb1-resistant and susecptible western corn rootworm (Coleoptera: Chrysomelidae) laboratory colonies. *J. Appl. Entomol.* **2012**, *136*, 730–740. [CrossRef]
72. Geisert, R.W.; Hibbard, B.E. Evaluation of potential fitness costs associated with eCry3.1Ab resistance in *Diabrotica virgifera virgifera* (Coleoptera: Chrysomelidae). *J. Econ. Entomol.* **2016**, *109*, 1853–1858. [CrossRef] [PubMed]
73. St. Clair, C.R.; Clifton, E.H.; Dunbar, M.W.; Masloski, K.E.; Paolino, A.R.; Shrestha, R.B.; Gassmann, A.J. Applying a selection experiment to test for fitness costs of Bt resistance in western corn rootworm and the effect of density on fitness costs. *J. Econ. Entomol.* **2020**, *113*, 2473–2479. [CrossRef]
74. Pereira, E.J.G.; Lang, B.A.; Storer, N.P.; Siegfried, B.D. Selection for Cry1F resistance in the European corn borer and cross-resistance to other Cry toxins. *Entomol. Exp. Appl.* **2008**, *126*, 115–121. [CrossRef]
75. Tabashnik, B.E.; Biggs, R.W.; Higginson, D.M.; Henderson, S.; Unnithan, D.C.; Unnithan, G.C.; Ellers-Kirk, C.; Sisterson, M.S.; Dennehy, T.J.; Carrière, Y.; et al. Association between resistance to Bt cotton and cadherin genotype in pink bollworm. *J. Econ. Entomol.* **2005**, *98*, 635–644. [CrossRef]
76. St. Clair, C.R.; Gassmann, A.J. Linking land use patters and pest outbreaks in Bt maize. *Ecol. Appl.* **2021**, e2295. [CrossRef]
77. Petzold-Maxwell, J.L.; Meinke, L.J.; Gray, M.E.; Estes, R.E.; Gassmann, A.J. Effect of Bt maize and soil insecticides on yield, injury, and rootworm survival: Implications for resistance management. *J. Econ. Entomol.* **2013**, *106*, 1941–1951. [CrossRef]
78. Shrestha, R.B.; Jakka, S.R.K.; French, B.W.; Gassmann, A.J. Field-based assessment of resistance to Bt corn by western corn rootworm (Coleoptera: Chrysomelidae). *J. Econ. Entomol.* **2016**, *109*, 1399–1409. [CrossRef]
79. Tabashnik, B.E.; Gould, F. Delaying corn rootworm resistance to Bt corn. *J. Econ. Entomol.* **2012**, *105*, 767–776. [CrossRef]
80. Cullen, E.M.; Gray, M.E.; Gassmann, A.J.; Hibbard, B.E. Resistance to Bt corn by western corn rootworm (Coleoptera: Chrysomelidae) in the U.S. Corn Belt. *J. Integ. Pest Manag.* **2013**, *4*, D1–D6. [CrossRef]
81. Gray, M.E.; Felsot, A.S.; Steffey, K.L.; Levine, E. Planting time application of soil insecticides and western corn rootworm (Coleoptera: Chrysomelidae) emergence: Implications for long-term management programs. *J. Econ. Entomol.* **1992**, *85*, 544–553. [CrossRef]
82. Martinez, J.C.; Caprio, M.A. IPM use with the deployment of a non-high dose Bt pyramid and mitigation of resistance for western corn rootworm (*Diabrotica virgifera virgifera*). *Environ. Entomol.* **2016**, *45*, 747–761. [CrossRef]
83. Carrière, Y.; Brown, Z.; Aglasanb, S.; Dutilleul, P.; Carroll, M.; Head, G.; Tabashnik, B.E.; Jørgensen, P.S.; Carroll, S.P. Crop rotation mitigates impacts of corn rootwormresistance to transgenic Bt corn. *Proc. Natl. Acad. Sci. USA* **2020**, *117*, 18385–18392. [CrossRef] [PubMed]
84. Baum, J.A.; Bogaert, T.; Clinton, W.; Heck, G.R.; Feldmann, P.; Ilagan, O.; Johnson, S.; Plaetinck, G.; Munyikwa, T.; Pleau, M.; et al. Control of coleopteran insect pests through RNA interference. *Nat. Biotechnol.* **2007**, *25*, 1322–1326. [CrossRef] [PubMed]
85. Bolognesi, R.; Ramaseshadri, R.; Anderson, J.; Bachman, P.; Clinton, W.; Flannagan, R.; Ilagan, O.; Lawrence, C.; Levine, S.; Moar, W.; et al. Characterizing the mechanism of action of double-stranded RNA activity against western corn rootworm (*Diabrotica virgifera virgifera* LeConte). *PLoS ONE* **2012**, *7*, e47534. [CrossRef] [PubMed]
86. Head, G.P.; Carroll, M.W.; Evans, S.P.; Rule, D.M.; Willse, A.R.; Clark, T.L.; Storer, N.P.; Flannagan, R.D.; Samuel, L.W.; Meinke, L.J. Evaluation of SmartStax and SmartStax PRO maize against western corn rootworm and northern corn rootworm: Efficacy and resistance management. *Pest Manag. Sci.* **2017**, *73*, 1883–1899. [CrossRef]
87. Khajuria, C.; Ivashuta, S.; Wiggins, E.; Flagel, L.; Moar, W.; Pleau, M.; Miller, K.; Zhang, Y.; Ramaseshadri, P.; Jiang, C.; et al. Development and characterization of the first dsRNA-resistant insect population from western corn rootworm, *Diabrotica virgifera virgifera* LeConte. *PLoS ONE* **2018**, *13*, e0197059. [CrossRef] [PubMed]
88. Schellenberger, U.; Oral, J.; Rosen, B.A.; Wei, J.-Z.; Zhu, G.; Xie, W.; McDonald, M.J.; Cerf, D.C.; Diehn, S.H.; Crane, V.C.; et al. A selective insecticidal protein from *Pseudomonas* for controlling corn rootworms. *Science* **2016**, *354*, 634–637. [CrossRef]
89. USDA [United States Department of Agriculture, Animal Plant Health Inspection Agency]. Pioneer Hi-Bred International, Inc.: Availability of a Petition for the Determination of Nonregulated Status for Insect Resistant and Herbicide-Tolerant Maize. 2020.

Available online: https://www.federalregister.gov/documents/2020/11/03/2020-24267/pioneer-hi-bred-international-inc-availability-of-a-petition-for-the-determination-of-nonregulated (accessed on 3 December 2020).
90. EPA [Environmental Protection Agency]. Framework to Delay Corn Rootworm Resistance. 2016. Available online: https://www.epa.gov/regulation-biotechnology-under-tsca-and-fifra/framework-delay-corn-rootworm-resistance (accessed on 28 January 2021).
91. Petzold-Maxwell, J.L.; Alves, A.P.; Estes, R.E.; Gray, M.E.; Meinke, L.J.; Shields, E.J.; Thompson, S.D.; Tinsley, N.A.; Gassmann, A.J. Applying an integrated refuge to manage western corn rootworm (Coleoptera: Chrysomelidae): Effects on survival, fitness and selection pressure. *J. Econ. Entomol.* **2013**, *106*, 2195–2207. [CrossRef]

Review

RNAi for Western Corn Rootworm Management: Lessons Learned, Challenges, and Future Directions

Molly Darlington [1], Jordan D. Reinders [1], Amit Sethi [2], Albert L. Lu [2], Partha Ramaseshadri [3], Joshua R. Fischer [3], Chad J. Boeckman [2], Jay S. Petrick [3], Jason M. Roper [2], Kenneth E. Narva [4] and Ana M. Vélez [1,*]

[1] Department of Entomology, University of Nebraska, Lincoln, NE 68583, USA; mndarlington@gmail.com (M.D.); jordan.reinders3@gmail.com (J.D.R.)
[2] Corteva Agriscience, Johnston, IA 50131, USA; amit.sethi@corteva.com (A.S.); albert.l.lu@corteva.com (A.L.L.); chad.boeckman@corteva.com (C.J.B.); jason.roper-1@corteva.com (J.M.R.)
[3] Bayer Crop Science, Chesterfield, MO 63017, USA; partha.ramaseshadri@bayer.com (P.R.); joshua.fischer@bayer.com (J.R.F.); jay.petrick@bayer.com (J.S.P.)
[4] GreenLight Biosciences, Research Triangle Park, NC 27709, USA; knarva@greenlightbio.com
*   Correspondence: avelezarango2@unl.edu; Tel.: +1-402-472-2152

Simple Summary: The western corn rootworm (WCR), *Diabrotica virgifera virgifera* LeConte, is an annual pest of maize in the United States Corn Belt. Larval feeding on the root system can promote significant yield loss through reduced water and nutrient uptake and decreased plant stability. Various management tactics, including crop rotation, insecticides, and transgenic crops expressing *Bacillus thuringiensis* Berliner proteins, have been used to manage WCR densities. However, resistance has evolved to each of these tactics in local areas, highlighting the need for new management strategies. The use of RNA interference (RNAi) technology for WCR management represents the next phase of species-specific pest management. This paper reviews the current knowledge of RNAi for WCR management. We present an overview of traits that have been explored and the accumulated knowledge acquired on mode of action, resistance, ecological risk assessment, and mammalian safety. We conclude by highlighting the challenges and future directions of this technology for WCR management.

Abstract: The western corn rootworm (WCR), *Diabrotica virgifera virgifera* LeConte, is considered one of the most economically important pests of maize (*Zea mays* L.) in the United States (U.S.) Corn Belt with costs of management and yield losses exceeding USD ~1–2 billion annually. WCR management has proven challenging given the ability of this insect to evolve resistance to multiple management strategies including synthetic insecticides, cultural practices, and plant-incorporated protectants, generating a constant need to develop new management tools. One of the most recent developments is maize expressing double-stranded hairpin RNA structures targeting housekeeping genes, which triggers an RNA interference (RNAi) response and eventually leads to insect death. Following the first description of *in planta* RNAi in 2007, traits targeting multiple genes have been explored. In June 2017, the U.S. Environmental Protection Agency approved the first *in planta* RNAi product against insects for commercial use. This product expresses a dsRNA targeting the WCR *snf7* gene in combination with *Bt* proteins (Cry3Bb1 and Cry34Ab1/Cry35Ab1) to improve trait durability and will be introduced for commercial use in 2022.

Keywords: RNAi; western corn rootworm; *Diabrotica virgifera virgifera*; plant-incorporated protectant; pyramid strategy; insect resistance management

Citation: Darlington, M.; Reinders, J.D.; Sethi, A.; Lu, A.L.; Ramaseshadri, P.; Fischer, J.R.; Boeckman, C.J.; Petrick, J.S.; Roper, J.M.; Narva, K.E.; et al. RNAi for Western Corn Rootworm Management: Lessons Learned, Challenges, and Future Directions. *Insects* 2022, 13, 57. https://doi.org/10.3390/insects13010057

Academic Editor: Joseph L. Spencer

Received: 18 October 2021
Accepted: 28 December 2021
Published: 5 January 2022

Publisher's Note: MDPI stays neutral with regard to jurisdictional claims in published maps and institutional affiliations.

1. Introduction

The western corn rootworm (WCR), *Diabrotica virgifera virgifera* LeConte (Coleoptera: Chrysomelidae), is one of the most destructive insect pests of maize (*Zea mays* L.) in the United States (U.S.) [1,2]. Native to Central America and first identified as a pest of cultivated maize in Colorado in 1909 [3,4], populations are now found throughout the midwestern U.S. and Europe [5,6]. Once achieving pest status, various control tactics have been used to reduce damage caused by WCR larval feeding including crop rotation, soil-applied insecticides, and maize hybrids expressing insecticidal proteins from the soil bacterium *Bacillus thuringiensis* Berliner (*Bt*) [2,7,8]. However, managing WCR has been historically challenging due to its remarkable ability to evolve resistance to all available management tactics throughout various local areas in the U.S. Corn Belt [2,7,9–14]. Economic analyses estimate costs associated with control strategies and yield loss exceed USD $2 billion annually [1,15]. Four insecticidal *Bt* proteins are currently available for WCR management: Cry3Bb1, mCry3A, eCry3.1Ab, and Cry34Ab1/Cry35Ab1 (now classified as Gpp34Ab1/Tpp35Ab1 [16]) [17–20]. Field-evolved resistance to Cry3Bb1 and mCry3A was first reported in 2011 [21] and has subsequently been confirmed in various areas of the U.S. Corn Belt [22–27]. Cross-resistance between Cry3Bb1, mCry3A, and eCry3.1Ab has been widely demonstrated [21,23,25,26,28]. More recently, resistance to Cry34Ab1/Cry35Ab1 was identified [27,29,30], highlighting the urgency for alternative approaches for WCR management. Recent review articles highlight the history, use of, and evolution of resistance to synthetic insecticides [7] and *Bt* traits [8].

In planta expression of double-stranded RNA (dsRNA) in the shape of a hairpin RNA triggers an RNA interference (RNAi) response within the insect, representing a new mode of action for WCR control [31,32]. In June 2017, the U.S. Environmental Protection Agency (EPA) registered the first transgenic maize product with an RNAi-based plant-incorporated protectant (PIP) for WCR management [33]. This product expresses three *Bt* proteins (Cry3Bb1 and Cry34Ab1/Cry35Ab1) and a dsRNA [32]. This review summarizes our current knowledge of RNAi for WCR management, including studies on the various RNAi traits explored to date, mode of action, susceptibility of field populations, resistance evolution, environmental risk assessment, and mammalian safety.

2. RNAi in the Western Corn Rootworm

RNAi is an inherent post-transcriptional gene silencing mechanism utilizing various non-coding RNAs (ncRNA) as substrates [34]. RNAi is thought to have evolved as a defense mechanism against invading nucleic acids such as viruses and transposable elements or anomalous endogenous transcription products [35,36]. RNAi was first described in *Caenorhabditis elegans* in 1998 and was subsequently shown to be conserved in most eukaryotic organisms [37,38].

Three RNAi pathways are present within eukaryote organisms, each triggered by distinct, short, duplexed, non-coding nucleic acids: microRNA (miRNA), piwi-interacting RNA (piRNA), and short interfering RNA (siRNA) [39] (reviewed by Zhu and Palli [40]). Each ncRNA embeds within an Argonaute protein and acts as a guide for the direct targeting of a single-stranded RNA molecule within the cell [41]. Despite overlap within the core machinery, each pathway differs in functionality. The RNAi pathway triggered by miRNA is involved in miRNA processing [42], regulation of nucleic acids through translation repression [43,44], and target degradation [45–48]. The piRNA pathway is involved in the silencing of transposable elements [49] and germline development [50]. Finally, the siRNA pathway responds to endogenous siRNAs [51] and long exogenous dsRNAs present during viral infections [52], which are then sliced into siRNAs within the cytoplasm. The siRNA pathway can be triggered exogenously through the application of synthetic dsRNA for specific silencing of inherent messenger RNA (mRNA) [34]. This pathway is commonly used for gene function studies and, more recently, for pest management [53,54]. Thus, this review focuses on the siRNA RNAi pathway in WCR.

In 2007, Baum et al. [55] demonstrated that oral exposure to *DvV-ATPase A* dsRNA elicited a silencing response in WCR larvae, leading to mortality ($LC_{50} < 0.52$ ng/cm^2). *In planta* expression conferred maize root protection (mean node injury score of 0.25), exhibiting the possibility of utilizing the natural RNAi response for pest management.

Another early study used RNAi in WCR to assess the function of WCR orthologues to *laccase 2* (*lac2*) and *chitin synthase 2* (*chs2*) by injecting dsRNA into second and third instars [56]. Injection of 200 ng/μL *DvLac2*-specific dsRNA resulted in a 95.8% reduction in *lac2* expression and the prevention of post-molt cuticular tanning, while injection of 200 ng/μL *DvCHS2*-specific dsRNA reduced chitin levels in midgut tissues by 78%. Both studies demonstrated that WCR has a robust RNAi response and systemic gene knockdown could be generated in WCR through both injection and feeding.

Silencing via the siRNA pathway involves three steps: dsRNA uptake, gene silencing, and systemic spread. Understanding the RNAi pathway in WCR is imperative to characterize potential resistance mechanisms. Much of the mode of action in WCR is understood and will be described in further detail in this review. However, knowledge gaps still exist and further examination in WCR is warranted.

2.1. dsRNA Uptake

To elicit an RNAi response, dsRNA must first enter the cell. In WCR, current evidence points to two putative uptake mechanisms: systemic RNA interference defective-like proteins (SID-like or SIL) and clathrin-mediated endocytosis (CME). In *C. elegans*, various SID proteins facilitate dsRNA movement across biological membranes [57–61]. Coleopteran homologs of *C. elegans sid* genes are labeled *sid-like* (*sil*) due to more substantial homology with *tag-130/chup-1*, a protein involved in cellular cholesterol import, than with SID proteins themselves [62]. At least two *sils* are found within the WCR genome (GenBank: PXJM00000000.2), *silA* (LOC114327414) and *silC* (LOC114340333). In WCR larvae, knockdown of each gene individually reduced the phenotype of a second dsRNA application (targeting *DvEbony*), implicating *silA* and *silC* as putative routes for dsRNA uptake in WCR [63]. However, in adult WCR, *silA* silencing did not affect *V-ATPase* knockdown after secondary exposure to *DvV-ATPase* dsRNA. Despite this, mortality levels typically associated with a *DvV-ATPase* dsRNA exposure were reduced in *silA* knockdown treatments, while silencing of *silC* alone and in combination with *silA* had no effect on *V-ATPase* silencing or mortality [64]. Due to these ambiguous results, more evidence is required to determine if *silA* and *silC* are indeed involved in dsRNA uptake in WCR. In *C. elegans*, *tag-130/chup-1*, the gene most similar to coleopteran *sils*, plays no role in systemic RNAi [62]; therefore, it is conceivable that *silA* and *silC* may have a function unrelated to dsRNA uptake. In addition, some insects with systemic RNAi responses lack SID homologs within their genome, suggesting the existence of alternative routes of cellular uptake [65].

The clearest evidence for dsRNA uptake in WCR is clathrin-mediated endocytosis (CME) [64]. CME is a form of endocytosis by which macromolecules such as proteins, lipids, and pathogens enter a cell in a clathrin-dependent manner [66,67]. CME requires cell surface receptors to recognize ligands and initiate internalization [68]. In WCR, silencing of key CME factors *clathrin-heavy chain* (*chc*), *vacuolar H⁺ ATPase 16 kDa subunit* (*vha16*), and *clathrin adaptor protein AP50* (*AP50*) led to a reduction in marker dsRNA effect, while knockdown of *ADP ribosylation factor-like 1* (*Arf72A*) and *small GTPase Rab7* did not produce the same effect [64]. Both AP50 and CHC are required for clathrin-coated pit formation, in which these proteins congregate around the intracellular portion of a receptor and allow for endosomal development [69]. Vha16 is also necessary for endosomal maturation by regulating vesicle acidification [70]. Interestingly, knockdown of *chc* only slightly reduced silencing caused by a marker dsRNA in adult WCR [64], while a similar assay using *T. castaneum* larvae found much a stronger effect on marker gene silencing [71]. The same experiment also identified Rab7 as an essential component in dsRNA uptake in *T. castaneum* [71], while in WCR, *rab7* knockdown did not show result in a reduction in the silencing of the marker gene [64]. Similarly, results from experiments with Colorado potato beetle, *Leptinotarsa decemlineata*, larvae mirrored those of *T. castaneum* [65]. Divergent results between coleopterans, *D. v. virgifera* (Chrysomelidae), *L. decemlineata* (Chrysomelidae), and *T. castaneum* (Tenebrionidae) could be due to physiological differences between species or from variable responses between life stages. Results suggesting involvement of *silA* and *silC* in the RNAi mechanism in *T. castaneum*, *L. decemlineata*, and WCR were found using the larval life stage. However, similar experiments carried out with WCR adults were not definitive. Expression of core RNAi genes was found to change based on the larval stage in *L. decemlineata*, affecting dsRNA efficacy [72] This finding provides an argument for life stage-dependent dsRNA uptake. However, further examinations of CME genes on WCR larvae are warranted to determine if these differences are due to physiological differences between insect species or life stages.

Despite conflicting results, the involvement of CME in dsRNA uptake in insects, and specifically in Coleoptera, is well established [65,70,71,73], suggesting that CME is the primary mechanism for dsRNA uptake in WCR (Figure 1a). Two key aspects of CME uptake, receptors responsible for dsRNA binding and genes involved in endosomal escape, have yet to be examined in WCR. Internalization of dsRNA is a significant barrier to RNAi efficacy in other insects [74]; therefore, understanding how dsRNA enters and is released from the endosome in WCR may illustrate possible deficiencies in other insects [75,76].

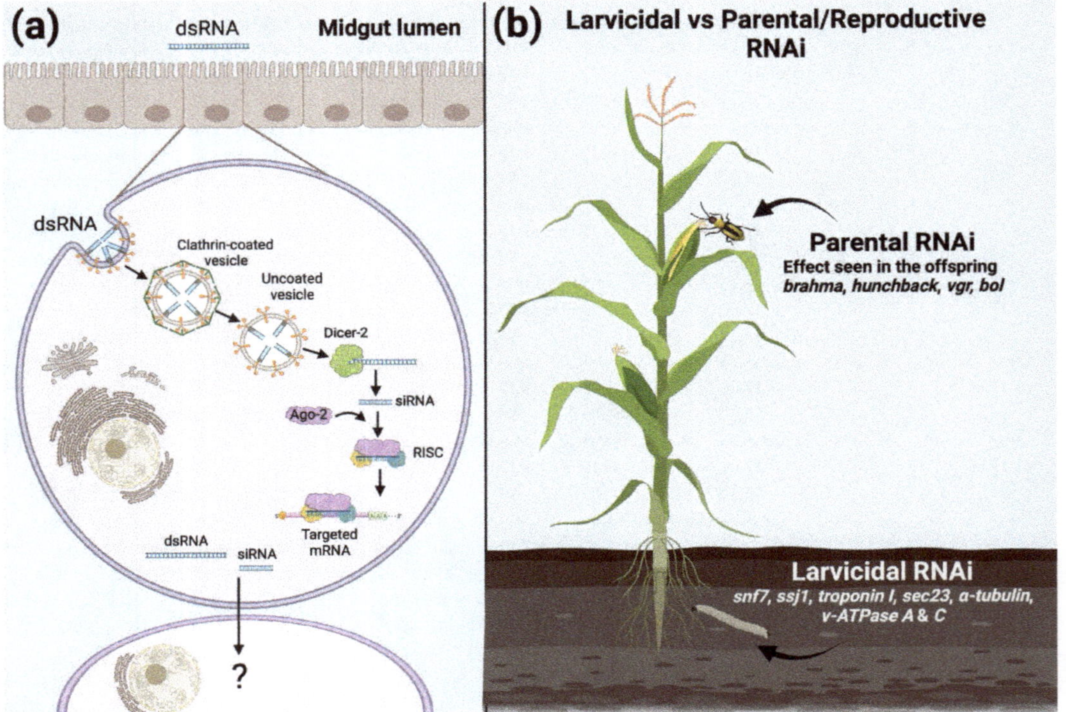

Figure 1. (**a**) Overview of the known RNAi uptake and processing mechanism in the western corn rootworm; (**b**) list of dsRNA gene targets active against the western corn rootworm subsequently transformed into maize.

2.2. Silencing Mechanism

The siRNA pathway is initiated when dsRNA within the cytoplasm is cleaved into multiple siRNAs by the RNase III family ribonuclease Dicer-2 (Dcr-2) [77–79]. Dcr-2 and the dsRNA-binding protein R2D2 form a complex to bind and cleave dsRNA [80]. siRNAs are 21–23 bp long duplexes with a 3′ two-nucleotide overhang [81–83]. A single copy of *dcr-2* has been annotated in the WCR genome (LOC114339627). Knockdown of *dcr-2* in WCR adults prevented the silencing of a marker dsRNA, indicating the requirement of Dcr-2 in the WCR RNAi response to exogenous dsRNA [84,85]. Knockdown of *dcr-2* in female WCR led to a reduction in oviposition and silencing during the third instar negatively affected adult emergence [86]. However, knockdown of *dcr-2* in neonates did not affect larval growth or development [87]. R2D2 not only plays a role in dsRNA binding but is also vital for loading siRNAs into the RNA-induced silencing complex (RISC) [88–90]. One copy of *r2d2*, expressing three isoforms, has been identified in WCR (LOC114342393), and silencing in WCR females led to a reduction in egg-laying capacity [86]. Based on a homology search within the WCR genome, one more *r2d2* gene may exist; however, functional evaluation is lacking.

After cleavage by Dcr-2, siRNAs are loaded onto the RISC by the RISC loading complex (RLC), comprised of Dcr-2, R2D2 [90,91], and the TATA-Box Binding Protein Associated Factor 11 (TAF11) [92]. Each siRNA is individually bound to an RNase H-like endonuclease, Argonaute-2 (Ago-2) [93,94], within the RISC. Argonaute proteins then separate siRNA duplexes into two single-stranded RNAs, a guide and a passenger strand, after which the passenger strand is degraded and the guide is retained [93,95]. Once the RISC/guide-siRNA complex locates a single-stranded RNA target through base-pair matching, silencing commences via mRNA hydrolysis [81,96,97]. Two silencing requirements include a catalytically active Argonaute protein, which does the actual degradation, and a sequence match between the guide siRNA and the target mRNA [98]. Data from *T. castaneum* suggest a greater than 80% sequence identify with target mRNA is required for an efficient RNAi response [97] and in *D. melanogaster*, 19 bp homology with mRNA led to effective target silencing [99]. After initial hydrolyzation of the mRNA target by Argonaute, the remaining nucleic acid is either degraded by endoribonucleases, exoribonucleases, or translated into incomplete proteins [100]. Multiple *ago2*s may be present in the WCR genome; however, current research has focused on one gene (LOC114327218). Knockdown of *ago2* prevented the downregulation of genes targeted by dsRNA, indicating the requirement of AGO2 for the RNAi response in WCR. Knockdown of *ago2* did not affect WCR development or survival [63,85].

The core RNAi silencing mechanism is relatively well conserved, and data from WCR have generally matched that of other insects. However, while key aspects of the pathway have been evaluated in WCR, many putative steps related to dsRNA and siRNA movement within the cytoplasm remain unstudied. For example, in *D. melanogaster*, the molecular chaperone complex Hsc70-Hsp90 facilitates the loading of siRNAs into Ago2 [101], and the small RNA methyltransferase HUA Enhancer 1 (HEN-1) activates the RISC through siRNA methylation [102]. Whether homologs exist with similar functions in WCR is unknown. Furthermore, crosstalk between the siRNA, miRNA, and piRNA RNAi pathways does exist. For example, silencing of *r2d2* and *ago2* negatively impacted miRNA processing in second instar WCR [86]. In *L. decemlineata*, Ago1 (miRNA) and Aubergine (piRNA) are involved in the dsRNA response [103], and the dsRNA binding protein Loquacious participates in both the siRNA and miRNA pathways in *D. melanogaster*, depending on the isoform expressed [104]. The extent to which crosstalk impacts the processing of exogenous dsRNA in WCR has not been determined. Further understanding of crosstalk between RNAi pathways may provide insight into why RNAi is efficient in Coleoptera and could be a cause of the differential response seen across Insecta. Lastly, the identification of coleopteran-specific *staufenC* in *L. decemlineata* opens up the possibility that other coleopterans have specific genes involved in the RNAi mechanisms [105], which may be responsible for enhanced efficacy that is characteristic of Coleoptera. Even though the main components of the dsRNA processing in WCR have been identified (Figure 1a), knowledge gaps remain. While verification of individual components identified in model organisms may be tedious, the strength of the WCR RNAi response is a valuable resource. Characterizing factors conferring the high susceptibility observed may lead to solutions to overcome barriers in other insects.

2.3. Systemic Spread

In organisms highly susceptible to RNAi, systemic spread of the silencing signal is an essential step for sufficient silencing. During a viral infection, cells initially exposed to dsRNA reverberate the signal to distal tissues, priming uninfected cells against the pathogen [106,107]. In *C. elegans*, RNA-directed RNA polymerases (RdRp) use primary siRNA and mRNA targets to form secondary siRNAs, amplifying the signal [108–110], while SID transmembrane proteins allow dsRNA to move from cell to cell [57,111]. However, putative homologs of SID proteins (SILs) in insects are not required for a systemic RNAi response and most insects lack RdRps entirely [65,112]. As previously stated, at least two homologs of *sid* genes can be found in the WCR genome; however, involvement in the RNAi mechanism is unclear [63,64]. In addition, no RdRp homolog is found in the WCR genome and small RNA sequencing after exposure to dsRNA found no evidence of secondary siRNA production in WCR, which suggests all siRNAs derive from the original dsRNA molecules [113]. However, this sequencing method would not have identified putative 5′ triphosphate modified siRNAs, small ncRNAs shown to be amplified in an AGO2-dependent manner from viral DNA in *D. melanogaster* hemocytes [106]. Whether this RdRp-independent amplification step is present in WCR or if it can be triggered through non-viral DNA is unknown and requires further examination.

Despite a lack of demonstrated causative elements, spread of the RNAi response in WCR has been documented, evidenced by transcript silencing in tissues distal to the exposure site [56,113–115]. Recent work in *T. castaneum* and *L. decemlineata* suggests that after the initial dsRNA uptake, siRNAs or dsRNAs travel cell to cell via extracellular vesicles [116,117]. Further work is required to illuminate the pathway of systemic spread in WCR.

3. RNAi Traits for Rootworm Control

3.1. RNAi as a New Mode of Action for Rootworm Control

Commercial traits used to control WCR have traditionally relied on insecticidal proteins identified from *B. thuringiensis* and other bacterial species [118–120]. When insecticidal proteins are expressed in transgenic maize, rootworm larvae exposed through root tissue feeding are significantly impacted in terms of mortality and stunting. Mortality is often the result of midgut epithelial cell disruption after receptor recognition results in surface membrane binding of the protein and a subsequent cascade of molecular events [121]. This effect in rootworm larvae translates directly into reduced root damage and yield protection.

As previously described, WCR is highly susceptible to environmental RNAi [122]. This has made it possible to target rootworm genes essential for survival through RNAi-mediated gene suppression as a new method to control rootworm damage complementary to insecticidal proteins. The next generation of rootworm-protected maize hybrids combine the MON87411 event, expressing Cry3Bb1 and *DvSnf7*, an RNAi trait, with the DAS-59122-7 event, expressing Cry34Ab1/Cry35Ab1, creating the first rootworm-active pyramid containing three modes of action [32]. A review of the efforts to identify efficient transgenic maize RNAi traits is provided below and a summary of the most successful genes evaluated in WCR is provided in Table 1.

Table 1. RNAi genes orally evaluated in western corn rootworm. Larvicidal genes were tested in neonates and the evaluated phenotype was mortality. Larvicidal genes reported by Baum et al. [55] include those with an $LC_{50} \leq 5.2$ ng/cm^2, while genes from Knorr et al. [123] include those with $\leq 60\%$ mortality in a nine-day bioassay. Parental and reproductive genes were tested in adults and the phenotype was evaluated in the offspring or adult fecundity, respectively.

Name	NCBI Accession No	Type of Exposure	Reference
Larvicidal Genes			
ESCRT III snf7 [1]	XM_028287710.1	Artificial diet, *in planta*	[31,55,124,125]
v-ATPase-A [1]	XM_028281990	Artificial diet, *in planta*	[55]
Actin [2]	XM_028292745.1	Artificial diet	[55]
apple ATPase [1]	XM_028281191.1	Artificial diet	[55]
alpha tubulin [1]	XM_028282553.1	Artificial diet	[55]
beta tubulin [1]	XM_028282553.1	Artificial diet	[55]
COPI coatomer subunit β [1]	XM_028291201.1	Artificial diet	[55]
ESCRT III_vps2 [1]	XM_028296669.1	Artificial diet	[55]
ESCRT I-Vps28 [1]	XM_028283797.1	Artificial diet	[55]
mov34 [1]	XM_028287237.1	Artificial diet	[55]
ribosomal protein L9 [1]	XM_028294395.1	Artificial diet	[55]
ribosomal protein L19 [1]	XM_028289442.1	Artificial diet	[55]
ribosomal protein S4 [1]	XM_028298505.1	Artificial diet	[55]
ribosomal protein rps-14 [2]	XM_028275245.1	Artificial diet	[55]
RNA polymerase II [1]	XM_028297193.1	Artificial diet	[55]
vATPase-D [1]	XM_028287428.1	Artificial diet	[55]
sec23 [1]	MK474471	Artificial diet, *in planta*	[126]
wupA/troponin I [1]	MH001576.1	Artificial diet, *in planta*	[127]
rop [3]	XM_028277045.1	Artificial diet, *in planta*	[123]
dre4 [3]	XM_028288745.1	Artificial diet, *in planta*	[123]
rpIII 140 [3]	XM_028297193.1	Artificial diet, *in planta*	[123]
ncm-1	XM_028276581.1	Artificial diet	[123]
Rpb7-1	XM_028299763.1	Artificial diet	[123]
smooth septate junction protein 1 (ssj1)	KU562965	Artificial diet, *in planta*	[114,128,129]
smooth septate junction protein 1 (ssj2)	KU562966	Artificial diet	[114]
proteasome subunit beta type-1-like (protb)	KU756279	Artificial diet	[114]
proteasome subunit alpha type-3-like (pat3)	KU756280	Artificial diet	[114]
ribosomal protein S10 (rps10)	KU756281	Artificial diet	[114]
Parental and Reproductive Genes			
hunchback [1]	XM_028272853.1	Artificial diet, *in planta*	[115,130]
brahma [1]	KR152260.1	Artificial diet, *in planta*	[115,130]
chd-1 [1]	KT364642	Artificial diet	[131]
iswi-1 [1]	KT364640	Artificial diet	[131]
mi-2 [1]	KT364639	Artificial diet	[131]
iswi-2 [1]	KT364641	Artificial diet	[131]
Vgr [1]	KY373243	Artificial diet, *in planta*	[132]
bol [1]	KY373244	Artificial diet, *in planta*	[132]

[1] Orthologs identified from *D. melanogaster*; [2] Orthologs identified from *C. elegans*; [3] Orthologs identified from *T. castaneum*.

3.2. Target Site Discovery

3.2.1. Larvicidal RNAi

Given the successful deployment of transgenic rootworm control traits to date, a primary focus for developing RNAi as another mode of action (MOA) for WCR control has been directed at identification and validation of genes where suppression of gene expression leads to rapid cessation of larval feeding, stunting, and lethality. The aspiration is to develop an RNAi-based trait with performance comparable to existing protein-based traits in overall lethality and speed of kill. An RNAi-based MOA may result in overall mortality comparable to an insecticidal protein; however, generally, the kill speed is slower due to the intrinsic multistep cascade process that leads to gene suppression by RNAi [35,55,133,134]. Consequently, selecting target gene(s) is an essential factor determining whether an RNAi-based MOA can be successfully developed as a control trait. In WCR, the RNAi targets do not have to be limited to the insect midgut since WCR mounts a strong systemic response against dsRNA [56,113]. Ideal targets are those that are involved in critical physiological processes and, therefore, are essential to insect survival. Gene targets that have been characterized as effective candidates for larvicidal RNAi suppress essential biological functions including protein trafficking/sorting and transport (*snf7*, *sec23* [31,126]), organ integrity (*ssj1* [114]), energy metabolism (*v-ATPase* [55]), and muscle function (*troponin I* [127]) (Figure 1b).

Baum et al. [55] conducted the first large-scale screen of RNAi targets in WCR. A total of 290 dsRNA molecules were provided to WCR neonate larvae in diet feeding assays and numerous gene targets exhibited significant levels of mortality or growth inhibition. Among the most effective targets were putative orthologues for *vacuolar ATPase subunits A* (*v-ATPase A*) and *D* (*v-ATPase D*), *COPI coatomer subunit β*, ribosomal proteins, and *ESCRT III snf7*. Ingestion of dsRNA targeting mRNA for these and other genes resulted in various levels of larval stunting and mortality. For *v-ATPase A*, a corresponding suppression of mRNA levels and generation of homologous siRNAs were demonstrated. v-ATPases are ATP-dependent proton pumps that function to transport protons across plasma membranes, acidify intracellular compartments, and play an essential role in membrane trafficking and protein degradation [55]. Significantly, the work by Baum et al. [55] provided the first demonstration of applying dsRNA technology for *in planta* insect resistance. Transgenic maize events expressing hairpin dsRNA targeting *v-ATPase A* conferred root protection against WCR feeding damage, demonstrating that artificial diet-based mortality could translate to transgenic plant-based reduction in root damage.

Further work on targets described by Baum et al. [55] led to the commercial development of dsRNA maize events targeting WCR *snf7* [31,124,125,135]. *DvSnf7* is a WCR ortholog of the *Drosophila vps32* or *shrub*, an essential component of the Endosomal Sorting Complex Required for Transport (ESCRT) pathway involved in intracellular protein trafficking [136]. Ultrastructural and histological studies showed that *snf7* suppression via exposure to *Dvsnf7* dsRNA in diet feeding assays led to progressive degeneration of midgut enterocytes, cell sloughing, cell lysis, and larval mortality [124,125]. RNAi-derived suppression of *snf7* represents the first RNAi-based MOA trait commercialized for insect pest control [32].

The successful engineering of *v-ATPase A* dsRNA for *in planta* control of WCR inspired RNAi target discovery research in WCR. Genome-wide screens utilizing high throughput target interrogation and knowledge-based approaches identified new RNAi-sensitive targets resulting in efficacious dsRNA maize events. Knorr et al. [123] tested 50 WCR genes selected based on genome-wide RNAi screens in *T. castaneum* [137]. From these, T0 maize plants expressing RNA hairpins for WCR orthologues *rop* (vesicular trafficking), *dre4* (transcription), or *rpII140* (transcription) showed protection from larval feeding damage. Knowledge-based approaches to dsRNA target selection have also been successful. Hu et al. [114] hypothesized, based on information from *Drosophila* genetic screens, that genes involved in insect midgut cell to cell septate junctions might be important to midgut integrity and function. Experimental evidence has shown that a class of WCR orthologous smooth septate junction (*ssj*) genes were sensitive to RNAi, resulting in insect mortality and plant protection. *DvSSJ1*, a homolog of the *snakeskin* (*ssk*) gene from *Drosophila*, prevents luminal content leakage across the midgut epithelial lining. Ultrastructural and histochemical studies performed by Hu et al. [114] showed that suppression of *DvSSJ1* destroys the integrity of SSJs and results in the collapse of midgut epithelial cells into the lumen, which disrupts midgut function, leading to feeding cessation and subsequent larval mortality [114,128]. Mortality directly correlated with a reduction in expression of *DvSSJ1* transcript and protein in WCR larvae.

Fishilevich et al. [127] examined a potentially haplolethal gene target, wings up A (*wup*A), which encodes Troponin I, an inhibitory protein of the Troponin-Tropomyosin complex involved in muscle contraction [138]. Due to the haplolethal phenotype, it was thought that RNAi of *wup*A might lead to greater dose sensitivity to dsRNA. RNAi of *wup*A resulted in a rapid onset of growth inhibition within two days. Knockdown of *wup*A resulted in significant food accumulation in the hindgut due to a loss of peristaltic motion of the alimentary canal. Finally, Vélez et al. [126] examined the lethal impact of dsRNA targeting *sec23*, a coatomer protein, and a component of the (COPII) complex that mediates ER-Golgi transport [139]. *sec23* was sensitive to RNAi in both larvae and adults. Interestingly, this study showed that 85% mRNA transcript knockdown resulted in 40% Sec23 protein knockdown, indicating that complete protein knockdown is not necessary to achieve insect mortality. Thus, it can be hypothesized that targeting important components of essential protein complexes could provide an effective route to insect control by disrupting protein complex stoichiometry. It should be noted that *sec23* dsRNA T0 maize events were protected against insect damage in greenhouse experiments and larval offspring of female adults that had been exposed to sublethal concentrations of *sec23* dsRNA in diet feeding assays displayed reduced weight and survival [126].

These examples illustrate that RNAi targets involved in essential processes can have larvicidal activity, translating to decreased root damage when hairpin RNAs are expressed. Although these genes are expressed throughout the larva, except for *DvSSJ1*, which is relegated to the midgut epithelia, it is unclear whether the larvicidal activity is primarily related to the loss of function of essential genes/gene products in the midgut/alimentary system cells compared to a more systemic effect on the larva.

3.2.2. Parental and Reproductive RNAi

Naturally, most RNAi trait targets have focused on those that result in mortality due to the need to protect maize roots from damage by WCR larval feeding. However, given the susceptibility of WCR to environmental RNAi and the ability of RNAi-mediated gene suppression to spread systemically within WCR after ingestion, additional complementary approaches to WCR management can be explored. Parental RNAi is a new concept where adult oral exposure to dsRNA targeting embryonic development genes leads to effects on fecundity or effects within the offspring [115,130,131]. Two genes, *brahma* and *hunchback*, significantly impacted egg viability after female beetles were exposed to corresponding target gene dsRNA in diet feeding assays [115] (Figure 1b). *Brahma* (*DvBrm*) encodes an SWI/SNF ATP-dependent chromatin remodeler and *hunchback* (*DvHb*) encodes a zinc finger transcription factor [115]. Larvae from unhatched eggs laid by *DvHb* dsRNA-treated adult females primarily displayed deformation of abdominal and thoracic segmentation. This phenotype is consistent with the role of *DvHb* as an essential regulatory gene in the anterior–posterior patterning of insect embryos. No embryo development was observed from unhatched eggs laid by female beetles treated with *DvBrm* dsRNA. Additional chromatin remodeling ATPases including *chd-1*, *iswi-1*, *mi-2*, and *iswi2* were tested and generated a similar genotype to that observed with *DvBrm* [131]. These results demonstrated that adult exposure to dsRNA can manifest in their offspring [115,130,131]. However, the results obtained from diet-based assays did not translate effectively to *in planta* experiments with adults feeding on maize vegetative tissue expressing *DvBrm* or *DvHb* (data not published).

Niu et al. [132] targeted two reproductive genes, *DvVgr* and *DvBol*, to determine the impact of suppression on insect fecundity (Figure 1b). While oral feeding of *DvVgr* and *DvBol* dsRNA in diet-based assays using WCR larvae resulted in a reduction in fecundity, this trend did not translate to *in planta* results, a discrepancy possibly explained by the reduced expression of dsRNA found in transgenic plants compared to diet-based assays. A significant reduction in fecundity was only observed when WCR larvae fed on roots of transgenic maize plants expressing *DvBol* hairpin RNA. *bol* encodes an RNA-binding protein whose homolog in *Drosophila* impacts male spermatogenesis. Unexposed adult female WCR showed a significant reduction in egg-lay number and percentage of egg hatch after mating with adult male WCR that fed on transgenic maize roots expressing *DvBol* hairpin RNA. This resulted in an overall net reduction in fecundity between 84–95%. However, only a modest reduction in the number of eggs laid per female was observed from mated WCR that emerged after feeding on *DvVgr* hairpin RNA-expressing maize roots during the larval stage. These results reveal that RNAi targeting reproductive genes in WCR adults can persist throughout the development cycle after initial exposure as larvae and significantly impact adult fecundity. This phenomenon could be explained by maternally transmitted dsRNAs [140] or RNA-directed DNA or histone methylation [141]. However, these studies were performed in *T. castaneum* and *C. elegans* and *D. melanogaster*, respectively, and further exploration is necessary to unveil the specific mechanism allowing parental and reproductive RNAi in WCR.

The development of parental and reproductive RNAi as potential RNAi trait MOA represents a truly novel RNAi-based approach to mitigating WCR damage in the field by reducing or suppressing rootworm populations (Figure 1b). Combining different RNAi trait MOAs may be explored to manage rootworm populations further and slow the development of resistance to RNAi and the pyramided insecticidal proteins to protect yields from this devastating maize pest.

3.3. Transgenic dsRNA Events

Identification of gene targets that generated a robust RNAi response in WCR, and most importantly larval mortality, was the first step in developing insect-specific dsRNA as a management tactic. Incorporation of dsRNA molecules into transgenic maize events was the most viable option to ensure both stability of dsRNA and activity against WCR larvae in field environments. Deployment of transgenic maize hybrids with WCR-active dsRNA in field studies allowed researchers to understand the impact of the RNAi response on WCR survival and feeding injury and evaluate potential non-target effects in the field.

RNA molecules directed at target genes are carefully selected to avoid putative siRNA sequences with homology to transcripts from non-target organisms or the crop of interest, in this case, maize. Bioinformatic workflows are used to focus on regions of target transcripts predicted to generate a high abundance of siRNA. In WCR, dsRNA needs to be at least 60 bp in length to trigger gene knockdown effectively, and a length of 200–400 bases has shown to be an adequate size to generate a robust RNAi response in WCR. Midgut uptake of long hairpin dsRNA, but not siRNA 21-mers, is efficient in WCR [31]. A broad genome-wide uptake of endogenous plant dsRNA and transgenic dsRNA and subsequent processing of long dsRNA into 21-nucleotide (nt) siRNA by WCR has been demonstrated [142]. When developing RNAi traits, it is a routine practice to express inverted repeats homologous to the mRNA target from a strong constitutive promoter. Inverted repeats are separated by a non-homologous neutral stuffer sequence or intron. Such dsRNA expression constructs are exemplified by those described in Baum et al. [55], Armstrong et al. [143], and Hu et al. [129]. Transgenic maize events are selected when they contain only one or two expression construct copies to avoid silencing of the insert by the plant and simplify the event molecular characterization. Selected events for advancement are ultimately based on root protection against WCR feeding using a root damage rating system (0–3 node-injury scale) [144]. Further characterization of selected events examines the amount of dsRNA produced. This work is essential for determining non-target organism exposure levels in the field. dsRNA levels expressed in maize events are most efficiently quantified using QuantiGene RNA technology, a nucleic acid hybridization assay that can be run on crude maize tissue lysates without the need for extensive RNA purification as is required for reverse-transcription polymerase chain reaction quantitation [143]. Greenhouse and field-based testing of transgenic plants expressing *DvSnf7* indicated that by itself this RNAi-based trait MOA did not provide a level of protection comparable to a commercial trait, Cry3Bb1. However, combining both MOAs enhanced root protection and resulted in a significant reduction in WCR adult emergence, which may delay resistance development to both traits [32,145].

4. Field Efficacy of RNAi for Insect Control, Insect Resistance Management, and RNAi Resistance

4.1. Field Efficacy of RNAi Traits

The U.S. EPA approved the first RNAi product for insect control in 2017 [146], representing the first new MOA for WCR control since the release of the Cry34Ab1/Cry35Ab1 binary protein in 2005 [147]. As previously indicated, this product (SmartStax® PRO) expresses three rootworm-active *Bt* proteins, Cry3Bb1 and Cry34Ab1/Cry35Ab1, as well as *DvSnf7* dsRNA [33]. SmartStax® PRO also contains three Lepidopteran-active *Bt* proteins (Cry1A.105/Cry2Ab2 and Cry1F) and genes for glyphosate tolerance [33]. Given reports of field-evolved resistance to Cry3Bb1 [21–23,26,30], an efficacy evaluation of this product on Cry3Bb1-resistant insects and field populations was performed [148]. Additionally, *DvSnf7* dsRNA concentration-response larval bioassays were conducted on artificial diet using field-derived populations, a susceptible laboratory population, and a Cry3Bb1-resistant population. Lastly, greenhouse experiments evaluating beetle emergence from plants expressing each trait individually (i.e., Cry3Bb1, Cry34Ab1/Cry35Ab1, and *DvSnf7*) and in combination were conducted. The Cry3Bb1-resistant population exhibited a significant 2.7-fold decrease in susceptibility to *DvSnf7* dsRNA compared to the Cry3Bb1-susceptible population. However, the Cry3Bb1-resistant population lowered susceptibility was similar to other WCR field populations in diet bioassays. Cry3Bb1-resistant and susceptible colonies had similar WCR adult emergence from plants expressing *DvSnf7* and Cry34Ab1/Cry35Ab1, and WCR adult emergence from plants expressing *DvSnf7* and *DvSnf7* + Cry3Bb1 was not significantly different when the Cry3Bb1-resistant colony was evaluated [148]. Collectively, diet assay and *in planta* experiments have demonstrated a lack of cross-resistance between Cry3Bb1, Cry34Ab1/Cry35Ab1, and *DvSnf7* [148]. Previous research has also documented significant variation in susceptibility of WCR larvae from field populations to *DvSnf7*, with an LC_{50} ranging from 4.07 to 40.51 ng/cm^2 [148]. This suggests that some populations might already exhibit higher tolerance to *DvSnf7* dsRNA and resistance monitoring will be essential to track susceptibility changes in field populations to promote trait durability.

Further field studies evaluated the efficacy of SmartStax® (Cry3Bb1 + Cry34Ab1/35Ab1 pyramid) and SmartStax® PRO maize (also expresses *DvSnf7* dsRNA) against western and northern corn rootworms [32]. Field trials conducted between 2013 and 2015 across the U.S. Corn Belt demonstrated that SmartStax® PRO could significantly reduce root damage ratings under high WCR larval densities, in areas of Cry3Bb1 resistance, and in areas with greater than expected injury to Cry3Bb1 or SmartStax®. SmartStax® PRO also significantly reduced root damage ratings relative to SmartStax® in some field trials, indicating that *DvSnf7* dsRNA can provide additional root protection when coupled with *Bt* proteins [32]. In addition to enhanced root protection, SmartStax® PRO also reduced adult emergence compared to SmartStax®, single-event Cry3Bb1 and Cry34Ab1/Cry35Ab1 hybrids, and the non-*Bt* control [32]. Collectively, the reduced root damage ratings and decreased adult emergence associated with SmartStax® PRO suggest that the addition of *DvSnf7* and other RNAi traits could serve as a valuable tool for insect resistance management (IRM) strategies [32]. Various resistance modeling scenarios indicated that inclusion of *DvSnf7* dsRNA could promote durability of the expressed *Bt* proteins and decrease the rate of resistance evolution relative to SmartStax®, even in areas with suspected resistance to Cry3Bb1 [32]. However, no work has been performed to evaluate the role of RNAi traits under scenarios with resistance to both Cry3Bb1 and Cry34Ab1/Cry35Ab1. This product and other products with RNAi traits will likely play an important role in WCR population management, especially in locations with high annual WCR densities and confirmed resistance to Cry3Bb1 or Cry34Ab1/Cry35Ab1 [32].

4.2. Insect Resistance Management

The U.S. EPA requires that registrants of PIPs, such as *Bt* proteins and RNAi traits, complete and submit an IRM plan for the target pest before registration [149]. IRM is the scientific approach to delay the development of resistance in pest populations. SmartStax® PRO follows the IRM pyramiding strategy and expresses three different MOAs targeting WCR (*DvSnf7* dsRNA, Cry3Bb1 and Cry34Ab1/Cry35Ab1), and stacks traits against Lepidopteran pests (Cry1F, Cry2Ab2, and Cry1A.105) and weeds (cp4/epsps (glyphosate resistance)). The IRM value of the pyramid is significantly reduced if individual components are deployed simultaneously, field-evolved resistance to one or more components is present, and/or cross-resistance between components is observed [150]. Confirmed WCR field-evolved resistance to Cry3Bb1 and/or Cry34Ab1/Cry35Ab1 has been reported in some populations [21,23,24,26–30], which could potentially impact the IRM value of this new pyramid. After this product is commercially available, it will be essential to monitor changes in susceptibility to the three traits, particularly Cry3Bb1 and Cry34Ab1/Cry35Ab1, to ensure durability of all traits in SmartStax® PRO.

An important consideration for IRM with RNA-based traits is that, in contrast to *Bt* proteins, larvicidal dsRNAs can cause mortality in adult WCR [75,126,151,152]. The adult WCR RNAi response is rapid and can persist throughout most of the life stage. For example, knockdown of *Lac2* generated 76% knockdown 10 hours after ingestion and 86% knockdown 20 days after ingestion [153]. Expression of *DvSnf7* dsRNA occurs throughout the plant (event MON87411), including two tissues commonly consumed by adults in the field (e.g., pollen, leaves). However, the concentrations expressed *in planta* are not sufficient to generate mortality in adults and will provide sublethal exposure (mean of 0.103 ng/g and 33.8 ng/g in fresh weight pollen and leaf tissue, respectively [154]; the LC_{50} of *DvSnf7* previously observed in WCR adults was 60.2 ng/cm^2 [151]). Therefore, adult sublethal exposure to *DvSnf7* may have implications for resistance management by adding selection pressure benefiting resistant individuals or individuals with resistance alleles [155,156]. Movement of adult WCR is common as maize phenology changes, with beetles searching for underdeveloped silks and pollinating maize as a primary food source [148]. Intrafield adult WCR movement could increase the likelihood of many individuals emerging from a single commercial maize field feeding on SmartStax® PRO tissue at some point during the life cycle, while interfield movement of adult WCR and subsequent feeding on these plants would increase the risk of exposure to sublethal concentrations of dsRNA [157]. No fitness costs were observed in WCR adults exposed to *DvSec23* dsRNA LC_{25}, suggesting that exposure to a sublethal concentration may not affect the fitness of exposed adults and their offspring [126]. However, due to the unique physiological effects of each dsRNA trait, future studies are important to determine the role sublethal exposure to *DvSnf7* field-relevant concentrations will have on adult WCR.

Reports of *Bt* trait field-evolved resistance suggested that the high dose refuge (HDR) approach as a standalone IRM strategy was ineffective at delaying *Bt* resistance in WCR. The most recent draft of the U.S. EPA's "Framework to Delay Corn Rootworm Resistance" requires incorporating integrated pest management (IPM) strategies into IRM plans to mitigate the spread of field-evolved resistance and minimize the risk of resistance evolution [158]. Suggested IPM strategies include crop rotation, rotation of PIP MOAs, adulticide application, and area-wide management [158]. Deploying this new dsRNA product within an IPM framework is necessary to increase trait durability and decrease the rate of resistance evolution [159].

4.3. Rootworm Resistance to RNAi

As in previous *Bt* PIPs, selection pressure from continuous exposure to dsRNA will eventually promote resistance evolution. Mechanisms of resistance to dsRNA were initially postulated to include degradation of dsRNA in the gut, reduced dsRNA uptake, alteration in proteins involved in dsRNA transport or formation of the RISC complex, loss of siRNA recognition by the RISC complex, mutation of the target gene, or systemic spread failure [76,134,160]. Identifying the resistance mechanism to *DvSnf7* dsRNA in WCR will inform effective IRM strategies to extend product durability if resistance is related to dsRNA processing, which may confer cross-resistance to other RNAi traits.

Efforts to evaluate the multigenerational effect of *DvSnf7* in field-collected insects resulted in the development of a resistant colony. Khajuria et al. [75] collected WCR adults emerging from areas planted with transgenic maize expressing *DvSnf7* dsRNA. Field-collected beetles were crossed with a non-diapausing WCR colony and exposed to *DvSnf7* dsRNA for eight generations, creating a population with ≥130-fold resistance to *DvSnf7* dsRNA [75]. Reciprocal crosses determined that *DvSnf7* dsRNA resistance was recessive, monogenic, and autosomal [75]. Resistance resulted from reduced uptake of dsRNA in gut cells and, interestingly, was found to be non-sequence specific. Cross-resistance to dsRNAs targeting *v-ATPase A*, *COPI B* (Coatomer Subunit beta), and *mov34* (26s proteasome) was identified, indicating adaptation occurred within the RNAi mechanism itself. Due to the development of cross-resistance to a variety of dsRNA targets in this study, it is possible dsRNA represents a single MOA in WCR [75]. However, it is yet unknown how insects will respond and adapt to dsRNA under field conditions. Further studies with other RNAi traits will provide a better understanding of potential resistance mechanisms that might exist in the field.

Because *DvSnf7* will not be released as a single trait product, it was important to identify the chromosomal location of resistance alleles and compare with current knowledge on the location of resistance alleles to the *Bt* traits in SmartStax® PRO. The Cry3Bb1 resistance gene(s) is located on linkage group 8 (LG8) [161] and the resistance locus for *DvSnf7* dsRNA is located on linkage group 4 (LG4) [75]. The fact that resistance causative elements are located on different chromosomes supports experimental evidence showing a lack of cross-resistance between *DvSnf7* dsRNA and Cry3Bb1 [75,148]. Therefore, resistance to both traits would have to develop independently and would have a lower probability of occurring than if a single trait or MOA were used.

Similar work in *L. decemlineata* furthers our understanding of how dsRNA resistance could develop in WCR. A laboratory-derived dsRNA-resistant *L. decemlineata* cell line and colony were recently established [105,162]. In both cases, cross-resistance to multiple dsRNAs was found, supporting results observed in WCR [75]. In dsRNA-resistant WCR, non-specific dsRNA uptake was disrupted, conferring resistance to all sequences tested. This fits with the transcriptional analysis of *L. decemlineata* cells, wherein genes related to uptake, *clathrin light chain*, *vha55*, *silA*, and the novel dsRNA binding protein *staufenC*, were downregulated in resistant cells [105]. However, while resistance in the WCR colony was narrowed to one locus in the genome, the *L. decemlineata* dsRNA-resistant colony displayed polygenic inheritance. This indicates that various mechanisms relating to dsRNA uptake could adapt to intense selection pressure. Given that resistance alleles have been detected in natural WCR field populations and those detected to date may confer cross-resistance to other dsRNA sequences, monitoring changes of susceptibility of field populations will be necessary to preserve RNAi technology. The deployment of *DvSnf7* with two additional functional MOAs will also assist in delaying RNAi trait resistance in populations that remain susceptible to Cry3Bb1 and Cry34Ab1/Cry35Ab1.

5. Environmental Risk Assessment

5.1. Environmental Risk Assessment Principles

Environmental Risk Assessment (ERA) is a three-step process that uses a science-based framework to evaluate whether a product of interest poses an unacceptable risk to the environment [163]. The framework is well established and broadly applicable to a wide array of substances [164], from traditional chemical molecules to traits used in the genetic modification of crops. The first step in the ERA process is problem formulation, which is a case-specific exercise that defines the scope and goals of the assessment, considers what is known about the substance, generates testable risk hypotheses in line with established protection goals, and formulates an approach for further data generation [165,166].

For example, the protection goals for non-target arthropods involve maintaining populations of beneficial arthropods that contribute to ecosystem services [167]. The most relevant populations evaluated include pollinators, parasitoids, and predators, as well as charismatic, protected, or endangered species. Once the species are selected, adequate surrogate species are selected to perform the testing in the next phase [168,169]. The flexibility and robustness of the ERA framework is driven by the case-specific nature of the problem formulation. Therefore, risk assessors must navigate the problem formulation phase before conducting thorough testing.

After problem formulation is complete, the analysis phase may begin. The analysis phase evaluates two key variables, exposure and hazard. The exposure assessment is conducted to identify routes of exposure between the substance of interest and valued non-target organisms and to characterize the magnitude and duration of such exposure [168]. The hazard characterization process evaluates the potency of the substance in a set of organisms from the problem formulation phase that serve as surrogate species for those that may be exposed in the environment [170,171]. Hazard or effects testing is often conducted in the context of a tiered framework, whereby early tier testing is conducted under highly conservative and controlled laboratory conditions [170]. For dsRNA, this can often be achieved by incorporating high concentrations of purified dsRNA, representing worse-case exposures, into an artificial diet to reach conservative test concentrations commonly used in Tier I testing. If no unacceptable effects are observed during this early tier testing, then the conclusion of minimal risk can be determined. If adverse effects are observed, higher tier tests may be conducted under more realistic conditions to determine whether adverse effects would likely occur under conditions more reflective of the environment [168]. The most common endpoint or effect evaluated is mortality, as it is unambiguous, easy to measure, and correlates with adverse effects to field populations. However, sublethal endpoints such as weight and developmental timing have also been used as additional endpoints in studies evaluating the risk of *Bt* crops (reviewed by Roberts et al. [172]).

In the third phase of the ERA process, risk characterization, exposure, and hazard assessment results are combined and conclusions are drawn about the likelihood or magnitude of risk based on introducing the substance into the environment. Risk is ultimately a function of both exposure and hazard; thus, if either exposure or hazard is deemed minimal, then the risk associated with that substance is low [173].

5.2. Environmental Risk Assessment for RNAi Traits

The use of RNAi as a new MOA to confer a particular desired phenotype in a genetically modified crop has increased over the past decade [154,174]. Given the flexibility and robustness of the existing ERA framework described above, this structure is well suited to characterize the environmental safety of transgenic crops expressing dsRNA [172]. To briefly illustrate that process, we consider a dsRNA expressed in maize intended to control WCR.

An evaluation of possible exposure pathways is needed to focus the assessment on species that may be exposed to the introduced dsRNA sequence under conditions of use for the trait. This includes identifying which species may be present in maize fields, maize tissue types to which non-target organisms of interest may be exposed, and the concentrations of dsRNA in those tissues. Only those species that contribute to the identified protection goals, as explained in the previous section, need to be assessed. Conservative estimates of maize-expressed dsRNA exposure to non-target organisms are frequently based on high-end expression level determinations. These conservative estimates inform exposure levels of an organism, including detritivores and pollinators that feed directly on plant tissues. Plant dsRNA expression information can also estimate exposure to organisms in other environmental compartments, including soil and aquatic environments, using established models [163,175,176]. Soil and aquatic dissipation studies may be used if needed to refine exposure quantification and evaluate the potential for environmental persistence of dsRNA. Studies to date indicate that dsRNA dissipates rapidly and is unlikely to persist in soil and aquatic environments [177–181] (reviewed by Bachman et al. [182]).

In addition to the exposure assessment, gathering baseline information about RNAi as a MOA to control WCR can help focus the risk assessment process. For instance, not all organisms appear to be sensitive to dsRNA, and those that are require sufficient sequence lengths and homology to the organismal mRNA for silencing to occur [31,129,183–185]. Once baseline information about general RNAi processes is gathered, understanding how conserved the gene mRNA sequence is among species related to WCR can provide testable hypotheses about the potential spectrum of activity for the introduced sequence [129]. By making use of ever-expanding bioinformatic databases and aligning known base pair sequence information, risk assessors can better identify which species may be sensitive to the introduced trait. An analysis plan can then be developed to incorporate these species, assuming they have relevant exposure pathways to the dsRNA, provide value to the risk assessment by contributing to the identified protection goals (i.e., pollination, predation, detritivory, etc.), and are available for testing in laboratory settings [171,186].

Developing an analysis plan for hazard testing involves an understanding of the ecosystem context for trait deployment (e.g., maize fields where WCR is present, hence non-target species in maize agroecosystems where WCR is present are considered). The analysis plan for hazard testing is also coupled with the exposure assessment (i.e., dsRNA concentration in relevant plant tissues and persistence in the environment) and the RNAi MOA and identified target gene, to generate hypotheses for which species may need to be tested in the laboratory. Increasing risk assessment certainty can be done by generating data to support or refute hypotheses on potentially sensitive species or characterizing the potential effects on organisms contributing to the identified protection goals. It is also important to consider endpoints or responses that are relevant to the risk assessment. Protection goals are typically framed around preserving ecosystem function based on the abundance of representative species. Therefore, endpoints such as mortality, weight, and time to development can be tied back to an effect on populations representing a specific ecosystem function and the overlying protection goals. A thorough understanding of the target gene is also helpful for designing studies and identifying relevant endpoints to observe in laboratory studies. The effects observed against the target pest, WCR, are informative for both the duration of laboratory studies and the endpoints that should be evaluated in those studies. Traditionally, mortality is the primary endpoint in risk assessment hazard studies with a threshold of at least 50% effect to trigger higher tier tests [168,172]. This is based on the notion that <50% effect at test concentrations in laboratory studies that exceed environmental concentrations are unlikely to result in adverse effects under field conditions. However, it should be noted that focusing on mortality for a non-essential gene may not provide relevant information and could lead to inaccurate conclusions on risk. In that way, understanding target gene function and the effects observed on the target pest from gene disruption can help risk assessors identify appropriate endpoints to observe in hazard studies. Similarly, careful observation of the timing of effects elicited on the target pest can help design laboratory hazard studies with beneficial insects important to protect. RNAi as a MOA relies upon protein suppression, a process that may take some time to lead to mortality. Therefore, it is important to ensure the duration of laboratory studies are long enough to measure the effect of the trait [154].

The case-specific and iterative nature of problem formation makes the ERA framework well suited to evaluate RNA-based traits. Collectively, the information gathered during problem formulation is used to generate specific and testable risk hypotheses relevant to a dsRNA-based trait that led to an analysis plan for exposure and hazard testing. Completing the analysis plan, detailing which data need to be generated, and producing those results mark the completion of problem formulation. Based on the analysis plan, data are generated during the analysis phase to characterize exposure and hazard posed by the trait of interest. After hazard and exposure data are collected, they are combined during risk characterization to ultimately make informed decisions about the likelihood the trait of interest poses an unacceptable risk to the environment or inform on problem formulation to guide additional testing.

5.3. Evaluation of Western Corn Rootworm RNAi Traits on Non-Target Organisms

Multiple studies have investigated the impact of WCR-active dsRNA constructs on non-target organisms. After the initial report that maize expressing *v-ATPase A* dsRNA could generate root protection from rootworm feeding damage, researchers started evaluating the potential effects of WCR *v-ATPase A* dsRNA on non-target arthropods, including a pollinator, *Apis mellifera* L. [187]; a herbivorous insect, *Danaus plexippus* (L.) larvae [188]; predators including the ladybeetle species, *Coccinella septempunctata* L., *Coleomegila maculata* (De Geer), *Hippodamia convergens* Guérin-Meneville, *Harmonia axyridis* (Pallas), and *Adalia bipunctata* (L.) [189,190]; and a decomposer, the collembolan *Sinella curviseta* Brook [191]. Results found that WCR *v-ATPase A* dsRNA has negligible effects on honey bee larvae and adults, monarch caterpillars, and *S. curviseta* [187,188,191]. Interestingly, species-specific *v-ATPase A* dsRNA experiments found that monarch larvae are not susceptible to dietary RNAi [188], and only a small downregulation from species-specific *v-ATPase A* dsRNA was observed in honey bees after 24 hours with gene expression recovering after 48 hours [187]. Furthermore, dsRNA was found bound to royal jelly, making it unavailable for honey bee larvae, suggesting environmental factors could render insecticidal dsRNAs unavailable for non-target organisms [187]. In contrast, two ladybeetles *H. axyridis* and *C. septempunctata*, experienced some gene knockdown and mortality when exposed to WCR *v-ATPase A* dsRNA [190]. In a different study, WCR *v-ATPase A* significantly reduced the development of *A. bipunctata* and *C. septempunctata* [189]. In these studies, ladybeetles with the strongest effects had more base pair matches between WCR *v-ATPase A* and their specific *v-ATPase A* sequence, highlighting the importance of bioinformatic analyses when designing insecticidal dsRNAs [189,190]. It is also important to consider that dsRNA concentrations tested in these studies were orders of magnitude higher than what is expected to occur in the field [189]; therefore, potential detrimental effects in a field scenario may be negligible.

Other studies have examined the impact of *DvSnf7* on non-target insects [154,192]. Bachman et al. [192] initially evaluated the lethal and sublethal effects of *DvSnf7* dsRNA on insects representing ten families and four orders. The *DvSnf7* dsRNA spectrum of activity was narrow, and activity was only observed in beetles within the Galerucinae subfamily of Chrysomelidae, whose homologous *snf7* sequence shared a 90% identity with WCR *Snf7* [192]. Tan et al. [193] evaluated *DvSnf7* dsRNA in honey bee larvae and adults and reported no adverse effects at high exposure levels. Bachman et al. [154] tested *DvSnf7* dsRNA activity at field concentrations in seven non-target insects, including honey bee larvae and adults; *C. maculata*; a rove beetle, *Aleochara bilineata* Gyllenhal; a ground beetle, *Poecilus chalcites* (Say); the green lacewing, *Chrysoperla carnea* (Stephens); the insidious flower bug, *Orius insidiosus* Say; and a parasitic wasp, *Pediobius foveolatus* (Crawford). Results from this study indicated that exposure to *DvSnf7* dsRNA at field-relevant exposure levels would not cause adverse effects on the non-target insect species described above [154]. These studies demonstrate the impact of *DvSnf7* dsRNA expressed in maize plants will be negligible on beneficial insects. Recent review papers provide more details regarding the problem formulation for off-target effects of dsRNA products for pest control [194], and sublethal endpoints for non-target organisms for insect-active transgenic crops, including RNAi [172].

6. RNAi Mammalian Safety

When evaluating human safety considerations for ingested RNA molecules associated with the use of RNA-mediated transgenic crops, one must consider the following: (1) the history of safe consumption of RNA; (2) biological barriers that limit internal exposure to exogenous RNAs; and (3) the weight of scientific evidence from published safety studies with RNA molecules. Collectively, these considerations enable proper hazard identification and, taken together with exposure assessment, can be leveraged to make an informed risk assessment decision on the use of transgenic maize expressing RNAi traits as a control tactic for corn rootworms.

6.1. RNA and Its History of Safe Consumption

RNAi is a ubiquitous process for gene expression modulation in all eukaryotic organisms. Therefore, longer dsRNA precursors and the small RNA molecules that this process leverages to regulate gene expression are also ubiquitous in widely consumed foods derived from plants and animals. Owing to the ubiquity of these RNA molecules, small RNAs with perfect sequence complementarity to transcripts of human and/or animal genes are evident in soybean, rice, maize, and fruits and vegetables [195–197]. These sequences include those important for key biological processes in mammals [196]. The presence of such sequences in staple food and feed crops demonstrates the safe consumption of small RNAs in the diet and supports the safety of RNAi for uses in transgenic crops, including those intended for control of agricultural pests such as corn rootworms.

Leveraging RNAi for insect control in a commercial transgenic crop represents a new application of this mechanism; however, RNAi has served as a natural process underlying plant phenotypes in domesticated crops and as a MOA in commercially approved transgenic traits (reviewed by Petrick et al. [198] and Sherman et al. [199]). One prominent example of RNA-mediated traits successfully leveraged in commercial transgenic products is the deployment of resistance to the papaya ringspot virus in papaya [200], a transgenic trait that played a vital role for Hawaiian growers amidst the devastation of the papaya crop. RNAi has also been leveraged commercially for the production of healthier oils in soybean [201,202], for reduction in browning in apples [203], and for reduction in acrylamide and blackspot bruising in potatoes [204,205]. Transgenic crops utilizing RNAi as a MOA have received regulatory approvals in several geographies. These include approvals for use in food; feed; and cultivation in crops such as maize, soybean, squash, potato, tomato, alfalfa, plum, apple, bean, and papaya [206]. The above information on RNA safe consumption in the diet and its safe use to date in commercialized transgenic crops with RNAi-based traits should also be applicable to those RNA-based traits intended for control of corn rootworms and other insects [55,129,145].

6.2. Biological Barriers to RNA Absorption

Owing to the ubiquitous presence of the RNAi mechanism in eukaryotic organisms, RNA molecules are ingested by vertebrates through the consumption of plants, animals, and fungi. Such RNA molecules include those with double-stranded regions that could initiate RNAi if they were to be absorbed from the diet and reach cells following consumption by the organism. As a further illustration of this point, exogenous dietary RNAs include those with sequence complementarity to vertebrate genes [195–197,207,208]. As vertebrates are constantly exposed to such RNA molecules, it is not surprising that there are a series of biological barriers that preclude these dietary components from serving in a regulatory capacity, and instead, they are harnessed for nutritional value. These biological barriers are reviewed in the scientific literature [198,199,209].

Ingested RNAs face an acidic digestive environment in the stomach following their initial exposure to nucleases in the saliva [210]. The efficacy of the low pH and hydrolysis in the stomach for nucleic acid degradation and the removal of bases from the nucleic acid backbone (e.g., depurination) has been well described and reviewed in the literature [198,209,211]. Further digestion of ingested nucleic acids occurs in the small intestine due to the presence of digestive enzymes and nucleases secreted by the pancreas [209]. Due to this extensive collection of barriers, nucleic acids from the diet are extensively degraded and do not undergo substantive absorption in an intact form, as demonstrated empirically with miRNAs in rodents [211,212], rhesus monkeys [213], and humans [214]. Thorough reviews of the safety considerations of plant expressed and externally applied RNA molecules for humans and other vertebrates have been published [198,215].

Biological barriers to the absorption and function of ingested nucleic acids expand beyond digestive barriers and include cellular membranes impermeable to RNAs that are both highly polar and large [198,216]. Each successive series of membrane barriers must be crossed for a dietary RNA to move from the intestinal tract lumen, across endothelial cells, and into the bloodstream. Once in the blood, RNAs are then subjected to nucleases [217–219] and rapid renal elimination [217,220]. To reach a putative target tissue, any RNA molecule escaping these nucleases would then have to cross through the endothelium and the epithelium within a target tissue to have the capacity to modulate tissue gene expression. Owing to the impermeability of these charged macromolecules across membrane barriers [198,216], any RNA surviving the intestinal tract would be unlikely to reach the bloodstream or a target tissue. Furthermore, RNAs reaching the cytoplasm of a cell in the consuming organism would subsequently be subjected to sequestration into endosomes that retain a vast majority of exogenous RNA molecules [221,222].

This array of physical, chemical, and biological barriers has made the development of pharmaceutical RNA drugs challenging, necessitating their chemical stabilization to limit degradation and specialized delivery formulations to elude barriers to exogenous RNA molecules [209,221,223,224]. Without such formulations, injection of these exogenous RNA molecules results in their rapid degradation and elimination [217,219,220,225]. Oral delivery of therapeutic macromolecules represents a desirable route of administration that is even more elusive to drug developers than systemically or locally administered therapeutics due to the aforementioned biological barriers [221,224]. These challenges are evident from literature reviews and studies demonstrating the need for specialized formulations and/or chemical modifications to facilitate limited delivery within proof-of-concept oral delivery evaluations of RNA therapeutics [209,223,226,227].

6.3. Evaluation of Potential Activity or Adverse Effects of Ingested RNA

The biological barriers described above collectively limit absorption of RNA molecules from the diet. Therefore, it is highly unlikely that ingested RNAs would have the capacity to regulate gene expression or induce adverse effects in a consuming mammalian organism. This concept stems primarily from a lack of significant RNA oral bioavailability [211,212], rendering the internal dose, e.g., the number of available copies of a given RNA molecule at the putative site of action, insufficient for regulatory function [214,219,228]. This is made especially challenging as uptake of RNA molecules in mammals is low (e.g., less than one copy per cell), and up to 1000 to 10,000 RNA copies per cell may be required for a functional RNAi response [229]. Further complicating the possibility of plant RNA activity in the diet is that small RNAs in plants are tightly bound to Argonaute proteins. Bound RNAs are not thought to dissociate from these complexes or exchange into Argonaute proteins in the consuming organism [228]. These principles collectively provide a solid biological basis for the history of safe dietary RNA consumption described above, even when ingested sequences possess sequence complementarity to transcripts in the consuming organisms.

The weight of evidence supporting the limited potential for functional activity ingested RNAs in mammals [230–233] has been called into question by several peer-reviewed publications alleging the opposite. However, a series of reviews and several laboratory studies on absorption and/or activity of ingested RNAs have been published since 2012, and collectively calls into question the potential for significant uptake and bioactivity of ingested RNAs [197,211–215,228,234–237]. For example, work challenging the concept of dietary RNA absorption/activity indicated that very low detection levels could have resulted from laboratory contamination or false-positive PCR results [213,238,239]. Furthermore, it is essential to ensure nutritionally balanced rodent diets when conducting rodent dietary studies. This is evidenced by one of the principal studies asserting uptake and activity of ingested RNAs in mammals [232]. Measured changes in blood cholesterol concomitant with dietary RNA detection in plasma were ultimately determined to result from dietary imbalances rather than ingested RNA activity [212]. The most comprehensive analysis of this phenomenon included an assessment of 800 human datasets, the results of which supported the conclusion that detection of small RNAs from exogenous sources in mammalian blood samples likely results from contamination as opposed to dietary uptake [235].

Toxicological evaluations of orally administered double-stranded RNAs have been conducted in mice to address the potential for oral activity and toxicity of these molecules. A 28-day repeated-dose oral toxicity study was conducted in mice with either a long dsRNA molecule or a pool of four 21 base pair siRNAs targeting the mouse *v-ATPase* gene. This proof of concept study for evaluating oral activity/toxicity of RNA was conducted utilizing a known gene target for corn rootworm control when expressed in plants [55] and to then construct a long dsRNA or a pool of predicted active siRNA sequences with 100% sequence complementarity to the mouse [237]. These mouse-targeting dsRNA sequences were then administered to mice by oral gavage for 28 consecutive days at doses of $\geq$48 mg/kg body weight, and traditional toxicology and target gene expression endpoints were evaluated. This study did not identify oral toxicity or suppression of *v-ATPase* gene expression in the gastrointestinal tract, kidney, liver, brain, or bone, demonstrating that biological barriers appear to preclude oral activity or toxicity of orally administered dsRNA molecules in mammals.

To further demonstrate RNA oral safety in a product-specific context, the *DvSnf7* dsRNA (240 base pair active dsRNA within a 968 nucleotide RNA sequence) was evaluated in a 28-day repeat-dose oral toxicity study [197]. *DvSnf7* dsRNA at oral doses of up to 100 mg/kg bodyweight for 28 consecutive days did not produce any treatment-related effects on weight, food consumption, clinical observations, clinical chemistry, hematology, gross pathology, or histopathology in mice [197]. Furthermore, the high dose utilized in this study (No-Observed Adverse-Effect-Level (NOAEL) of 100 mg/kg) was billions of times greater than highly conservative estimates of mean per capita human exposure to *DvSnf7* dsRNA in Europe, the U.S., Mexico, China, Japan, and Korea. This study demonstrates the safety of a dsRNA molecule used to control insects, specifically corn rootworms, illustrating further that transgenic crops using dsRNA for insect control do not pose unique risks to the health of consuming mammals, including humans.

7. Conclusions

The efficacy of dsRNA as a control tactic against WCR, along with commercial viability and regulatory approval, has spearheaded the use of RNAi technology for insect pest management and initiated a new phase of highly species-specific insecticides. The ERA and safety studies performed with dsRNA active against WCR have shown that the effects on non-target organisms, including humans and the environment, will be negligible.

The commercial release of dsRNA for corn rootworm control represents a milestone for WCR management, as it is the first new MOA to be deployed since 2005 [147]. However, the presence of dsRNA resistance alleles in natural populations exhibiting cross-resistance to other dsRNAs suggests RNAi may represent a single MOA in WCR. Further studies with other RNAi traits and WCR populations will determine how resistance may develop in the field and the uncertainty that presents for other RNAi traits. Furthermore, in contrast to historically commercialized *Bt* proteins, dsRNAs are more potent to adults at high concentrations, suggesting the likelihood of exposure to sublethal concentrations in the field upon feeding. Hence, future studies evaluating the effect of adult exposure to field-realistic dsRNA concentrations are important to characterize the potential risk of sublethal exposure on resistance evolution and the impact on WCR population dynamics. Resistance monitoring for the three MOAs expressed in SmartStax® PRO and management of WCR within an IPM framework will be fundamental to guarantee the durability of these traits, ensuring available WCR control measures in upcoming years.

Finally, continued exploration into the WCR RNAi mechanisms (i.e., uptake gene silencing, and systemic spread) will benefit the search for putative routes of resistance, inform mitigation strategies, and perhaps lead to the development of efficient monitoring strategies utilizing bioinformatics and molecular approaches. In addition, an understanding of WCR RNAi mechanisms could provide insights into the efficiency (or lack thereof) in other insect pests, potentially providing paths for innovation. The development and validation of parental and reproductive RNAi have demonstrated the flexibility that RNAi could provide for insect pest management, representing a new approach to mitigating insect damage and a complementary approach to larvicidal RNAi.

The specificity of RNAi technologies continues to move agriculture towards more environmentally friendly insecticides. The lessons learned with WCR, the first insect to be targeted with an RNAi technology, can serve as an example for the development of novel RNAi insecticides in other insect pests, providing a safe and efficient MOA.

Author Contributions: Conceptualization and writing project administration, A.M.V.; writing, A.M.V., M.D., J.D.R., A.S., A.L.L., P.R., C.J.B., J.R.F., J.M.R., J.S.P. and K.E.N.; reviewing and editing, A.M.V., M.D., J.D.R., A.S., J.S.P., C.J.B., J.M.R. and K.E.N. All authors have read and agreed to the published version of the manuscript.

Funding: This research received no external funding.

Institutional Review Board Statement: Not applicable.

Data Availability Statement: No new data were created or analyzed in this study. Data sharing is not applicable to this article.

Conflicts of Interest: A.M.V. has received research funding related to RNAi in western corn rootworm from Dow AgroSciences, DuPont Pioneer, Monsanto Company, Corteva Agriscience, Bayer Crop Science, and GreenLight Biosciences; some of these results are reported in this review. She also holds six RNAi patents. Graduate research on RNAi in western corn rootworm conducted by J.D.R. was funded by Bayer Crop Science. A.S., A.L.L., C.J.B. and J.M.R. work at Corteva Agriscience. P.R., J.R.F. and J.S.P. work at Bayer Crop Science. K.E.N. works at Greenlight Biosciences and holds multiple RNAi patents. The information and views represented in this review are those of the authors as individuals and experts in the field and do not necessarily represent those of their organizations. The mention of a product does not constitute an endorsement or a recommendation for its use.

References

1. Sappington, T.W.; Siegfried, B.D.; Guillemaud, T. Coordinated *Diabrotica* genetics research: Accelerating progress on an urgent insect pest problem. *Am. Entomol.* **2006**, *52*, 90–97. [CrossRef]
2. Gray, M.E.; Sappington, T.W.; Miller, N.J.; Moeser, J.; Bohn, M.O. Adaptation and invasiveness of western corn rootworm: Intensifying research on a worsening pest. *Annu. Rev. Entomol.* **2009**, *54*, 303–321. [CrossRef] [PubMed]
3. Gillette, C.P. *Diabrotica virgifera* LeConte as a corn rootworm. *J. Econ. Entomol.* **1912**, *54*, 364–366. [CrossRef]
4. Melhus, I.E.; Painter, R.H.; Smith, F.O. A search for resistance to the injury caused by species of *Diabrotica* in the corns of Guatemala. *Iowa State Coll. J. Sci.* **1954**, *29*, 75–94.
5. Baca, F. New member of the harmful entomofauna of Yugoslavia *Diabrotica virgifera virgifera* LeConte (Coleoptera: Chrysomelidae). *Zast Bilja* **1994**, *45*, 125–131.
6. Kiss, J.; Edwards, C.R.; Berger, H.K.; Cate, P.; Cean, M.; Cheek, S.; Derron, J.; Festic, H.; Furlan, L.; Iger-Barcic, J.; et al. Monitoring of western corn rootworm (*Diabrotica virgifera virgifera* LeConte) in Europe 1992–2003. In *Western Corn Rootworm: Ecology and Management*; Vidal, S., Kuhlman, U., Eds.; C.R. Edwards, CABI International: Wallingford, UK, 2005; Volume 2, pp. 29–39.
7. Meinke, L.J.; Souza, D.; Siegfried, B.D. The use of insecticides to manage the western corn rootworm, *Diabrotica virgifera virgifera*, LeConte: History, field-evolved resistance, and associated mechanisms. *Insects* **2021**, *12*, 112. [CrossRef] [PubMed]
8. Gassmann, A.J. Resistance to Bt maize by western corn rootworm: Effects of pest biology, the pest–crop interaction and the agricultural landscape on resistance. *Insects* **2021**, *12*, 136. [CrossRef]
9. Metcalf, R.L. Implications and prognosis of resistance to insecticides. In *Pest Resistance to Pesticides*; Georghiou, G.P., Saito, T., Eds.; Plenum Press: New York, NY, USA, 1983; pp. 769–792.
10. Meinke, L.J.; Siegfried, B.D.; Wright, R.J.; Chandler, L.D. Adult susceptibility of Nebraska populations of western corn rootworm to selected insecticides. *J. Econ. Entomol.* **1998**, *91*, 594–600. [CrossRef]
11. Levine, E.; Spencer, J.L.; Isard, S.A.; Onstad, D.W.; Gray, M.E. Adaptation of the western corn rootworm to crop rotation: Evolution of a new strain in response to a cultural management practice. *Am. Entomol.* **2002**, *48*, 94–107. [CrossRef]
12. Pereira, A.E.; Wang, H.; Zukoff, S.N.; Meinke, L.J.; French, B.W.; Siegfried, B.D. Evidence of field-evolved resistance to bifenthrin in western corn rootworm (*Diabrotica virgifera virgifera* LeConte) populations in western Nebraska and Kansas. *PLoS ONE* **2015**, *10*, e0142299. [CrossRef] [PubMed]
13. Souza, D.; Vieira, B.C.; Fritz, B.K.; Hoffmann, W.C.; Peterson, J.A.; Kruger, G.R.; Meinke, L.J. Western corn rootworm pyrethroid resistance confirmed by aerial application simulations of commercial insecticides. *Sci. Rep.* **2019**, *9*, 6713. [CrossRef]

14. Gassmann, A.J.; Petzold-Maxwell, J.L.; Clifton, E.H.; Dunbar, M.W.; Hoffmann, A.M.; Ingber, D.A.; Keweshan, R.S. Field-evolved resistance by western corn rootworm to multiple *Bacillus thuringiensis* toxins in transgenic maize. *Proc. Natl. Acad. Sci. USA* **2014**, *111*, 5141–5146. [CrossRef]
15. Wechsler, S.; Smith, D. Has resistance taken root in U.S. corn fields? Demand for insect control. *Am. J. Agric. Econ.* **2018**, *100*, 1136–1150. [CrossRef]
16. Crickmore, N.; Berry, C.; Panneerselvam, S.; Mishra, R.; Connor, T.R.; Bonning, B.C. A structure-based nomenclature for *Bacillus thuringiensis* and other bacteria-derived pesticidal proteins. *J. Invertebr. Pathol.* **2020**, 107438. [CrossRef] [PubMed]
17. USEPA [U.S. Environmental Protection Agency]. Pesticide Product Label, Corn Event MON 863: Corn Rootworm-Protected Corn. 2003. Available online: https://www3.epa.gov/pesticides/chem_search/ppls/000524-00528-20030224.pdf (accessed on 27 December 2021).
18. USEPA [U.S. Environmental Protection Agency]. Pesticide Product Label, Agrisure™ RW Rootworm-Protected Com. 2006. Available online: https://www3.epa.gov/pesticides/chem_search/ppls/067979-00005-20061003.pdf (accessed on 27 December 2021).
19. USEPA [U.S. Environmental Protection Agency]. Pesticide Product Label, 5307 Corn. 2012. Available online: https://www3.epa.gov/pesticides/chem_search/ppls/067979-00022-20120731.pdf (accessed on 27 December 2021).
20. USEPA [U.S. Environmental Protection Agency]. Pesticide Product Label, Herculex Xtra Insect Protection. 2005. Available online: https://www3.epa.gov/pesticides/chem_search/ppls/029964-00005-20051027.pdf (accessed on 27 December 2021).
21. Gassmann, A.J.; Petzold-Maxwell, J.L.; Keweshan, R.S.; Dunbar, M.W. Field-evolved resistance to Bt maize by western corn rootworm. *PLoS ONE* **2011**, *6*, e22629. [CrossRef] [PubMed]
22. Schrader, P.M.; Estes, R.E.; Tinsley, N.A.; Gassmann, A.J.; Gray, M.E. Evaluation of adult emergence and larval root injury for Cry3Bb1-resistant populations of the western corn rootworm. *J. Appl. Entomol.* **2017**, *141*, 41–52. [CrossRef]
23. Reinders, J.D.; Hitt, B.D.; Stroup, W.W.; French, B.W.; Meinke, L.J. Spatial variation in western corn rootworm (Coleoptera: Chrysomelidae) susceptibility to Cry3 toxins in Nebraska. *PLoS ONE* **2018**, *13*, e0208266. [CrossRef] [PubMed]
24. Calles-Torrez, V.; Knodel, J.J.; Boetel, M.A.; French, B.W.; Fuller, B.W.; Ransom, J.K. Field-evolved resistance of northern and western corn rootworm (Coleoptera: Chrysomelidae) populations to corn hybrids expressing single and pyramided Cry3Bb1 and Cry34/35Ab1 Bt proteins in North Dakota. *J. Econ. Entomol.* **2019**, *112*, 1875–1886. [CrossRef] [PubMed]
25. Jakka, S.R.; Shrestha, R.B.; Gassmann, A.J. Broad-spectrum resistance to *Bacillus thuringiensis* toxins by western corn rootworm (*Diabrotica virgifera virgifera*). *Sci. Rep.* **2016**, *6*, 27860. [CrossRef]
26. Wangila, D.S.; Gassmann, A.J.; Petzold-Maxwell, J.L.; French, B.W.; Meinke, L.J. Susceptibility of Nebraska western corn rootworm (Coleoptera: Chrysomelidae) populations to Bt corn events. *J. Econ. Entomol.* **2015**, *108*, 742–751. [CrossRef] [PubMed]
27. Gassmann, A.J.; Shrestha, R.B.; Kropf, A.L.; St Clair, C.R.; Brenizer, B.D. Field-evolved resistance by western corn rootworm to Cry34/35Ab1 and other *Bacillus thuringiensis* traits in transgenic maize. *Pest Manag. Sci.* **2020**, *76*, 268–276. [CrossRef] [PubMed]
28. Zukoff, S.N.; Ostlie, K.R.; Potter, B.; Meihls, L.N.; Zukoff, A.L.; French, L.; Ellersieck, M.R.; Wade French, B.; Hibbard, B.E. Multiple assays indicate varying levels of cross resistance in Cry3Bb1-selected field populations of the western corn rootworm to mCry3A, eCry3.1Ab, and Cry34/35Ab1. *J. Econ. Entomol.* **2016**, *109*, 1387–1398. [CrossRef] [PubMed]
29. Gassmann, A.J.; Shrestha, R.B.; Jakka, S.R.; Dunbar, M.W.; Clifton, E.H.; Paolino, A.R.; Ingber, D.A.; French, B.W.; Masloski, K.E.; Dounda, J.W.; et al. Evidence of resistance to Cry34/35Ab1 corn by western corn rootworm (Coleoptera: Chrysomelidae): Root injury in the field and larval survival in plant-based bioassays. *J. Econ. Entomol.* **2016**, *109*, 1872–1880. [CrossRef]
30. Ludwick, D.C.; Meihls, L.N.; Ostlie, K.R.; Potter, B.D.; French, L.; Hibbard, B.E. Minnesota field population of western corn rootworm (Coleoptera: Chrysomelidae) shows incomplete resistance to Cry34Ab1/Cry35Ab1 and Cry3Bb1. *J. Appl. Entomol.* **2017**, *141*, 28–40. [CrossRef]
31. Bolognesi, R.; Ramaseshadri, P.; Anderson, J.; Bachman, P.; Clinton, W.; Flannagan, R.; Ilagan, O.; Lawrence, C.; Levine, S.; Moar, W.; et al. Characterizing the mechanism of action of double-stranded RNA activity against western corn rootworm (*Diabrotica virgifera virgifera* LeConte). *PLoS ONE* **2012**, *7*, e47534. [CrossRef]
32. Head, G.P.; Carroll, M.W.; Evans, S.P.; Rule, D.M.; Willse, A.R.; Clark, T.L.; Storer, N.P.; Flannagan, R.D.; Samuel, L.W.; Meinke, L.J. Evaluation of SmartStax and SmartStax PRO maize against western corn rootworm and northern corn rootworm: Efficacy and resistance management. *Pest Manag. Sci.* **2017**, *73*, 1883–1899. [CrossRef] [PubMed]
33. USEPA [U.S. Environmental Protection Agency]. Pesticide Product Label, SmartStax® PRO Enlist™ Refuge Advanced®. 2017. Available online: https://www3.epa.gov/pesticides/chem_search/ppls/062719-00707-20170608.pdf (accessed on 27 December 2021).
34. Fire, A.; Xu, S.; Montgomery, M.K.; Kostas, S.A.; Driver, S.E.; Mello, C.C. Potent and specific genetic interference by double-stranded RNA in *Caenorhabditis elegans*. *Nature* **1998**, *391*, 806–811. [CrossRef] [PubMed]
35. Agrawal, N.; Dasaradhi, P.V.; Mohmmed, A.; Malhotra, P.; Bhatnagar, R.K.; Mukherjee, S.K. RNA interference: Biology, mechanism, and applications. *Microbiol. Mol. Biol. Rev.* **2003**, *67*, 657–685. [CrossRef] [PubMed]
36. Timmons, L.; Fire, A. Specific interference by ingested dsRNA. *Nature* **1998**, *395*, 854. [CrossRef] [PubMed]
37. Ghildiyal, M.; Zamore, P.D. Small silencing RNAs: An expanding universe. *Nat. Rev. Genet.* **2009**, *10*, 94–108. [CrossRef] [PubMed]
38. Shabalina, S.A.; Koonin, E.V. Origins and evolution of eukaryotic RNA interference. *Trends Ecol. Evol.* **2008**, *23*, 578–587. [CrossRef]

39. Mukherjee, K.; Campos, H.; Kolaczkowski, B. Evolution of animal and plant dicers: Early parallel duplications and recurrent adaptation of antiviral RNA binding in plants. *Mol. Biol. Evol.* **2013**, *30*, 627–641. [CrossRef]

40. Zhu, K.Y.; Palli, S.R. Mechanisms, Applications, and challenges of insect RNA interference. *Annu. Rev. Entomol.* **2020**, *65*, 293–311. [CrossRef]

41. Ender, C.; Meister, G. Argonaute proteins at a glance. *J. Cell Sci.* **2010**, *123*, 1819–1823. [CrossRef]

42. Tang, R.; Li, L.; Zhu, D.; Hou, D.; Cao, T.; Gu, H.; Zhang, J.; Chen, J.; Zhang, C.Y.; Zen, K. Mouse miRNA-709 directly regulates miRNA-15a/16-1 biogenesis at the posttranscriptional level in the nucleus: Evidence for a microRNA hierarchy system. *Cell Res.* **2012**, *22*, 504–515. [CrossRef]

43. Olsen, P.H.; Ambros, V. The lin-4 regulatory RNA controls developmental timing in *Caenorhabditis elegans* by blocking LIN-14 protein synthesis after the initiation of translation. *Dev. Biol.* **1999**, *216*, 671–680. [CrossRef]

44. Seggerson, K.; Tang, L.; Moss, E.G. Two genetic circuits repress the *Caenorhabditis elegans* heterochronic gene lin-28 after translation initiation. *Dev. Biol.* **2002**, *243*, 215–225. [CrossRef] [PubMed]

45. Lim, L.P.; Lau, N.C.; Garrett-Engele, P.; Grimson, A.; Schelter, J.M.; Castle, J.; Bartel, D.P.; Linsley, P.S.; Johnson, J.M. Microarray analysis shows that some microRNAs downregulate large numbers of target mRNAs. *Nature* **2005**, *433*, 769–773. [CrossRef] [PubMed]

46. Krützfeldt, J.; Rajewsky, N.; Braich, R.; Rajeev, K.G.; Tuschl, T.; Manoharan, M.; Stoffel, M. Silencing of microRNAs in vivo with 'antagomirs'. *Nature* **2005**, *438*, 685–689. [CrossRef]

47. Bagga, S.; Bracht, J.; Hunter, S.; Massirer, K.; Holtz, J.; Eachus, R.; Pasquinelli, A.E. Regulation by let-7 and lin-4 miRNAs results in target mRNA degradation. *Cell* **2005**, *122*, 553–563. [CrossRef]

48. Wu, L.; Belasco, J.G. Micro-RNA regulation of the mammalian lin-28 gene during neuronal differentiation of embryonal carcinoma cells. *Mol. Cell. Biol.* **2005**, *25*, 9198–9208. [CrossRef]

49. Aravin, A.A.; Naumova, N.M.; Tulin, A.V.; Vagin, V.V.; Rozovsky, Y.M.; Gvozdev, V.A. Double-stranded RNA-mediated silencing of genomic tandem repeats and transposable elements in the *D. melanogaster* germline. *Curr. Biol.* **2001**, *11*, 1017–1027. [CrossRef]

50. Cox, D.N.; Chao, A.; Baker, J.; Chang, L.; Qiao, D.; Lin, H. A novel class of evolutionarily conserved genes defined by piwi are essential for stem cell self-renewal. *Genes Dev.* **1998**, *12*, 3715–3727. [CrossRef] [PubMed]

51. Yang, N.; Kazazian, H.H. L1 retrotransposition is suppressed by endogenously encoded small interfering RNAs in human cultured cells. *Nat. Struct. Mol. Biol.* **2006**, *13*, 763–771. [CrossRef] [PubMed]

52. Jacobs, B.L.; Langland, J.O. When two strands are better than one: The mediators and modulators of the cellular responses to double-stranded RNA. *Virology* **1996**, *219*, 339–349. [CrossRef] [PubMed]

53. Bellés, X. Beyond *Drosophila*: RNAi in vivo and functional genomics in insects. *Annu. Rev. Entomol.* **2010**, *55*, 111–128. [CrossRef]

54. Price, D.R.; Gatehouse, J.A. RNAi-mediated crop protection against insects. *Trends Biotechnol.* **2008**, *26*, 393–400. [CrossRef]

55. Baum, J.A.; Bogaert, T.; Clinton, W.; Heck, G.R.; Feldmann, P.; Ilagan, O.; Johnson, S.; Plaetinck, G.; Munyikwa, T.; Pleau, M.; et al. Control of coleopteran insect pests through RNA interference. *Nat. Biotechnol.* **2007**, *25*, 1322–1326. [CrossRef]

56. Alves, A.P.; Lorenzen, M.D.; Beeman, R.W.; Foster, J.E.; Siegfried, B.D. RNA interference as a method for target-site screening in the Western corn rootworm, *Diabrotica virgifera virgifera*. *J. Insect Sci.* **2010**, *10*, 162. [CrossRef] [PubMed]

57. Winston, W.M.; Molodowitch, C.; Hunter, C.P. Systemic RNAi in *C. elegans* requires the putative transmembrane protein SID-1. *Science* **2002**, *295*, 2456–2459. [CrossRef]

58. Feinberg, E.H.; Hunter, C.P. Transport of dsRNA into cells by the transmembrane protein SID-1. *Science* **2003**, *301*, 1545–1547. [CrossRef] [PubMed]

59. McEwan, D.L.; Weisman, A.S.; Hunter, C.P. Uptake of extracellular double-stranded RNA by SID-2. *Mol. Cell* **2012**, *47*, 746–754. [CrossRef] [PubMed]

60. Jose, A.M.; Kim, Y.A.; Leal-Ekman, S.; Hunter, C.P. Conserved tyrosine kinase promotes the import of silencing RNA into *Caenorhabditis elegans* cells. *Proc. Natl. Acad. Sci. USA* **2012**, *109*, 14520–14525. [CrossRef] [PubMed]

61. Hinas, A.; Wright, A.J.; Hunter, C.P. SID-5 is an endosome-associated protein required for efficient systemic RNAi in *C. elegans*. *Curr. Biol.* **2012**, *22*, 1938–1943. [CrossRef]

62. Tomoyasu, Y.; Miller, S.C.; Tomita, S.; Schoppmeier, M.; Grossmann, D.; Bucher, G. Exploring systemic RNA interference in insects: A genome-wide survey for RNAi genes in *Tribolium*. *Genome Biol.* **2008**, *9*, R10. [CrossRef] [PubMed]

63. Miyata, K.; Ramaseshadri, P.; Zhang, Y.; Segers, G.; Bolognesi, R.; Tomoyasu, Y. Establishing an in vivo assay system to identify components involved in environmental RNA interference in the western corn rootworm. *PLoS ONE* **2014**, *9*, e101661. [CrossRef]

64. Pinheiro, D.H.; Vélez, A.M.; Fishilevich, E.; Wang, H.; Carneiro, N.P.; Valencia-Jiménez, A.; Valicente, F.H.; Narva, K.E.; Siegfried, B.D. Clathrin-dependent endocytosis is associated with RNAi response in the western corn rootworm, *Diabrotica virgifera virgifera* LeConte. *PLoS ONE* **2018**, *13*, e0201849. [CrossRef] [PubMed]

65. Cappelle, K.; de Oliveira, C.F.; Van Eynde, B.; Christiaens, O.; Smagghe, G. The involvement of clathrin-mediated endocytosis and two Sid-1-like transmembrane proteins in double-stranded RNA uptake in the Colorado potato beetle midgut. *Insect Mol. Biol.* **2016**, *25*, 315–323. [CrossRef] [PubMed]

66. Gruenberg, J. The endocytic pathway: A mosaic of domains. *Nat. Rev. Mol. Cell Biol.* **2001**, *2*, 721–730. [CrossRef]

67. Marsh, M.; Helenius, A. Virus entry: Open sesame. *Cell* **2006**, *124*, 729–740. [CrossRef] [PubMed]

68. Sorkin, A. Cargo recognition during clathrin-mediated endocytosis: A team effort. *Curr. Opin. Cell Biol.* **2004**, *16*, 392–399. [CrossRef] [PubMed]

69. McMahon, H.T.; Boucrot, E. Molecular mechanism and physiological functions of clathrin-mediated endocytosis. *Nat. Rev. Mol. Cell Biol.* **2011**, *12*, 517–533. [CrossRef] [PubMed]
70. Saleh, M.C.; van Rij, R.P.; Hekele, A.; Gillis, A.; Foley, E.; O'Farrell, P.H.; Andino, R. The endocytic pathway mediates cell entry of dsRNA to induce RNAi silencing. *Nat. Cell Biol.* **2006**, *8*, 793–802. [CrossRef]
71. Xiao, D.; Gao, X.; Xu, J.; Liang, X.; Li, Q.; Yao, J.; Zhu, K.Y. Clathrin-dependent endocytosis plays a predominant role in cellular uptake of double-stranded RNA in the red flour beetle. *Insect Biochem. Mol. Biol.* **2015**, *60*, 68–77. [CrossRef] [PubMed]
72. Guo, W.C.; Fu, K.Y.; Yang, S.; Li, X.X.; Li, G.Q. Instar-dependent systemic RNA interference response in *Leptinotarsa decemlineata* larvae. *Pestic Biochem. Physiol.* **2015**, *123*, 64–73. [CrossRef]
73. Ulvila, J.; Parikka, M.; Kleino, A.; Sormunen, R.; Ezekowitz, R.A.; Kocks, C.; Rämet, M. Double-stranded RNA is internalized by scavenger receptor-mediated endocytosis in *Drosophila* S2 cells. *J. Biol. Chem.* **2006**, *281*, 14370–14375. [CrossRef]
74. Vélez, A.M.; Fishilevich, E. The mysteries of insect RNAi: A focus on dsRNA uptake and transport. *Pestic. Biochem. Physiol.* **2018**, *151*, 25–31. [CrossRef] [PubMed]
75. Khajuria, C.; Ivashuta, S.; Wiggins, E.; Flagel, L.; Moar, W.; Pleau, M.; Miller, K.; Zhang, Y.; Ramaseshadri, P.; Jiang, C.; et al. Development and characterization of the first dsRNA-resistant insect population from western corn rootworm, *Diabrotica virgifera virgifera* LeConte. *PLoS ONE* **2018**, *13*, e0197059. [CrossRef]
76. Cooper, A.M.; Silver, K.; Zhang, J.; Park, Y.; Zhu, K.Y. Molecular mechanisms influencing efficiency of RNA interference in insects. *Pest Manag. Sci.* **2019**, *75*, 18–28. [CrossRef] [PubMed]
77. Grishok, A.; Pasquinelli, A.E.; Conte, D.; Li, N.; Parrish, S.; Ha, I.; Baillie, D.L.; Fire, A.; Ruvkun, G.; Mello, C.C. Genes and mechanisms related to RNA interference regulate expression of the small temporal RNAs that control *C. elegans* developmental timing. *Cell* **2001**, *106*, 23–34. [CrossRef]
78. Bernstein, E.; Caudy, A.A.; Hammond, S.M.; Hannon, G.J. Role for a bidentate ribonuclease in the initiation step of RNA interference. *Nature* **2001**, *409*, 363–366. [CrossRef]
79. Lee, Y.S.; Nakahara, K.; Pham, J.W.; Kim, K.; He, Z.; Sontheimer, E.J.; Carthew, R.W. Distinct roles for *Drosophila* Dicer-1 and Dicer-2 in the siRNA/miRNA silencing pathways. *Cell* **2004**, *117*, 69–81. [CrossRef]
80. Liu, Q.; Rand, T.A.; Kalidas, S.; Du, F.; Kim, H.E.; Smith, D.P.; Wang, X. R2D2, a bridge between the initiation and effector steps of the *Drosophila* RNAi pathway. *Science* **2003**, *301*, 1921–1925. [CrossRef] [PubMed]
81. Zamore, P.D.; Tuschl, T.; Sharp, P.A.; Bartel, D.P. RNAi: Double-stranded RNA directs the ATP-dependent cleavage of mRNA at 21 to 23 nucleotide intervals. *Cell* **2000**, *101*, 25–33. [CrossRef]
82. Elbashir, S.M.; Martinez, J.; Patkaniowska, A.; Lendeckel, W.; Tuschl, T. Functional anatomy of siRNAs for mediating efficient RNAi in *Drosophila melanogaster* embryo lysate. *EMBO J.* **2001**, *20*, 6877–6888. [CrossRef] [PubMed]
83. Elbashir, S.M.; Lendeckel, W.; Tuschl, T. RNA interference is mediated by 21- and 22-nucleotide RNAs. *Genes Dev.* **2001**, *15*, 188–200. [CrossRef]
84. Vélez, A.M.; Khajuria, C.; Wang, H.; Narva, K.E.; Siegfried, B.D. Knockdown of RNA interference pathway genes in western corn rootworms (*Diabrotica virgifera virgifera* Le Conte) demonstrates a possible mechanism of resistance to lethal dsRNA. *PLoS ONE* **2016**, *11*, e0157520. [CrossRef] [PubMed]
85. Wu, K.; Camargo, C.; Fishilevich, E.; Narva, K.E.; Chen, X.; Taylor, C.E.; Siegfried, B.D. Distinct fitness costs associated with the knockdown of RNAi pathway genes in western corn rootworm adults. *PLoS ONE* **2017**, *12*, e0190208. [CrossRef]
86. Davis-Vogel, C.; Ortiz, A.; Procyk, L.; Robeson, J.; Kassa, A.; Wang, Y.; Huang, E.; Walker, C.; Sethi, A.; Nelson, M.E.; et al. Knockdown of RNA interference pathway genes impacts the fitness of western corn rootworm. *Sci. Rep.* **2018**, *8*, 7858. [CrossRef] [PubMed]
87. Camargo, C.; Wu, K.; Fishilevich, E.; Narva, K.E.; Siegfried, B.D. Knockdown of RNA interference pathway genes in western corn rootworm, *Diabrotica virgifera virgifera*, identifies no fitness costs associated with Argonaute 2 or Dicer-2. *Pestic Biochem. Physiol.* **2018**, *148*, 103–110. [CrossRef] [PubMed]
88. Pham, J.W.; Pellino, J.L.; Lee, Y.S.; Carthew, R.W.; Sontheimer, E.J. A Dicer-2-dependent 80s complex cleaves targeted mRNAs during RNAi in *Drosophila*. *Cell* **2004**, *117*, 83–94. [CrossRef]
89. Tomari, Y.; Du, T.; Haley, B.; Schwarz, D.S.; Bennett, R.; Cook, H.A.; Koppetsch, B.S.; Theurkauf, W.E.; Zamore, P.D. RISC assembly defects in the *Drosophila* RNAi mutant armitage. *Cell* **2004**, *116*, 831–841. [CrossRef]
90. Liu, X.; Jiang, F.; Kalidas, S.; Smith, D.; Liu, Q. Dicer-2 and R2D2 coordinately bind siRNA to promote assembly of the siRISC complexes. *RNA* **2006**, *12*, 1514–1520. [CrossRef]
91. Pham, J.W.; Sontheimer, E.J. Molecular requirements for RNA-induced silencing complex assembly in the *Drosophila* RNA interference pathway. *J. Biol. Chem.* **2005**, *280*, 39278–39283. [CrossRef]
92. Liang, C.; Wang, Y.; Murota, Y.; Liu, X.; Smith, D.; Siomi, M.C.; Liu, Q. TAF11 Assembles the RISC loading complex to enhance RNAi efficiency. *Mol. Cell* **2015**, *59*, 807–818. [CrossRef] [PubMed]
93. Okamura, K.; Ishizuka, A.; Siomi, H.; Siomi, M.C. Distinct roles for Argonaute proteins in small RNA-directed RNA cleavage pathways. *Genes Dev.* **2004**, *18*, 1655–1666. [CrossRef]
94. Hammond, S.M.; Boettcher, S.; Caudy, A.A.; Kobayashi, R.; Hannon, G.J. Argonaute2, a link between genetic and biochemical analyses of RNAi. *Science* **2001**, *293*, 1146–1150. [CrossRef]
95. Matranga, C.; Tomari, Y.; Shin, C.; Bartel, D.P.; Zamore, P.D. Passenger-strand cleavage facilitates assembly of siRNA into Ago2-containing RNAi enzyme complexes. *Cell* **2005**, *123*, 607–620. [CrossRef]

96. Hammond, S.M.; Bernstein, E.; Beach, D.; Hannon, G.J. An RNA-directed nuclease mediates post-transcriptional gene silencing in *Drosophila* cells. *Nature* **2000**, *404*, 293–296. [CrossRef]

97. Chen, J.; Peng, Y.; Zhang, H.; Wang, K.; Zhao, C.; Zhu, G.; Reddy Palli, S.; Han, Z. Off-target effects of RNAi correlate with the mismatch rate between dsRNA and non-target mRNA. *RNA Biol.* **2021**, *18*, 1747–1759. [CrossRef]

98. Rivas, F.V.; Tolia, N.H.; Song, J.J.; Aragon, J.P.; Liu, J.; Hannon, G.J.; Joshua-Tor, L. Purified Argonaute2 and an siRNA form recombinant human RISC. *Nat. Struct. Mol. Biol.* **2005**, *12*, 340–349. [CrossRef]

99. Kulkarni, M.M.; Booker, M.; Silver, S.J.; Friedman, A.; Hong, P.; Perrimon, N.; Mathey-Prevot, B. Evidence of off-target effects associated with long dsRNAs in *Drosophila melanogaster* cell-based assays. *Nat. Methods* **2006**, *3*, 833–838. [CrossRef]

100. Pratt, A.J.; MacRae, I.J. The RNA-induced silencing complex: A versatile gene-silencing machine. *J. Biol. Chem.* **2009**, *284*, 17897–17901. [CrossRef] [PubMed]

101. Iwasaki, S.; Kobayashi, M.; Yoda, M.; Sakaguchi, Y.; Katsuma, S.; Suzuki, T.; Tomari, Y. Hsc70/Hsp90 chaperone machinery mediates ATP-dependent RISC loading of small RNA duplexes. *Mol. Cell* **2010**, *39*, 292–299. [CrossRef] [PubMed]

102. Ji, L.; Chen, X. Regulation of small RNA stability: Methylation and beyond. *Cell Res.* **2012**, *22*, 624–636. [CrossRef]

103. Yoon, J.S.; Shukla, J.N.; Gong, Z.J.; Mogilicherla, K.; Palli, S.R. RNA interference in the Colorado potato beetle, *Leptinotarsa decemlineata*: Identification of key contributors. *Insect Biochem. Mol. Biol.* **2016**, *78*, 78–88. [CrossRef]

104. Zhou, R.; Czech, B.; Brennecke, J.; Sachidanandam, R.; Wohlschlegel, J.A.; Perrimon, N.; Hannon, G.J. Processing of *Drosophila* endo-siRNAs depends on a specific Loquacious isoform. *RNA* **2009**, *15*, 1886–1895. [CrossRef]

105. Yoon, J.S.; Mogilicherla, K.; Gurusamy, D.; Chen, X.; Chereddy, S.C.R.R.; Palli, S.R. Double-stranded RNA binding protein, Staufen, is required for the initiation of RNAi in coleopteran insects. *Proc. Natl. Acad. Sci. USA* **2018**, *115*, 8334–8339. [CrossRef]

106. Tassetto, M.; Kunitomi, M.; Andino, R. Circulating immune cells mediate a systemic RNAi-based adaptive antiviral response in *Drosophila*. *Cell* **2017**, *169*, 314–325.e313. [CrossRef]

107. Karlikow, M.; Goic, B.; Saleh, M.C. RNAi and antiviral defense in *Drosophila*: Setting up a systemic immune response. *Dev. Comp. Immunol.* **2014**, *42*, 85–92. [CrossRef]

108. Pak, J.; Fire, A. Distinct populations of primary and secondary effectors during RNAi in *C. elegans*. *Science* **2007**, *315*, 241–244. [CrossRef] [PubMed]

109. Gu, W.; Shirayama, M.; Conte, D.; Vasale, J.; Batista, P.J.; Claycomb, J.M.; Moresco, J.J.; Youngman, E.M.; Keys, J.; Stoltz, M.J.; et al. Distinct argonaute-mediated 22G-RNA pathways direct genome surveillance in the *C. elegans* germline. *Mol. Cell* **2009**, *36*, 231–244. [CrossRef]

110. Aoki, K.; Moriguchi, H.; Yoshioka, T.; Okawa, K.; Tabara, H. *In vitro* analyses of the production and activity of secondary small interfering RNAs in *C. elegans*. *EMBO J.* **2007**, *26*, 5007–5019. [CrossRef]

111. Shih, J.D.; Hunter, C.P. SID-1 is a dsRNA-selective dsRNA-gated channel. *RNA* **2011**, *17*, 1057–1065. [CrossRef] [PubMed]

112. Gordon, K.H.; Waterhouse, P.M. RNAi for insect-proof plants. *Nat. Biotechnol.* **2007**, *25*, 1231–1232. [CrossRef]

113. Li, H.R.; Bowling, A.J.; Gandra, P.; Rangasamy, M.; Pence, H.E.; McEwan, R.E.; Khajuria, C.; Siegfried, B.D.; Narva, K.E. Systemic RNAi in western corn rootworm, *Diabrotica virgifera virgifera*, does not involve transitive pathways. *Insect Sci.* **2018**, *25*, 45–56. [CrossRef]

114. Hu, X.; Richtman, N.M.; Zhao, J.Z.; Duncan, K.E.; Niu, X.; Procyk, L.A.; Oneal, M.A.; Kernodle, B.M.; Steimel, J.P.; Crane, V.C.; et al. Discovery of midgut genes for the RNA interference control of corn rootworm. *Sci. Rep.* **2016**, *6*, 30542. [CrossRef]

115. Khajuria, C.; Velez, A.M.; Rangasamy, M.; Wang, H.; Fishilevich, E.; Frey, M.L.; Carneiro, N.P.; Gandra, P.; Narva, K.E.; Siegfried, B.D. Parental RNA interference of genes involved in embryonic development of the western corn rootworm, *Diabrotica virgifera virgifera* LeConte. *Insect Biochem. Mol. Biol.* **2015**, *63*, 54–62. [CrossRef] [PubMed]

116. Mingels, L.; Wynant, N.; Santos, D.; Peeters, P.; Gansemans, Y.; Billen, J.; Van Nieuwerburgh, F.; Vanden Broeck, J. Extracellular vesicles spread the RNA interference signal of *Tribolium castaneum* TcA cells. *Insect Biochem. Mol. Biol.* **2020**, *122*, 103377. [CrossRef]

117. Yoon, J.S.; Kim, K.; Palli, S.R. Double-stranded RNA in exosomes: Potential systemic RNA interference pathway in the Colorado potato beetle, *Leptinotarsa decemlineata*. *J. Asia-Pac. Entomol.* **2020**, *23*, 1160–1164. [CrossRef]

118. Vaughn, T. A method of controlling corn rootworm feeding using a *Bacillus thuringiensis* protein expressed in transgenic maize. *Crop. Sci.* **2005**, *45*, 931–938. [CrossRef]

119. Deitloff, J.; Dunbar, M.W.; Ingber, D.A.; Hibbard, B.E.; Gassmann, A.J. Effects of refuges on the evolution of resistance to transgenic corn by the western corn rootworm, *Diabrotica virgifera virgifera* LeConte. *Pest Manag. Sci.* **2016**, *72*, 190–198. [CrossRef]

120. Schellenberger, U.; Oral, J.; Rosen, B.A.; Wei, J.Z.; Zhu, G.; Xie, W.; McDonald, M.J.; Cerf, D.C.; Diehn, S.H.; Crane, V.C.; et al. A selective insecticidal protein from *Pseudomonas* for controlling corn rootworms. *Science* **2016**, *354*, 634–637. [CrossRef]

121. Domínguez-Arrizabalaga, M.; Villanueva, M.; Escriche, B.; Ancín-Azpilicueta, C.; Caballero, P. Insecticidal activity of *Bacillus thuringiensis* proteins against coleopteran pests. *Toxins* **2020**, *12*, 430. [CrossRef]

122. Joga, M.R.; Zotti, M.J.; Smagghe, G.; Christiaens, O. RNAi Efficiency, systemic properties, and novel delivery methods for pest insect control: What we know so far. *Front. Physiol.* **2016**, *7*, 553. [CrossRef]

123. Knorr, E.; Fishilevich, E.; Tenbusch, L.; Frey, M.L.F.; Rangasamy, M.; Billion, A.; Worden, S.E.; Gandra, P.; Arora, K.; Lo, W.; et al. Gene silencing in *Tribolium castaneum* as a tool for the targeted identification of candidate RNAi targets in crop pests. *Sci. Rep.* **2018**, *8*, 2061. [CrossRef]

124. Koci, J.; Ramaseshadri, P.; Bolognesi, R.; Segers, G.; Flannagan, R.; Park, Y. Ultrastructural changes caused by *Snf7* RNAi in larval enterocytes of western corn rootworm (*Diabrotica virgifera virgifera* Le Conte). *PLoS ONE* **2014**, *9*, e83985. [CrossRef]

125. Ramaseshadri, P.; Segers, G.; Flannagan, R.; Wiggins, E.; Clinton, W.; Ilagan, O.; McNulty, B.; Clark, T.; Bolognesi, R. Physiological and cellular responses caused by RNAi- mediated suppression of *Snf7* orthologue in western corn rootworm (*Diabrotica virgifera virgifera*) larvae. *PLoS ONE* **2013**, *8*, e54270. [CrossRef]

126. Vélez, A.M.; Fishilevich, E.; Rangasamy, M.; Khajuria, C.; McCaskill, D.G.; Pereira, A.E.; Gandra, P.; Frey, M.L.; Worden, S.E.; Whitlock, S.L.; et al. Control of western corn rootworm via RNAi traits in maize: Lethal and sublethal effects of *Sec23* dsRNA. *Pest Manag. Sci.* **2020**, *76*, 1500–1512. [CrossRef]

127. Fishilevich, E.; Bowling, A.J.; Frey, M.L.F.; Wang, P.H.; Lo, W.; Rangasamy, M.; Worden, S.E.; Pence, H.E.; Gandra, P.; Whitlock, S.L.; et al. RNAi targeting of rootworm *Troponin I* transcripts confers root protection in maize. *Insect Biochem. Mol. Biol.* **2019**, *104*, 20–29. [CrossRef]

128. Hu, X.; Steimel, J.P.; Kapka-Kitzman, D.M.; Davis-Vogel, C.; Richtman, N.M.; Mathis, J.P.; Nelson, M.E.; Lu, A.L.; Wu, G. Molecular characterization of the insecticidal activity of double-stranded RNA targeting the smooth septate junction of western corn rootworm (*Diabrotica virgifera virgifera*). *PLoS ONE* **2019**, *14*, e0210491. [CrossRef]

129. Hu, X.; Boeckman, C.J.; Cong, B.; Steimel, J.P.; Richtman, N.M.; Sturtz, K.; Wang, Y.; Walker, C.L.; Yin, J.; Unger, A.; et al. Characterization of *DvSSJ1* transcripts targeting the smooth septate junction (SSJ) of western corn rootworm (*Diabrotica virgifera virgifera*). *Sci. Rep.* **2020**, *10*, 11139. [CrossRef]

130. Velez, A.M.; Fishilevich, E.; Matz, N.; Storer, N.P.; Narva, K.E.; Siegfried, B.D. Parameters for successful parental RNAi as an insect pest management tool in western corn rootworm, *Diabrotica virgifera virgifera*. *Genes* **2016**, *8*, 7. [CrossRef]

131. Fishilevich, E.; Vélez, A.M.; Khajuria, C.; Frey, M.L.; Hamm, R.L.; Wang, H.; Schulenberg, G.A.; Bowling, A.J.; Pence, H.E.; Gandra, P.; et al. Use of chromatin remodeling ATPases as RNAi targets for parental control of western corn rootworm (*Diabrotica virgifera virgifera*) and Neotropical brown stink bug (*Euschistus heros*). *Insect Biochem. Mol. Biol.* **2016**, *71*, 58–71. [CrossRef]

132. Niu, X.; Kassa, A.; Hu, X.; Robeson, J.; McMahon, M.; Richtman, N.M.; Steimel, J.P.; Kernodle, B.M.; Crane, V.C.; Sandahl, G.; et al. Control of western corn rootworm (*Diabrotica virgifera virgifera*) reproduction through plant-mediated RNA interference. *Sci. Rep.* **2017**, *7*, 12591. [CrossRef]

133. Siomi, M.C.; Sato, K.; Pezic, D.; Aravin, A.A. PIWI-interacting small RNAs: The vanguard of genome defence. *Nat. Rev. Mol. Cell Biol.* **2011**, *12*, 246–258. [CrossRef]

134. Fishilevich, E.; Vélez, A.M.; Storer, N.P.; Li, H.; Bowling, A.J.; Rangasamy, M.; Worden, S.E.; Narva, K.E.; Siegfried, B.D. RNAi as a management tool for the western corn rootworm, *Diabrotica virgifera virgifera*. *Pest Manag. Sci.* **2016**, *72*, 1652–1663. [CrossRef]

135. Levine, S.L.; Tan, J.; Mueller, G.M.; Bachman, P.M.; Jensen, P.D.; Uffman, J.P. Independent action between DvSnf7 RNA and Cry3Bb1 protein in southern corn rootworm, *Diabrotica undecimpunctata howardi* and Colorado potato beetle, *Leptinotarsa decemlineata*. *PLoS ONE* **2015**, *10*, e0118622. [CrossRef]

136. Vaccari, T.; Rusten, T.E.; Menut, L.; Nezis, I.P.; Brech, A.; Stenmark, H.; Bilder, D. Comparative analysis of ESCRT-I, ESCRT-II and ESCRT-III function in *Drosophila* by efficient isolation of ESCRT mutants. *J. Cell Sci.* **2009**, *122*, 2413–2423. [CrossRef] [PubMed]

137. Ulrich, J.; Dao, V.A.; Majumdar, U.; Schmitt-Engel, C.; Schwirz, J.; Schultheis, D.; Strohlein, N.; Troelenberg, N.; Grossmann, D.; Richter, T.; et al. Large scale RNAi screen in *Tribolium* reveals novel target genes for pest control and the proteasome as prime target. *BMC Genom.* **2015**, *16*, 674. [CrossRef] [PubMed]

138. Lehman, W.; Galińska-Rakoczy, A.; Hatch, V.; Tobacman, L.S.; Craig, R. Structural basis for the activation of muscle contraction by troponin and tropomyosin. *J. Mol. Biol.* **2009**, *388*, 673–681. [CrossRef]

139. Lord, C.; Ferro-Novick, S.; Miller, E.A. The highly conserved COPII coat complex sorts cargo from the endoplasmic reticulum and targets it to the golgi. *Cold Spring Harb. Perspect Biol.* **2013**, *5*. [CrossRef]

140. Horn, T.; Narov, K.D.; Panfilio, K.A. Persistent parental RNAi in the beetle *Tribolium castaneum* involves maternal transmission of long double-stranded RNA. *bioRxiv* **2021**. [CrossRef]

141. Holoch, D.; Moazed, D. RNA-mediated epigenetic regulation of gene expression. *Nat. Rev. Genet.* **2015**, *16*, 71–84. [CrossRef] [PubMed]

142. Ivashuta, S.; Zhang, Y.; Wiggins, B.E.; Ramaseshadri, P.; Segers, G.C.; Johnson, S.; Meyer, S.E.; Kerstetter, R.A.; McNulty, B.C.; Bolognesi, R.; et al. Environmental RNAi in herbivorous insects. *RNA* **2015**, *21*, 840–850. [CrossRef]

143. Armstrong, T.A.; Chen, H.; Ziegler, T.E.; Iyadurai, K.R.; Gao, A.G.; Wang, Y.; Song, Z.; Tian, Q.; Zhang, Q.; Ward, J.M.; et al. Quantification of transgene-derived double-stranded RNA in plants using the QuantiGene nucleic acid detection platform. *J. Agric. Food Chem.* **2013**, *61*, 12557–12564. [CrossRef]

144. Oleson, J.D.; Park, Y.-L.; Nowatzki, T.M.; Tollefson, J.J. Node-injury scale to evaluate root injury by corn rootworms (Coleoptera: Chrysomelidae). *J. Econ. Entomol.* **2005**, *98*, 1–8. [CrossRef] [PubMed]

145. Baum, J.A.; Roberts, J.K. Progress toward RNAi-mediated insect pest management. *Adv. Insect Phys.* **2014**, *47*, 249–295.

146. USEPA [U.S. Environmental Protection Agency]. EPA Registers Innovative Tool to Control Corn Rootworm. 2017. Available online: https://archive.epa.gov/epa/newsreleases/epa-registers-innovative-tool-control-corn-rootworm.html (accessed on 27 December 2021).

147. USEPA [U.S. Environmental Protection Agency]. Current and Previously Registered Section 3 Plant-Incorporated Protectant (PIP) Registrations. 2015. Available online: https://archive.epa.gov/pesticides/reregistration/web/html/current-previously-registered-section-3-plant-incorporated.html (accessed on 27 December 2021).

148. Moar, W.; Khajuria, C.; Pleau, M.; Ilagan, O.; Chen, M.; Jiang, C.; Price, P.; McNulty, B.; Clark, T.; Head, G. Cry3Bb1-resistant western corn rootworm, *Diabrotica virgifera virgifera* (LeConte) does not exhibit cross-resistance to *DvSnf7* dsRNA. *PLoS ONE* **2017**, *12*, e0169175. [CrossRef] [PubMed]
149. USEPA [U.S. Environmental Protection Agency]. Pesticide Product Label, Bt11 × MIR162 × MIR604 × TC1507 × 5307 Corn. 2012. Available online: https://www3.epa.gov/pesticides/chem_search/ppls/067979-00023-20120731.pdf (accessed on 27 December 2021).
150. Carrière, Y.; Fabrick, J.A.; Tabashnik, B.E. Can pyramids and seed mixtures delay resistance to Bt crops? *Trends Biotechnol.* **2016**, *34*, 291–302. [CrossRef]
151. Pereira, A.E.; Carneiro, N.P.; Siegfried, B.D. Comparative susceptibility of southern and western corn rootworm adults and larvae to v*ATPase-A* and *Snf7* dsRNAs. *J. RNAi Gene Silenc.* **2016**, *12*, 528–535.
152. Rangasamy, M.; Siegfried, B.D. Validation of RNA interference in western corn rootworm *Diabrotica virgifera virgifera* LeConte (Coleoptera: Chrysomelidae) adults. *Pest Manag. Sci.* **2012**, *68*, 587–591. [CrossRef] [PubMed]
153. Wu, K.; Taylor, C.E.; Fishilevich, E.; Narva, K.E.; Siegfried, B.D. Rapid and persistent RNAi response in western corn rootworm adults. *Pestic. Biochem. Physiol.* **2018**, *150*, 66–70. [CrossRef] [PubMed]
154. Bachman, P.M.; Huizinga, K.M.; Jensen, P.D.; Mueller, G.; Tan, J.; Uffman, J.P.; Levine, S.L. Ecological risk assessment for *DvSnf7* RNA: A plant-incorporated protectant with targeted activity against western corn rootworm. *Regul. Toxicol. Pharmacol.* **2016**, *81*, 77–88. [CrossRef] [PubMed]
155. Tabashnik, B.E.; Brévault, T.; Carrière, Y. Insect resistance to Bt crops: Lessons from the first billion acres. *Nat. Biotechnol.* **2013**, *31*, 510–521. [CrossRef] [PubMed]
156. Tabashnik, B.E.; Gould, F.; Carrière, Y. Delaying evolution of insect resistance to transgenic crops by decreasing dominance and heritability. *J. Evol. Biol.* **2004**, *17*, 904–912. [CrossRef] [PubMed]
157. Spencer, J.L.; Hibbard, B.E.; Moeser, J.; Onstad, D.W. Behaviour and ecology of the western corn rootworm (*Diabrotica virgifera virgifera* LeConte). *Agric. For. Entomol.* **2009**, *11*, 9–27. [CrossRef]
158. USEPA [U.S. Environmental Protection Agency]. Framework to Delay Corn Rootworm Resistance. 2016. Available online: https://www.epa.gov/regulation-biotechnology-under-tsca-and-fifra/framework-delay-corn-rootworm-resistance (accessed on 27 December 2021).
159. Martinez, J.C.; Caprio, M.A. IPM use with the deployment of a non-high dose Bt pyramid and mitigation of resistance for western corn rootworm (*Diabrotica virgifera virgifera*). *Environ. Entomol.* **2016**, *45*, 747–761. [CrossRef]
160. Palli, S.R. RNA interference in Colorado potato beetle: Steps toward development of dsRNA as a commercial insecticide. *Curr. Opin. Insect Sci.* **2014**, *6*, 1–8. [CrossRef]
161. Flagel, L.E.; Bansal, R.; Kerstetter, R.A.; Chen, M.; Carroll, M.; Flannagan, R.; Clark, T.; Goldman, B.S.; Michel, A.P. Western corn rootworm (*Diabrotica virgifera virgifera*) transcriptome assembly and genomic analysis of population structure. *BMC Genom.* **2014**, *15*, 195. [CrossRef]
162. Mishra, S.; Dee, J.; Moar, W.; Dufner-Beattie, J.; Baum, J.; Dias, N.P.; Alyokhin, A.; Buzza, A.; Rondon, S.I.; Clough, M.; et al. Selection for high levels of resistance to double-stranded RNA (dsRNA) in Colorado potato beetle (*Leptinotarsa decemlineata* Say) using non-transgenic foliar delivery. *Sci. Rep.* **2021**, *11*, 6523. [CrossRef] [PubMed]
163. Carstens, K.L.; Hayter, K.; Layton, R.J. A perspective on problem formulation and exposure assessment of transgenic crops. In Proceedings of the Fourth Meeting on Ecological Impact of Genetically Modified Organisms, GMOs in Integrated Plant Production, Rostock, Germany, 14–16 May 2009; Volume 52, pp. 23–30. Available online: https://www.iobc-wprs.org/members/shop_en.cfm?mod_Shop_detail_produkte=59 (accessed on 27 December 2021).
164. USEPA [U.S. Environmental Protection Agency]. Guidelines for Ecological Risk Assessment, EPA/630/R-95-002F. 1998. Available online: https://www.epa.gov/sites/default/files/2014-11/documents/eco_risk_assessment1998.pdf (accessed on 27 December 2021).
165. Raybould, A. Problem formulation and hypothesis testing for environmental risk assessments of genetically modified crops. *Environ. Biosaf. Res.* **2006**, *5*, 119–125. [CrossRef]
166. Wolt, J.D.; Peterson, R.K. Prospective formulation of environmental risk assessments: Probabilistic screening for Cry1A(b) maize risk to aquatic insects. *Ecotoxicol. Environ. Saf.* **2010**, *73*, 1182–1188. [CrossRef]
167. Devos, Y.; Romeis, J.; Luttik, R.; Maggiore, A.; Perry, J.N.; Schoonjans, R.; Streissl, F.; Tarazona, J.V.; Brock, T.C. Optimising environmental risk assessments: Accounting for ecosystem services helps to translate broad policy protection goals into specific operational ones for environmental risk assessments. *EMBO Rep.* **2015**, *16*, 1060–1063. [CrossRef] [PubMed]
168. Rose, R.I. White Paper on Tier-Based Testing for the Effects of Proteinaceous Insecticidal Plant-Incorporated Protectants on Non-Target Invertebrates for Regulatory Risk Assessment. 2007. Available online: https://www.epa.gov/sites/default/files/2015-09/documents/tier-based-testing.pdf (accessed on 27 December 2021).
169. EFSA. Guidance on the environmental risk assessment of genetically modified plants. *EFSA J.* **2010**, *8*, 1879. [CrossRef]
170. Romeis, J.; Hellmich, R.L.; Candolfi, M.P.; Carstens, K.; De Schrijver, A.; Gatehouse, A.M.; Herman, R.A.; Huesing, J.E.; McLean, M.A.; Raybould, A.; et al. Recommendations for the design of laboratory studies on non-target arthropods for risk assessment of genetically engineered plants. *Transgenic Res.* **2011**, *20*, 1–22. [CrossRef]

171. Romeis, J.; Raybould, A.; Bigler, F.; Candolfi, M.P.; Hellmich, R.L.; Huesing, J.E.; Shelton, A.M. Deriving criteria to select arthropod species for laboratory tests to assess the ecological risks from cultivating arthropod-resistant genetically engineered crops. *Chemosphere* **2013**, *90*, 901–909. [CrossRef]
172. Roberts, A.; Boeckman, C.J.; Mühl, M.; Romeis, J.; Teem, J.L.; Valicente, F.H.; Brown, J.K.; Edwards, M.G.; Levine, S.L.; Melnick, R.L.; et al. Sublethal endpoints in non-target organism testing for insect-active GE crops. *Front. Bioeng. Biotechnol.* **2020**, *8*, 556. [CrossRef]
173. Garcia-Alonso, M.; Jacobs, E.; Raybould, A.; Nickson, T.E.; Sowig, P.; Willekens, H.; Van der Kouwe, P.; Layton, R.; Amijee, F.; Fuentes, A.M.; et al. A tiered system for assessing the risk of genetically modified plants to non-target organisms. *Environ. Biosaf. Res.* **2006**, *5*, 57–65. [CrossRef]
174. Kim, Y.H.; Soumaila Issa, M.; Cooper, A.M.; Zhu, K.Y. RNA interference: Applications and advances in insect toxicology and insect pest management. *Pestic. Biochem. Physiol.* **2015**, *120*, 109–117. [CrossRef]
175. Raybould, A.; Vlachos, D. Non-target organism effects tests on Vip3A and their application to the ecological risk assessment for cultivation of MIR162 maize. *Transgenic Res.* **2011**, *20*, 599–611. [CrossRef]
176. Fischer, J.R.; MacQuarrie, G.R.; Malven, M.; Song, Z.; Rogan, G. Dissipation of *DvSnf7* RNA from late-season maize tissue in aquatic microcosms. *Environ. Toxicol. Chem.* **2020**, *39*, 1032–1040. [CrossRef]
177. Dubelman, S.; Fischer, J.; Zapata, F.; Huizinga, K.; Jiang, C.; Uffman, J.; Levine, S.; Carson, D. Environmental fate of double-stranded RNA in agricultural soils. *PLoS ONE* **2014**, *9*, e93155. [CrossRef]
178. Parker, K.M.; Barragán Borrero, V.; van Leeuwen, D.M.; Lever, M.A.; Mateescu, B.; Sander, M. Environmental fate of RNA interference pesticides: Adsorption and degradation of double-stranded RNA molecules in agricultural soils. *Environ. Sci. Technol.* **2019**, *53*, 3027–3036. [CrossRef]
179. Fischer, J.R.; Zapata, F.; Dubelman, S.; Mueller, G.M.; Uffman, J.P.; Jiang, C.; Jensen, P.D.; Levine, S.L. Aquatic fate of a double-stranded RNA in a sediment water system following an over-water application. *Environ. Toxicol. Chem.* **2017**, *36*, 727–734. [CrossRef]
180. Albright, V.C.; Wong, C.R.; Hellmich, R.L.; Coats, J.R. Dissipation of double-stranded RNA in aquatic microcosms. *Environ. Toxicol. Chem.* **2017**, *36*, 1249–1253. [CrossRef]
181. Fischer, J.R.; Zapata, F.; Dubelman, S.; Mueller, G.M.; Jensen, P.D.; Levine, S.L. Characterizing a novel and sensitive method to measure dsRNA in soil. *Chemosphere* **2016**, *161*, 319–324. [CrossRef]
182. Bachman, P.; Fischer, J.; Song, Z.; Urbanczyk-Wochniak, E.; Watson, G. Environmental fate and dissipation of applied dsRNA in soil, aquatic systems, and plants. *Front. Plant. Sci.* **2020**, *11*, 21. [CrossRef]
183. Li, H.; Khajuria, C.; Rangasamy, M.; Gandra, P.; Fitter, M.; Geng, C.; Woosely, A.; Hasler, J.; Schulenberg, G.; Worden, S.; et al. Long dsRNA but not siRNA initiates RNAi in western corn rootworm larvae and adults. *J. Appl. Entomol.* **2015**, *139*, 432–445. [CrossRef]
184. Paces, J.; Nic, M.; Novotny, T.; Svoboda, P. Literature review of baseline information on RNAi to support the environmental risk assessment of RNAi-based GM plants. *EFSA J.* **2017**, *15*, 1424E. [CrossRef]
185. Christiaens, O.; Dzhambazova, T.; Kostov, K.; Arpaia, S.; Joga, M.R.; Urru, I.; Sweet, J.; Smagghe, G. Literature review of baseline information on RNAi to support the environmental risk assessment of RNAi-based GM plants. *EFSA J.* **2018**, *15*, 1424E. [CrossRef]
186. Romeis, J.; Bartsch, D.; Bigler, F.; Candolfi, M.P.; Gielkens, M.M.; Hartley, S.E.; Hellmich, R.L.; Huesing, J.E.; Jepson, P.C.; Layton, R.; et al. Assessment of risk of insect-resistant transgenic crops to nontarget arthropods. *Nat. Biotechnol.* **2008**, *26*, 203–208. [CrossRef]
187. Vélez, A.M.; Jurzenski, J.; Matz, N.; Zhou, X.; Wang, H.; Ellis, M.; Siegfried, B.D. Developing an in vivo toxicity assay for RNAi risk assessment in honey bees, *Apis mellifera* L. *Chemosphere* **2016**, *144*, 1083–1090. [CrossRef]
188. Pan, H.; Yang, X.; Bidne, K.; Hellmich, R.L.; Siegfried, B.D.; Zhou, X. Dietary risk assessment of *v-ATPase A* dsRNAs on monarch butterfly larvae. *Front. Plant. Sci.* **2017**, *8*, 242. [CrossRef]
189. Haller, S.; Widmer, F.; Siegfried, B.D.; Zhuo, X.; Romeis, J. Responses of two ladybird beetle species (Coleoptera: Coccinellidae) to dietary RNAi. *Pest Manag. Sci.* **2019**, *75*, 2652–2662. [CrossRef]
190. Pan, H.; Yang, X.; Romeis, J.; Siegfried, B.D.; Zhou, X. Dietary RNAi toxicity assay exhibits differential responses to ingested dsRNAs among lady beetles. *Pest Manag. Sci.* **2020**, *76*, 3606–3614. [CrossRef]
191. Pan, H.; Xu, L.; Noland, J.E.; Li, H.; Siegfried, B.D.; Zhou, X. Assessment of potential risks of dietary RNAi to a soil micro-arthropod, *Sinella curviseta* Brook (Collembola: Entomobryidae). *Front. Plant. Sci.* **2016**, *7*, 1028. [CrossRef]
192. Bachman, P.M.; Bolognesi, R.; Moar, W.J.; Mueller, G.M.; Paradise, M.S.; Ramaseshadri, P.; Tan, J.; Uffman, J.P.; Warren, J.; Wiggins, B.E.; et al. Characterization of the spectrum of insecticidal activity of a double-stranded RNA with targeted activity against western corn rootworm (*Diabrotica virgifera virgifera* LeConte). *Transgenic Res.* **2013**, *22*, 1207–1222. [CrossRef]
193. Tan, J.; Levine, S.L.; Bachman, P.M.; Jensen, P.D.; Mueller, G.M.; Uffman, J.P.; Meng, C.; Song, Z.; Richards, K.B.; Beevers, M.H. No impact of *DvSnf7* RNA on honey bee (*Apis mellifera* L.) adults and larvae in dietary feeding tests. *Environ. Toxicol. Chem.* **2016**, *35*, 287–294. [CrossRef]
194. Raybould, A.; Burns, A. Problem formulation for off-target effects of externally applied double-stranded RNA-based products for pest control. *Front. Plant. Sci.* **2020**, *11*, 424. [CrossRef]
195. Frizzi, A.; Zhang, Y.; Kao, J.; Hagen, C.; Huang, S. Small RNA profiles from virus-infected fresh market vegetables. *J. Agric. Food Chem.* **2014**, *62*, 12067–12074. [CrossRef]

196. Ivashuta, S.I.; Petrick, J.S.; Heisel, S.E.; Zhang, Y.; Guo, L.; Reynolds, T.L.; Rice, J.F.; Allen, E.; Roberts, J.K. Endogenous small RNAs in grain: Semi-quantification and sequence homology to human and animal genes. *Food Chem. Toxicol.* **2009**, *47*, 353–360. [CrossRef] [PubMed]

197. Petrick, J.S.; Frierdich, G.E.; Carleton, S.M.; Kessenich, C.R.; Silvanovich, A.; Zhang, Y.; Koch, M.S. Corn rootworm-active RNA DvSnf7: Repeat dose oral toxicology assessment in support of human and mammalian safety. *Regul. Toxicol. Pharmacol.* **2016**, *81*, 57–68, Erratum in *Regul. Toxicol. Pharmacol.* **2016**, *82*, 191. [CrossRef]

198. Petrick, J.S.; Brower-Toland, B.; Jackson, A.L.; Kier, L.D. Safety assessment of food and feed from biotechnology-derived crops employing RNA-mediated gene regulation to achieve desired traits: A scientific review. *Regul. Toxicol. Pharmacol.* **2013**, *66*, 167–176. [CrossRef]

199. Sherman, J.H.; Munyikwa, T.; Chan, S.Y.; Petrick, J.S.; Witwer, K.W.; Choudhuri, S. RNAi technologies in agricultural biotechnology: The Toxicology Forum 40th Annual Summer Meeting. *Regul. Toxicol. Pharmacol.* **2015**, *73*, 671–680. [CrossRef]

200. Tripathi, S.; Suzuki, J.Y.; Ferreira, S.A.; Gonsalves, D. Papaya ringspot virus-P: Characteristics, pathogenicity, sequence variability and control. *Mol. Plant. Pathol.* **2008**, *9*, 269–280. [CrossRef]

201. Wagner, N.; Mroczka, A.; Roberts, P.D.; Schreckengost, W.; Voelker, T. RNAi trigger fragment truncation attenuates soybean FAD2-1 transcript suppression and yields intermediate oil phenotypes. *Plant. Biotechnol. J.* **2011**, *9*, 723–728. [CrossRef] [PubMed]

202. Yang, J.; Xing, G.; Niu, L.; He, H.; Guo, D.; Du, Q.; Qian, X.; Yao, Y.; Li, H.; Zhong, X.; et al. Improved oil quality in transgenic soybean seeds by RNAi-mediated knockdown of GmFAD2-1B. *Transgenic Res.* **2018**, *27*, 155–166. [CrossRef]

203. Waltz, E. Nonbrowning GM apple cleared for market. *Nat. Biotechnol.* **2015**, *33*, 326–327. [CrossRef]

204. Pence, M.; Spence, R.; Rood, T.; Habig, J.; Collinge, S. Petition for Extension of Nonregulated Status for X17 Ranger Russet and Y9 Atlantic Potatoes with Late Blight Resistance, Low Acrylamide Potential, Lowered Reducing Sugars, and Reduced Black Spot, Petition No. 16-064-01P. 2016. Available online: https://www.aphis.usda.gov/brs/aphisdocs/16_06401p.pdf (accessed on 27 December 2021).

205. Tran, N.L.; Barraj, L.M.; Collinge, S. Reduction in dietary acrylamide exposure-impact of potatoes with low acrylamide potential. *Risk Anal.* **2017**, *37*, 1754–1767. [CrossRef]

206. ISAAA [International Service for the Acquisition of Agri-Biotech Applications]. GM Crop Events List. 2021. Available online: https://www.isaaa.org/gmapprovaldatabase/cropslist/default.asp (accessed on 27 December 2021).

207. Heisel, S.E.; Zhang, Y.; Allen, E.; Guo, L.; Reynolds, T.L.; Yang, X.; Kovalic, D.; Roberts, J.K. Characterization of unique small RNA populations from rice grain. *PLoS ONE* **2008**, *3*, e2871. [CrossRef] [PubMed]

208. Jensen, P.D.; Zhang, Y.; Wiggins, B.E.; Petrick, J.S.; Zhu, J.; Kerstetter, R.A.; Heck, G.R.; Ivashuta, S.I. Computational sequence analysis of predicted long dsRNA transcriptomes of major crops reveals sequence complementarity with human genes. *GM Crops Food* **2013**, *4*, 90–97. [CrossRef]

209. O'Neill, M.J.; Bourre, L.; Melgar, S.; O'Driscoll, C.M. Intestinal delivery of non-viral gene therapeutics: Physiological barriers and preclinical models. *Drug Discov. Today* **2011**, *16*, 203–218. [CrossRef]

210. Park, N.J.; Li, Y.; Yu, T.; Brinkman, B.M.; Wong, D.T. Characterization of RNA in saliva. *Clin. Chem.* **2006**, *52*, 988–994. [CrossRef]

211. Huang, H.; Davis, C.D.; Wang, T.T.Y. Extensive degradation and low bioavailability of orally consumed corn miRNAs in mice. *Nutrients* **2018**, *10*, 215. [CrossRef]

212. Dickinson, B.; Zhang, Y.; Petrick, J.S.; Heck, G.; Ivashuta, S.; Marshall, W.S. Lack of detectable oral bioavailability of plant microRNAs after feeding in mice. *Nat. Biotechnol.* **2013**, *31*, 965–967. [CrossRef]

213. Witwer, K.W.; McAlexander, M.A.; Queen, S.E.; Adams, R.J. Real-time quantitative PCR and droplet digital PCR for plant miRNAs in mammalian blood provide little evidence for general uptake of dietary miRNAs: Limited evidence for general uptake of dietary plant xenomiRs. *RNA Biol.* **2013**, *10*, 1080–1086. [CrossRef]

214. Snow, J.W.; Hale, A.E.; Isaacs, S.K.; Baggish, A.L.; Chan, S.Y. Ineffective delivery of diet-derived microRNAs to recipient animal organisms. *RNA Biol.* **2013**, *10*, 1107–1116. [CrossRef]

215. Rodrigues, T.B.; Petrick, J.S. Safety considerations for humans and other vertebrates regarding agricultural uses of externally applied RNA molecules. *Front. Plant. Sci.* **2020**, *11*, 407. [CrossRef]

216. Gilmore, I.R.; Fox, S.P.; Hollins, A.J.; Sohail, M.; Akhtar, S. The design and exogenous delivery of siRNA for post-transcriptional gene silencing. *J. Drug Target.* **2004**, *12*, 315–340. [CrossRef]

217. Christensen, J.; Litherland, K.; Faller, T.; van de Kerkhof, E.; Natt, F.; Hunziker, J.; Krauser, J.; Swart, P. Metabolism studies of unformulated internally [3H]-labeled short interfering RNAs in mice. *Drug Metab. Dispos.* **2013**, *41*, 1211–1219. [CrossRef] [PubMed]

218. Layzer, J.M.; McCaffrey, A.P.; Tanner, A.K.; Huang, Z.; Kay, M.A.; Sullenger, B.A. In vivo activity of nuclease-resistant siRNAs. *RNA* **2004**, *10*, 766–771. [CrossRef]

219. White, P.J. Barriers to successful delivery of short interfering RNA after systemic administration. *Clin. Exp. Pharmacol. Physiol.* **2008**, *35*, 1371–1376. [CrossRef] [PubMed]

220. Thompson, J.D.; Kornbrust, D.J.; Foy, J.W.; Solano, E.C.; Schneider, D.J.; Feinstein, E.; Molitoris, B.A.; Erlich, S. Toxicological and pharmacokinetic properties of chemically modified siRNAs targeting p53 RNA following intravenous administration. *Nucleic Acid Ther.* **2012**, *22*, 255–264. [CrossRef] [PubMed]

221. Forbes, D.C.; Peppas, N.A. Oral delivery of small RNA and DNA. *J. Control Release* **2012**, *162*, 438–445. [CrossRef]

222. Gilleron, J.; Querbes, W.; Zeigerer, A.; Borodovsky, A.; Marsico, G.; Schubert, U.; Manygoats, K.; Seifert, S.; Andree, C.; Stöter, M.; et al. Image-based analysis of lipid nanoparticle-mediated siRNA delivery, intracellular trafficking and endosomal escape. *Nat. Biotechnol.* **2013**, *31*, 638–646. [CrossRef]
223. Juliano, R.L. The delivery of therapeutic oligonucleotides. *Nucleic Acids Res.* **2016**, *44*, 6518–6548. [CrossRef]
224. Moroz, E.; Matoori, S.; Leroux, J.C. Oral delivery of macromolecular drugs: Where we are after almost 100 years of attempts. *Adv. Drug Deliv. Rev.* **2016**, *101*, 108–121. [CrossRef]
225. Molitoris, B.A.; Dagher, P.C.; Sandoval, R.M.; Campos, S.B.; Ashush, H.; Fridman, E.; Brafman, A.; Faerman, A.; Atkinson, S.J.; Thompson, J.D.; et al. siRNA targeted to p53 attenuates ischemic and cisplatin-induced acute kidney injury. *J. Am. Soc. Nephrol.* **2009**, *20*, 1754–1764. [CrossRef]
226. Ballarín-González, B.; Dagnaes-Hansen, F.; Fenton, R.A.; Gao, S.; Hein, S.; Dong, M.; Kjems, J.; Howard, K.A. Protection and systemic translocation of siRNA following oral administration of Chitosan/siRNA nanoparticles. *Mol. Ther. Nucleic Acids* **2013**, *2*, e76. [CrossRef]
227. Tillman, L.G.; Geary, R.S.; Hardee, G.E. Oral delivery of antisense oligonucleotides in man. *J. Pharm. Sci.* **2008**, *97*, 225–236. [CrossRef]
228. Witwer, K.W. Hypothetical plant-mammal RNA communication: Packaging and stoichiometry. In *Non-Coding RNAs and Inter.-Kingdom Communication*; Leitao, A.L., Enguita, F.J., Eds.; Springer International Publishing: Basel, Switzerland, 2016; pp. 161–176.
229. Title, A.C.; Denzler, R.; Stoffel, M. Uptake and function studies of maternal milk-derived microRNAs. *J. Biol. Chem.* **2015**, *290*, 23680–23691. [CrossRef] [PubMed]
230. Li, M.; Chen, T.; Wang, R.; Luo, J.Y.; He, J.J.; Ye, R.S.; Xie, M.Y.; Xi, Q.Y.; Jiang, Q.Y.; Sun, J.J.; et al. Plant MIR156 regulates intestinal growth in mammals by targeting the Wnt/β-catenin pathway. *Am. J. Physiol. Cell Physiol.* **2019**, *317*, C434–C448. [CrossRef] [PubMed]
231. Mlotshwa, S.; Pruss, G.J.; MacArthur, J.L.; Endres, M.W.; Davis, C.; Hofseth, L.J.; Peña, M.M.; Vance, V. A novel chemopreventive strategy based on therapeutic microRNAs produced in plants. *Cell Res.* **2015**, *25*, 521–524. [CrossRef] [PubMed]
232. Zhang, L.; Hou, D.; Chen, X.; Li, D.; Zhu, L.; Zhang, Y.; Li, J.; Bian, Z.; Liang, X.; Cai, X.; et al. Exogenous plant MIR168a specifically targets mammalian LDLRAP1: Evidence of cross-kingdom regulation by microRNA. *Cell Res.* **2012**, *22*, 107–126. [CrossRef]
233. Zhou, Z.; Li, X.; Liu, J.; Dong, L.; Chen, Q.; Kong, H.; Zhang, Q.; Qi, X.; Hou, D.; Zhang, L.; et al. Honeysuckle-encoded atypical microRNA2911 directly targets influenza A viruses. *Cell Res.* **2015**, *25*, 39–49. [CrossRef]
234. Chan, S.Y.; Snow, J.W. Formidable challenges to the notion of biologically important roles for dietary small RNAs in ingesting mammals. *Genes Nutr.* **2017**, *12*, 13. [CrossRef]
235. Kang, W.; Bang-Berthelsen, C.H.; Holm, A.; Houben, A.J.; Müller, A.H.; Thymann, T.; Pociot, F.; Estivill, X.; Friedländer, M.R. Survey of 800+ data sets from human tissue and body fluid reveals xenomiRs are likely artifacts. *RNA* **2017**, *23*, 433–445. [CrossRef]
236. Witwer, K.W.; Halushka, M.K. Toward the promise of microRNAs—Enhancing reproducibility and rigor in microRNA research. *RNA Biol.* **2016**, *13*, 1103–1116. [CrossRef]
237. Petrick, J.S.; Moore, W.M.; Heydens, W.F.; Koch, M.S.; Sherman, J.H.; Lemke, S.L. A 28-day oral toxicity evaluation of small interfering RNAs and a long double-stranded RNA targeting vacuolar ATPase in mice. *Regul. Toxicol. Pharmacol.* **2015**, *71*, 8–23. [CrossRef]
238. Lusk, R.W. Diverse and widespread contamination evident in the unmapped depths of high throughput sequencing data. *PLoS ONE* **2014**, *9*, e110808. [CrossRef]
239. Tosar, J.P.; Rovira, C.; Naya, H.; Cayota, A. Mining of public sequencing databases supports a non-dietary origin for putative foreign miRNAs: Underestimated effects of contamination in NGS. *RNA* **2014**, *20*, 754–757. [CrossRef]

Review

Western Corn Rootworm (*Diabrotica virgifera virgifera* LeConte) in Europe: Current Status and Sustainable Pest Management

Renata Bažok [1], Darija Lemić [1], Francesca Chiarini [2] and Lorenzo Furlan [2,*]

[1]   Department for Agricultural Zoology, Faculty of Agriculture, University of Zagreb, Svetosimunska 25, 10000 Zagreb, Croatia; rbazok@agr.hr (R.B.); dlemic@agr.hr (D.L.)

[2]   Veneto Agricoltura, Agricultural Research Department, 35020 Legnaro, PD, Italy; francesca.chiarini@venetoagricoltura.org

*   Correspondence: lorenzo.furlan@venetoagricoltura.org; Tel.: +39-049-829-3879

Simple Summary: *Diabrotica virgifera virgifera*, also known as western corn rootworm (WCR), is a maize-specific pest that has been a serious threat in Europe since the mid-1990s. Between 1995 and 2010, European countries were involved in international projects to plan pest control strategies. However, since 2011, collaborative efforts have declined and the overview of knowledge on WCR is in great need of updating. Therefore, a review of scientific papers published between 2008 and 2020, in addition to direct interviews with experts responsible for WCR management in several European countries, was conducted to (1) summarize the research conducted over the last 12 years and (2) describe the current WCR distribution and population in the EU, and the management strategies implemented. A considerable amount of new knowledge has been gained over the last 12 years, which has contributed to the development of pest management strategies applicable in EU agricultural systems. There is no EU country reporting economic damage on a large scale. In many countries, solutions based on crop rotation are regularly implemented, avoiding insecticide use. Therefore, WCR has not become as serious a pest as was expected when it was discovered in much of Europe.

Citation: Bažok, R.; Lemić, D.; Chiarini, F.; Furlan, L. Western Corn Rootworm (*Diabrotica virgifera virgifera* LeConte) in Europe: Current Status and Sustainable Pest Management. *Insects* **2021**, *12*, 195. https://doi.org/10.3390/insects12030195

Academic Editor: Kent M. Daane

Received: 1 January 2021
Accepted: 21 February 2021
Published: 25 February 2021

Publisher's Note: MDPI stays neutral with regard to jurisdictional claims in published maps and institutional affiliations.

Abstract: Western corn rootworm (WCR), or *Diabrotica virgifera virgifera* LeConte, became a very serious quarantine maize pest in Europe in the mid-1990s. Between 1995 and 2010, European countries were involved in international projects to share information and plan common research for integrated pest management (IPM) implementation. Since 2011, however, common efforts have declined, and an overview of WCR population spread, density, and research is in serious need of update. Therefore, we retained that it was necessary to (1) summarize the research activities carried out in the last 12 years in various countries and the research topics addressed, and analyze how these activities have contributed to IPM for WCR and (2) present the current distribution of WCR in the EU and analyze the current population levels in different European countries, focusing on different management strategies. A review of scientific papers published from 2008 to 2020, in addition to direct interviews with experts in charge of WCR management in a range of European countries, was conducted. Over the past 12 years, scientists in Europe have continued their research activities to investigate various aspects of WCR management by implementing several approaches to WCR control. A considerable amount of new knowledge has been produced, contributing to the development of pest management strategies applicable in EU farming systems. Among the 10 EU countries analyzed, there is no country reporting economic damage on a large scale. Thanks to intensive research leading to specific agricultural practices and the EU Common Agricultural Policy, there are crop-rotation-based solutions that can adequately control this pest avoiding insecticide use.

Keywords: IPM; WCR distribution; WCR damage; crop-rotation-based solutions; agronomic alternative; eradication; containment

1. Introduction

The most severe maize pest in North America, the western corn rootworm (*D. virgifera virgifera* LeConte) (WCR), known as the billion-dollar beetle [1], was first discovered in Europe in 1992 near Belgrade, Serbia [2].

Immediately after the news of this pest arrival had spread among scientists, it became clear that international action against the WCR would be necessary. The International Working Group on *Ostrinia* and other Maize Pests (IWGO), which was established in the mid-20th century as part of the International Organization for Biological Control (IOBC) to study the European corn borer (ECB) and other corn pests, included the WCR in its program activities. Since 1995, IWGO, in collaboration with the European and Mediterranean Plant Protection Organization (EPPO) and Food and Agriculture Organization (FAO) of the United Nations (UN), has organized annual meetings to share new information among scientists on pest distribution and the damage caused. WCR is still on the agenda on regular biannual IWGO meetings. This made the WCR the only pest in the world to be monitored using the same method in most countries, and its spread was determined in detail every year. The threatened Eastern European countries were particularly active in this sense. The monitoring of adult WCR by European countries enabled the rapid detection and determination of the spread of this invasive pest species since its first observation [3]. Permanent monitoring stations were established by each network partner. These stations enabled the measurement of population fluctuations over the years.

Since the 1980s, there have been three phases of WCR invasion in Europe [4]—the first phase was from accidental introduction until the pest was first identified in the maize field in Serbia. Accidental introduction in Europe took place, according to Szalai et al. [5], between 1979 and 1984, i.e., 8 to 13 years before this species was discovered and started to damage maize fields. The second phase was the spread and establishment of WCR in the Eastern European countries that surround Serbia (i.e., Hungary, Croatia, Romania, Bulgaria, and Bosnia and Herzegovina). The third phase of invasion (2001–2018) was a dispersal phase in which WCR spread across most European countries (EPPO) [6].

According to WCR population genetics studies by Miller et al. [7] and Ciosi et al. [8], WCR was introduced into Serbia, with the population source probably being Pennsylvania. From Serbia, the pest spread naturally to most of the countries of central and southeastern Europe, in addition to the Italian region of Friuli. Later on, there were four other introductions of WCR into several other European countries: Italy (Lombardy region), France (Alsace region), France (Paris region), and the United Kingdom [8].

A significant bulk of new WCR knowledge has been created by Europe's scientific communities in the past 25 years. At the very beginning, European research focused on WCR monitoring and spread [3,9–18], predictions of its further spread and damage [19–26], tools for monitoring [27,28], ecology [13,29–35], and damage [36], and on control methods [37] and tools, including biological control [38–40].

This paper aims to analyze what has happened since the end of FAO and EU monitoring and research projects in different areas of the EU, and what the current WCR situation is in Europe in terms of research, spread, population levels, damage, and control measures. Our aims were to (1) summarize the research activities carried out in the last 12 years in various countries and the research topics addressed, and analyze how these activities have contributed to IPM for WCR and (2) present the current distribution of WCR in the EU and analyze the current population levels in different European countries, focusing on different management strategies.

2. History of WCR Management in Europe

The first international project "Development and Implementation of Containment and Control of the Western Corn Rootworm in Europe" was implemented from 1997 to 2000. It was founded by FAO. In 2003, a new FAO project "Integrated Pest Management for Western Corn Rootworm in Central and Eastern Europe (GTFS/RER/017/ITA)" was launched in the seven most endangered countries (Hungary, Croatia, Serbia, Bosnia and

Herzegovina, Slovakia, Romania, and Bulgaria) and implemented until 2008. Participatory research activities, field studies, and field-training sessions implemented in each country have demonstrated successful management approaches for controlling WCR in a range of agroecological and socioeconomic conditions. The introduction of farmer field schools (FFSs) and student field schools (SFSs) provided an innovative model of working with farmers, and for collaboration among farmers. The original focus on WCR risk management has also widened and led to a better understanding of local agrobiodiversity. The involvement of additional institutions (secondary schools, local and regional administrations) increased WCR awareness and led to new approaches to agricultural extension. Regional networking contributed to reaching a common understanding in all participating countries, from training activities to WCR monitoring and research [41].

However, only countries from east and central Europe were involved in the project. In many countries, the project was carried out by quarantine officials and not by scientists. The scientists involved very often had weak links with scientists from more developed EU countries and with better research infrastructure. Due to the quarantine status and weak research infrastructure, no laboratory colonies were available and all research activities were conducted in field conditions depending on the fluctuating WCR population, from very low to very high. As a result, the research results were not widely disseminated in the scientific community and had a limited impact on further research within the EU.

The first EU-funded scientific project on WCR, "Threat to European Maize Production by Invasive Quarantine Pest, Western Corn Rootworm (*Diabrotica virgifera virgifera*): A New Sustainable Crop Management Approach", conducted from 2000 to 2003, focused on eradication and containment measures. As a result, measures including crop management, plant-insect interactions, natural enemy assessment, risk management, and biotechnological control were investigated and developed [42]. The project results contributed to the creation of an EU strategy to contain and/or eradicate the pest, and in 2003, the Commission of the European Union issued emergency measures (Commission Decision 2003/766/EC of 24 October 2003 on emergency measures to prevent the spread of *Diabrotica virgifera virgifera* LeConte within the community) [43]. In 2006, the 2003 regulation was supplemented by Decision 2006/564/EC [44], which introduced additional requirements for the containment of WCR in the infested zones to limit the further spread of the pest. European Commission (EC) Recommendation 2006/565/EC [45] made it possible to switch from an eradication policy to a containment policy. However, from 2000 to 2009, WCR spread extensively over non-EU territory, and over some EU countries [4].

As the result of the second EU-funded project "Harmonise the Strategies for Fighting *Diabrotica virgifera virgifera*" implemented between 2006 and 2008 [46], several control strategies for WCR management were explored, including biological control, utilization of plant resistance traits, plus the adaptation of biotechnological approaches and cultural techniques. All of the explored measures had to carry a minimum impact on biodiversity and the environment. Additionally, a database was constructed containing all available literature on WCR ecology and current research activities, and a comprehensive review was written of past research focusing on WCR biology. Experts in maize agriculture and WCR ecology were brought together to develop biological WCR control strategies that could integrate with established control options used against other maize pests. Researchers also looked into the possibility of enhancing and maintaining various natural WCR enemies to reduce pest outbreaks.

Additional research has also been financed by national sources. Among the many national programs, the most comprehensive was the German *Diabrotica* research program financed by Germany's Ministry of Food, Agriculture, and Consumer Protection and implemented from 2008 to 2012. This program consisted of 11 research activities carried out at the federal level and 12 research activities carried out locally, with the region of Bavaria being the most endangered region in Germany [47].

In the past 10 years, WCR has not been considered a new pest in many EU countries because it became a regular part of entomofauna. Moreover, research and/or monitoring activities were organized at a national (or even local) level. WCR was removed from the EU quarantine list in 2014 [48]. However, it remains on the A1 or A2 list of some European countries that are non-EU members, such as Azerbaijan (on A1 list since 2007), Georgia (on A1 list since 2018), Moldova (on A1 list since 2006), Russia (on A1 list since 2014), Turkey (on A1 list since 2016), and Ukraine (on A2 list since 2019).

In the EU, solutions to manage WCR damage in maize must comply with current legislation requiring the implementation of the principles of integrated pest management (IPM), as described in Annex III of Directive 2009/128/EC [49]. The first IPM step is prevention, i.e., the implementation of a set of agronomic measures such as crop rotation (the first measure listed in Annex III) and, where appropriate, the use of resistant/tolerant varieties, which create the conditions for reducing the risk of pest outbreaks and thus the need for plant protection measures.

Because the most effective strategy against WCR is rotation [3,50,51], its implementation is made mandatory by the aforementioned legislation, but this may be a problem for livestock farms that have to maintain forage production at the best level in terms of yield and quality.

3. Research Activities on WCR in Europe and Topics Investigated

Google Scholar literature review was queried using the following keyword combinations: "western corn rootworm in Europe," "*Diabrotica virgifera virgifera* in Europe," etc. The search was limited to scientific articles or communications published in English, Croatian, Serbian, and German from 2008 to 2020 and researches conducted on the European area. We aimed to create an overview of all European research groups, a list of research topics, and a reference set of published articles. In Figure 1, the main research areas and number of published papers by each area have been presented.

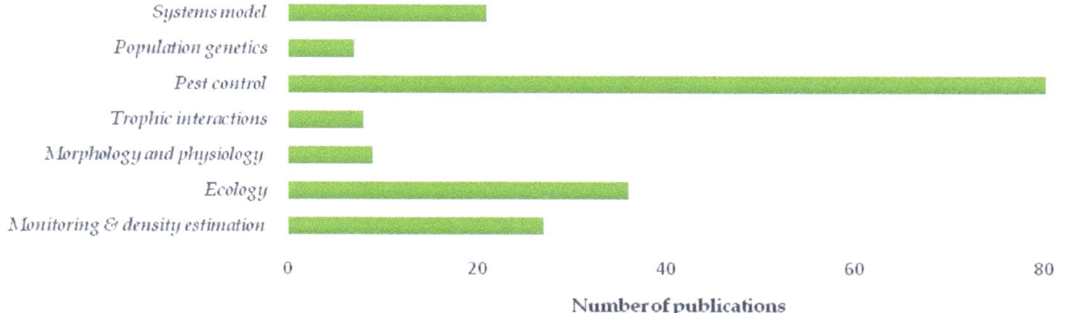

Figure 1. Western corn rootworm (WCR) research area in Europe and published papers accordingly.

Our review of research activities was composed of 187 relevant references covering seven WCR research areas in Europe over the last 12 years. All research areas are divided into sub-themes describing the focus of the research conducted. The main findings of each research area have been listed. The country of affiliation of the authors in the overview of research activities is also listed. The reference list of papers relevant to each defined subtopic is given in Table 1, and a brief description of all research areas is presented here with the main methods and results.

Table 1. Overview of the WCR research topics with their main findings and research groups from Europe in the period 2008–2020.

Research Area		Main Findings	Researches' Group Affiliation Country and Reference
Monitoring and density estimation	**First occurrence and spreading**	the pheromone monitoring is used in two directions: continuous monitoring on territories populated by pest; spreading monitoring—new hearth depistation on a non-populated area	Romania, Slovenia, Poland, Greece, Germany, Russia, Moldova, Italy, Croatia	[52–62]
	Population level	the highest abundance was quantified in semi-early and semi-late hybrids	Croatia, Germany, Romania, Ukraine, Slovakia, Poland	[55,63–69]
	Monitoring methods/techniques/designs	traditional monitoring can be effectively used to predict population abundance, and modern monitoring procedures can be used to estimate inter and intra-population variation	Germany, Croatia, Austria, Serbia, Hungary	[6,70–74]
	Area-wide monitoring	significant relationship of WCR flight dynamic with the weather and geographical conditions	Slovenia, Romania, Serbia, Hungary	[75–78]
Ecology	**Hosts**	cereals and oil pumpkin plants were not suitable as host plants for larval development while *Miscanthus* sp., are good hosts for WCR	Romania, Poland, Hungary, Austria, Germany, Switzerland, Serbia	[52,79–92]
	Individual movement	distance between maize fields and the phenological status of maize influenced inter-field movements	Slovakia, Romania, Hungary	[93–96]
	Climatic influence on WCR caches	physiological limit as a result of climate change might increase the strength of outbreaks at higher latitudes	Romania, Bulgaria, Croatia, Serbia, Spain, Finland	[97–105]
	Soil activities	plowing and disking had diminishing effects on WCR	Bosnia and Herzegovina, Romania	[106–109]
	Attractants	the most effective in catching WCR were traps with sex attractant	Poland, Croatia, Hungary, Germany	[110–114]
Morphology and physiology	**Sexual dimorphism**	sexual dimorphism may be modulated by natural selection	Romania, Germany	[115,116]
	Wings morphology	changes in hind wing shape and size are related to identifiable invasion processes	Croatia, Italy, Hungary, Serbia, Austria	[117–121]
	Enzyme activity	increase in the esterase activity after pesticide exposure was followed by a significant decrease in AChE	Romania, Poland	[122,123]
Trophic interaction	**Influence on root feeding**	larval behaviour respond to root volatiles	Germany, Italy	[124–126]
	Plant signal	biological control can be improved by manipulating the production of and responsiveness to a plant signal	Switzerland	[127,128]
	WCR as a disease vector	WCR may be an important vector of maize fungal diseases	Germany, Croatia, Poland	[129–131]

Table 1. *Cont.*

Research Area		Main Findings	Researches' Group Affiliation Country and Reference	
Pest control	Risk assessment	international research cooperation is the most important key to successfully manage WCR; presented global zones of climatic favourability and invasion risk for the WCR	Spain, Czech Republic, Germany, France	[132–135]
	Forecast	identification of the reproductive potential and longevity of the females of WCR under different rearing conditions	Croatia, Romania, Serbia, Hungary	[65,101,136–143]
	Crop rotation	low WCR population in maize fields managed by crop rotation	Serbia, Germany, Romania	[144–147]
	Host plant resistance	significant differences were found in the tolerance levels of the hybrids	Croatia, Germany, Hungary, Italy	[148–156]
	Attract and kill	host-specific compounds, combined with a CO_2 source, make attract and kill a feasible management option against WCR	Germany	[157–161]
	Bt	proteolytic processing of Bt toxins by WCR midgut juice was examined, no degradation of any of these toxins was observed	Belgium, Hungary, Germany	[162,163]
	Eradication	buffer zones large enough to allow eradication are economically unpalatable	UK, Hungary	[164]
	Chemical control of adults and larvae	insecticide application led to a significant reduction in the WCR larval density	Netherlands, Poland, Italy, Slovakia, France, Switzerland	[165–171]
	Entomopathogenic fungi	fungal strains significantly influenced the mortality of WCR larvae	Austria, Switzerland, Germany, Slovakia	[172–179]
	Entomopathogenic nematodes	nematodes appeared as effective as, or better than standard pesticides at reducing WCR populations	Hungary, Germany, Switzerland, Austria, Serbia, Romania, Italy, Croatia, Belgium, Slovenia	[173,175,178,180–198]
	Natural enemies	natural enemies can be useful elements of a strategic approach to the control of WCR	Switzerland	[199–201]
	Biopesticides	best results with bioproducts applied to the seed	Hungary, Romania, Slovenia, Germany	[202–207]
	Alternatives and benefits	annual welfare gain of ca. €190 million from biocontrol of WCR	Italy, France, Italy Spain, Germany, Austria, Romania, Belgium	[208–213]

Table 1. *Cont.*

Research Area		Main Findings	Researches' Group Affiliation Country and Reference	
Population genetics	**Dispersal**	large European outbreak was expanding by stratified dispersal, involving continuous diffusion and discontinuous long-distance dispersal	France, Italy, UK, Germany, Serbia, Hungary, Austria, Croatia, Slovenia	[214–216]
	Genetic monitoring	temporal genetic monitoring allowed a deeper understanding of the population genetics of WCR (multiple introductions, admixture, etc.)	Croatia, Serbia, Hungary, Italy	[51,217–222]
Systems model	**IPM**	management provides a basis for analyzing impacts of climate change and crop rotation on the spread, abundance, and damage of WCR	Austria, Germany, Hungary, Italy, Slovenia, Netherlands, Serbia	[223–226]
	Farmers education	through Farmer Field Schools, farmers were educated on WCR risk assessment	Serbia, Austria	[227,228]
	Simulations	development of a mechanistic understanding of the maize-pest system	Netherlands, Hungary, Germany, Sweden, Romania, Austria, Italy	[5,229–239]
	Remote sensing	the methodology that identifies WCR larval damage efficiently	Hungary, France, UK	[240–242]

We organized the evaluation of WCR research areas in Europe into seven main categories, depending on the primary type of interest, namely, (1) monitoring and density estimation; (2) ecology; (3) morphology and physiology; (4) trophic interactions; (5) pest control; (6) population genetics; and (7) systems modeling.

From the collected set of references, we first extracted those specifically dealing with "monitoring" procedures in Europe. Thus, we included all papers describing the first WCR occurrence, studies of population levels during the first WCR invasion process, and recorded movements of WCR individuals at short and long distances. The monitoring-based work described in detail the three phases of the WCR invasion process—introduction, establishment and spread, and the influences of various biotic and abiotic factors on WCR (e.g., weather, host plants). A major sub-theme of this research area is the description of different monitoring techniques and procedures that are integrated with the standard monitoring process in Europe. It was important to include "density estimation" because it is a research area closely related to monitoring, as the listed papers deal with different methods, measurements, and estimations of WCR population density, and with climatic conditions and climate change impacts on WCR in monitored areas in Europe. We divided this common research area into four main categories, depending on the type of monitoring and the tool used to describe population density, namely, (1) initial occurrence and spread; (2) population-level; (3) monitoring methods/techniques and designs; and (4) area-wide monitoring.

International collaborations in the field of WCR "ecology" over the past 12 years have addressed climate change and its influence on expanding areas of WCR invasion. Moreover, dozens of studies have been conducted on various WCR host opportunities, ground preferences, and movements between and within areas. Particular attention has been paid to larval behavior in response to root escapes and to the identification of various attractants that provide communication channels for reproduction and feeding, respectively.

"Morphology and Physiology" of WCR is the third research area and is represented by three sub-themes—(1) dimorphism, (2) wing morphology, and (3) enzyme activity. Researchers were concerned with specific changes in WCR explained in the context of natural selection, flight maneuverability, invasion process, resistance evolution, etc.

"Trophic interactions" were the subject of a small research area dealing with (1) WCR-maize-root interactions, (2) plant signals, and (3) WCR vectoring abilities. The main focus was on exploring ways to enhance biological control by manipulating the production of and responsiveness to plant signals. Smaller groups addressed bacterial and fungal community shifts in response to larval feeding and the identification of WCR larvae as potential disease vectors on maize plants.

Research addressing the "population genetics" of WCR in Europe focused mainly on the genetic basis of WCR dispersal and the temporal and spatial genetic monitoring that allows a deeper understanding of the changes that WCR populations have undergone as a result of the replacement of their original habitat in the USA with a new one in Europe.

The largest research area in Europe was "pest control" as a topic of interest to researchers from all WCR-infested countries. Special attention was given to risk assessment and prediction of WCR in all maize-growing areas in Europe. Crop rotation was explored as the most effective control measure against WCR. In contrast to America, few scientific groups in Europe have focused on WCR resistance. Furthermore, conventional genetic research on finding tolerant maize hybrids through intensive breeding programs for native resistance (e.g., extended root systems, ion concentrations in roots) has been very limited. Some papers have identified chemical insecticide control (either in-furrow microgranules or seed coating) as a factor that reduces root and yield damage. Even trials with the most effective insecticides show that pesticide protection is partial; they also note that when significantly lower than untreated plots, root damage in treated plots remains appreciable. No experiment made clear whether WCR root damage, usually still present in treated plots, leaded to yield reductions or whether it leaded to a yield comparable to that of plants with uninfested root systems in maize rotation fields. When WCR populations exceed thresholds, first-year maize fields in the rotation are the only ones suffering negligible root damage. Various non-chemical control measures were explored, with entomopathogenic fungi and nematodes identified as having a high potential to reduce WCR larvae in most European soils. All this led to commercial mass production of these environmentally safe control agents. Moreover, biopesticides and natural WCR enemies were identified as useful elements for a strategic approach to WCR pest control.

We found that over the past 12 years, many groups from Europe have been involved in predicting and assessing the further spread of WCR, its evolution, and the achievement of thresholds. This has led to a mechanistic understanding of the maize–WCR system and the development of remote sensing models that identify WCR larval damage, aspects of the invasion, growth rate, hatching prediction, effects of climate change, and crop rotation on spread and occurrence, etc. In European countries, the need for knowledge transfer, training for farmers, and the need for regulations for the introduction of alternative pest control options have been identified. Long-term and proactive coordination is required for the implementation of collective WCR control measures that meet the needs of individual farmers. Through farmer field schools, farmers were educated on WCR risk assessment and integrated pest management, resulting in successful WCR control in Europe.

4. Current Status of the Pest in Europe

4.1. Pest Distribution

To determine the current distribution of the pest in Europe, we consulted the EPPO Global Database (2020) [48] and the available literature sources found by searching Web of Science, SCOPUS, CAB direct, and Google Scholar databases. Based on the collected information, we created a map of WCR distribution in Europe.

According to the EPPO Global Database [48] and other literature data [52,58,62,69,243], WCR is currently distributed across 21 European countries (Figure 2). In the United

Kingdom and the Netherlands, WCR has been successfully eradicated, and in Belgium, the pest no longer exists even though the eradication has not been carried out. In Denmark, Estonia, and Spain, the absence of the pest is confirmed by surveys. In Finland, no pest has been recorded. The degree of WCR distribution varies in each European country—in some, the pest is widespread (the dark red color on the map); in others, it is distributed across a limited area, and this limited area of pest distribution corresponds to the area suitable for maize cultivation (e.g., Austria, Croatia, Italy, Montenegro). In several countries, including Germany and France, the restricted area is limited to restricted regions (orange color on the map). Here, the WCR population is monitored and suppressed by containment measures after the first introduction. Several countries are in the process of eradicating the pest (e.g., Switzerland).

Figure 2. The distribution map of western corn rootworm in Europe—countries are categorized according to pest presence and population level in seven categories (1–7) based on EPPO Global Database [48] and other literature data [53,58,62,69,239], in addition to data reported by Bieńkowski and Orlova-Bienkowskaja, [62] Modič et al. [53], Raileanu and Odobesky, [239] Voineac et al. [63], and Voineac et al. [69].

4.2. WCR Population Level, Damage, and Management Practices

WCR population level, damage, and management practices were analyzed in 10 European countries: Austria, Croatia, France, Germany, Hungary, Italy, Serbia, Slovenia, Switzerland, and Romania. Based on the EPPO Global Database [48], the selected countries were divided into the following three categories according to their pest distribution status:

Category I. Countries infested with WCR either in the whole maize-growing area or in the whole territory—Austria, Croatia, Hungary, Italy, Serbia, Slovenia, and Romania reported damage in the 2000s.

Category II. Countries partially infested—Germany and France.

Category III. Countries where containment measures have been moderately successful—Switzerland.

To obtain information on WCR population levels, damage, and control measures in various EU countries, we reviewed available literature sources regarding the period between 2008 and 2020. We used the overview explained in Section 2. All abstracts of publications were screened, and literature references were chosen for their relevance to the countries in question.

In addition, we obtained information from expert scientists and pest monitoring in the selected countries to obtain up-to-date information on surveillance activities, pest status, and damage. We asked them to send us articles in local languages and to share links for websites where official data on WCR monitoring can be viewed. We also asked them to send us their data on the percentage of infested cropland, an area with economic damage, average yield losses, and average insecticide use (where available).

- Category I (Austria, Croatia, Hungary, Italy, Serbia, Slovenia, and Romania)

Because WCR has already spread over the entire territory of Category I countries, there was no need to monitor its distribution at the country level over the last few years. Therefore, we did not find many articles published in the last 10 to 12 years reporting on the national spread, population density, and WCR damage for these countries.

Research initiatives in the observed countries aimed to identify factors affecting adult and larval populations and damage at field level in Croatia [64,66,72,102,137,139], Hungary [78,162,231], Serbia [66,79,82,104,145,148,224], Slovenia, [53,76] and Romania [52, 56,61,68,85,93,97–99,140].

The articles dealing with population monitoring in Croatia describe monitoring techniques [6,72] rather than population levels and damage. In contrast, the article by Falkner et al. (2019) [234] is based on data collected during WCR monitoring in Austria from 2002 to 2015. In the paper, the authors reported the highest WCR population level in Styria, the province where WCR has occurred since 2002. Based on the collected data, the authors developed a spatial zero-inflation Poisson mixture model (ZIP) to relate WCR counts to climatic conditions and maize proportions and to account for zero inflation and spatial correlation in the counting data. The developed model provides a scientifically sound basis for analyzing the effects of future climate change scenarios and maize rotation restrictions on WCR distribution and abundance.

Models for efficient WCR management have been developed in Austria [236,238,239], Hungary [233], and Italy [244], showing that crop rotation restrictions can help to reduce WCR spread and abundance regardless of the climate change scenario considered. Therefore, the impact of climate change is limited compared with the impact of crop rotation restriction measures. The developed models show that legislation requiring 100% crop rotation to control WCR seems too strict. Using the meta-models developed in Hungary, one can easily estimate the percentage of maize fields that would promote the increase of a pest population above the threshold. The results can be used by regional or national agricultural policy decision makers, and for integrated pest management. These models are also recommended for Italy with some adaptations [244] and could be useful for decision makers and farmers when planning flexible rotation with arable crops. These plans may allow farmers to plant the maximal maize percentage for a cultivated area, even including some continuous maize fields, which is a move that would also prevent the establishment of an economic damage WCR population.

Of the seven countries belonging to Category I, official pest monitoring is still carried out in Austria [245] and Serbia [246]. The other five monitored countries (Hungary, Croatia, Slovenia, Italy, and Romania) are no longer conducting official monitoring [247–249]. Due to this situation, it was difficult to collect extensive and reliable information on population density and damage.

Based on the data from WCR monitoring in Austria [245], it is clear that the pest is distributed throughout the maize-growing area in this country. Monitoring activities were carried out throughout Austria's maize-growing area, except for the Alps. Calculations by Feusthuber et al [236] show a large spatial variability in WCR damage potential on gross markets. They indicated that a large WCR population in a field combined with adverse weather conditions can result in total maize yield loss. As reported by Falkner et al. (2019) [236], there are no official data on damage and yield loss. As the spatial variability in the economic damage potential of maize yield losses corresponds to the regional maize density, the frequency of maize cultivation is already regulated by law in Styria, where the WCR population is high. A maximum maize share of 75% in crop rotations was legally allowed until 2016 and was reduced to 66% in 2017 [239]. In addition to crop rotation, there are chemical insecticides and entomopathogenic nematodes on the Austrian market that are approved for WCR control [228]. Kropf et al. [224] investigated farmers' behavior concerning individual and collective WCR control measures, and the results suggest that new forms of knowledge transfer are needed to facilitate the proactive implementation of individual and collective WCR control measures before triggering events, such as severe WCR damage.

In Croatia, official WCR monitoring was discontinued in 2012. The area of Croatia where WCR population density reaches the economic threshold is located in the northern part of Croatia on the border with Hungary and part of Slovenia (Međimurje and Podravina region) next to the River Drava (on approximately 10,000 km^2 [55]. In this area, maize is grown on approximately 110,000 ha and the share of maize cultivation in agricultural land is over 60% because farmers depend on maize due to intensive livestock farming. To prevent damage, crop rotation is practiced in all fields where the adult WCR population reaches the economic threshold. As a reliable tool enabling farmers to select the most suitable field for continuous management, Kos et al. [139] suggested using Pherocon AM traps (Trece Inc., Adair, OK, USA) between the 29th and 32nd weeks. The estimated WCR adult catch that could cause significant larval infestation is $\geq$22 adults/trap in the 29th week. Because IPM is mandatory for all Croatian farmers receiving income support through direct payments, crop rotation is also mandatory as one of the elements of IPM. A granulated insecticide (tefluthrin) is approved for larvae control. WCR damage is not officially recorded.

In Hungary, monitoring activities were stopped because WCR has spread over the whole area. The national survey conducted in the early 2000s showed that root damage occurred in 22.9% of heavily infected continuous maize fields (more than 10 adults/plant/day). At that time, the greatest damage was measured in Tolna, Baranya, Békés, Bács-Kiskun, and Csongrád counties. In the early 2000s, the pest conquered the best maize-growing areas in Hungary [250]. As in all EU countries, the plant protection regulation in Hungary requires the implementation of the principles of integrated pest management (IPM), as described in Annex III of Directive 2009/128/EC [49], where crop rotation is the first mandatory measure. In the mid-2010s, WCR was deregulated and placed under the general IPM management approach; its population is now managed by the mandatory minimum number of non-host crops in the rotation, and by other components of EU greening. According to recent information collected by plant protection specialists in some regions of Hungary [247], WCR is still an important pest and there are outbreaks of the pest in some regions from year to year, as was the case in 2012 [251] and 2017 [252]. Moreover, Gyeraj et al. [141] reported silking damage caused by adult WCR on sweet corn and the need to determine the economic threshold to justify the area application of insecticides.

After WCR was successfully eradicated in Europe's first focus area, which was identified in Italy, around Venice Marco Polo airport in 1998 [63,253], WCR spread to all Italian maize-growing areas. The spread started with new focus sites in Lombardy [254] and Friuli Venezia Giulia [63,255]. After the first WCR crop damage was observed in 2002 in Lombardy [256], new eradication/containment programs were established in Italy [257]. Despite these programs, WCR populations continued their spread, albeit slowly, and increased year by year. Therefore, official national WCR monitoring was carried out until 2012 [255,258–262] when WCR completed its spread to all the major maize-growing areas, including central Italy [261]. From that point, it was clear that economic threshold levels of WCR populations would be reached wherever continuous maize cultivation was the prevalent practice. Therefore, after emergency eradication, large-scale IPM measures were established regularly [262] and were supported by an innovative insurance tool to cover the risk of IPM implementation [244,263].

Official monitoring observed that maize crop-damage cases occurred in newly infested areas after 4–5 years from the first beetle captures, with subsequent population stabilization and crop-damage decrease or disappearance [255]. We can hypothesize that this trend was due to the implementation of containment measures with the interruption of continuous maize together with growing farmer awareness of the WCR problem, once farmers had seen WCR damage symptoms directly. The lack of damaged maize fields in 2011 and 2012 might be considered a confirmation of this hypothesis [260,261].

Later, during 2016–2017, area-wide (hundreds of hectares) WCR management strategies were assessed in northeast Italy [244]. A strategy based on chemical control (high presence of continuous maize plots with adult treatments and/or seed treatments) was compared to a flexible rotation approach (continuous corn for two or more years, interrupted when WCR populations exceeded the damage threshold of six beetles/Ph AM trap per day averaged over 42 days). WCR beetle levels were found significantly higher in the chemical control scenario than in the flexible rotation scenario, confirming that crop rotation is the most effective strategy for maintaining WCR populations permanently below the damage threshold without insecticide applications, even with a flexible approach. Flexible crop rotation may imply a higher risk of local sporadic damage because it means allowing the beetle's population to approach its damage threshold; to avoid an economic loss for farmers, this risk has been successfully managed with the introduction of insurance instruments, such as mutual funds [244,263].

In Serbia, organized monitoring collected data on damage to over 140,000 ha of maize until 1999. After 2000 and 2003, WCR population density and the number of damaged maize fields decreased significantly due to the massive application of crop rotations. Due to WCR presence and severe damage in the early phase of the invasion, the proportion of continuous maize in Serbia, once as high as 30%, was reduced to almost 0%. Sivčev et al. studied 794 maize fields from 2002 to 2006, and under the conditions of their study, the ratio of maize to non-maize fields was about 50:50 [145]. Rotation proved to be very effective because 87.8% of the fields lacked an economically harmful WCR population. Although WCR is considered a well-established pest that can be effectively controlled through diversified crop rotation, official WCR monitoring continues. We analyzed data collected between 2013 and 2020 by official monitoring activities available on the official website [264]. In the last eight years, the level of WCR population caught in pheromone traps in Serbia has fluctuated (Table 2), with a peak in 2016 and a significant decrease in the last three years (2018–2020).

Table 2. The analysis of the WCR monitoring activities in Serbia in the period from 2013 until 2020 (data available at Portal izveštajno prognozne službe zaštite bilja [264]).

	2013	2014	2015	2016	2017	2018	2019	2020
Number of monitored fields	22	24	30	27	29	25	24	25
Percent of fields with the capture	91	87.5	86.7	85.2	86.2	84	75	80
Lowes and highest maximal daily capture	1–186	2–536	1–236	2–250	1–350	1–121	1–442	1–237
Number of fields with maximal daily capture ≥100	3	4	5	12	6	2	1	1

For monitoring purposes, pheromone-baited traps were used. Therefore, the maximal daily captures were very high in some fields. Based on the collected data on pheromone-baited traps is not possible to conclude whether or not the WCR population has reached the economic threshold on the monitored field. No official data on yield losses caused by WCR are available but, based on the discussion with experts [246], farmers are paying much attention to crop rotation, thus economic damages are limited to continuous maize fields in the third or fourth year of continuous sowing. As it was reported by Filipović et al., farmers' education through farmer field schools contributed to a better understanding of why crop rotation plays such an important role in reducing pest damage even though maize has the highest gross margin when compared to soybean and wheat [227].

The status of WCR in Slovenia is the same as in Hungary and Croatia—it is not considered a quarantine pest since it spread over the whole territory of Slovenia in 2009 [248]. According to Razinger [248] crop rotation is the predominant (alternative) control tool for WCR and it is performed on around 80% of maize planted-surface in Slovenia. However, the percentage of monoculture maize production differs a lot among the regions (3% in Pomurje region and almost 1/3 in Gorenjska region). Approximately 30% of maize is treated with 13.3 kg/ha Force (tefluthrin) at sowing against WCR; approximately 60% of maize seed is treated with Sonido (thiacloprid) against wireworms; approximately 10–25% of maize is treated with both, tefluthrin and thiacloprid, at sowing. Often, crop rotation and insecticides (seed coating—Sonido; and granular—Force) are used together. Generally, in Slovenia, economic damage occurs every year on less than 10 ha and yield loss is negligible, i.e., 0–1% every year. To estimate hatching time and adult emergence, a decision-support model based on degree-days is practiced in Slovenia [230]. In addition, the use of entomopathogenic nematodes and pheromone-based mating confusion are under investigation as ecologically acceptable tools for WCR control.

Until 2010, WCR was a real danger for maize plants in the western part of Romania and the speed of spreading was astounding [123]. After 2010, at the national level, no extensive monitoring research or official data centralization was performed. At that time, it was estimated that the pest was present only in the western half of the country [249]. After 2010, WCR continued to expand slower, probably due to the Carpathian Mountains that prevented the flight of adult forms [93]. Even though adults have not migrated to other parts of the country, they have grown to size in the western populations (where they first appeared) [123]. According to Grozea [249], WCR is still present in Romania, especially in the west of the country (in Timis county), and continues to expand eastward. In 2019, it was also reported in the east of the country (Neamt County). The presence of the species and its extension were probably favored by the cultivation of corn on an appreciable area of approximately 2.5 million hectares and by the practice of monoculture on large areas. The highest population level is by far in the western area (especially in Timis County) where corn is cultivated on 189,000 hectares, the plain area, and there are still farmers who practice monoculture (20%). Studies conducted in this county in the period 2015–2018 showed that the insect population was located at a fairly high level, but still not as in the period 1997–2010 when there were economic losses.

- Category II (Germany and France)

The first occurrence of WCR in Germany was recorded in 2007, in Baden-Württemberg, and Bavaria. Beetles were also caught in North Rhine-Westphalia in 2010 and Hesse and Rhineland-Palatinate have also been affected since 2011. WCR is not quarantine pest anymore and WCR monitoring is performed by each of the federal states separately, but data are also sent to JKI, Institute for Plant Protection in Field Crops and Grassland [264]. The federal states Mecklenburg-Pomerania and Schleswig-Holstein are not participating in the WCR monitoring because the species is not a quarantine pest anymore. The traps used in the monitoring are PAL traps (Csalomon® Pheromone Traps) with pheromone dispensers. Most federal states use the products from Trifolio-M, but Csalomon products are also used. According to collected data from monitoring activities [265,266], WCR is spread in six out of ten federal states involved in monitoring in Germany. The federal states that are not infested are Hesse, Lower Saxony, North Rhine-Westfalia, and Thuringia. In the federal states Brandenburg, Rhineland-Palatinate, Saxony-Anhalt, and Saxony, the population level is very low. The regions Baden-Wurttemburg and Bavaria have the highest population level. The number of trapping locations varies markedly between the federal states with heavy infestation such as Baden-Wurttemberg (700 sites in 2019) and Bavaria (243 sites in 2019) and those that have no infestations such as Lower Saxony (90 sites in 2019) or Thuringia (20 sites in 2019). Bayern and Baden-Wurttemberg have had rapid increases in WCR during the last six years. A slight decrease in 2020 is recorded in Bavaria [261]. In contrast to Bavaria in which the beetles came mainly from the southeast spreading northwest and west [267], the main focus of the infestation in Baden-Württemberg is the upper Rhine valley. Here, also the adjacent regions in France are infested. In Saxony, the infestation is likely spreading from the Czech Republic and Poland, and in Rhineland-Palatinate, it is coming from Baden-Wurttemberg. There are about 600,000 hectares of maize in Bavaria, of which about 420,000 hectares are infested [268]. Despite WCR is still spreading and has increased in numbers in recent years, so far, no economic damage has been observed in Germany, even in the most heavily infested federal states. Only occasional damages of plants are visible to the trained eye in parts of the fields. German IPM measure is to have maize at maximum in two of three years on the same fields in the regions where WCR occurs. It can be stated that if farmers grow maize in the same plot only every two or three years, they do not need to fear considerable damage. Even if an infestation is established, no control measures are necessary. Transferred to the situation of Bavarian maize cultivation this implies that of 420,000 ha cultivated maize area, about 30,000 ha can be considered as highly endangered. Southern Bavaria is concerned almost exclusively [269].

After the first WCR detections in 2002 in France (near Paris Charles de Gaulle airport) [270], multiple regulatory changes until the deregulation of the species in 2014 occurred. However, official monitoring activities are carried out by Arvalis [270]. WCR is significantly present in only two areas of France—Alsace and Rhône-Alpes. In 2019, for the first time, insects were easily observable (without traps) in certain plots of these two regions. The first damage attributable in part to the WCR was even observed in an area in the Alpine valley in Rhône-Alpes (Grésivaudan valley) where the first captures had taken place 10 years earlier. Moreover, the outbreaks of WCR are multiplying in the great southwest of France and in particular in New Aquitaine. In Alsace, 100% of the 91 pheromone traps revealed the presence of WCR, which confirms once again that the insect is present throughout the region. Given the high intensities of WCR catches in 2018, some parts were monitored using unbaited chromotropic traps (free of sexual pheromone) in 2019. A plot even accounted for 4.7 insects/trap/day, approaching the established economic thresholds of 5–6 beetles/trap/day (i.e., 40 beetles/trap/week). At the end of 2019, the damages were observed on less than 10 hectares (and it was the first time damages were observed in France) [271]. In Alsace and Rhône-Alpes, there are 300,000 hectares of maize (grain and silage), which represents 10% of maize in France (3 million hectares of maize grain + silage). It is estimated [271] that in this area, the WCR population is reaching the economic threshold of less than 0.01 % (i.e., 30 hectares). The insecticides against WCR are not used and crop rotation is recommended. The recommendation to farmers is to consider the establishment of other crops one year out of six when possible, prioritizing their implementation first on plots where the highest levels of catches were observed in 2019.

- Category III (Switzerland)

In Switzerland, WCR is under eradication. Due to the crop rotation system, the pest was unable to establish itself in Switzerland [272]. Because the beetle has not spread in Switzerland, WCR is still a quarantine organism in this country and is therefore regulated by phytosanitary law. Any suspected infestation must be reported to the cantonal plant protection service immediately.

In south Switzerland, near Ticino, WCR has been caught every year since 2000, as the beetles fly in regularly from Italy. In the Canton of Ticino since 2004, the cultivation of maize on maize has been generally prohibited in the entire area. Official monitoring activities have been taking place since 2003 [272]. The WCR situation in Switzerland has been monitored annually using around 150 pheromone traps each year. Traps are set up mainly in the maize growing areas and in places where the beetle was caught the previous year. Until 2019, special attention was paid to the traffic axes and airports. Since 2020, the traps have been distributed in a grid pattern across the entire Swiss maize-growing area because an increasing number of adult corn rootworms have flown in from the surrounding countries. If corn rootworms are caught, it is compulsory to follow a crop rotation restriction (cultivation of maize on maize is prohibited) within 10 kilometers (demarcated area) of the trap location [272]. Probably because the WCR population level is still very low and that crop rotation is regulated, the damages caused by larval feeding have not been reported until now.

The no-control scenario of potential damage cost of WCR infestation in Europe published by Wesseler and Fall foresaw that the economic benefits of WCR control would range between EUR 143 million and EUR 1.739 million. In the analysis, they included 18 countries (at that time 16 EU member countries plus Croatia and Switzerland) [230]. In our research, the current situation in 10 countries (eight EU countries plus Serbia and Switzerland) is analyzed. The analysis shows that WCR has established its population in all countries belonging to category I and, in some regions of countries, belonging to category II. The current situation with WCR leads us to the conclusion that quarantine and agricultural experts in EU countries involved in this analysis correctly approached the WCR problem. The countries involved in this analysis reported that the frequency of continuous maize was from 20 % in Croatia to 35 % in Hungary and Romania.

Among analyzed countries, there is no country reporting economic damage on a large scale. In addition, the insecticide application for reducing WCR larval damage is a regular practice in Romania and Slovenia. All involved countries are requesting crop rotation as the main measure against WCR. Taking into account that the level of maize dependency of different farms varies, some countries are advocating breaking continuous maize growing in one out of three years (Germany) up to one in six years (France). In Croatia and Serbia, the education of farmers took place to carry out a risk assessment on maize fields destined for repeated sowing [66,139,227]. The need for farmer education is also highlighted by Kropf et al. [228]. Crop rotation is not considered necessary on the fields with a WCR population lower than the economic threshold level established by monitoring. A decision support system for WCR is developed in Slovenia [225]. In Austria, Hungary, and Italy, crop rotation is also advocated, suggesting modulation of its intensity with the use of different models developed by scientists [229,234,240].

5. Conclusions

Over the past 12 years, WCR has continued to multiply and spread throughout Europe, infesting new countries and increasing population density. During the same period, scientists in Europe continued their research activities to investigate different aspects of WCR management by implementing a range of approaches to WCR control. The research topics are very diverse, resulting in a considerable amount of new knowledge that can contribute to the development of pest control strategies applicable in EU agricultural systems, which are completely different from agricultural systems in the USA region, where this pest causes the most problems. No major economic damage was observed in any of the countries surveyed. This status confirms that EU countries surveyed are applying appropriate WCR management measures by implementing research findings. Research findings are confirmed by the situation in EU countries and crop rotation has proven itself to be the most effective strategy for maintaining WCR populations permanently below the damage threshold. To maintain forage production on livestock farms at the best level in terms of yield and quality, different models of crop rotation could be applied. The application of different models could enable farmers to grow maize for two or more continuous years, while other crops could be planted according to a flexible scheme when the WCR population increases significantly, whenever the threshold is exceeded. Implementation of available models of crop rotation prevents any application of pesticides following current European legislation (Directive 128/2009/EU).

Now that the European agricultural system has coexisted with WCR for 35 years, we can conclude that WCR is a potentially serious threat that can be effectively contrasted in all EU countries. Due to intensive research and professional activities leading to specific agricultural practices and the EU's Common Agricultural Policy, there are crop-rotation rotation-based solutions that can manage this pest properly with negligible impacts on farmers and the environment. In many countries, these solutions are regularly implemented either by policymakers, extension services, or farmers themselves. Therefore, WCR has not become as serious a pest as expected when it was discovered on the majority of European territories.

Author Contributions: Conceptualization, R.B. and L.F.; methodology, R.B., D.L., F.C., and L.F.; software, R.B., D.L., F.C., and L.F.; validation, R.B., D.L., F.C., and L.F.; formal analysis, R.B., D.L., F.C., and L.F.; investigation, R.B., D.L., F.C., and L.F.; resources, R.B. and L.F.; data curation, R.B., D.L., F.C., and L.F.; writing—original draft preparation, R.B., D.L., F.C., and L.F.; writing—review and editing, R.B., D.L., F.C., and L.F.; visualization, R.B.; supervision, L.F.; project administration, R.B.; funding acquisition, R.B. All authors have read and agreed to the published version of the manuscript.

Funding: This research was supported by the Croatian Science Foundation through the project MONPERES (2016-06-7458) "Monitoring of Insect Pest Resistance: Novel Approach for Detection and Effective Resistance Management Strategies".

Institutional Review Board Statement: Not applicable.

Informed Consent Statement: Not applicable.

Acknowledgments: The authors thank all of the experts who provided the data on WCR distribution in their countries: Peter Baufeld, Joern Lehmnus, Michael Zellner, Stefan Topfer, Jozsef Kiss, Geza Vörös, Geza Ripka, Ioanna Grozea, Slađan Stanković, Jaka Razinger, and Jean Baptiste Thibord.

Conflicts of Interest: The authors declare no conflict of interest.

References

1. Metcalf, R.L. Foreword. In *Methods for the Study of Pest Diabrotica*; Springer-Verlag: New York, NY, USA, 1986; pp. vii–xvi.
2. Baca, F. New member of the harmful entomofauna of Yugoslavia, *Diabrotica virgifera virgifera* LeConte (Coleoptera, Chrysomelidae). *Zast. Bilja* **1994**, *45*, 125–131.
3. Kiss, J.; Edwards, C.R.; Berger, H.K.; Cate, P.; Cean, M.; Cheek, S.; Derron, J.; Festic, H.; Furlan, L.; Igrc Barčić, J.; et al. Monitoring of western corn rootworm (*Diabrotica virgifera virgifera* LeConte) in Europe 1992–2003. In *Western Corn Rootworm: Ecology and Management*; CABI Publishing: Wallingford, UK, 2005; pp. 29–39.
4. EPPO Situation of *Diabrotica virgifera virgifera* in the EPPO Region. 2012. Available online: https://www.eppo.int/ACTIVITIES/plant_quarantine/shortnotes_qps/diabrotica_virgifera (accessed on 20 December 2020).
5. Szalai, M.; Komáromi, J.P.; Bažok, R.; Igrc-Barčić, J.; Kiss, J.; Toepfer, S. The growth rate of *Diabrotica virgifera virgifera* populations in Europe. *J. Pest. Sci.* **2010**, *84*, 133–142. [CrossRef]
6. Mrganić, M.; Bažok, R.; Mikac, K.M.; Benitez, H.A.; Lemic, D. Two Decades of Invasive Western Corn Rootworm Population Monitoring in Croatia. *Insects* **2018**, *9*, 160. [CrossRef]
7. Miller, N.; Estoup, A.; Toepfer, S.; Bourguet, D.; Lapchin, L.; Derridj, S. Multiple transatlantic introductions of the western corn rootworm. *Science* **2005**, *310*, 992. [CrossRef]
8. Ciosi, M.; Miller, N.J.; Kim, K.S.; Giordano, R.; Estoup, A.; Guillemaud, T. Invasion of Europe by the western corn rootworm, *Diabrotica virgifera virgifera*: Multiple transatlantic introductions with various reductions of genetic diversity. *Mol. Ecol.* **2008**, *17*, 3614–3627. [CrossRef] [PubMed]
9. Prinzinger, G. Monitoring of western corn rootworm (*Diabrotica virgifera virgifera* LeConte) in Hungary in 1995. *IWGO News Lett.* **1996**, *16*, 7–11.
10. Igrc Barčić, J.; Maceljski, M. Kukuruzna zlatica (*Diabrotica virgifera virgifera* LeConte-Col.: Chrysomelidae)-novi štetnik u hrvatskom podunavlju. *Agron. Glas.* **1997**, *5–6*, 429–443.
11. Edwards, C.R.; Barcic, J.I.; Berberovic, H.; Berger, H.K.; Festic, H.; Kiss, J.; Princzinger, G.; Schulten, G.G.M.; Vonica, I. Results of the 1997-1998 multi-country FAO activity on containment and control of the western corn rootworm. *Diabrotica virgifera virgifera* LeConte, in Central Europe. *Acta Phytopathol. Èntomol. Hung.* **1999**, *34*, 373–386.
12. Berger, H.K. The western corn rootworm (*Diabrotica virgifera virgifera*): A new maize pest threatening Europe. *EPPO Bull.* **2001**, *31*, 411–414. [CrossRef]
13. Dobrinčić, R. An Investigation of the Biology and Ecology of *Diabrotica virgifera virgifera* LeConte, a New Member of the Entomofauna of Croatia. Ph.D. Thesis, University of Zagreb, Zagreb, Croatia, 10 February 2001.
14. Vilsan, D.; Vonica, I. Results of Monitoring *Diabrotica virgifera virgifera* LeConte in Romania. *Acta Phytopathol. Èntomol. Hung.* **2002**, *37*, 175–182. [CrossRef]
15. Igrc Barčić, J.; Dobrinčić, R. Results of Monitoring *Diabrotica virgifera virgifera* LeConte in Croatia. *Acta Phytopathol. Èntomol. Hung.* **2002**, *37*, 137–144. [CrossRef]

16. Berger, H.K.; Baufeld, P.; Pajmon, A.; Reynauld, P.; Sivicek, P.; Ulvee, C.C.; Urek, G. 1998–1999 Detection Survey Results of *Diabrotica virgifera virgifera* LeConte in Non-infested European Countries. *Acta Phytopathol. Èntomol. Hung.* **2002**, *37*, 183–192. [CrossRef]
17. Festić, H.; Karić, N.; Berberović, H.; Huremović, H. The 1998 Monitoring Results of *Diabrotica virgifera virgifera* LeConte (Coleoptera: Chrysomelidae) in Bosnia and Herzegovina. *Acta Phytopathol. Èntomol. Hung.* **2002**, *37*, 159–162. [CrossRef]
18. Urek, G.; Modič, S. Occurrence of the western corn rootworm (*Diabrotica virgifera virgifera* Le Conte) in Slovenia. *Acta Agric. Slov.* **2004**, *83*, 5–13.
19. Maceljski, M.; Igrc Barčić, J. Significance of *Diabrotica virgifera virgifera* LeConte (Coleoptera: Chrysomelidae) for Croatia. *Poljopr. Znan. Smotra* **1994**, *59*, 413–423.
20. Sivčev, I.; Manojlovic, B.; Krnjajic, S.; Dimic, N.; Draganic, M. Distribution and harmfulness of *Diabrotica virgifera virgifera* LeConte (Coleoptera, Chrysomelidae), a new maize pest in Yugoslavia. *Zaštita Bilja* **1994**, *45*, 19–26.
21. Baufeld, P.; Enzian, S. Maize growing, maize high-risk areas and potential yield losses due to western corn rootworm (*Diabrotica virgifera virgifera* LeConte) damage in selected European countries. In *Western Corn Rootworm: Ecology and Management*; CABI Publishing: Wallingford, UK, 2005; pp. 285–302.
22. Hummel, H.E. Introduction of *Diabrotica virgifera virgifera* into the Old World and its consequences: A recently acquired invasive alien pest species on *Zea mays* from North America. *Commun. Agric. Appl. Biol. Sci.* **2003**, *68*, 45–57. [PubMed]
23. Hemerik, L.; Busstra, C.; Mols, P. Predicting the temperature-dependent natural population expansion of the western corn rootworm, *Diabrotica virgifera*. *Entomol. Exp. Appl.* **2004**, *111*, 59–69. [CrossRef]
24. Wudtke, A.; Hummel, H.; Ulrichs, C. The western corn rootworm *Diabrotica virgifera virgifera* LeConte en route to Germany. *Gesunde Pflanz.* **2005**, *57*, 73–80. [CrossRef]
25. Hummel, H.E.; Bertossa, M.A.; Hein, D.F.; Wudtke, A.; Urek, G.R.; Modic, S.P.; Ulrichs, C.H. The western corn rootworm *Diabrotica virgifera virgifera* en route to Germany. *Commun. Agric. Appl. Biol. Sci.* **2005**, *70*, 677–686.
26. Hummel, H.E.; Urek, G.; Modic, S.; Hein, D.F. Monitoring presence and advance of the alien invasive western corn rootworm beetle in eastern Slovenia with highly sensitive Metcalf traps. *Commun. Agric. Appl. Biol. Sci.* **2005**, *70*, 687–692. [PubMed]
27. Tóth, M.; Sivcev, I.; Ujváry, I.; Tomasek, I.; Imrei, Z.; Horváth, P.; Szarukán, I. Development of Trapping Tools for Detection and Monitoring of *Diabrotica v. virgifera* in Europe. *Acta Phytopathol. Èntomol. Hung.* **2003**, *38*, 307–322. [CrossRef]
28. Toepfer, S.; Gueldenzoph, C.; Ehlers, R.-U.; Kuhlmann, U. Screening of entomopathogenic nematodes for virulence against the invasive western corn rootworm, *Diabrotica virgifera virgifera* (Coleoptera: Chrysomelidae) in Europe. *Bull. Entomol. Res.* **2005**, *95*, 473–482. [CrossRef]
29. Igrc Barčić, J.; Bažok, R.; Maceljski, M. Research on the western corn rootworm (*Diabrotica virgifera virgifera* LeConte, Coleoptera: Chrysomelidae) in Croatia (1994–2003). *Entomol. Croat.* **2003**, *7*, 63–83.
30. Tóth, F.; Horváth, L.; Komáromi, J.; Kiss, J.; Széll, E. Field Data on the Presence of Spiders Preying on Western Corn Rootworm (*Diabrotica virgifera virgifera* LeConte) in Szeged Region, Hungary. *Acta Phytopathol. Èntomol. Hung.* **2002**, *37*, 163–168. [CrossRef]
31. Igrc Barčić, J.; Bažok, R. The influence of different food sources on the life parameters of western corn rootworm (*Diabrotica virgifera virgifera* LeConte, Coleoptera: Chrysomelidae). *Razpr. IV. Razreda SAZU* **2004**, *XLV-1*, 75–85.
32. Moeser, J.; Vidal, S. Do Alternative Host Plants Enhance the Invasion of the Maize Pest *Diabrotica virgifera virgifera* (Coleoptera: Chrysomelidae, Galerucinae) in Europe? *Env. Entomol.* **2004**, *33*, 1169–1177. [CrossRef]
33. Toepfer, S.; Kuhlmann, U. Survey for natural enemies of the invasive alien chrysomelid, *Diabrotica virgifera virgifera*, in Central Europe. *Biocontrol* **2004**, *49*, 385–395. [CrossRef]
34. Bažok, R.; Igrc-Barčić, J.; Edwards, C.R. Effects of proteinase inhibitors on western corn rootworm life parameters. *J. Appl. Èntomol.* **2005**, *129*, 185–190. [CrossRef]
35. Moeser, J.; Vidal, S. Nutritional resources used by the invasive maize pest *Diabrotica virgifera virgifera* in its new South-east-European distribution range. *Entomol. Exp. Appl.* **2005**, *114*, 55–63. [CrossRef]
36. Dobrinčić, R.; Igrc Barčić, J.; Edwards, R.C. (Determining of the injuriousness of the larvae of western corn rootworm (*Diabrotica virgifera virgifera* LeConte) in Croatian conditions. *Agr. Consp. Sc.* **2002**, *67*, 1–9.
37. Szell, E.; Zseller, I.; Ripka, G.; Kiss, J.; Prinzinger, G. Strategies for controlling western corn rootworm (*Diabrotica virgifera virgifera*. *Acta Agron Hung.* **2005**, *53*, 71–79. [CrossRef]
38. Zhang, F.; Toepfer, S.; Kuhlmann, U. Basic biology and small-scale rearing of *Celatoria compressa* (Diptera: Tachinidae), a parasitoid of *Diabrotica virgifera virgifera* (Coleoptera: Chrysomelidae). *Bull. Entomol. Res.* **2003**, *93*, 569–575. [CrossRef]
39. Toepfer, S.; Levay, N.; Kiss, J. Suitability of different fluorescent powders for mass-marking the Chrysomelid, *Diabrotica virgifera virgifera* LeConte. *J. Appl. Entomol.* **2005**, *129*, 456–464. [CrossRef]
40. Rasmann, S.; Köllner, T.G.; Degenhardt, J.; Hiltpold, I.; Toepfer, S.; Kuhlmann, U.; Gershenzon, J.; Turlings, T.C.J. Recruitment of entomopathogenic nematodes by insect-damaged maize roots. *Nature* **2005**, *434*, 732–737. [CrossRef] [PubMed]
41. FAO. Evaluation of Integrated Pest Management for Western Corn Rootworm (WCR) in Central and Eastern Europe (GTFS/RER/017/ITA). Available online: http://www.fao.org/fileadmin/user_upload/oed/docs/GTFSRER017ITA_2008_ER.pdf (accessed on 19 December 2020).
42. CORDIS. Threat to European Maize Production by Invasive Quarantine Pest, Western corn Rootworm (Diabrotica virgifera virgifera): A new Sustainable Crop Management Approach. Available online: https://cordis.europa.eu/article/id/81870-modelling-the-spread-of-western-corn-rootworm (accessed on 19 December 2020).

43. EUR-Lex. Commission Decision 2003/766/EC on Emergency Measures to Prevent the Spread within the Community of *Diabrotica virgifera* Le Conte. Available online: https://eur-lex.europa.eu/legal-content/GA/TXT/?uri=CELEX:32003D0766 (accessed on 19 December 2020).

44. EUR-Lex. 2006/564/EC: Commission Decision of 11 August 2006 Amending Decision 2003/766/EC on Emergency Measures to Prevent the Spread within the Community of *Diabrotica virgifera* Le Conte (notified under document number C(2006) 3582). Available online: https://eur-lex.europa.eu/legal-content/EN/TXT/?uri=CELEX%3A32006D0564 (accessed on 19 December 2020).

45. EUR-Lex. 2006/565/EC: Commission Recommendation of 11 August 2006 on Containment Programmes to Limit the Further Spread of Diabrotica Virgifera Le Conte in Community areas Where its Presence is Confirmed. Available online: https://eur-lex.europa.eu/legal-content/EN/TXT/?uri=CELEX%3A32006H0565 (accessed on 21 December 2020).

46. CORDIS. DIABR-ACT—Harmonise the Strategies for Fighting *Diabrotica virgifera virgifera*. Available online: https://cordis.europa.eu/docs/results/22/22623/123869751-6_en.pdf (accessed on 19 December 2020).

47. Diabrotica. Julius Kuehn. *Diabrotica virgifera virgifera* LeConte. Available online: https://diabrotica.julius-kuehn.de/index.php?menuid=1 (accessed on 19 December 2020).

48. EPPO Global Database. *Diabrotica virgifera virgifera* LeConte. Available online: https://gd.eppo.int/taxon/DIABVI/distribution (accessed on 19 December 2020).

49. EUR-Lex. Directive 2009/128/EC of the European Parliament and of the Council of 21 October 2009 Establishing a Framework for Community Action to Achieve the Sustainable Use of Pesticides. Available online: https://eur-lex.europa.eu/legal-content/EN/ALL/?uri=celex%3A32009L0128 (accessed on 21 December 2020).

50. Furlan, L.; Benvegnu, I.; Cecchin, A.; Chiarini, F.; Fracasso, F.; Sartori, A.; Manfredi, V.; Frigimelica, G.; Davanzo, M.; Canzi, S.; et al. Difesa integrata del mais: Come applicarla in campo. *L'Informatore Agrar.* **2014**, *9*, 11–14.

51. Meinke, L.J.; Sappington, T.W.; Onstad, D.W.; Guillemaud, T.; Miller, N.J.; Komáromi, J.; Levay, N.; Furlan, L.; Kiss, J.; Toth, F. Western corn rootworm (*Diabrotica virgifera virgifera* LeConte) population dynamics. *Agric. For. Entomol.* **2009**, *11*, 29–46. [CrossRef]

52. Antonie, V.I.; Tanase, M.; Neagu, M. The within control of the populations of *Diabrotica virgifera virgifera* LeConte in the Mureş county. *Acta Univ. Cibiniensis Agric. Sci.* **2008**, *1*, 20–28.

53. Modic, Š.; Knapič, M.; Urek, G. Širjenje koruznega hrošča *Diabrotica v. virgifera* v Sloveniji v obdobju 2003–2007. *Acta Agric. Slov.* **2008**, *91*, 259–270.

54. Konefal, T.; Beres, P.K. *Diabrotica virgifera* LeConte in Poland in 2005–2007 and regulations in the control of the pest in 2008. *J. Plant. Prot. Res.* **2009**, *49*, 129–134. [CrossRef]

55. Bažok, R.; Igrc-Barčić, J. *Pheromone Applications in Maize Pest. Control.*, 1st ed.; Novascience Publishers: Haupauge, NY, USA, 2010; pp. 23–35.

56. Grozea, I.; Stef, R.; Virteiu, A.M.; Carabet, A. Development of partial maps of WCR spreading in accordance with environmental factors. *Res. J. Agric. Sci.* **2010**, *42*, 44–49.

57. Michaelakis, A.N.; Papdopoulos, N.T.; Antonatos, S.A.; Zarpas, K. First data on the occurrence of *Diabrotica virgifera virgifera* Le Conte (Coleoptera: Chrysomelidae) in Greece. *Hell. Plant. Prot. J.* **2010**, *3*, 29–32.

58. Voineac, V.; Volosciuc, L.; Babidorich, M.; Rosca, G.; Odobescu, V.; Patrasc, T. Western Corn Rootworm (*Diabrotica virgifera virgifera* Le Conte) Population Monitoring With Help Of Sex Pheromone. In *Actual Problems of Protection and Sustainable Use of the Animal World Diversity. International Conference of Zoologists Dedicated to the 50th Anniversary from the Foundation of Institute of Zoology of ASM*; Academy of sciences of Moldova Department of nature and life sciences Institute of zoology: Chişinău, Moldova, 2011; pp. 152–153.

59. Schaub, L.; Furlan, L.; Toth, M.; Steinger, T.; Carrasco, L.R.; Toepfer, S. Efficiency of pheromone traps for monitoring *Diabrotica virgifera virgifera* LeConte. *Bull. OEPP* **2011**, *41*, 189–194. [CrossRef]

60. Dicke, D.; Martinez, O.; Frosch, M.; Lenz, M.; Jung, J.; Willig, W.; Jostock, M.; Kerber, M. First occurrence of western corn root worm beetles in the federal states Hesse and Rhineland-Palatinate (Germany), 2011. *Jul. Kühn Arch.* **2014**, *444*, 17–19.

61. Manole, T.; Chireceanu, C.; Teodpru, A. Current Status of *Diabrotica virgifera virgifera* LeConte, 1868 (Coleoptera: Chrysomelidae) in Romania. *Acta Zool. Bulg.* **2017**, *9*, 143–148.

62. Bieńkowski, A.O.; Orlova-Bienkowskaja, M.J. Alien leaf beetles (Coleoptera, Chrysomelidae) of European Russia and some general tendencies of leaf beetle invasions. *PLoS ONE* **2018**, *13*, e0203561. [CrossRef]

63. De Luigi, V.; Furlan, L.; Palmieri, S.; Vettorazzo, M.; Zanini, G.; Edwards, C.R.; Burgio, G. Results of WCR monitoring plans and evaluation of an eradication programme using GIS and Indicator Kriging. *J. Appl. Entomol.* **2011**, *135*, 38–46. [CrossRef]

64. Lemić, D.; Bažok, R. Risk assessmenet against western corn rootworm *Diabrotica virgifera virgifera* LeConte in the Moslavina region. *Agron. Glas.* **2009**, *5–6*, 337–346.

65. Beres, P.K.; Sionek, R. Study on the fecundity, egg collection technique and longevity of *Diabrotica virgifera* LeConte females under laboratory and field conditions. *J. Plant. Prot. Res.* **2010**, *50*, 429–437. [CrossRef]

66. Bažok, R.; Sivčev, I.; Kos, T.; Igrc-Barčić, J.; Kiss, J.; Janković, S. Pherocon AM trapping and the "Whole plant count" method—A comparison of two sampling techniques to estimate the WCR adult densities in Central Europe. *Cereal Res. Commun.* **2011**, *39*, 298–305. [CrossRef]

67. Wilstermann, A. A Regional View of an Impending Invasion: Western Corn Rootworm Development in Northern Germany. Ph.D. Thesis, Der Georg-August-Universität Göttingen, Gottingen, Germany, 2012.

68. Grozea, I.; Trusca, R.; Stef, R.; Molnar, L.; Fericean, M.; Prunar, S.; Mazare, V.; Dobrin, I. Is *Diabrotica virgifera virgifera* Still Considered a Dangerous Pest From Crops Of Romania? *Bull. UASVM Hortic.* **2014**, *71*, 351–352. [CrossRef]

69. Voineac, V.; Elisovetcaia, D.; Cristman, D.; Babidorici, M.; Tulgara, E. The results of pheromone monitoring of invasive pest *Diabrotica virgifera virgifera* LeConte in Transcarpathian region of Ukraine. *Agron. Ser. Sci. Res./Lucr. Stiint. Ser. Agron.* **2015**, *58*, 75–78.

70. Cagáň, L.; Števo, J.; Peťovská, K. Development of the western corn rootworm, *Diabrotica virgifera virgifera* in soil. *J. Cent. Eur. Agric.* **2016**, *17*, 1050–1069. [CrossRef]

71. Acker, M.; Zintel, A.; Benker, U. Western corn rootworm: Experiments on the improvement of monitoring at low population densities. *Jul. Kühn Arch.* **2014**, *444*, 33–38.

72. Lemic, D.; Mikac, K.M.; Kozina, A.; Benitez, H.A.; McLean, C.M.; Bažok, R. Monitoring techniques of the western corn rootworm are the precursor to effective IPM strategies. *Pest. Manag. Sci.* **2016**, *72*, 405–417. [CrossRef] [PubMed]

73. Rauch, H.; Zelger, R.; Strasser, H. Highly Efficient Field Emergence Trap for Quantitative Adult Western Corn Rootworm Monitoring. *J. Kans. Entomol. Soc.* **2016**, *89*, 256–266. [CrossRef]

74. Marković, D.; Ranđić, S.; Tanasković, S.; Gvozdenac, S. Possibility of monitoring *D. v. virgifera* flight by processing image of phero-traps using Raspberry Pi based devices. *Acta Agric. Serb.* **2017**, *22*, 207–217. [CrossRef]

75. Tóth, Z.; Tóth, M.; Jósvai, J.K.; Tóth, F.; Flórián, N.; Gergócs, V.; Dombos, M. Automatic Field Detection of Western Corn Rootworm (*Diabrotica virgifera virgifera*; Coleoptera: Chrysomelidae) with a New Probe. *Insects* **2020**, *11*, 486. [CrossRef]

76. Knapič, M.; Urek, G.; Modic, Š. GIS Analysis of the Spread and Population Density of *Diabrotica virgifera virgifera* LeConte and its Impact on Agricultural Practice in Slovenia during the Period from 2003 to 2007. *Cereal Res. Commun.* **2009**, *37*, 227–236. [CrossRef]

77. Cagáň, L.; Rosca, I. Seasonal dispersal of the western corn rootworm (*Diabrotica virgifera virgifera*) adults in Bt and non-Bt maize fields. *Plant. Protect. Sci.* **2012**, *48*, S36–S42. [CrossRef]

78. Levay, N.; Terpo, I.; Kiss, J.; Toepfer, S. Quantifying Inter-field Movements of the Western Corn Rootworm (*Diabrotica virgifera virgifera* LeConte)—A Central European Field Study. *Cereal Res. Commun.* **2015**, *43*, 155–165. [CrossRef]

79. Tanasković, S.; Popović, B.; Gvozdena, S.A.; Karpáti, Z.; Bógnar, C.; Erb, M. Level of larval atack on maize roots as a consequence of artificial infestation with western corn rootworm eggs. *AGROFOR* **2017**, *2*, 2490–3442. [CrossRef]

80. Floarea, A.; Grozea, I.; Popescu, G.; Jurca, D. Setting attack frequency produced by the larvae of *Diabrotica virgifera virgifera* LeConte in the Arad area. *Res. J. Agric. Sci.* **2008**, *39*, 494.

81. Marton, C.L.; Szoke, C.; Pinter, J. *Studies of the Tolerance of Maize Hybrids to Corn Rootworm in Hungary*. 59; Tagung der Vereinigung der Pflanzenzüchter und Saatgutkaufleute Österreichs: Raumberg-Gumpenstein, Austria, 2008; pp. 77–80.

82. Kadličko, S.R.; Tollefson, J.J.; Prasifka, J.R.; Bača, F.; Stanković, G.; Delić, N. Evaluation of Serbian commercial maize hybrid tolerance to feeding by larval western corn rootworm (*Diabrotica virgifera virgifera* LeConte) using the novel 'difference approach'. *Maydica* **2010**, *55*, 179–185.

83. Gloyna, K.; Thieme, T.; Zellner, M. Miscanthus, a host for larvae of a European population of *Diabrotica v. virgifera*. *J. Appl. Entomol.* **2011**, *135*, 780–785. [CrossRef]

84. Toepfer, S.; Zellner, M.; Kuhlamann, U. Food and oviposition preferences of *Diabrotica v. virgifera* in multiple-choice crop habitat situations. *Entomologia* **2013**, *1*, 60–68. [CrossRef]

85. Fora, C.G.; Lauer, K.F. Host plants for the western corn rootworm *Diabrotica virgifera virgifera* (Coleoptera: Chrysomelidae). *Rom. Agric. Res.* **2014**, *30*, 291–295.

86. Foltin, K.; Robier, J. Host plant specificity studies of the western corn rootworm—experiments in isolation cages. *Jul. Kühn Arch.* **2014**, *444*, 144–146.

87. Grabenweger, G.; Zellner, M. Winter wheat and volunteer cereals as host plants for the western corn rootworm in Europe. *Jul. Kühn Arch.* **2014**, *444*, 133.

88. Gloyna, K.; Thieme, T.; Zellner, M. Sorghum, Miscanthus & Co: Energy crops as potential host plants of western corn rootworm larvae. *Jul. Kühn Arch.* **2014**, *444*, 134–143.

89. Grozea, I.; Stef, R.; Virteiu, A.M.; Molnar, L.; Carabet, A.; Puia, C.; Dobrin, I. Feeding Behaviour of *Diabrotica virgifera virgifera* Adults on Corn Crops. *Bull. UASVM Hortic.* **2015**, *72*, 463–464. [CrossRef]

90. Guzik, J.; Nakonieczny, M.; Tarnawska, M.; Bereš, P.K.; Drzewiecki, J.; Migula, P. The Glycolytic Enzymes Activity in the Midgut of *Diabrotica virgifera virgifera* (Coleoptera: Chrysomelidae) adult and their Seasonal Changes. *J. Insect Sci.* **2015**, *15*, 56. [CrossRef]

91. Toepfer, S.; Zellner, M.; Szalai, M.; Kuhlmann, U. Field survival analyses of adult *Diabrotica virgifera virgifera* (Coleoptera: Chrysomelidae). *J. Pest. Sci.* **2015**, *88*, 25–35. [CrossRef]

92. Grozea, I.; Trusca, R.; Virteiu, A.M.; Stef, R.; Butnariu, M. Interaction Between *Diabrotica virgifera virgifera* and Host Plants Determined By Feeding Behavior And Chemical Composition. *Rom. Agric. Res.* **2017**, *34*, 329–337.

93. Manole, T.; Chireceanu, C.; Teodpru, A. The broadening of distribution of the invasive species *Diabrotica virgifera virgifera* Leconte in the area of Muntenia region under specific climatic and trophic conditions. *Sci. Pap. Ser. A Agron.* **2017**, *60*, 495–499.

94. Stanikova, K.; Rosca, I.; Cagan, L. Movement of the western corn rootworm (*Diabrotica virgifera virgifera*) adults in a trial with Bt and non-Bt maize plots. *Insect Pathog. Insect Parasit. Nematodes IOBC/wprs Bull.* **2009**, *45*, 453–456.

95. Grozea, I. Western Corn Rootworm (WCR), *Diabrotica virgifera virgifera* Le Conte—Several Years of Research in Western Part of Romania. *Bull. USAMV Agric.* **2010**, *67*, 122–129.

96. Surek, G.; Nádor, G.; Fényes, D.; Vasas, L. Monitoring of western corn rootworm damage in maize fields by using integrated radar (ALOS PALSAR) and optical (IRS LISS, AWiFS) satellite data. *Geocarto Int.* **2013**, *28*, 63–79. [CrossRef]

97. Grozea, I.; Stef, R.; Carabet, A.; Virteiu, A.M.; Dinnesen, S.; Chris, C.; Molnar, L. Te influence of weather and geographical conditions on flight dynamics of WCR adults. *Comm. Appl. Biol. Sci. Ghent Univ.* **2009**, *75*, 1–9.

98. Grozea, I.; Carabet, A.; Stef, R.; Virteiu, A.M.; Chis, C.; Dinnesen, S. Analysis of correlations between WCR adults recorded at different altitudes and climate factors. *Res. J. Agric. Sci.* **2011**, *43*, 44–50.

99. Ciobanu, C.; Ciobanu, G.; Domuta, C.; Sandor, M.; Domuta, C.; Albu, R.; Vuscan, A.; Popov, C. The influence of ecological factors from northwestern part of romania on *Diabrotica virgifera virgifera* LeConte (western corn rootworm) species. *Nat. Resour. Sustain. Dev.* **2011**, *1*, 89–96.

100. Agargon, P.; Lobo, J.M. Predicted effect of climate change on the invasibility and distribution of the western corn root-worm. *Agric. For. Entomol.* **2012**, *14*, 13–18. [CrossRef]

101. Fora, C.G. On the influence of different soil cultivation practices in autumn and spring on the population development of the western corn rootworm *Diabrotica virgifera virgifera* LeConte (Col.: Chrysomelidae). *Jul. Kühn Arch.* **2014**, *444*, 105–111.

102. Kos, T.; Bažok, R.; Lemić, D.; Igrc Barčić, J. Forecasting of root damage, plant lodging and yield loss caused by western corn root worm larval feeding based on larval population density. *Jul. Kühn Arch.* **2014**, *444*, 40.

103. Lindström, L.; Lehmann, P. Climate Change Effects on Agricultural Insect Pests in Europe. In *Climate Change and Insect Pests*; CABI Climate Change Series (7); CABI: Delémont, Switzerland, 2015.

104. Popović, B.; Tanasković, S.; Gvozdenac, S.; Karpati, Z.; Bognar, C.; Erb, M. Population Dynamics of WCR and ECB in Maize Field In Bečej, Vojvodina Provance. *XXI Savetov. Biotehnol. Zb. Rad.* **2016**, *21*, 341–346.

105. Toshova, T.B.; Velchev, D.I.; Abaev, V.D.; Subchev, M.A.; Atanasova, D.Y.; Tóth, M. Detection and Monitoring of *Diabrotica virgifera virgifera* LeConte, 1868 (Coleoptera: Chrysomelidae) by KLP+ Traps with Dual (Pheromone and Floral) Lures in Bulgaria. *Acta Zool. Bulg.* **2017**, *9*, 247–254.

106. Karić, N.; Festić, H. The effect of soil type on the western corn root worm (*Diabrotica virgifera virgifera* Le Conte) population density. *Works Fac. Agric. Food Sci. Univ. Sarajevo* **2011**, *105*, 75–86.

107. Rancov, I.; Carciu, G. Impact of soil works on the dynamics of the population of *Diabrotica virgifera virgifera* Le Conte. *J. Hortic. For.* **2011**, *15*, 55–59.

108. Rancov, I.; Carciu, G. Impact of soil works on dynamics and spread of *Diabrotica virgifera virgifera* LeConte in the conditions of Western Romania. *J. Hortic. For.* **2012**, *16*, 142–144.

109. Rancov, I.P.; Cârciu, G.; Lăzureanu, A.; Cristea, T.; Alda, S.; Molnar, L. Influence of Soil Works on the Damage by the Western Corn Rootworm (*Diabrotica virgifera virgifera* LeConte) in Grain Maize in the Banat's Plain. *ProEnvironment* **2015**, *8*, 234–239.

110. Tóth, M.; Törôcsik, G.; Imrei, Z.; Vörös, G. Diel Rhythmicity of Field Responses to Synthetic Pheromonal or Floral Lures in the Western Corn Rootworm *Diabrotica v. virgifera*. *Acta Phytopathol. Entomol. Hung.* **2010**, *45*, 323–328. [CrossRef]

111. Toth, M.; Ujvary, I.; Imrei, Z. 8-Methyldecan-2-yl acetate inhibits response to the pheromone in the western corn rootworm *Diabrotica v. virgifera*. *J. Appl. Entomol.* **2010**, *134*, 462–466. [CrossRef]

112. Beres, P.K.; Sionek, R. Catches of western corn rootworm (*Diabrotica v. virgifera* Le Conte) beetles by pheromone traps type PAL and feeding traps type PALs in Krzeczowice in 2009–2011. *Prog. Plant. Prot./Post. Ochr. Roślin* **2012**, *52*, 14–19.

113. Bereś, P.K.; Drzewiecki, S.; Nakonieczny, M.; Tarawska, M.; Guzik, J.; Migula, P. Population dynamics of western corn rootworm beetles on different varieties of maize identified using pheromone and floral baited traps. *J. Agric. Sci.* **2015**, *153*, 1479–1490. [CrossRef]

114. Schumann, M.; Ladin, Z.S.; Beatens, J.M.; Hiltpold, I. Navigating on a chemical radar: Usage of root exudates by foraging *Diabrotica virgifera virgifera* larvae. *J. Appl. Entomol.* **2017**, *142*, 911–920. [CrossRef]

115. Florian, T.; Oltean, I.; Brunescu, H.; Todoran, F.C.; Varga, M. Reserach Regarding the External Morphology of *Diabrotica virgifera* Le Conte, Western Corn Root Worm. *ProEnvironment* **2011**, *4*, 233–236.

116. Gloyna, K.; Thieme, T.; Gorb, S.; Voigt, D. Sexual differences in tarsal adhesive setae of *Diabrotica virgifera virgifera*. *Eur. J. Environ. Sci.* **2014**, *4*, 97–101.

117. Benitez, H.A.; Lemic, D.; Bažok, R.; Bravi, R.; Buketa, M.; Puschel, T. Morphological integration and modularity in *Diabrotica virgifera virgifera* LeConte (Coleoptera: Chrysomelidae) hind wings. *Zool. Anz.* **2014**, *253*, 461–468. [CrossRef]

118. Benitez, H.A.; Lemic, D.; Bažok, R.; Gallardo-Araya, C.M.; Mikac, K.M. Evolutionary directional asymmetry and shape variation in *Diabrotica virgifera virgifera* (Coleoptera: Chrysomelidae): An example using hind wings. *Biol. J. Linn. Soc.* **2014**, *111*, 110–118. [CrossRef]

119. Lemic, D.; Benitez, H.A.; Bažok, R. Intercontinental effect on sexual shape dimorphism and allometric relationships in the beetle pest *Diabrotica virgifera virgifera* LeConte (Coleoptera: Chrysomelidae). *Zool. Anz.* **2014**, *253*, 203–206. [CrossRef]

120. Mikac, K.M.; Lemic, D.; Bažok, R.; Benitez, H.A. Wing shape changes: A morphological view of the *Diabrotica virgifera virgifera* European invasion. *Biol. Invasions* **2016**, *18*, 3401–3407. [CrossRef]

121. Mikac, K.M.; Lemic, D.; Benitez, H.A.; Bažok, R. Changes in corn rootworm wing morphology are related to resistance development. *J. Pest. Sci.* **2019**, *92*, 443–451. [CrossRef]

122. Tarnawska, M.; Bereś, P.K.; Drzewiecki, S.; Guzik, J.; Migula, P.; Brom, K.R.; Brzozowska-Wojoczek, K.; Doleżych, B.; Nakonieczny, M. Esterase activity and heat shock protein levels in western corn rootworm beetles (*Diabrotica virgifera virgifera* LeConte, Coleoptera: Chrysomelidae) after pesticide exposure. *Int. J. Pest. Manag.* **2019**. [CrossRef]

123. Horgoș, H.; Grozea, I. The current assessment of the structure of *Diabrotica virgifera* (Coleoptera: Chrysomelidae) populations and the possible correlation of adult coloristic with the type and composition of ingested maize plants. *Rom. Agric. Res.* **2020**, *37*, 197–210.

124. Bahar, M.H.; Vidal, S.; Möser, J. Tritrophic interaction between the invasive insect pest *Diabrotica virgifera virgifera*, its host plant maize and endophytic fungi, Acremonium strictum. *Khulna Univ. Stud.* **2006**, Special Issue (1st Research Cell Conference). 47–50.

125. Dematheis, F.; Zimmerling, U.; Flocco, C.; Kurtz, B.; Vidal, S.; Kropf, S.; Smalla, K. Multitrophic Interaction in the Rhizosphere of Maize: Root Feeding of Western Corn Rootworm Larvae Alters the Microbial Community Composition. *PLoS ONE* **2012**, *7*, e37288. [CrossRef] [PubMed]

126. Dematheis, F.; Kurtz, B.; Vidal, S.; Smalla, K. Multitrophic interactions among western corn rootworm, *Glomus* intraradices and microbial communities in the rhizosphere and endorhiza of maize. *Front. Microbiol.* **2013**, *4*, 357. [CrossRef] [PubMed]

127. Hiltpold, I. Manipulation of Tritrophic Interactions: A Key for Belowground Biological Control? Ph.D. Thesis, University of Neuchâtel, Neuchatel, Switzerland, 2008.

128. Hartings, H.; Lanzanova, C.; Lazzaroni, N.; Balconi, C. How maize (Zea mays L) tackles the *Diabrotica v. virgifera* attack. *Maydica* **2016**, *61*, 1–8.

129. Dematheis, F.; Kurtz, B.; Vidal, S.; Smalla, K. Microbial Communities Associated with the Larval Gut and Eggs of the Western Corn Rootworm. *PLoS ONE* **2012**, *7*, e44685. [CrossRef] [PubMed]

130. Sever, Z.; Kos, T.; Miličević, T.; Bažok, R. Western Corn Rootworm (*Diabrotica vigrifera vigrifera* LeConte) as potential vector of phytopathogenic fungi on maize. In Proceedings of the 49th Croatian & 9th International Symposium on Agriculture, Dubrovnik, Croatia, 16–21 February 2014; pp. 416–419.

131. Krawczyk, K.; Foryś, J.; Nakonieczny, M.; Tarnawska, M.; Bereś, P.K. Transmission of Pantoea ananatis, the causal agent of leaf spot disease of maize (*Zea mays*), by western corn rootworm (*Diabrotica virgifera virgifera* LeConte). *Crop. Prot.* **2020**, 105431. [CrossRef]

132. Moeser, J.; Guillemaud, T. International cooperation on western corn rootworm ecology research: State-of-the-art and future research. *Agric. For. Entomol.* **2009**, *11*, 3–7. [CrossRef]

133. Agargon, P.; Baselga, A.; Lobo, J.M. Global estimation of invasion risk zones for the western corn rootworm *Diabrotica virgifera virgifera*: Integrating distribution models and physiological thresholds to assess climatic favorability. *J. Appl. Ecol.* **2010**, *47*, 1026–1035. [CrossRef]

134. Gray, M.E.; Sappington, T.W.; Miller, N.J.; Moeser, J.; Bohn, M.O. Adaptation and Invasiveness of Western Corn Rootworm: Intensifying Research on a Worsening Pest. *Annu. Rev. Entomol.* **2009**, *54*, 303–321. [CrossRef]

135. Spencer, J.L.; Hibbard, B.E.; Moeser, J.; Onstad, D.W. Behavior and ecology of the western corn rootworm (*Diabrotica virgifera virgifera* LeConte) (Coleoptera: Chrysomelidae). *Agric. Forest Entomol.* **2009**, *11*, 9–27. [CrossRef]

136. Středa, T.; Vahala, O.; Středová, H. Prediction of adult western corn rootworm (*Diabrotica virgifera virgifera* LeConte) emergence. *Plant. Protect. Sci.* **2012**, *49*, 89–97. [CrossRef]

137. Kos, T.; Bažok, R.; Varga, B.; Igrc Barčić, J.; Kozina, A. Estimation of western corn rootworm (*Diabrotica virgifera virgifera* LeConte) egg abundance based on the previous year adult capture. *J. Cent. Eur. Agric.* **2013**, *14*, 1488–1501. [CrossRef]

138. Trusca, R.; Grozea, I.; Stef, R. Attractiveness and injury levels of adults by *Diabrotica virgifera virgifera* (Le Conte) on different host plant. *J. Food Agric. Environ.* **2013**, *11*, 773–776.

139. Kos, T.; Gunjača, J.; Igrc Barčić, J. Western corn rootworm adult captures as a tool for the larval damage prediction in continuous maize. *J. Appl. Entomol.* **2014**, *138*, 173–182. [CrossRef]

140. Grozea, I.; Horgos, H.; Stef, R.; Carabet, A.; Virteiu, A.; Butnariu, M.; Molnar, L. Assessment of population density of insect species called "species problem", in lots with different maize hybrids. *Res. J. Agric. Sci.* **2019**, *51*, 132–137.

141. Popović, B.; Tanasković, S.; Gvozdenac, S. Root damages and root mass under conditions of artificial infestation with western corn rootworm eggs. *Res. J. Agric. Sci.* **2017**, *49*, 261–268.

142. Popović, B.; Tanasković, S.; Gvozdenac, S. Effects of the low-level western corn rootworm egg infestation on maize plants in the field. *Contemp. Agric.* **2019**, *68*, 28–33. [CrossRef]

143. Gyeraj, A.; Szalai, M.; Pálinkás, Z.; Edwards, C.R.; Kiss, J. Effects of adult western corn rootworm (*Diabrotica virgifera virgifera* LeConte, Coleoptera: Chrysomelidae) silk feeding on yield parameters of sweet maize. *Crop. Prot.* **2020**. [CrossRef]

144. Illes, A.; Bojtor, C.; Mohammad, S.; Mousavi, N.; Marton, L.C.; Ragan, P.; Nagy, J. Maize hybrid and nutrient specific evaluation of the population dynamics and damage of the western corn rootworm (*Diabrotica virgifera virgifera* LeConte) in a long-term field experiment. *Prog. Agric. Eng. Sci.* **2020**, *16*, 11–24.

145. Sivčev, I.; Stanković, S.; Kostic, M.; Lakic, N.; Popovic, Z. Population density of *Diabrotica virgifera virgifera* LeConte beetles in Serbian first year and continuous maize fields. *J. Appl. Entomol.* **2009**, *133*, 430–437. [CrossRef]

146. Horgoș, H.; Grozea, I. The occurrence dynamics and sex ratio of *Diabrotica virgifera virgifera* adults on different corn hybrids in western Romania. *Lucr. Științifice* **2017**, *60*, 127–130.

147. Schumann, M.; Tappe, B.; French, W.; Vidal, S. Semi field trials to evaluate undersowings in maize for management of western corn rootworm larvae. *Bull. Insectol.* **2017**, *70*, 63–68.

148. Gošić-Dondo, S.; Srdić, J.; Popović, Ž.; Tancik, J. The population level of western corn rootworm adults in the period 2005–2009. *Sel. Semen.* **2018**, *24*, 39–48. [CrossRef]

149. Ivezić, M.; Raspudić, E.; Brmež, M.; Majić, I.; Džoić, D.; Brkić, A. Maize Tolerance to Western Corn Rootworm Larval Feeding: Screening through Five Years of Investigation. *Agric. Conspec. Sci.* **2009**, *74*, 291–295.

150. Kaiser-Alexnat, R. Protease activities in the midgut of western corn rootworm (*Diabrotica virgifera virgifera* LeConte). *J. Invertebr. Pathol.* **2009**, *100*, 169–174. [CrossRef]

151. Marton, L.C.; Szoke, C.; Pinter, J.; Bodnar, E. Studies on the tolerance of maize hybrids to western corn rootworm (*Diabrotica virgifera virgifera* LeConte). *Maydica* **2009**, *54*, 217–220.

152. Brkić, I.; Brkić, A.; Ivezić, M.; Ledenčan, T.; Jambrović, A.; Zdunić, Z.; Brkić, J.; Raspudić, E.; Šimić, D. Resource allocation in a maize breeding program for native resistance to western corn rootworm. *Poljoprivreda* **2012**, *18*, 3–7.

153. Lanzanova, C.; Hartings, H.; Balconi, C. Analysis of genetic variability for selection of tolerance to western corn rootworm damage in maize. *Maydica* **2014**, *59*, 329–335.

154. Brkić, A.; Raspudić, E.; Brmež, M.; Brkić, J.; Šimić, D. Relations among western corn rootworm resistance traits and elements concentration in maize germplasm roots. *Poljoprivreda* **2015**, *21*, 3–7. [CrossRef]

155. Brkić, A.; Brkić, I.; Jambrović, A.; Ivezić, M.; Raspudić, E.; Brmež, M.; Zdunić, Z.; Ledenčan, T.; Brkić, J.; Marković, M.; et al. Maize germplasm of eastern Croatia with native resistance to western corn rootworm (*Diabrotica virgifera virgifera* LeConte). *Genetika* **2017**, *49*, 1023–1034. [CrossRef]

156. Brkić, A.; Šimić, D.; Jambrović, A.; Zdunić, Z.; Ledenčan, T.; Raspudić, E.; Brmež, M.; Brkić, J.; Mazur, M.; Galić, V. QTL analysis of western corn rootworm resistance traits in maize ibm population grown in continuous maize. *Genetika* **2020**, *52*, 137–148. [CrossRef]

157. Schumann, M. Development of an Attract & Kill"Strategy for the Control of Western Corn Rootworm Larvae. Ph.D. Thesis, der Georg-August-Universität Göttingen, Göttingen, Germany, 2012.

158. Schumann, M.; Patel, A.; Vemmer, M.; Vidal, S. The role of carbon dioxide as an orientation cue for western corn rootworm larvae within the maize root system: Implications for an attract-and-kill approach. *Pest. Manag. Sci.* **2013**, *70*, 642–650. [CrossRef]

159. Schumann, M.; Patel, A.; Vidal, S. Evaluation of an attract and kill strategy for western corn rootworm larvae. *Appl. Soil Ecol.* **2013**, *64*, 178–189. [CrossRef]

160. Schumann, M.; Patel, A.; Vidal, S. Soil Application of an Encapsulated CO_2 Source and Its Potential for Management of Western Corn Rootworm Larvae. *J. Econ. Entomol.* **2014**, *107*, 230–239. [CrossRef]

161. Schumann, M.; Toepfer, S.; Vemmer, M.; Patel, A.; Kuhlmann, U.; Vidal, S. Field evaluation of an attract and kill strategy against western corn rootworm larvae. *J. Pest. Sci.* **2014**, *87*, 259–271. [CrossRef]

162. Kaiser-Alexnat, R.; Buchs, W.; Huber, J. Studies on the proteolytic processing and binding of Bt toxins Cry3Bb1 and Cry34Ab1/Cry35Ab1 in the midgut of western corn rootworm (*Diabrotica virgifera virgifera* LeConte). *Insect Pathog. Insect Parasit. Nematodes IOBC/wprs Bull.* **2009**, *45*, 235–238.

163. Dillen, K.; Mitchell, P.D.; Tollens, E. On the competitiveness of *Diabrotica virgifera virgifera* damage abatement strategies in Hungary: A bio-economic approach. *J. Appl. Entomol.* **2010**, *134*, 395–408. [CrossRef]

164. Carrasco, L.R.; Harwood, T.D.; Toepfer, S.; MacLeod, A.; Levay, N.; Kiss, J.; Baker, R.H.A.; Mumford, J.D.; Knight, J.D. Dispersal kernels of the invasive alien western corn rootworm and the effectiveness of buffer zones in eradication programmes in Europe. *Ann. Appl. Biol.* **2010**, *156*, 63–77. [CrossRef]

165. Ciosi, M.; Toepfer, S.; Li, H.; Haye, T.; Kuhlmann, U.; Wang, H.; Siegfried, B.; Guillemaud, T. European populations of *Diabrotica virgifera virgifera* are resistant to aldrin, but not to methyl-parathion. *J. Appl. Entomol.* **2009**, *133*, 307–314. [CrossRef]

166. Van Rozen, K.; Ester, A. Chemical control of *Diabrotica virgifera virgifera* LeConte. *J. Appl. Entomol.* **2010**, *134*, 376–384. [CrossRef]

167. Čerevkova, A.; Cagan, L. The influence of western corn rootworm seed coating and granular insecticides on the seasonal fluctuations of soil nematode communities in a maize field. *Helminthologia* **2013**, *50*, 205–214. [CrossRef]

168. Bereś, P. Effects of chemical control of *Diabrotica virgifera virgifera* LeConte larvae on maize using methiocarb and thiacloprid with deltamethrin. *Prog. Plant. Prot.* **2016**, *56*, 219–224.

169. Drzewiecki, S.; Pietryga, J.; Bereś, P.K. The Effects of Chemical Control Used Against Males of *Diabrotica virgifera* LeConte In Maize in Southern Poland in 2011–2012. *Nauka Przyr. Technol.* **2016**, *10*, 7. [CrossRef]

170. Blandino, M.; Ferracini, C.; Rigamonti, I.; Testa, G.; Saladini, M.A.; Jucker, C.; Agosti, M.; Alma, A.; Reyneri, A. Control of western corn rootworm damage by application of soil insecticides at different maize planting times. *Crop. Prot.* **2017**, *93*, 19–27. [CrossRef]

171. Ferracini, C.; Blandino, M.; Rigamonti, I.E.; Jucker, C.; Busato, E.; Saladini, M.A.; Reyneri, A.; Alma, A. Chemical-based strategies to control the western corn rootworm, *Diabrotica virgifera virgifera* LeConte. *Crop. Prot.* **2021**, *139*, 105306. [CrossRef]

172. Pilz, C. Biological Control of the Invasive Maize Pest *Diabrotica virgifera virgifera* by the Entomopathogenic Fungus Metarhizium Anisopliae. Ph.D. Thesis, Universiät für Bodenkultur Wien, Vienna, Austria, 2008.

173. Pilz, C.; Wegensteiner, R.; Keller, S. Natural occurrence of insect pathogenic fungi and insect parasitic nematodes in *Diabrotica virgifera virgifera* populations. *BioControl* **2008**, *53*, 353–359. [CrossRef]

174. Meissle, M.; Pilz, C.; Romeis, J. Susceptibility of *Diabrotica virgifera virgifera* (Coleoptera: Chrysomelidae) to the Entomopathogenic Fungus *Metarhizium anisopliae* When Feeding on *Bacillus thuringiensis* Cry3Bb1-Expressing Maize. *Appl. Environ. Microbiol.* **2009**, *75*, 3937–3943. [CrossRef] [PubMed]

175. Pilz, C.; Keller, S.; Kuhlmann, U.; Toepfer, S. Comparative efficacy assessment of fungi, nematodes and insecticides to control western corn rootworm larvae in maize. *BioControl* **2009**, *54*, 671–684. [CrossRef]

176. Kurtz, B. Interaction of Maize Root Associated Fungi and the Western Corn Rootworm. Ph.D. Thesis, der Georg-August-Universität Göttingen, Gottingen, Germany, 2010.

177. Kurtz, B.; Karlovsky, P.; Vidal, S. Interaction Between Western Corn Rootworm (Coleoptera: Chrysomelidae) Larvae and Root-Infecting *Fusarium verticillioides*. *Environm. Entomol.* **2010**, *39*, 1532–1538. [CrossRef]

178. Rauch, H.; Steinwender, B.M.; Mayerhofer, J.; Sigsgaard, L.; Eilenberg, J.; Enkerli, J.; Zelger, R.; Strasser, H. Field efficacy of *Heterorhabditis bacteriophora* (Nematoda: Heterorhabditidae), *Metarhizium brunneum* (Hypocreales: Clavicipitaceae), and chemical insecticide combinations for *Diabrotica virgifera virgifera* larval management. *Biol. Contr.* **2017**, *107*, 1–10. [CrossRef]

179. Cagáň, L.; Števo, J.; Gašparovič, K.; Matušíková, S. Mortality of the western corn rootworm, *Diabrotica virgifera virgifera* larvae caused by entomopathogenic fungi. *J. Cent. Eur. Agric.* **2019**, *20*, 678–685. [CrossRef]

180. Ehlers, R.-U.; Hiltpold, I.; Kuhlmann, U.; Toepfer, S. Field results on the use of *Heterorhabditis bacteriophora* against the invasive maize pest *Diabrotica virgifera virgifera*. *Insect Pathog. Insect Parasit. Nematodes IOBC/wprs Bull.* **2008**, *31*, 332–335.

181. Toepfer, S.; Ehlers, R.-U.; Burger, R.; Peters, A.; Kuhlmann, U. Biological control of western corn rootworm larvae using nematodes. *Insect Pathog. Insect Parasit. Nematodes IOBC/wprs Bull.* **2009**, *45*, 390.

182. Toepfer, S.; Enkerli, J.; Kuhlamann, U. Research needs and promising approaches for the biological control of *Diabrotica* and other emerging soil insect pests with pathogens or nematodes. *Insect Pathog. Insect Parasit. Nematodes IOBC/wprs Bull.* **2009**, *45*, 47–58.

183. Kurtz, B.; Hiltpold, I.; Turlings, T.C.J.; Kuhlmann, U.; Toepfer, S. Comparative susceptibility of larval instars and pupae of the western corn rootworm to infection by three entomopathogenic nematodes. *BioControl* **2009**, *54*, 255–262. [CrossRef]

184. Hiltpold, I.; Toepfer, S.; Kuhlmann, U.; Turlings, T.C.J. How maize root volatiles affect the efficacy of entomopathogenic nematodes in controlling the western corn rootworm? *Chemoecology* **2010**, *20*, 155–162. [CrossRef]

185. Toepfer, S.; Burger, R.; Ehlerst, R.-U.; Peters, A.; Kuhlmann, U. Controlling western corn rootworm larvae with entomopathogenic nematodes: Effect of application techniques on plant-scale efficacy. *J. Appl. Entomol.* **2010**, *134*, 467–480. [CrossRef]

186. Toepfer, S.; Hatala-Zseller, I.; Ehler, R.-U.; Peters, A.; Kuhlmann, U. The effect of application techniques on field-scale efficacy: Can the use of entomopathogenic nematodes reduce damage by western corn rootworm larvae? *Agric. For. Entomol.* **2010**, *12*, 389–402. [CrossRef]

187. Toepfer, S.; Kurtz, B.; Kuhlmann, U. Influence of soil on the efficacy of entomopathogenic nematodes in reducing *Diabrotica virgifera virgifera* in maize. *J. Pest. Sci.* **2010**, *83*, 257–264. [CrossRef] [PubMed]

188. Anbesse, S.; Ehlers, R.-U. *Heterorhabditis* sp. not attracted to synthetic (E)-bcaryophyllene, a volatile emitted by roots upon feeding by corn rootworm. *J. Appl. Entomol.* **2013**, *137*, 88–96. [CrossRef]

189. Grubišić, D.; Vladić, M.; Gotlin Čuljak, T.; Benković Lačić, T. Primjena entomopatogenih nematoda u suzbijanju kukuruzne zlatice. *Glas. Biljn. Zaštite* **2013**, *13*, 223–231.

190. Kahrer, A.; Pilz, C.; Heimbach, U.; Grabenweger, G. Biological control of the western corn rootworm (*Diabrotica virgifera virgifera*) by entomoparasitic nematodes. *Jul. Kühn Arch.* **2014**, *444*, 54–58.

191. Pilz, C.; Toepfer, S.; Knuth, P.; Strimitzer, T.; Heimbach, U.; Grabnweger, G. Persistence of the entomoparasitic nematode *Heterorhabditis bacteriophora* in maize fields. *J. Appl. Entomol.* **2014**, *138*, 202–212. [CrossRef]

192. Toepfer, S.; Knuth, P.; Glas, M.; Kuhlmann, U. Successful application of entomopathogenic nematodes for the biological control of western corn rootworm larvae in Europe—A mini review. *Jul. Kühn Arch.* **2014**, *444*, 59–66.

193. Babendreier, D.; Jeanneret, P.; Pilz, C.; Toepfer, S. Non-target effects of insecticides, entomopathogenic fungi and nematodes applied against western corn rootworm larvae in maize. *J. Appl. Entomol.* **2015**, *139*, 457–467. [CrossRef]

194. Carrera, A.C. Control of the Western Corn Rootworm in Maize with *Heterorhabditis bacteriophora*: Influence of Field Persistence and Infectivity. Master Thesis, University of Ghent, Ghent, Belgium, 2015.

195. Machado, R.A.R.; Thönen, L.; Arce, C.C.M.; Theepan, V.; Prada, F.; Wutrich, D.; Robert, C.A.M.; Vogiatzki, E.; Shi, Y.-M.; Schaeren, O.P.; et al. Engineering bacterial symbionts of nematodes improves their biocontrol potential to counter the western corn rootworm. *Nat. Biotechnol.* **2020**, *38*, 600–608. [CrossRef]

196. Modič, Š.; Žigon, P.; Kolomanič, A.; Trdan, S.; Razinger, J. Evaluation of the Field Efficacy of *Heterorhabditis bacteriophora* Poinar (Rhabditida: Heterorhabditidae) and Synthetic Insecticides for the Control of Western Corn Rootworm Larvae. *Insects* **2020**, *11*, 202. [CrossRef]

197. Toepfer, S.; Toth, S. Entomopathogenic nematode application against root-damaging *Diabrotica* larvae in maize: What, when, and how? *Microb. Nematode Control. Invertebr. Pests IOBC-WPRS Bull.* **2020**, *150*, 185–188.

198. Toth, S.; Szalai, M.; Kiss, J.; Toepfer, S. Missing temporal efects of soil insecticides and entomopathogenic nematodes in reducing the maize pest *Diabrotica virgifera virgifera*. *J. Pest. Sci.* **2020**, *93*, 767–781. [CrossRef]

199. Toepfer, S.; Cabrera Walsh, G.; Eben, A.; Alvarez-Zagoya, R.; Haye, T.; Zhang, F.; Kuhlmann, U. A critical evaluation of host ranges of parasitoids of the subtribe Diabroticina (Coleoptera: Chrysomelidae: Galerucinae: Luperini) using field and laboratory host records. *Biocontrol Sci. Technol.* **2008**, *18*, 483–504. [CrossRef]

200. Toepfer, S.; Haye, T.; Erlandson, M.; Goettel, M.; Lundgren, J.G.; Kleespies, R.G.; Weber, D.C.; Cabrera Walsh, G.; Peters, A.; Ehlers, R.-U.; et al. A review of the natural enemies of beetles in the subtribe Diabroticina (Coleoptera: Chrysomelidae): Implications for sustainable pest management. *Biocontrol Sci. Technol.* **2009**, *19*, 1–65. [CrossRef]

201. Toepfer, S.; Zhang, F.; Kuhlmann, U. Assessing host specificity of a classical biological control agent against western corn rootworm with a recently developed testing protocol. *Biol. Control.* **2009**, *51*, 26–33. [CrossRef]

202. Florian, T.; Oltean, I.; Brunescu, H.; Todoran, F.C.; Florian, C.V. Results Obtained in the Biological Control of Western Corn Root Worm. *ProEnvironment* **2010**, *3*, 305–308.

203. Florian, T.; Oltean, I.; Brunescu, H.; Florian, C.V.; Todoran, F.C.; Bodis, I. Results obtained in control of *Diabrotica virgifera virgifera* larvae. *Res. J. Agric. Sci.* **2011**, *43*, 40–43.

204. Balog, E.; Hung, B.H.; Kiss, J.; Turoczi, G. Efficacy of biological control agents for the control of western corn rootworm. *Insect Pathog. Entomoparasit. Nematodes IOBC-WPRS Bull.* **2013**, *90*, 33–36.

205. Florian, T.; Oltean, I.; Bunescu, H.; Todoran, F.C.; Florian, V. Results obtained in the biological control of western corn root worm, *Diabrotica virgifera virgifera* LeConte (2007–2010). *J. Food Agric. Environ.* **2013**, *11*, 306–308.

206. Brandl, M.A.; Schumann, M.; French, B.W. Screening of Botanical Extracts for Repellence against Western Corn Rootworm Larvae. *J. Insect Behav.* **2016**, *29*, 395–414. [CrossRef]

207. Panevska, A.; Hodnik, V.; Skočaj, M.; Novak, M.; Modic, Š.; Pavlic, I.; Podržaj, S.; Zarić, M.; Resnik, N.; Maček, P.; et al. Pore-forming protein complexes from *Pleurotus* mushrooms kill western corn rootworm and Colorado potato beetle through targeting membrane ceramide phosphoethanolamine. *Sci. Rep.* **2019**, *9*, 5073. [CrossRef]

208. Dillen, K.; Mitchell, P.D.; Van Looy, T.; Tollens, E. The western corn rootworm, a new threat to European agriculture: Opportunities for biotechnology? *Pest. Manag. Sci.* **2010**, *66*, 956–966. [CrossRef]

209. Kehlenbeck, H. Assessment of economic impacts of the western corn rootworm (*Diabrotica virgifera virgifera*) in Germany. *Jul. Kühn Arch.* **2014**, *444*, 198–201.

210. Feusthuber, E.; Schönhart, M.; Schmid, E. Spatial analysis of maize cropping systems to relieve crop pest pressure in Austria. In Proceedings of the 150th EAAE Seminar, Edinburgh, UK, 22–23 October 2015.

211. Furlan, L.; Kreutzweiser, D. Alternatives to neonicotinoid insecticides for pest control: Case studies in agriculture and forestry. *Environ. Sci. Pollut. Res.* **2015**, *22*, 135–147. [CrossRef] [PubMed]

212. Benjamin, E.O.; Wesseler, J.H.H. A socioeconomic analysis of biocontrol in integrated pest management: A review of the effects of uncertainty, irreversibility and flexibility. *NJAS-Wagen. J. Life Sc.* **2016**, *77*, 53–60. [CrossRef]

213. Benjamin, E.O.; Grabenweger, G.; Strasser, H.; Wesseler, J. The socioeconomic benefits of biological control of western corn rootworm *Diabrotica virgifera virgifera* and wireworms *Agriotes* spp. in maize and potatoes for selected European countries. *J. Plant. Dis. Prot.* **2018**, *125*, 273–285. [CrossRef]

214. Ciosi, M.; Miller, N.J.; Toepfer, S.; Estoup, A.; Guillemaud, T. Stratified dispersal and increasing genetic variation during the invasion of Central Europe by the western corn rootworm, *Diabrotica virgifera virgifera*. *Evol. Appl.* **2011**, *4*, 54–70. [CrossRef]

215. Bermond, G.; Ciosi, M.; Lombaert, E.; Blin, A.; Boriani, M.; Furlan, L.; Toepfer, S.; Guillemaud, T. Secondary Contact and Admixture between Independently Invading Populations of the Western Corn Rootworm, *Diabrotica virgifera virgifera* in Europe. *PLoS ONE* **2012**, *7*, e50129. [CrossRef] [PubMed]

216. Bermond, G.; Blin, A.; Vercken, E.; Ravignem, V.; Rieux, A.; Mallez, S.; Morel-Journel, T.; Guillemaud, T. Estimation of the dispersal of a major pest of maize by cline analysis of a temporary contact zone between two invasive outbreaks. *Mol. Ecol.* **2013**, *22*, 5368–5381. [CrossRef] [PubMed]

217. Lemic, D.; Mikac, K.M.; Ivkosic, S.A.; Bažok, R. The Temporal and Spatial Invasion Genetics of the Western Corn Rootworm (Coleoptera: Chrysomelidae) in Southern Europe. *PLoS ONE* **2015**, *10*, e0138796. [CrossRef] [PubMed]

218. Lemic, D.; Mikac, K.M.; Bažok, R. Historical and Contemporary Population Genetics of the Invasive Western Corn Rootworm (Coleoptera: Chrysomelidae) in Croatia. *Environ. Entomol.* **2013**, *42*, 811–819. [CrossRef]

219. Bermond, G.; Cavigliasso, F.; Mallez, S.; Spencer, J.; Guillemaud, T. No Clear Effect of Admixture between Two European Invading Outbreaks of *Diabrotica virgifera virgifera* in Natura. *PLoS ONE* **2014**, *9*, e106139. [CrossRef] [PubMed]

220. Ivkosic, S.A.; Gorman, J.; Lemic, D.; Mikac, K.M. Genetic Monitoring of Western Corn Rootworm (Coleoptera: Chrysomelidae) Populations on a Microgeographic Scale. *Environ. Entomol.* **2014**, *43*, 804–818. [CrossRef]

221. Lemić, D. Temporal and Spatial Influence on the Genetic Variability of Western Corn Rootworm (*Diabrotica virgifera virgifera* LeConte Coleoptera: Chrysomelidae) Populations. Ph.D. Thesis, University of Zagreb, Zagreb, Croatia, 2014.

222. Miller, N.J.; Guillemaud, T.; Giordano, R.; Siegfried, B.D.; Gray, M.E.; Meinke, L.J.; Sappington, T.W. Genes, gene flow and adaptation of *Diabrotica virgifera virgifera*. *Agric. Forest Entomol.* **2009**, *11*, 47–60. [CrossRef]

223. Egartner, A.; Heimbach, U.; Grabenweger, G. A new method for efficacy testing of control measures against adult western corn rootworm in maize. *J. Kult.* **2012**, *64*, 342–347.

224. Sivčev, I.; Kljajić, P.; Kostić, M.; Sivčev, L.; Stanković, S. Management of Western Corn Rootworm (*Diabrotica virgifera virgifera*). *Pestic. Phytomed.* **2012**, *27*, 189–201. [CrossRef]

225. Furlan, L.; Vasileiadis, V.P.; Chiarini, F.; Huiting, H.; Leskovšek, R.; Razinger, J.; Holb, I.J.; Sartori, E.; Urek, G.; Verschwele, A.; et al. Risk assessment of soil-pest damage to grain maize in Europe within the framework of Integrated Pest Management. *Crop. Prot.* **2017**, *97*, 52–59. [CrossRef]

226. Veres, A.; Wyckhuys, K.A.G.; Kiss, J.; Tóth, F.; Burgio, G.; Pons, X.; Avilla, C.; Vidal, S.; Razinger, J.; Bazok, R.; et al. An update of the Worldwide Integrated Assessment (WIA) on systemic pesticides. Part 4: Alternatives in major cropping systems. *Environ. Sci. Pollut. Res.* **2020**, *27*, 29867–29899. [CrossRef] [PubMed]

227. Filipović, J.; Stanković, S.; Ceranić, S. Gross margin as an indicator of the significance of farmer education on the WCR risk assessment in repeated sowing. *Econ. Agric.* **2015**, *62*, 137–153. [CrossRef]

228. Kropf, B.; Schmid, E.; Schönhart, M.; Mitter, H. Exploring farmers' behavior toward individual and collective measures of western corn rootworm control—A case study in south-east Austria. *J. Environ. Manag.* **2020**, *264*, 110431. [CrossRef] [PubMed]

229. Hemerik, L.; van Nes, E.H. A new release of INSIM: A temperature dependent model for insect development. *Proc. Neth. Entomol. Soc. Meet.* **2008**, *19*, 147–155.

230. Wesseler, J.; Fall, E.H. Potential damage costs of *Diabrotica virgifera virgifera* infestation in Europe–the 'no control' scenario. *J. Appl. Entomol.* **2010**, *134*, 385–394. [CrossRef]

231. Szalai, M. Modelling the Population Dynamics of Western Corn Rootworm (*Diabrotica virgifera virgifera* LeConte) on Landscape Level. Ph.D. Thesis, Szent István University, Godollo, Hungary, 2012.

232. Wilstermann, A.; Vidal, S. Western corn rootworm egg hatch and larval development under constant and varying temperatures. *J. Pest. Sci.* **2013**, *86*, 419–428. [CrossRef]

233. Szalai, M.; Kiss, J.; Kover, S.; Toepfer, S. Simulating crop rotation strategies with a spatiotemporal lattice model to improve legislation for the management of the maize pest *Diabrotica virgifera virgifera*. *Agric. Syst.* **2014**, *124*, 39–50. [CrossRef]

234. Svystun, T. Modeling the Potential Impact of Climate Change on the Distribution of Western Corn Rootworm in Europe. Master's Thesis, Lund University, Lund, Sweden, 2015.

235. Grozea, I.; Chris, C.; Carabet, A.; Virteiu, A.M.; Grozea, A.; Stef, R.; Corcionivoschi, N. Mathematical model to analyze the population changes of *Diabrotica virgifera* in terms of geographical coordinates and climatic factors. *Rom. Biotechnol. Lett.* **2017**, *22*, 12630–12642.

236. Feusthuber, E.; Mitter, H.; Schönhart, M.; Schmid, E. Integrated modelling of efficient crop management strategies in response to economic damage potentials of the western corn rootworm in Austria. *Agric. Syst.* **2017**, *157*, 93–106. [CrossRef]

237. Marchi, S.; Guidotti, D.; Riccioloni, M.; Petacchi, R. Validating spatiotemporal predictions of western corn rootworm at the regional scale (Tuscany, central Italy). *Ital. J. Agrometeorol.* **2017**, *3*, 13–24.

238. Falkner, K.; Mitter, H.; Moltchanova, E.; Schmid, E. A zero-inflated Poisson mixture model to analyse spread and abundance of the western corn rootworm in Austria. *Agric. Syst.* **2019**, *174*, 105–116. [CrossRef]

239. Falkner, K.; Mitter, H.; Moltchanova, E.; Schmid, E. Modelling crop rotation regulations to control western corn rootworm infestation under climate change in Styria. *Austrian J. Agric. Econ. Rural Stud.* **2019**, *28*, 47–53.

240. Nador, G.; Fenyes, D.; Vasas, L.; Surek, G. Monitoring of maize damage caused by western corn rootworm by remote sensing. In Proceedings of the 4th Int. Workshop on Science and Applications of SAR Polarimetry and Polarimetric, Frascati, Italy, 26–30 January 2009.

241. Dupin, M.; Reynaud, P.; Jarosık, V.; Baker, R.; Brunel, S.; Eyre, D.; Pergl, J.; Makowski, D. Effects of the Training Dataset Characteristics on the Performance of Nine Species Distribution Models: Application to *Diabrotica virgifera virgifera*. *PLoS ONE* **2011**, *6*, e20957. [CrossRef] [PubMed]

242. Agatz, A.; Ashauer, R.; Sweeney, P.; Brown, C.D. A knowledge-based approach to designing control strategies for agricultural pests. *Agric. Syst.* **2020**, *183*, 102865. [CrossRef]

243. Raileanu, N.; Odobesky, V. Monitoring of the Western Corn Beetle in the Republic of Moldova (Мониторниг западного кукурузного жука в Республике Молдова). In Proceedings of the Protecţia Plantelor-Realizări şi Perspective, Кишинев, Moldova, 27–28 October 2020; pp. 79–83.

244. Furlan, L.; Cossalter, S.; Chiarini, F.; Signori, A.; Bincoletto, S.; Faraon, F.; Codato, F. Strategie di difesa integrata dalla diabrotica del mais. *L'Informatore Agrar.* **2018**, *10*, 74–77.

245. AGES. Maiswurzelbohrer *Diabrotica Virgifera Virgifer*a. Available online: https://www.ages.at/en/topics/harmful-organisms/maiswurzelbohrer/verbreitung/ (accessed on 20 December 2020).

246. Stanković, S.; Institute for Science Application in Agriculture, Belgrade, Serbia. Personal communication, 2020.

247. Kiss, J.; Szent Istvan University, Godollo, Hungary. Personal communication, 2020.

248. Razinger, J.; Agricultural Institute of Slovenia, Ljubljana, SIovenia. Personal communication, 2020.

249. Grozea, I.; Banat's University of Agricultural Sciences and Veterinary Medicine, Temisvar, Romania. Personal communication, 2020.

250. Hegyi, T.; Tóth, B.; Hataláné Zsellér, I.; Szeőke, K.; Ciklin, M.; Vörös, G. Helyzetkép a kukoricabogárról 2003 őszén. *Agrofórum* **2003**, *13*, 16–24.

251. Vörös, G. Az amerikai kukoricabogár 2012. évi biológiai ciklusa, jövő évi kártételi veszélye. *Növényorvosi* **2012**, *14*, 93–94.

252. Vörös, G. Töretlenül támad a kukoricabogár! *Növényorvosi* **2017**, *12*, 105–107.

253. Furlan, L.; Di Bernardo, A.; Girolami, V.; Vettorazzo, M.; Piccolo, A.M.; Santamaria, G.; Donantoni, L.; Funes, V. *Diabrotica virgifera virgifera* eradication containment programme in Veneto: Year 2001: Distribution, population level and what has to be done. In Proceedings of the XXI IWGO Conference, Padova, Italy, 27 October–3 November 2001; pp. 47–51.

254. Boriani, M.; Gervasini, E. La diabrotica del mais è arrivata in Lombardia. *L'Informatore Agrar.* **2000**, *39*, 75.

255. Boriani, M.; Bettoni, D.; Notarangelo, N. Primi danni da diabrotica su mais in Italia. *L'Informatore Agrar.* **2002**, *31*, 61–62.

256. Furlan, L.; Faraglia, B.; Zanini, G.; Vettorazzo, M.; Palmieri, S.; Martini, G.; Governatori, G.; Frausin, C.; Gremo, F.; Bariselli, M.; et al. *Diabrotica*: Risultati 2008 della presenza in Italia. *L'Informatore Agrar.* **2009**, *5*, 47–49.

257. Furlan, L.; Vettorazzo, M.; Frausin, C. *Diabrotica virgifera virgifera* LeConte: What has been done and what will be done in Italy. *Acta Phytopathol. Entomol. Hung.* **2002**, *37*, 169–173. [CrossRef]

258. VV.AA. Diabrotica: Aggiornamento della situazione al 2009. *L'Informatore Agrar.* **2010**, *5*, 50–52.

259. Governatori, G.; Furlan, L.; Bariselli, M.; Boriani, M.; Cavicchini, R.; Faraglia, B.; Franchi, R.; Giovanelli, P.; Gremo, F.; Luppino, M.; et al. Il 2010 della diabrotica: Danni e diffusione contenuti. *L'Informatore Agrar.* **2011**, *5*, 49–51.
260. Governatori, G.; Furlan, L.; Bariselli, M.; Boriani, M.; Cavicchini, R.; Faraglia, B.; Franchi, R.; Giovanelli, P.; Gremo, F.; Luppino, M.; et al. Nel 2011 la diabrotica colpisce senza fare danni. *L'Informatore Agrar.* **2012**, *4*, 43–45.
261. Governatori, G.; Bariselli, M.; Boriani, M.; Carli, M.; Cavicchini, R.; Franchi, R.; Furlan, L.; Giovanelli, P.; Gremo, F.; Luppino, M.; et al. Il monitoraggio della diabrotica in Italia nel 2012. *L'Informatore Agrar.* **2013**, *8*, 8–10.
262. Furlan, L.; Capellari, C.; Porrini, C.; Radeghieri, P.; Ferrari, R.; Pozzati, M.; Davanzo, M.; Canzi, S.; Saladini, M.A.; Alma, A.; et al. Difesa integrata del mais: Come effettuarla nelle prime fasi. *L'Informatore Agrar.* **2011**, *7*, 15–19.
263. Furlan, L.; Pozzebon, A.; Duso, C.; Simon-Delso, N.; Sànchez-Bayo, F.; Marchand, P.A.; Codato, F.; Bijleveld van Lexmond, M.; Bonmatin, J.M. An update of the Worldwide Integrated Assessment (WIA) on systemic insecticides. Part 3: Alternatives to systemic insecticides. *Environ. Sci. Pollut. Res.* **2018**. [CrossRef] [PubMed]
264. Portal Izveštajno Prognozne Službe Zaštite Bilja. Alpha 1.1. Available online: http://www.pissrbija.com:8888/ISPIS/Grafikoni/BrojeviJedinki (accessed on 23 December 2020).
265. Lehmhus, J.; Juhlius Kuehn Institute, Quedlinburg, Germany. Personal communication, 2020.
266. LFL Pflanzenschutz. Available online: https://www.lfl.bayern.de/ips/blattfruechte/086682/index.php (accessed on 23 December 2020).
267. Zellner, M.; Bayerische Landesanstalt für Landwirtschaft, Freising, Germany. Personal communication, 2020.
268. LFL Pflanzenschutz. Western Corn Rootworm—a Dangerous Pest in Maize Cultivation. Available online: https://www.lfl.bayern.de/ips/blattfruechte/052077/index.php (accessed on 23 December 2020).
269. CABI. *DIABROTICA Virgifera Virgifera* (Western Corn Rootworm). Available online: https://www.cabi.org/isc/datasheet/18637#REF-DDB-179771 (accessed on 20 December 2020).
270. Thibord, J.B.; Arvalis–Institut du végétal, Montardon, France. Personal communication, 2020.
271. Thibord, J.B. La chrysomèle du maïs franchit encore un nouveau cap en 2019. *Choisir Maïs Ravag.* **2019**, V20191220-V4 WCR.
272. Agroscope. Maiswurzelbohrer, Diabrotica Virgifera Virgifera. Available online: https://www.agroscope.admin.ch/agroscope/de/home/themen/pflanzenbau/pflanzenschutz/agroscope-pflanzenschutzdienst/geregelte-schadorgnismen/quarantaeneorganismen/diabrotica.html (accessed on 23 December 2020).

Review

Biology and Management of Pest *Diabrotica* Species in South America

Guillermo Cabrera Walsh [1,*], Crébio J. Ávila [2], Nora Cabrera [3], Dori E. Nava [4], Alexandre de Sene Pinto [5] and Donald C. Weber [6]

[1] ARS-SABCL/FuEDEI (Foundation for the Study of Invasive Species), Hurlingham B1686EFA, Argentina
[2] EMBRAPA Agropecuaria Oeste, Dourados, Mato Grosso de Sul Caixa-postal 449, Brazil; crebio.avila@embrapa.br
[3] Facultad de Ciencias Naturales y Museo, Universidad Nacional de La Plata, La Plata B1900FWA, Argentina; ncabrera@fcnym.unlp.edu.ar
[4] EMBRAPA Clima Temperado, Pelotas, Rio Grande do Sul Caixa-Postal 403, Brazil; dori.edson-nava@embrapa.br
[5] Centro Universitario Moura Lacerda, Ribeirão Preto, São Paulo 14076-510, Brazil; aspinn@uol.com.br
[6] USDA-ARS Invasive Insect Biocontrol & Behavior Laboratory, Baltimore Avenue, Beltsville, MD 10300, USA; don.weber@usda.gov
*   Correspondence: gcabrera@fuedei.org; Tel.: +54-11-4452-4838

Received: 27 May 2020; Accepted: 4 July 2020; Published: 8 July 2020

Abstract: The genus *Diabrotica* has over 400 described species, the majority of them neotropical. However, only three species of neotropical *Diabrotica* are considered agricultural pests: *D. speciosa*, *D. balteata*, and *D. viridula*. *D. speciosa* and *D. balteata* are polyphagous both as adults and during the larval stage. *D. viridula* are stenophagous during the larval stage, feeding essentially on maize roots, and polyphagous as adults. The larvae of the three species are pests on maize, but *D. speciosa* larvae also feed on potatoes and peanuts, while *D. balteata* larvae feed on beans and peanuts. None of these species express a winter/dry season egg diapause, displaying instead several continuous, latitude-mediated generations per year. This hinders the use of crop rotation as a management tool, although early planting can help in the temperate regions of the distribution of *D. speciosa*. The parasitoids of adults, *Celatoria bosqi* and *Centistes gasseni*, do not exert much control on *Diabrotica* populations, or show potential for inundative biocontrol plans. Management options are limited to insecticide applications and Bt genetically modified (GM) maize. Other techniques that show promise are products using *Beauveria bassiana* and *Heterorhabditis bacteriophora*, semiochemical attractants for monitoring purposes or as toxic baits, and plant resistance.

Keywords: *Diabrotica speciosa*; *Diabrotica balteata*; *Diabrotica viridula*; rootworm management; maize pests; cucurbitacins; semiochemicals

---

1. General Biology of South American Pest *Diabrotica*

The genus *Diabrotica* has over 400 described species [1], the majority of them neotropical, but only 7 species, plus six subspecies, are considered agricultural pests in the Americas [2]. Of these, only three species are considered agricultural pests in South America: *D. speciosa* (Germar) with subspecies *speciosa* and *vigens*, *D. balteata* (LeConte), and *D. viridula* (F.) (Figure 1). The genus *Diabrotica* is divided into three species groups: *virgifera*, *fucata*, and *signifera* [3,4]. However, studies on South American *virgifera* group species suggest that these groups are not as well defined as previously thought [5,6]. *D. speciosa* and *D. balteata* are in the *fucata* group, which is the group with the largest number of species. The species in this group that have been studied are polyphagous both as adults and during the larval stage. Another characteristic of the North American pest *Diabrotica* of the *fucata* species group is that they overwinter as adults and lack resistant stages to deal with harsh climatic conditions [2]. *D. viridula* is in the *virgifera* group, the same clade of the Northern, Western, and Mexican corn rootworms (*Diabrotica barberi*, *Diabrotica virgifera virgifera*, and *Diabrotica virgifera zeae*, respectively). The larvae of the North American species in the *virgifera* group feed exclusively on Poaceae [7], although the host range has been observed or tested for only a few of the species in the group [8]. The North American pest species in the *virgifera* species group are univoltine, or sometimes semivoltine, and possess diapausing eggs that allow them to overwinter in temperate climates or survive dry seasons in the subtropics [9,10], both situations during which the adult cannot find sustenance or survive the extreme conditions.

Diabrotica balteata (photo Stephen Cresswell)

Diabrotica speciosa adult feeding on a bean leaf (photo Dirceu N. Gassen)

Diabrotica viridula

Figure 1. Photographs of the adult of the three species of pest *Diabrotica* from South America.

D. speciosa is distributed throughout South America, from agricultural patches in the temperate Patagonian steppes to the tropics, with the exception of Chile, and up to altitudes of over 2500 m above sea level [2,11] (Figure 2). It is the best studied *Diabrotica* species in South America due to its impact on many crops. The adult has over 132 recorded host species, in 24 different plant families [11, and literature therein]. Larval hosts are not as well known, but *D. speciosa* has at least five confirmed larval hosts: maize (*Zea mays* L.), wheat (*Triticum* spp.), Johnsongrass (*Sorghum halepense* Persoon), peanut (*Arachis hypogaea* L.), and potato (*Solanum tuberosum* L.). Another four plant species hosted full development in the laboratory [11–15]. However, the fact that larvae can develop on plant species in four families of three different orders suggests that there could be many more larval hosts that simply have not been discovered because of the hypogeous habit of the larva.

D. speciosa is documented in most crops in South America, but is considered mainly a horticultural pest as an adult, and a pest of potato, maize, and peanuts as larva [11,13,16]. Yet these generalizations are not without exceptions. In Brazil, this species is considered a pest of maize as a larva, and a minor pest as an adult as well [17,18]. It is also regarded as an important pest of potato during both the adult and larval stages, although this depends heavily on the cultivar [19]. In addition, the adult is also regarded as an important pest of seedlings and young plants of some extensive crops, such as soybeans, beans (*Phaseolus vulgaris*), cotton, sunflower, maize, tobacco, wheat, and canola [20–22], and, curiously, of table grapes [23] (Table 1).

Figure 2. Distribution of *Diabrotica speciosa* in South America (crosshatched area).

Table 1. Main crops attacked by the South American pest *Diabrotica* species, and current and potential control methods.

	D. balteata		*D. speciosa*		*D. viridula*		**Control Methods**		**Promising Control Methods**	
Host Crop	**Adults**	**Larvae**	**Adults**	**Larvae**	**Adults**	**Larvae**	**Adults**	**Larvae**	**Adults**	**Larvae**
beans	x	x	x		x		Cb, Op, Nn, Py [1]		intercropping plant resistance	
cucurbits	x		x				Cb, Op, Nn, Py		cucurbitacin baits	
maize		x	x	x		x	Cb, Op, Nn, Py	Bt maize seed treatment (Nn, Cb, Di) [1]	silicon cucurbitacin baits	IGR [1] seed treatment with fungi, plant resistance, nematodes
peanuts		x	x	x			Cb, Op, Nn, Py			
potatoes	x		x	x			Nn		plant resistance	plant resistance, nematodes
soybeans			x				Cb, Op, Nn, Py			
tobacco			x				Cb, Op, Nn, Py			

[1] Cb, carbamates; Op, organophosphates; Nn, neonicotinoids; Py, phenylpyrazole; Di, diamides; IGR, insect growth regulators

D. balteata is found from subtropical North America through Central America and Caribbean islands including Cuba, Hispaniola, and Puerto Rico, to South America, although its distribution in South America is limited to Venezuela and Colombia [2,24], where it can occur at altitudes ranging from 0 to 2000 m [25]. However, there is insufficient data to infer species distribution patterns in either country. The adult of *D. balteata* also has an extremely wide range of host plants, as it has been documented on over 140 plant species [26]. There is a more conservative estimate of 50 species in 23 families, with a preference for plants in the Cucurbitaceae, Rosaceae, Fabaceae, and Brassicaceae [27]. The *D. balteata* adult is considered a pest on squash (*Cucurbita* spp., Cucurbitaceae), several bean species (*P. vulgaris*, *Glycine max*, *Mucuna pruriens*, and *Vigna unguiculata*, Fabaceae), lettuce (*Lactuca sativa*, Asteraceae), sugar cane (*Saccharum officinarum*, Poaceae), and potato [28]. Adults are also implicated in the transmission of the tomato brown rugose fruit virus (Tobamovirus, ToBRFV) to *P. vulgaris* [29], and other viruses of *P. vulgaris* and calapo (*Calopogonium mucunoides* Desv.) [30,31]. Larval damage is reported only from Colombia, where this species is known to attack beans, but as considered a minor problem [32], maize, on which it can be locally problematic [33,34], and peanuts, on which it is considered among the 10–12 worst pests in Colombia [35] (Table 1). The larva has also been reported to attack sweet potato in the USA [36], although not in South America. Yet, the fact that these hosts are also from three families in three orders suggests that there could be many more larval hosts as well. In addition, phylogenetic studies indicate *D. speciosa* and *D. balteata* are sister clades [37].

D. viridula is distributed from Mexico to northern Argentina, and apparently absent in Uruguay and Chile, except on Easter Island, where it was introduced [2,13,38] (Figure 3). Like *D. balteata*, its distribution is primarily tropical and subtropical. The *D. viridula* adult is considered a minor pest of beans in Peru [39], while the larva is considered locally important on maize in Central America and Peru [40]. In greenhouse tests, both the larvae and the adults of this species were able to transmit maize chlorotic mottle virus (MCMV) to maize, and they are assumed to be one of its vectors in the field [41]. *D. viridula* is also assumed to be an important, albeit new, pest of maize roots in Argentina, Paraguay, and Brazil [11,42], but its damage cannot be differentiated from that of *D. speciosa*. Studies to clarify what proportion of the damage is owed to each species (e.g., collections of larvae directly in the field) have not been done. The larva has been found feeding on maize roots only, and in the laboratory, it developed successfully on wheat as well, but not on any of the species tested from outside the Poaceae, suggesting it is stenophagous during the larval stage [2,43]. As an adult it is polyphagous, albeit reduced to fewer hosts than *D. speciosa* and *D. balteata*, as it has been recorded only on 21 plant species in the Poacae, Cucurbitaceae, and Asteraceae [11] (Table 1). Yet, similarities with the North American species in the *virgifera* group end here, as *D. viridula* eggs do not diapause. This species was reared in the laboratory for many generations, and the eggs never expressed any delay in hatching at optimal developmental temperature (8 ± 1 days at 25 ± 1 °C), regardless of previous photoperiod and temperature conditions (0 ± 1, 5 ± 1, 13 ± 1 °C; 10:14, 12:12, 14:10 h (L:D)) [13,43,44]. Eggs from field-collected adults, including overwintering adults, expressed no delay in hatching either [44].

Figure 3. Distribution of *Diabrotica viridula* in South America (stippled area).

Evidence indicates that the three South American pest *Diabrotica* overwinter as adults, are multivoltine, and do not have diapausing eggs. A reproductive diapause has been observed for *D. speciosa*, at least for the populations from the temperate and higher subtropical areas, but the fact that it could be overridden by manipulating temperature and light hours suggests it may not exist in the lower latitudes [44].

2. Control of South American *Diabrotica*

As the North American corn rootworms in the *virgifera* species group overwinter as diapausing eggs, are univoltine, and have a narrow larval host range limited to maize and a few grasses, their life cycle is tightly coupled to the phenology of one or very few annual host species. This provides opportunities for the use of different management strategies to reduce damage levels on susceptible crops, such as crop rotation and manipulation of sowing dates [45,46], expected density functions based on preceding density data [47], and anticipation of adult appearance through degree-day models [48]. Also, as the eggs are found anywhere in the soil from before the crop is planted, different tillage techniques could be applied to hinder the larvae from reaching the roots, for instance, compacting the soil between rows, thus affecting neonate larval movement [49]. Furthermore, factors behind the recommencement and completion of embryonic development after winter in univoltine *Diabrotica* are fairly well understood, so it is possible to estimate a "fixed point" (or interval) for the conclusion of embryonic development of the egg bank laid during the previous season in any given area [50]. However, none of these options have been developed for multivoltine species.

The field biology of the multivoltine species of the North American pest *Diabrotica* is also relatively well understood. Yet, in contrast to the univoltine species, predicting the incidence of the multivoltine species is not easily achieved. The only predictive tool of which the authors are aware has been used to calculate the probable damage of *Diabrotica undecimpunctata howardi* Barber on peanuts. This index used data such as soil texture, soil drainage class, planting date, cultivar resistance, and field history of rootworm damage to determine when to apply soil insecticides. Although the index recommended insecticide applications for 98.5% of the fields that actually needed insecticide treatment, it also recommended treatment for over 50% of fields that did not need it [51].

Although it is certain that the South American pest *Diabrotica* are multivoltine, seasonal reproductive patterns are not well known for these species. Soil and air temperatures were used in a linear degree-day model in laboratory and greenhouse experiments, to predict the occurrence of adults of *D. speciosa* [52]. The authors found that soil and air temperatures provided a significantly different prediction of insect occurrence than those observed experimentally. However, the prediction of occurrence based on soil temperature was more accurate than when the air temperature was used. One study in Argentina based on teneral collections in different regions suggests that the single most important determinant for the emergence of *D. speciosa* adults was weekly average temperatures above 13 °C. Due to this, in the temperate distribution areas of *D. speciosa*, there could be around three generations a year, and in subtropical regions, no fewer than five. However, no obvious or discrete voltinism pattern could be observed, expressing, to all practical effects, continuous generations [53]. What is known of the reproductive biology of the other two pest species suggests the same may be expected for them. Under the circumstances, it may be feasible to predict the appearance of a first generation after winter, in the areas where larval development might be temperature-limited, but such prediction may not be accurate enough to calculate planting dates, and certainly not apt to determine predictable cohorts. The practical implications of this study were that the life history pattern of this pest seems to leave few management alternatives. In the temperate regions of this species' distribution, early planting of maize could ensure that the first generations of larvae encounter more mature, and thus less susceptible stages of the crop. Other than this, the seasonal dispersion and unpredictability of *D. speciosa* outbreaks suggest that the only pre-emptive action available to protect maize crops from this pest is to plant Bt maize [53].

As mentioned above, the damage on maize from *D. speciosa* larval feeding cannot be differentiated from that of *D. viridula*, so control measures implemented for the control of *D. speciosa* larvae apply to *D. viridula* as well (Figure 4). In addition, the vast majority of references to research on *Diabrotica* spp. control in South America apply to *D. speciosa*, or are general for several agricultural pests.

Figure 4. Top, typical damage on maize roots and lodging caused by *D. speciosa* and *D. viridula* larvae. (photos by Dirceu N. Gassen); below, *D. speciosa* larva on potato with typical pinprick damage (photo by Pablo Lanzetta).

2.1. Chemical Control

Most control efforts in agriculture in South America are aimed at foliar pests and stem borers. There are published recommendations for treatment thresholds based on adult *Diabrotica* sampling protocols and foliar damage rates for beans and soybeans, respectively [54,55]. Yet, some control measures for root-feeders have been attempted, mainly seed treatments, in-furrow spraying, and granular pesticide applications [56,57]. There are no published calculations of the input of pesticides used for maize, beans, and potato, but they are generally considered to be high [58]. In Brazil there are 129 pesticides registered for *D. speciosa* in maize, potatoes, and beans, including foliar sprays, in-furrow, seed treatments, and four biological products based on *Beauveria bassiana* and one based on *Heterorhabditis bacteriophora* [58] (Table 1).

References for chemical control of *Diabrotica* in Argentina, Peru, and Uruguay follow more or less the same tendency of recommending several broad spectrum pesticides for adult control: chlorpyrifos, methomyl, other carbamates, fenitrothion, and several pyrethroids [59,60]. We have not found references to chemical control of larvae, and in fact concern for larval damage from *Diabrotica* is relatively recent, and all root-damaging insects are combined insofar as treatment actions are concerned. Their control has been trusted essentially to seed treatments with carbamates, neonicotinoids such as clothianidin, thiamethoxam, and imidacloprid, recently combined with diamides (cyantraniliprole and chlorantraniliprole), and genetically modified (GM) maize [61,62] (Table 1). However, seed treatments have been reported to be inefficient ways of controlling *D. speciosa* larvae on maize in Brazil [63]. Several authors reported that the most effective treatments are liquid in-furrow applications with organophosphates and phenylpyrazole insecticides in maize [64,65], and neonicotinoids for potatoes [66]. Granular applications also showed promise, but are not recommended due to technical limitations related to the cost and efficiency of granular applicators, and toxicity risks [67]. Finally, silicon applications have been reported to help decrease adult damage from *D. speciosa* and *Liriomyza* spp. (Diptera: Agromyzidae), leaf miners in organic potatoes [68].

Insecticides that interfere with the development of immature forms of insects (insect growth regulators (IGR)) can also cause a sterilizing effect on adult Coleoptera, affecting their fecundity and egg viability [69,70]. *D. speciosa* adults fed bean leaves treated with the IGR lufenuron showed a significant reduction in fertility and egg viability [71,72]. This deleterious effect on the progeny might reduce their biotic potential in the field, without using soil treatments (Table 1), although this has yet to be confirmed.

References to the evolution of insecticide resistance in South American *Diabrotica* are absent in the literature. However, this does not mean that it does not occur, but perhaps that it has not been studied.

2.2. Genetically Modified Crops

GM crops are one of the most widespread options for insect management in South America. GM maize, cotton, and soya are widely planted in Brazil and Argentina, the second and third countries with the largest productions of GM crops in the world, respectively, after the USA [73]. GM maize containing the Cry3Bb1 gene has been available in both countries since 2010 [57]. Up to 90% of the maize sown in Brazil is GM [57], and 96% in Argentina [73], mostly for control of Lepidoptera. Field tests showed that root damage levels were, without exception, lower than economic threshold, while yield was 2 to 5% higher than that of susceptible maize of the same variety [57]. Several lines of maize containing the Cry3Bb1 and the Cry1Ab genes were tested in greenhouse feeding tests with *D. speciosa* in Argentina in 2004. A 15-stage rating system was applied, which revealed that both events afforded some protection from larval damage compared to that seen in their conventional near-isolines. In the tests, however, the lines with the Cry3Bb1 gene suffered significantly lower damage levels (Cabrera Walsh, unpublished). Other countries in South America show a similar pattern, such as Paraguay (virtually 100% of its maize, [74]), and Uruguay, where there are no official data, but the area cultivated with GM maize is estimated at 86% [75]. This situation is not observed in Colombia, with only 31% of its maize crop being GM [76], Peru, where there is a moratorium on GM crops until 2021 [77], or Bolivia, where GM maize has recently been approved for planting, but its level of adoption remains unreported [78] (Table 1).

A new Bt protein, aimed especially for the control of *D. speciosa* larvae, was made available to maize growers during the 2013–2014 season, especially in south-central Brazil. This transgenic cultivar contained two Bt proteins expressed in the aerial parts aimed at caterpillars, and another specific protein (Cry3Bb1) for the control of *D. speciosa* larvae. Silva et al. [79] evaluated the efficiency of the Cry3Bb1 protein present in maize for the control of *D. speciosa* larvae, confirming higher productivity than that of the susceptible maize, and fewer larvae in the rhizosphere. Gallo [80] also evaluated the efficacy of corn genotypes that express the Cry3Bb1 protein for the control of *D. speciosa* larvae, and reported that both genotypes tested were effective in reducing corn root damage compared to that of other genotypes free of this toxin.

Potatoes expressing both the Cry3A and Cry1Ia1 genes were developed, field tested, and deemed to be effective to control *D. speciosa* [81]. However, these potato varieties were never commercialized.

2.3. Plant Resistance

Damage of *D. speciosa* on potatoes can be locally severe, both from adult damage to the aerial parts, and larval damage to the roots and tubers [82]. Work has been done to promote natural resistance in potato. This can come from chemical defenses, such as leptins (which are insecticidal peptides) and natural glycoalkaloids, which can confer resistance to both adults and larvae. Furthermore, the density and type of trichomes expressed by the plant can influence adult feeding behavior. These defense mechanisms can be selected from different cultivars, or incorporated from different species of wild potatoes [82–84].

In South Carolina (USA), sweet potatoes have been evaluated for *D. balteata* resistance [85]. In Florida (USA), where *D. balteata* is a key pest of lettuce, resistance has been evaluated based on the effective expression of latex upon injury [86,87]. Beans can also be selected for trichome expression to confer defoliation resistance to many pests, not only *Diabrotica* spp. [88,89].

Native resistance in maize to South American *Diabrotica* has not been tested, but it should be explored given the high number of native maize varieties in South America. Experiments in the US indicate that some maize genotypes expressed native antibiosis that reduced *D. virgifera virgifera* feeding significantly, as compared to that in the more susceptible genotypes. Damage was still higher than for a control GM maize, but larval development was not significantly different between the GM control and the more resistant maize genotypes [90] (Table 1).

Although not actually a form of plant resistance, intercropping shows some promise as a management option as well. There is some evidence of reduced incidence and damage from several bean pests, including *Diabrotica* sp., on *P. vulgaris*, based on intercropping with sugar cane in Colombia [91]. Intercropping beans with banana, maize, and other crops has shown mixed, although often favorable results in Central America [92,93] (Table 1).

2.4. Biological Control

In spite of the large number of species in the *Diabrotica* genus, and how widespread several of them are, only five species of parasitoids are known for the whole genus [94,95]. This is not the result of a lack of survey efforts, since many entomologists have surveyed for parasitoids and pathogens for many years throughout the Americas, and only one new species was detected in 60 years ([94], and literature therein). The scarcity of parasitoids of adults in the genus has been hypothesized to be due to the accumulation of cucurbitacins in fatty tissues [96,97]. These triterpenes are frequent in the Cucurbitaceae, common feeding hosts of adults in the genus, and are known to have antifeedant properties, but act as feeding stimulants for *Diabrotica* spp. [98,99]. There are no references of predators or parasitoids of larvae of South American species of *Diabrotica* [94]. However, based on the wide range of predators detected for *D. virgifera virgifera* in North America [100,101], it is to be expected that there are egg and larval predators of South American *Diabrotica* as well, which are yet to be discovered. *Diabrotica virgifera virgifera* larvae were found to have potent hemolymph defenses against predators [102,103], which may also be present in other *Diabrotica* spp.

Two adult parasitoid species, *Centistes gasseni* (Hymenoptera: Braconidae) and *Celotoria bosqi* (Diptera: Tachinidae), are known to parasitize *D. speciosa* and *D. viridula*, but with extremely low incidences in the latter. *Celatoria compressa* (Diptera: Tachinidae) is known to parasitize *D. balteata* in North and Central America, with no records for South America [104–107]. Other than these, at least 10 generalist predators have been recorded for adult *D. speciosa* [108].

Natural parasitism levels in *D. speciosa* have been reported between 1 and 28%, and on rare occasions over 30% [105,106]. Furthermore, the higher levels of parasitoidism are always recorded toward the end of the growing season, when most of the crop damage is done, suggesting that natural control levels are of minor importance to pest management [108]. It seems unlikely that biological control with macro-organisms will provide any significant relief to agriculture, or to have much potential at this stage for inundative biocontrol plans, given their low reproductive rate, comparatively long development, and dependence on laboratory-reared adults. However, new advances in parasitoid rearing could change this situation in the future [109].

Biological control with pathogens and nematodes offers a different outlook, with several promising laboratory and greenhouse results. Several strains of *Beauveria bassiana*, *B. brongniartii* (Hypocreales: Cordycipitaceae), and *Metarhizium anisopliae* (Hypocreales: Clavicipitaceae) were effective in controlling *Diabrotica virgifera virgifera* larvae for up to 21 days after application [110]. Similar results have been obtained for South American species. In Brazil, the microbial control of *D. speciosa* larvae with entomopathogenic fungi or nematodes is considered to have great potential because the soil is a relatively stable environment in terms of temperature and humidity, especially in no-till farming [111]. Argentine strains of *M. anisopliae* and *B. bassiana* killed third instars of *D. speciosa* in the laboratory [112]. Brazilian strains of *Isaria fumosorosea* (Hypocreales: Clavicipitaceae) and *Purpureocillium lilacinum* (Hypocreales: Ophiocordycipitaceae) killed eggs of these species, also in the laboratory [113]. Twenty strains of entomopathogic fungi (*B. bassiana*, *M. anisopliae*, and *P. lilacinum*) were colonized as endophytes in tobacco from northern Argentina. However, feeding tests on *D. speciosa* adults with the treated plants showed no significant differences with endophyte-free plants [114] (Table 1).

A few studies have also been translated to field conditions for biological control of *D. speciosa* in production systems [113,115]. Promising results were obtained with the strain of *B. bassiana* ESALQ PL63, used in seed treatments, which decreased the defoliation caused by *D. speciosa* adults in beans for more than three weeks after seeding [116]. Similar results were obtained in maize when the soil was treated with *Pseudomonas* (Pseudomonadales: Pseudomonadaceae) [117] and *Bacillus pumilus* [118].

Rhabditid nematodes (Steinernematidae and Heterorhabditidae) have been studied to control corn rootworms for decades, often with promising results. In the field, *Heterorhabditis bacteriophora* Poinar (Rhabditida: Heterorhabditidae) was as effective as tefluthrin in controlling *Diabrotica virgifera virgifera* in corn crop [119] and with a long residual action in the soil [120,121]. Seventeen native and exotic entomopathogenic nematode isolates (Steinernematidae and Heterorhabditidae) were tested against *D. speciosa* under laboratory and greenhouse conditions in Brazil on eggs, third (last) instars, and pupae. High mortality rates were obtained with *Heterorhabditis* sp. RSC01 and JPM04, *Steinernema glaseri*, and *Heterorhabditis amazonensis* on larvae and pupae, while eggs were unaffected [121]. These nematodes are considered to have great potential to control *D. speciosa* in irrigated maize and potatoes [122] (Table 1).

Maize roots attract entomopathogenic nematodes with (E)-β-caryophyllene when fed upon by *D. balteata* and other *Diabrotica*, and production of this chemical is enhanced by certain root-colonizing bacteria [123]. Furthermore, Jaffuel et al. [120] have shown that *Heterorhabditis bacteriophora*, encapsulated in durable alginate-based beads, effectively controlled *D. balteata* larvae in greenhouse tests.

Mermithidae have been cited quite often from *D. speciosa* adults [95,124,125] as well as *D. balteata* [126], but they are generally considered to be too difficult to mass rear, so are probably not feasible biocontrol agents [122].

2.5. Semiochemicals

D. speciosa females exhibited calling behavior similar to that described for *Diabrotica virgifera virgifera* [127,128]. Nardi [129] studied the sexual behavior of *D. speciosa*, observing mating from the third day after the emergence of the females. Mating was concentrated from 6 p.m. to midnight. Based on these studies, it became evident that the sexual behavior of *D. speciosa* was well defined, and that sexual attraction was probably mediated by a sexual pheromone produced by females. Yellow plastic cups coated with an adhesive and baited with females, especially virgin females, attracted males. Males of different age or reproductive state enclosed in the same cups did not attract females nor males [130]. Y-tube olfactometer and GC-EAG tests showed that males of *D. speciosa* were attracted by volatile compounds emitted by females. However, this compound has not been identified yet. Male volatiles were not attractive to either sex [128].

The female-produced sex pheromone for *D. balteata* is (R,R) 6,12-dimethylpentadecan-2-one [131,132]. Although stereospecific syntheses have been published [133–135], the racemic mixture is attractive, based on the single active stereoisomer [132]. It was attractive to males in the field in South Carolina (USA) and potentially useful for monitoring and management [136], and is commercially available in the USA [137].

The floral compound 1,2-dimethoxybenzene, one of the main floral volatiles of *Cucurbita maxima*, was found attractive to *D. speciosa* adults. Traps baited with TIC (1,2,4-trimethoxybenzene + indole + trans-cinnamaldehyde) and VIP (veratrole + indole + phenylacetaldehyde) also attracted *D. speciosa* adults, but less effectively [138]. Although 1,2-dimethoxybenzene is a very abundant and well-known floral component, it had not been reported as an attractant for *Diabrotica* spp. before, suggesting *D. speciosa* has a unique response pattern for floral volatiles [130]. Ensuing studies showed that the attractiveness of this compound was quite specific, as none of the analogs tested were attractive to adults [139].

Olfactometer tests with seedlings have shown that CO_2 and unidentified host specific root compounds from maize and oat seedlings were attractive to *D. speciosa* larvae. Wheat, beans, and soybean seedlings also elicited a response, albeit less vigorous [140]. Johnson and Gregory [141] reported that CO_2 is involved in general orientation, while specific compounds are involved in fine orientation toward the host plant roots. In any case, Nardi [129] argued that *D. speciosa* larvae have a very limited capacity for movement and host location, and it is the gravid female that chooses the host plants, suggesting there may not be much of a future for *D. speciosa* management in larval attractants. Regardless, this information could be useful in future research on chemical communication and development of management techniques for this species [142], but to date, no pheromones or floral attractants have been synthesized for practical uses.

As mentioned above, there are many references to the attractant and/or arrestant effects of cucurbitacins to adult *Diabrotica* spp. Several pest management tactics have been implemented based on the phagostimulatory effect of cucurbitacins on diabroticine beetles. These include lacing bitter cucurbit roots or fruit with an insecticide [128,143,144], using the roots or fruits in traps for monitoring and collecting Luperini [10,145–148], bitter cucurbit juice formulations combined with fungal pathogens [149], and in toxic baits [150–155]. Cucurbitacins have also been included as baits in traps for monitoring purposes [10,145,147,156,157].

Although it is clear that cucurbitacins are phagostimulants, there were contradictory reports as to them being volatile kairomones as well (see [147] for a full discussion on the subject). The difference is that volatile kairomones have the power to attract the recipient from a distance, whereas arrestants cause the recipient to remain only after the individual has made contact with the compound. These characteristics potentially provide different applications, because whereas an arrestant in a toxic bait can drive the target insect to ingest the insecticide, it will not attract it from a distance, precluding its use in traps. Kairomones, on the other hand can serve both purposes if they are phagostimulants as well, as is the case with cucurbitacins. Field experiments in Argentina showed that only males of *D. speciosa* were attracted from a distance to cucurbitacins (*ca.* 20 m), whereas for females these compounds acted only as arrestants, and to a lesser degree than for males [11,148]. This indicates that control or monitoring devices reliant on distance attraction to bitter cucurbit extracts would function exclusively on *D. speciosa* males. However, the wide dispersal of a toxic bait based on cucurbitacins promoted encounters and control of both sexes within the treated area, without any significant non-target effects [155,158] (Table 1).

3. Conclusions

Diabrotica management in South America has been stagnated for several years. Apart from insecticide applications, the major innovation of applicable use of the last 30 years has been the introduction of GM maize. However, other techniques that show promise must continue to be explored, such as the use of toxic baits with semiochemical attractants to suppress adult populations and for monitoring purposes, IGR insecticides aimed at adults to reduce their progeny, development of plant resistance, and biological control using *Heterorhabditis* nematodes and entomopathogenic fungus against larvae. Insecticide + cucurbitacin baits also deserve a special mention, because this combination has proved to be an effective technique that probably warrants further development.

Pest *Diabrotica* in South America are widely regarded as important, but usually are not differentiated from other foliar pests or root-feeders when it comes to management. Farmers do not identify them among the worst pests, and seldom deploy specific control measures for these beetles, except for potatoes in Brazil, where producers consider *D. speciosa* to be the main pest. Yet, the actual impact of the larvae of *D. speciosa* and *D. viridula*, especially on maize, may not be properly assessed, and until that is done, we cannot be sure of the real importance of these pests.

Author Contributions: Writing—original draft preparation and conceptualization, G.C.W.; validation, investigation, data curation, writing—review and editing, G.C.W., C.J.Á., N.C., D.E.N., A.d.S.P. and D.C.W. All authors have read and agreed to the published version of the manuscript.

Funding: This research received no external funding.

Acknowledgments: We wish to thank Joseph Spencer and Lance Meinke for inviting us to prepare this review and Paulo Lanzetta, Dirceu Gassen, and Stephen Cresswell for letting us use their photographs.

Conflicts of Interest: The authors declare no conflict of interest.

References

1. Derunkov, A.; Konstantinov, A. Taxonomic changes in the genus *Diabrotica* Chevrolat (Coleoptera: Chrysomelidae: Galerucinae): Results of a synopsis of North and Central America *Diabrotica* species. *Zootaxa* **2013**, *3686*, 301–325. [CrossRef]
2. Krysan, J.L. Introduction: Biology, distribution, and identification of pest *Diabrotica*. In *Methods for the Study of Pest Diabrotica*, 1st ed.; Krysan, J.L., Miller, T.A., Eds.; Springer: New York, NY, USA, 1986; pp. 1–23.
3. Wilcox, J.A. Chrysomelidae: Galerucinae: Luperini: Diabroticina; Pars. 78, Fasc. 2. In *Coleopterum Catalogus Supplementa*, 1st ed.; Wilcox, J.A., Ed.; Uitgeverij Dr. W. Junk's: Gravenhage, The Netherlands, 1972; pp. 296–343.
4. Krysan, J.L.; Smith, R.F. Systematics of the *virgifera* species group of *Diabrotica* (Coleoptera: Chrysomelidae: Galerucinae). *Entomography* **1987**, *5*, 375–484.
5. Cabrera, N.; Sosa Gómez, D.; Micheli, A. A morphological and molecular characterization of a new species of *Diabrotica* (Coeloptera: Chrysomelidae: Galerucinae). *Zootaxa* **2008**, *1922*, 33–46. [CrossRef]
6. Cabrera, N.; Cabrera Walsh, G. *Diabrotica collicola* (Coleoptera: Chrysomelidae), a new species of leaf beetle from Argentina. Discussion and key to some similar species of the *Diabrotica virgifera* group. *Zootaxa* **2010**, *2683*, 45–55. [CrossRef]
7. Branson, T.F.; Krysan, J.L. Feeding and oviposition behavior and life cycle strategies of *Diabrotica*: An evolutionary view with implications for pest management. *Environ. Entomol.* **1981**, *10*, 826–831. [CrossRef]
8. Clark, T.L.; Hibbard, B.E. Comparison of nonmaize hosts to support western corn rootworm (Coleoptera: Chrysomelidae) larval biology. *Environ. Entomol.* **2004**, *33*, 681–689. [CrossRef]
9. Krysan, J.L. Diapause in the nearctic species of the *virgifera* group of *Diabrotica*: Evidence for tropical origin and temperate adaptations. *Ann. Entomol. Soc. Am.* **1982**, *75*, 136–142. [CrossRef]
10. Krysan, J.L.; Branson, T.F.; Díaz Castro, G. Diapause in *Diabrotica virgifera* (Coleoptera: Chrysomelidae): A comparison of eggs from temperate and subtropical climates. *Entomol. Exp. Appl.* **1977**, *22*, 81–89. [CrossRef]
11. Cabrera Walsh, G.; Cabrera, N. Distribution and hosts of the pestiferous and other common Diabroticites from Argentina and Southern South America: A geographic and systematic view. In *New Developments in the Biology of Chrysomelidae*; Jolivet, P.H., Santiago-Blay, J.A., Schmitt, M., Eds.; SPB Academic Publishers: The Hague, The Netherlands, 2004; pp. 333–350.
12. Ávila, C.J.; Parra, J.R.P. Desenvolvimento de *Diabrotica speciosa* (Germar) (Coleoptera: Chrysomelidae) em diferentes hospedeiros. *Cienc. Rural* **2002**, *32*, 739–743. [CrossRef]
13. Cabrera Walsh, G. Host range and reproductive traits of *Diabrotica speciosa* (Germar) and *Diabrotica viridula* (F.) (Coleoptera: Chrysomelidae), two species of South American pest rootworms, with notes on other species of Diabroticina. *Environ. Entomol.* **2003**, *32*, 276–285. [CrossRef]
14. Cabrera Walsh, G. Sorghum halepense (L.) Persoon (Poaceae), a new larval host for the South American corn rootworm *Diabrotica speciosa* (Germar) (Coleoptera: Chrysomelidae). *Coleopt. Bull.* **2007**, *61*, 83–84. [CrossRef]
15. Ávila, C.J.; Bitencourt, D.R.; Silva, I.F. Biology, reproductive capacity, and foliar consumption of *Diabrotica speciosa* (Germar) (Coleoptera: Chrysomelidae) in different host plants. *J. Agric. Sci.* **2019**, *11*, 1–9. [CrossRef]
16. Marques, G.B.C.; Ávila, C.J.; Parra, J.R.P. Danos causados por larvas e adultos de *Diabrotica speciosa* (Coleoptera: Chrysomelidae) em milho. *Pesqui. Agropecu. Bras.* **1999**, *34*, 1983–1986. [CrossRef]
17. Gassen, D.N. *Insetos Subterráneos Perjudiciais às Culturas no Sul de Brasil Documentos, 13*; Embrapa-CNPT: Passo Fundo, Brazil, 1989; pp. 32–33.

18. Ávila, C.J.; Milanez, J.M. Larva alfinete. In *Pragas de Solo no Brasil*; Salvadori, J.R., Ávila, C.J., Silva, M.T.B., Eds.; Fundacep-Fecotrigo: Passo Fundo/Dourados/Cruz Alta, Brazil, 2004; pp. 345–378.

19. Salles, L.A. Incidência de danos de *Diabrotica speciosa* en cultivares e linhagens de batata. *Cienc. Rural* **2000**, *30*, 205–209. [CrossRef]

20. Haji, N.F.P. Biologia, dano e controle do adulto de *Diabrotica speciosa* (Germar, 1824) (Coleoptera: Chrysomelidae na cultura da batatinha (*Solanum tuberosum* L.). Ph.D. Thesis, Escola Superior de Agricultura "Luiz de Queiroz", Piracicaba, Brazil, 1981.

21. Ávila, C.J. Principais pragas e seu controle. In *A Cultura do Feijoeiro em Mato Grosso do Sul, Circular Tecnica 17*; Embrapa-UEPAE: Dourados, Brazil, 1990; pp. 54–56.

22. Ávila, C.J.; Santana, A.G. Cap. 4: Danos causados às culturas por adultos e larvas de *Diabrotica speciosa*. In *Diabrotica speciosa*, 1st ed.; Nava, D.E., Ávila, C.J., Pinto, A.S., Eds.; Occasio Editora: Piracicaba/São Paulo, Brasil, 2016; pp. 59–67.

23. Roberto, S.R.; Genta, W.; Ventura, M.U. *Diabrotica speciosa* (Ger.) (Coleoptera: Chrysomelidae): New pest in table grape orchards. *Neotrop. Entomol.* **2001**, *30*, 721–722. [CrossRef]

24. Segarra-Carmona, A.E.; Flores-López, L.; Cabrera-Asencio, I. New report of a leaf beetle pest from North America in Puerto Rico: *Diabrotica balteata* Le Conte (Coleoptera: Chrysomelidae) and its chemical control. *J. Agric. Univ. Puerto Rico* **2008**, *92*, 119–122.

25. Gonzalez, R.; Cardona, C.; Schoonhoven, A.V. Morfología y biología de los crisomélidos *Diabrotica balteata* LeConte y *Cerotoma facialis* Erickson como plagas del frijol común. *Turrialba* **1982**, *32*, 257–264.

26. Clark, S.M.; LeDoux, D.G.; Seeno, T.N.; Riley, E.G.; Gilbert, A.J.; Sullivan, J.M. *Host Plants of Leaf Beetle Species Occurring in the United States and Canada (Coleoptera: Megalopodidae, Orsodacnidae, Chrysomelidae, Excluding Bruchinae), Special Publication No. 2*; Coleopterists Society: Sacramento, CA, USA, 2004; pp. 86–87.

27. Saba, F. Host plant spectrum and temperature limitations of *Diabrotica balteata*. *Can. Entomol.* **1970**, *102*, 684–691. [CrossRef]

28. Agrosavia. Available online: https://www.agrosavia.co/ctni/ctc/coleoptera/chrysomelidae/diabrotica/diabrotica-balteata (accessed on 16 April 2020).

29. Morales, F.; Gámez, R. Beetle-transmitted viruses. In *Bean Production Problems in the Tropics*, 2nd ed.; Schwartz, H.F., Pastor Corrales, M.A., Eds.; CIAT: Cali, Colombia, 1989; pp. 363–378.

30. Cano Piedrahíta, C.A. Evaluación de tres Extractos Vegetales para el Control de Plagas en el Cultivo de Frijol Arbustivo *Phaseolus vulgaris* L. Master's Thesis, Universidad de Manizales, Caldas, Colombia, 2016.

31. Morales, F.J.; Castano, M.; Arroyave, J.A.; Ospina, M.D.; Calvert, L.A. A sobemovirus hindering the utilization of *Calopogonium mucunoides* as a forage legume in the lowland tropics. *Plant Dis.* **1995**, *79*, 1220–1224. [CrossRef]

32. Cardona, C.; Gonzalez, R.; Schoonhoven, A.V. Evaluation of damage to common beans by larvae and adults of *Diabrotica balteata* and *Cerotoma facialis*. *J. Econ. Entomol.* **1982**, *75*, 324–327. [CrossRef]

33. Bandas, L.D.C.; Corredor, D.; Corredor, S. Efecto de la asociación patilla (*Citrullus lanatus*) con maíz (*Zea mays*) sobre la población y daño causado por tres insectos plaga y el rendimiento de estos cultivos en la Ciénaga Grande de Lorica, Córdoba. *Rev. Colomb. Entomol.* **2004**, *30*, 161–169.

34. Rodríguez Chalarca, J.; Valencia, S.J. Daño por larvas de Diabrotica balteata (Coleoptera: Chrysomelidae) en raíces de maíz en condiciones controladas. In Proceedings of the 39 Congreso de la Sociedad Colombiana de Entomología, Ibagué, Universidad Cooperativa de Colombia, Bogota, Colombia, 11–13 June 2012; p. 93.

35. Tobar, J.A. *Manejo Integrado de Insectos Plaga en el Cultivo de la Mani (Arachis hypogaea L.)*; Facultad de Ciencias Agrícolas, Universidad de Nariño: Nariño, Colombia, 1990; p. 21.

36. Pitre, H.N., Jr.; Kantack, E.J. Biology of the banded cucumber beetle, *Diabrotica balteata*, in Louisiana. *J. Econ. Entomol.* **1962**, *55*, 904–906. [CrossRef]

37. Clark, T.L.; Meinke, L.J.; Foster, J.E. Molecular phylogeny of *Diabrotica* beetles (Coleoptera: Chrysomelidae) inferred from analysis of combined mitochondrial and nuclear DNA sequences. *Insect Mol. Biol.* **2001**, *10*, 303–314. [CrossRef] [PubMed]

38. Olalquiaga, F.G. Aspectos fitosanitarios de la Isla de Pascua. *Rev. Chil. Entomol.* **1980**, *10*, 101–102.

39. Anteparra, M.; Velásquez, J. Revisión de la familia Chrysomelidae asociada a leguminosas de grano en el trópico sudamericano. *Invest. Amazonía* **2015**, *4*, 62–69.

40. King, A.B.S.; Saunders, J.L. *The Invertebrate Pests of Annual Food Crops in Central America*, 1st ed.; Overseas Development Administration: London, UK, 1984; pp. 44–45.

41. Reyes, H.E.; Castillo, L.J. Transmisión del virus del moteado clorótico del maíz (maize chlorotic mottle virus -MCMV) por dos especies del género *Diabrotica*, familia Chrysomelidae. *Fitopatología* **1988**, *23*, 65–73.

42. Waquil, J.M.; Mendes, S.M.; Marucci, R.C. *Comunicado Técnico 178: Ocorrência de Especies de Diabrotica em milho no Brasil: Qual a Predominante, Diabrotica Speciosa ou Diabrotica Viridula*; Embrapa Milho e Sorgo: Sete Lagoas/Minas Gerais, Brazil, 2010; pp. 1–6.

43. Cabrera Walsh, G. Laboratory rearing and vital statistics of *Diabrotica speciosa* (Germar) and *Diabrotica viridula* (F.) (Coleoptera: Chrysomelidae), two species of South American pest rootworms. *Rev. Soc. Entomol. Argent.* **2001**, *60*, 239–248.

44. Cabrera Walsh, G. *Crisomélidos Diabroticinos Americanos: Hospederos y Enemigos Naturales. Biología y Factibilidad de Manejo de las Especies Plaga*, 1st ed.; Lap Lambert Academic Publishing GmbH & Co.: Saarbrücken, Germany, 2012; pp. 42–60.

45. Levine, E.; Oloumi-Sadeghi, H. Management of diabroticite rootworms in corn. *Annu. Rev. Entomol.* **1991**, *36*, 229–255. [CrossRef]

46. Spencer, J.L.; Hibbard, B.E.; Moeser, J.; Onstad, D.W. Behaviour and ecology of the western corn rootworm (*Diabrotica virgifera virgifera* LeConte). *Agric. For. Entomol.* **2009**, *11*, 9–27. [CrossRef]

47. Schaafsma, A.W.; Whitfield, G.H.; Ellis, C.R. A temperature-dependent model of egg development of the western corn rootworm, *Diabrotica virgifera virgifera* Leconte (Coleoptera: Chrysomelidae). *Can. Entomol.* **1991**, *123*, 1183–1197. [CrossRef]

48. Stevenson, D.E.; Michels, G.J.; Bible, J.B.; Jackman, J.A.; Harris, M.K. Physiological time model for predicting adult emergence of western corn rootworm (Coleoptera: Chrysomelidae) in the Texas High Plains. *J. Econ. Entomol.* **2008**, *101*, 1584–1593. [CrossRef]

49. Park, Y.; Tollefson, J.J. Spatial prediction of corn rootworm (Coleoptera: Chrysomelidae) adult emergence in Iowa cornfields. *J. Econ. Entomol.* **2005**, *98*, 121–128. [CrossRef]

50. Meinke, L.J.; Sappington, T.W.; Onstad, D.W.; Guillemaud, T.; Miller, N.J.; Komáromi, J.; Levay, N.; Furlan, L.; Kiss, J.; Toth, F. Western corn rootworm (*Diabrotica virgifera virgifera* LeConte) population dynamics. *Agric. For. Entomol.* **2009**, *11*, 29–46. [CrossRef]

51. Herbert, D.A., Jr.; Malone, S.; Brandenburg, R.L.; Royals, B.M. Evaluation of the peanut southern corn rootworm advisory. *Peanut Sci.* **2004**, *31*, 28–32. [CrossRef]

52. Ávila, C.J.; Milanez, J.M.; Parra, J.R.P. Previsão de ocorrência de *Diabrotica speciosa* utilizando o modelo de graus-dia de laboratório. *Pesqui. Agropecu. Bras.* **2002**, *37*, 427–432. [CrossRef]

53. Cabrera Walsh, G.; Sacco, J.; Mattioli, F. Voltinism of *Diabrotica speciosa* (Coleoptera: Chrysomelidae) in Argentina: Latitudinal clines and implications for damage anticipation. *Pest Manag. Sci.* **2013**, *69*, 1272–1279.

54. Hoffmann-Campo, C.B.; Moscardi, F.; Corrêa-Ferreira, B.S.; Oliveira, L.J.; Sosa-Gómez, D.R.; Panizzi, A.R.; Corso, I.C.; Gazzoni, D.L.; Oliveira, E.B. *Pragas da Soja no Brasil e seu Manejo Integrado, Circular Técnica 30*; Embrapa Soja: Londrina, Brazil, 2000; pp. 16–17.

55. Silva, C.C.; Peloso, M.J.D. *Informações técnica para o cultivo do feijoeiro comum na região central-brasileira 2005–2007*; Embrapa arroz e feijão: Santo Antônio de Goiás, Brazil, 2006; pp. 124–136.

56. Ávila, C.J. Eficiência do inseticida terbufós no controle de larvas de vaquinha (*Diabrotica speciosa*) em milho (*Zea mays* L.). In Proceedings of the 15 Congresso Brasileiro de Entomologia, Universidade Federal de Lavras, Lavras, Brazil, 12–17 March 1995; p. 467.

57. Carvalho, R.A.; Dourado, P.M.; Oliveira Junio, J.A.; Martinelli, S. Cap. 6: Plants transgênicas no controle de *Diabrotica* spp. In *Diabrotica Speciosa*, 1st ed.; Nava, D.E., Ávila, C.J., Pinto, A.S., Eds.; Occasio Editora: Piracicaba/São Paulo, Brasil, 2016; pp. 85–103.

58. Ávila, C.J.; Santana, A.G. Cap. 9: Controle químico de *Diabrotica speciosa*. In *Diabrotica speciosa*, 1st ed.; Nava, D.E., Ávila, C.J., Pinto, A.S., Eds.; Occasio Editora: Piracicaba/São Paulo, Brasil, 2016; pp. 139–149.

59. AGROFIT. Available online: http://agrofit.agricultura.gov.br/agrofit_cons/principal_agrofit_cons (accessed on 6 May 2020).

60. Programa de Hortalizas. 2020. Available online: http://www.lamolina.edu.pe/hortalizas (accessed on 30 March 2020).

61. INTA. Manejo de Plagas de Maíz. Available online: https://inta.gob.ar/sites/default/files/script-tmp-inta-manejo_de_plagas_en_el_cultivo_de_maz.pdf (accessed on 22 April 2020).

62. On24. Available online: https://www.on24.com.ar/negocios/agro/a-la-vanguardia-en-tratamientos-de-semillas/ (accessed on 22 April 2020).

63. Ávila, C.J.; Gomez, S.A. Diagnóstico de pragas de solo no Estado de Mato Grosso do Sul. In Proceedings of the 9 Reunião Sul-Brasileira de Pragas de solo, EPAGRI, Estação Experimental de Itajaí, Camboriú, Brazil, 3–5 September 2005; pp. 30–34.

64. Ávila, C.J.; Gomez, S.A. Controle químico de larvas de *Diabrotica speciosa* Coleoptera: Chrysomelidae) na cultura do milho. In Proceedings of the 8 Reunião sul Brasileira de Pragas do Solo, Londrina, Brazil, 26–27 September 2001; Embrapa Soja: Londrina, Brazil, 2001; pp. 254–257.

65. Viana, P.A.; Marochi, A.I. Controle químico da larva de *Diabrotica* spp. na cultura do milho em sistema de plantio direto. *Rev. Bras. Milho Sorgo* **2002**, *1*, 1–11. [CrossRef]

66. Salles, L.A. Eficiência do inseticida thiamethoxam (actara) no controle das pragas de solo da batata, *Diabrotica speciosa* (Col., Chrysomelidae) e *Heteroderes* spp. (Col., Elateridae). *Rev. Bras. Agrociencia* **2000**, *6*, 149–151.

67. Ávila, C.J.; Botton, M. *Aplicação de Inseticidas no Solo*; FEALQ: Piracicaba, Brazil, 2000; pp. 24–26.

68. Gomes, F.B.; Moraes, J.C.; Ner, D.K.P. Adubação com silício como fator de resistência a insetos-praga e promotor de produtividade em cultura de batata inglesa em sistema orgânico. *Cienc. Agrotec.* **2009**, *33*, 18–23. [CrossRef]

69. Lovestrand, S.G.; Beavers, J.B. Effect of diflubenzuron on four species of weevil attacking citrus in Florida. *Fla. Entomol.* **1980**, *63*, 112–115. [CrossRef]

70. Elek, J.A.; Longstaff, B.C. Effect of chitin-synthesis inhibitors on stored-products beetles. *Pestic. Sci.* **1994**, *40*, 225–230. [CrossRef]

71. Ávila, C.J.; Nakano, O.; Chagas, M.C.M. Efeito do regulador de crescimento de insetos lufenuron na fecundidade e viabilidade dos ovos de *Diabrotica speciosa* (Germar), 1924 (Coleoptera: Chrysomelidae). *Rev. Agric.* **1998**, *73*, 69–78.

72. Ávila, C.J.; Nakano, O. Efeito do regulador de crescimento de insetos lufenuron na reprodução de *Diabrotica speciosa* (Germar) (Coleoptera: Chrysomelidae). *An. Soc. Entomol. Bras.* **1999**, *28*, 293–299. [CrossRef]

73. ArgenBio. Available online: http://www.argenbio.org/cultivos-transgenicos (accessed on 22 April 2020).

74. INBIO. Available online: https://inbio.org.py/wp-content/uploads/maiz-soja-zafri%C3%B1a-2019-INBIO-para-web-1-1.pdf (accessed on 22 April 2020).

75. ISAAA. *ISAAA Brief No. 53: Global Status of Commercialized Biotech/GM Crops in 2017: Biotech Crop Adoption Surges as Economic Benefits Accumulate in 22 Years*; ISAAA: Ithaca, NY, USA, 2017; pp. 53–55.

76. Cultivos Transgénicos en Colombia. Available online: https://www.semillas.org.co/apc-aa-files/5d99b14191 c59782eab3da99d8f95126/informe-pais-ogm-2018_web.pdf (accessed on 15 April 2020).

77. Delgado Gutiérrez, D. *Regulación de los transgénicos en el Perú*; Sociedad Peruana de Derecho Ambiental: Lima, Peru, 2015; pp. 56–61.

78. Hernández, X. Bolivia abandona su política anti transgénicos y se suma al mercado de los OGM. Available online: https://www.infocampo.com.ar/bolivia-abandona-su-politica-anti-transgenicos-y-se-suma-al-mercado-de-los-ogm/ (accessed on 16 April 2020).

79. Silva, J.R.; Feldmann, N.A.; Muhl, F.R.; Rhoden, A.C.; Blabinot, M.; Asolin, L.; Pava, D. Avaliação da eficiência da biotecnologia no controle da larva-alfinete (*Diabrotica speciosa*) na cultura do milho. *Rev. Cienc. Agrovet. Aliment.* **2016**, *1*, 1–11.

80. Gallo, P. Avaliação da eficácia do evento MON88017 (Cry3bb1) na redução do dano da larva de *Diabrotica speciosa* (Germar, 1824) (Coleoptera: Chrysomelidae) na raiz do milho. Master's Thesis, Universidade Estadual de Ponta Grossa, Ponta Grossa, Brazil, 2012.

81. Afonso da Rosa, A.P.S.; Castro, C.M.; Pereira, A.S.; Lourenção, A.L. Cap. 5. Resistência de plantas a *Diabrotica speciosa*. In *Diabrotica speciosa*, 1st ed.; Nava, D.E., Ávila, C.J., Pinto, A.S., Eds.; Occasio Editora: Piracicaba/São Paulo, Brasil, 2016; pp. 71–82.

82. Lara, F.M.; Scaranello, A.L.; Baldin, E.L.L.; Bolça Junior, A.L.; Lourenção, A.L. Resistência de genótipos de batata a larvas e adultos de *Diabrotica speciosa*. *Hortic. Bras.* **2004**, *22*, 761–765. [CrossRef]

83. Lara, F.M.; Poletti, M.; Barbosa, J.C. Resistência de genótipos de batata (*Solanum* spp.) a *Diabrotica speciosa* (Germar, 1824) (Coleoptera: Chrysomelidae). *Cienc. Rural* **2000**, *30*, 927–931. [CrossRef]

84. Teodoro, J.S.; Martins, J.F.S.; Rosa, A.P.; Castro, C.M. Characterization of potato genotypes for resistance to *Diabrotica speciosa*. *Hortic. Bras.* **2014**, *32*, 440–445. [CrossRef]

85. Jackson, D.M.; Bohac, J.R. Resistance of sweetpotato genotypes to adult *Diabrotica* beetles. *J. Econ. Entomol.* **2014**, *100*, 566–572. [CrossRef]

86. Lu, H.; Wright, A.L.; Sui, D. Responses of lettuce cultivars to insect pests in southern Florida. *Horttechnology* **2011**, *21*, 773–778. [CrossRef]

87. Sethi, A.; Alborn, H.T.; McAuslane, H.J.; Nuessly, G.S.; Nagata, R.T. Banded cucumber beetle (Coleoptera: Chrysomelidae) resistance in romaine lettuce: Understanding latex chemistry. *Arthropod Plant Interact.* **2012**, *6*, 269–281. [CrossRef]

88. Heyer, W.; Cruz, B.; Chiang-Lok, M.L. Comportamiento y preferencia de los adultos de *Diabrotica balteata*, *Andrector ruficornis*, *Systena basalis* (Coleoptera: Chrysomelidae) y *Empoasca fabae* (Homoptera: Cicadellidae), en frijol. *Cienc. Agric.* **1986**, *27*, 61–76.

89. Vieira, C.; Borém, A.; Ramalho, M.A.P. Melhoramento do feijão. In *Melhoramento de Espécies Cultivadas*; Borém, A., Ed.; UFV: Viçosa, Brazil, 2005; pp. 301–391.

90. El Khishen, A.A.; Bohn, M.O.; Prischmann-Voldseth, D.A.; Dashiel, K.E.; French, B.W.; Hibbard, B.E. Native resistance to western corn rootworm (Coleoptera: Chrysomelidae) larval feeding: Characterization and mechanisms. *J. Econ. Entomol.* **2009**, *102*, 2350–2359. [CrossRef]

91. García, J.; Cardona, C.; Raigosa, J. Evaluación de poblaciones de insectos plaga en la asociación caña de azúcar–fríjol y su relación con los rendimientos. *Rev. Colomb. Entomol.* **1979**, *5*, 17–24.

92. Risch, S. The population dynamics of several herbivorous beetles in a tropical agroecosystem: The effect of intercropping corn, beans and squash in Costa Rica. *J. Appl. Ecol.* **1980**, *17*, 593–611. [CrossRef]

93. Cardona, C. Effect of intercropping on insect populations: The case of beans. In *Proceedings, Workshop on Research Methods for Cereal/Legume Intercropping in Eastern and Southern Africa (Lilongwe, Malawi)*; Waddington, S.R., Palmer, A.F.E., Edje, O.T., Eds.; CIMMYT: Mexico City, Mexico, 1989; pp. 56–61.

94. Toepfer, S.; Cabrera-Walsh, G.; Eben, A.; Alvarez Zagoya, R.; Haye, T.; Zhang, F.; Kuhlmann, U. A critical evaluation of host ranges of parasitoids of the subtribe Diabroticina (Coleoptera: Chrysomelidae: Galerucinae: Luperini) using field and laboratory host records. *Biocontrol Sci. Technol.* **2008**, *18*, 485–508. [CrossRef]

95. Toepfer, S.; Haye, T.; Erlandson, M.; Goettel, M.; Lundgren, J.G.; Kleespies, R.G.; Weber, D.C.; Cabrera Walsh, G.; Peters, A.; Ehlers, R.-U.; et al. A review of the natural enemies of beetles in the subtribe Diabroticina (Coleoptera: Chrysomelidae): Implications for sustainable pest management. *Biocontrol Sci. Technol.* **2009**, *19*, 1–65. [CrossRef]

96. Metcalf, R.L. Chemical ecology of Diabroticites. In *Novel Aspects of the Biology of Chrysomelidae, Series Entomologica*, 1st ed.; Jolivet, P.H., Cox, M.L., Petitpierre, E., Eds.; Springer: Dordrecht, The Netherlands, 1994; Volume 50, pp. 153–169.

97. Tallamy, D.W.; Stull, J.; Ehresman, N.P.; Gorski, P.M.; Mason, C.E. Cucurbitacins as feeding and oviposition deterrents to insects. *Environ. Entomol.* **1997**, *26*, 678–683. [CrossRef]

98. Contardi, H.G. Estudios genéticos en *Cucurbita* y consideraciones agronómicas. *Physis* **1939**, *18*, 332–347.

99. Howe, W.L.; Sanborn, J.R.; Rhodes, A.M. Western corn rootworms and spotted cucumber beetle associations with *Cucurbita* and cucurbitacin. *Environ. Entomol.* **1976**, *5*, 1043–1048. [CrossRef]

100. Lundgren, J.G.; Fergen, J.K. Predator community structure and trophic linkage strength to a focal prey. *Mol. Ecol.* **2014**, *23*, 3790–3798. [CrossRef]

101. Lundgren, J.G.; McDonald, T.; Rand, T.A.; Fausti, S.W. Spatial and numerical relationships of arthropod communities associated with key pests of maize. *J. Appl. Entomol.* **2015**, *139*, 446–456. [CrossRef]

102. Lundgren, J.G.; Haye, T.; Toepfer, S.; Kuhlmann, U. A multifaceted hemolymph defense against predation in *Diabrotica virgifera virgifera* larvae. *Biocontrol Sci. Technol.* **2009**, *19*, 871–880. [CrossRef]

103. Lundgren, J.G.; Toepfer, S.; Haye, T.; Kuhlmann, U. Haemolymph defence of an invasive herbivore: Its breadth of effectiveness against predators. *J. Appl. Entomol.* **2010**, *134*, 439–448. [CrossRef]

104. Eben, A.; Barbercheck, M.E. Field observations on host plant associations enemies of diabroticite beetles (Chrisomelidae: Luperini) in Veracruz, Mexico. *Acta Zool. Mex.* **1996**, *67*, 47–65.

105. Heineck-Leonel, M.A.; Salles, L.A.B. Incidência de parasitóides e patógenos em adultos de *Diabrotica speciosa* (Germar, 1824) (Col., Chrysomelidae) na região de Pelotas, RS. *Ann. Soc. Entomol. Bras.* **1997**, *26*, 81–85. [CrossRef]

106. Cabrera Walsh, G. Distribution, host specificity, and overwintering of *Celatoria bosqi* Blanchard (Diptera: Tachinidae), a South American parasitoid of *Diabrotica* spp. (Coleoptera: Chrysomelidae: Galerucinae). *Biol. Control* **2004**, *29*, 427–434. [CrossRef]

107. Cabrera Walsh, G.; Athanas, M.M.; Salles, L.A.B.; Schroder, R.F.W. Distribution, host range, and climatic constraints on *Centistes gasseni* (Hymenoptera: Braconidae), a South American parasitoid of cucumber beetles, *Diabrotica* spp. (Coleoptera: Chrysomelidae). *Bull. Entomol. Res.* **2004**, *93*, 561–567. [CrossRef]

108. Cabrera Walsh, G.; Pinto, A.S.; Nava, D.E. Cap. 7: Controle biológico de *Diabrotica speciosa*: Parasitoides e predadores. In *Diabrotica Speciosa*, 1st ed.; Nava, D.E., Ávila, C.J., Pinto, A.S., Eds.; Occasio Editora: Piracicaba/São Paulo, Brasil, 2016; pp. 107–117.

109. Pinto, A.d.S.; Parra, J.R.P. Liberação de inimigos naturais. In *Controle Biológico No Brasil: Parasitóides e Predadores*, 1st ed.; Parra, J.R.P., Botelho, P.S.M., Corrêa-Ferreira, B.S., Bento, J.M.S., Eds.; Manole: São Paulo, Brasil, 2002; pp. 325–342.

110. Cagan, L.; Stevo, J.; Gasparovic, K.; Matusikova, S. Mortality of the Western corn rootworm, *Diabrotica virgifera virgifera* larvae caused by entomopathogenic fungi. *J. Cent. Eur. Agric.* **2019**, *20*, 678–685. [CrossRef]

111. Santos, V.; Moino Junior, A.; Andaló, V.; Moreira, C.C.; Olinda, R.A. Virulence of entomopathogenic nematodes (Rhabditida: Steinernematidae and Heterorhabditidae) for the control of *Diabrotica speciosa* Germar (Coleoptera: Chrysomelidae). *Cienc. Agrotec.* **2011**, *35*, 1149–1156. [CrossRef]

112. Consolo, V.; Salerno, G.; Beron, C. Pathogenicity, formulation and storage of insect pathogenic hyphomycetous fungi tested against *Diabrotica speciosa*. *BioControl* **2003**, *48*, 705–712. [CrossRef]

113. Tigano-Milani, M.S.; Carneiro, R.G.; Faria, M.R.; Frazão, H.S.; McCoy, C.W. Isozyme characterization and pathogenicity of *Paecilomyces fumosoroseus* and *P. lilacinus* to *Diabrotica speciosa* (Coleoptera: Chrysomelidae) and *Meloidogyne javanica* (Nematoda: Tylenchidae). *Biol. Control* **1995**, *5*, 378–382. [CrossRef]

114. Vianna, M.F. Capacidad biocida de hongos entomopatógenos para el control de plagas del tabaco (*Nicotiana tabacum* L.) en la provincia de Jujuy, República Argentina. Ph.D. Thesis, Universidad de La Plata, La Plata, Argentina, 2019.

115. Silva-Werneck, J.O.; de Faria, M.R.; Abreu Neto, B.P.; Magalhães, B.P.; Schimidt, F.G.V. Técnica de criação de *Diabrotica speciosa* (Germ.) (Coleoptera: Chrysomelidae) para bioensaios com bacilos e fungos entomopatogênicos. *An. Soc. Entomol. Bras.* **1995**, *24*, 45–52.

116. Pinto, A.d.S.; Hernandes, A.J.; Miyazaki, M.J.; Miralha, V.R.; Rodrigues, L.R.; de Sousa, E.N. Tratamento de sementes de feijoeiro com *Beauveria bassiana* e *Metarhizium anisopliae* visando ao manejo de pragas de folhas. In Proceedings of the 16 Simpósio de Controle Biológico, Londrina, Brazil, 11–15 August 2019; EMBRAPA Soja: Londrina, Brazil, 2019; p. 53.

117. Jaffuel, G.; Imperiali, N.; Shelby, K.; Campos-Herrera, R.; Geisert, R.; Maurhofer, M.; Loper, J.; Keel, C.; Turlings, T.C.J.; Hibbard, B.E. Protecting maize from rootworm damage with the combined application of arbuscular mycorrhizal fungi, *Pseudomonas* bacteria and entomopathogenic nematodes. *Sci. Rep.* **2019**, *9*, 3127. [CrossRef] [PubMed]

118. Disi, J.O.; Kloepper, J.W.; Fadamiro, H.Y. Seed treatment of maize with *Bacillus pumilus* strain INR-7 affects host location and feeding by Western corn rootworm, *Diabrotica virgifera virgifera*. *J. Pest Sci.* **2018**, *91*, 515–522. [CrossRef]

119. Modic, S.; Zigon, P.; Kolmanic, A.; Trdan, S.; Razinger, J. Evaluation of the field efficacy of *Heterorhabditis bacteriophora* Poinar (Rhabditida: Heterorhabditidae) and synthetic insecticides for the control of Western Corn Rootworm Larvae. *Insects* **2020**, *11*, 202. [CrossRef]

120. Jaffuel, G.; Sbaiti, I.; Turlings, T.C. Encapsulated entomopathogenic nematodes can protect maize plants from *Diabrotica balteata* larvae. *Insects* **2020**, *11*, 27. [CrossRef]

121. Toth, S.; Szalai, M.; Kiss, J.; Toepfer, S. Missing temporal effects of soil insecticides and entomopathogenic nematodes in reducing the maize pest *Diabrotica virgifera virgifera*. *J. Pest Sci.* **2020**, *93*, 767–781. [CrossRef]

122. Santos, V.; Leite, L.G.; Moino Junior, A. Cap. 8. Controle de *Diabrotica speciosa* com entomopatógenos. In *Diabrotica Speciosa*, 1st ed.; Nava, D.E., Ávila, C.J., Pinto, A.S., Eds.; Occasio Editora: Piracicaba/São Paulo, Brasil, 2016; pp. 121–136.

123. Chiriboga, X.; Guo, H.; Campos-Herrera, R.; Röder, G.; Imperiali, N.; Keel, C.; Maurhofer, M.; Turlings, T.C. Root-colonizing bacteria enhance the levels of (E)-β-caryophyllene produced by maize roots in response to rootworm feeding. *Oecologia* **2018**, *187*, 459–468. [CrossRef]

124. Nickle, W.R.; Schroder, R.F.W.; Krysan, J.L. A new Peruvian *Hexamermis* sp. (Nematoda: Mermithidae) parasite of corn rootworms, *Diabrotica* spp. *Proc. Helminthol. Soc. Wash.* **1984**, *51*, 212–216.

125. Gassen, D.N. *Circular Técnica, 1. Parasitos, Patógenos e Predadores de Insetos Associados à Cultura do Trigo*, 2nd ed.; EMBRAPA-CNPT: Passo Fundo, Brazil, 1986; pp. 32–33.

126. Creighton, C.S.; Fassuliotis, G. Infectivity and suppression of the banded cucumber beetle (Coleoptera: Chrysomelidae) by the mermithid nematode *Filipjevimermis leipsandra* (Mermithida: Mermithidae). *J. Econ. Entomol.* **1983**, *76*, 615–618. [CrossRef]

127. Hammack, L. Calling behavior in female western corn rootworm beetles (Coleoptera: Chrysomelidae). *Ann. Entomol. Soc. Am.* **1995**, *88*, 562–569. [CrossRef]

128. Nardi, C.; Ventura, M.U.; Santos, F.; Bento, J.M.S. Cap. 10: Comportamento e ecología química de *Diabrotica speciosa*. In *Diabrotica speciosa*, 1st ed.; Nava, D.E., Ávila, C.J., Pinto, A.S., Eds.; Occasio Editora: Piracicaba/São Paulo, Brasil, 2016; pp. 153–184.

129. Nardi, C. Estímulos Olfativos Envolvidos no Comportamento Sexual e na Seleção Hospedeira de *Diabrotica speciosa* (Germar) (Coleoptera: Crysomelidae). Ph.D. Thesis, Escola Superior de Agricultura "Luiz de Queiroz", Universidade de São Paulo, Piracicaba, Brazil, 2010.

130. Ventura, M.U.; Mello, E.P.; Oliveira, A.R.M.; Simonelli, F.; Marques, F.A.; Zarbin, P.H.G. Males are attracted by female traps: A new perspective for management of *Diabrotica speciosa* (Germar) (Coleoptera: Chrysomelidae) using sexual pheromone. *Neotrop. Entomol.* **2001**, *30*, 361–364. [CrossRef]

131. Chuman, T.; Guss, P.L.; Doolittle, R.E.; McLaughlin, J.R.; Krysan, J.L.; Schalk, J.M.; Tumlinson, J.H. Identification of female-produced sex pheromone from banded cucumber beetle, *Diabrotica balteata* LeConte (Coleoptera: Chrysomelidae). *J. Chem. Ecol.* **1987**, *13*, 1601–1616. [CrossRef] [PubMed]

132. McLaughlin, J.R.; Tumlinson, J.H.; Mori, K. Responses of male *Diabrotica balteata* (Coleoptera: Chrysomelidae) to stereoisomers of the sex pheromone 6,12-dimethylpentadecan-2-one. *J. Econ. Entomol.* **1991**, *84*, 99–102. [CrossRef]

133. Mori, K.; Igarashi, Y. Synthesis of the four stereoisomers of 6,12-dimethyl-2-pentadecanone, the sex pheromone of *Diabrotica balteata* LeConte. *Liebigs Ann. Chem.* **1988**, *7*, 717–720. [CrossRef]

134. Enders, D.; Jandeleit, B.; Prokopenko, O.F. Convergent synthesis of (R,R)-6,12-dimethylpentadecan-2-one, the female sex pheromone of the banded cucumber beetle by iron mediated chirality transfer. *Tetrahedron* **1995**, *51*, 6273–6284. [CrossRef]

135. Shen, W.; Hao, X.; Shi, Y.; Tian, W.S. Synthesis of (6R,12R)-6,12-dimethylpentadecan-2-one, the female-produced sex pheromone from banded cucumber beetle *Diabrotica balteata*, based on a chiron approach. *Nat. Prod. Commun.* **2015**, *10*, 2155–2160. [CrossRef]

136. Schalk, J.M.; McLaughlin, J.R.; Tumlinson, J.H. Field response of feral male banded cucumber beetles to the sex pheromone 6,12-dimethylpentadecan-2-one. *Fla. Entomol.* **1990**, *73*, 292–297. [CrossRef]

137. Evergreen Growers Supply. Available online: www.evergreengrowers.com/banded-cucumber-beetle-lure-group-diabal.html (accessed on 6 May 2020).

138. Ventura, M.U.; Martins, M.C.; Pasini, A. Responses of *Diabrotica speciosa* and *Cerotoma arcuata tingomariana* (Coleoptera: Chrysomelidae) to volatile attractants. *Fla. Entomol.* **2000**, *83*, 403–410. [CrossRef]

139. Marques, F.A.; Wendler, E.P.; Macedo, A.; Wosch, C.L.; Maia, B.H.S.; Mikami, A.Y.; Arruda-Gatt, I.C.; Pissina, A.; Mingotte, F.L.C.; Alves, A.; et al. Response of *Diabrotica speciosa* (Coleoptera: Chrysomelidae) to 1,4-Dimethoxybenzene and analogs in common bean crop. *Braz. Arch. Biol. Technol.* **2009**, *52*, 1333–1340. [CrossRef]

140. Pereira, T.; Ventura, M.U.; Marques, M.A. Comportamento de larvas de *Diabrotica speciosa* (Coleoptera: Chrysomelidae) em resposta ao CO_2 e a plântulas de espécies cultivadas. *Cienc. Rural* **2005**, *35*, 981–985. [CrossRef]

141. Johnson, S.N.; Gregory, P.J. Chemically-mediated host-plant location and selection by root-feeding insects. *Physiol. Entomol.* **2006**, *31*, 1–13. [CrossRef]

142. Nardi, C.; Luvizotto, R.A.; Parra, J.R.P.; Bento, J.M.S. Mating behavior of *Diabrotica speciosa* (Coleoptera: Chrysomelidae). *Environ. Entomol.* **2012**, *41*, 562–570. [CrossRef] [PubMed]

143. Lorenzato, D. Controle integrado de *Diabrotica speciosa* (Germar 1824) em frutiferas de clima temperado com cairomonio encontrado em raizes de plantas nativas da familia Cucurbitaceae. In Proceedings of the 7 Congresso Brasileiro de Fruticultura, Florianópolis, Brazil, 25–26 July 1983; Empresa de Pesquisa Agropecuária e Extensão Rural: Florianópolis, Brazil, 1984; pp. 347–355.

144. Hamerschmidt, I. Uso do tajujá e purungo como atraentes de vaquinha em olericultura. *Hortic. Bras.* **1985**, *3*, 45.

145. Shaw, J.T.; Ruesink, W.G.; Briggs, S.P.; Luckmann, W.H. Monitoring populations of corn rootworm beetles (Coleoptera: Chrysomelidae) with a trap baited with cucurbitacins. *J. Econ. Entomol.* **1984**, *77*, 1495–1499. [CrossRef]

146. Ventura, M.U.; Ito, M.; Montalván, R. An attractive trap to capture *Diabrotica speciosa* (Ger.) and *Cerotoma arcuata tingomariana* Bechyné. *An. Soc. Entomol. Bras.* **1996**, *25*, 529–535.

147. Cabrera Walsh, G.; Weber, D.C.; Mattioli, F.M.; Heck, G. Qualitative and quantitative responses of Diabroticina (Coleoptera: Chrysomelidae) to cucurbit extracts linked to species, sex, weather, and deployment method. *J. Appl. Entomol.* **2008**, *132*, 205–215. [CrossRef]

148. Cabrera Walsh, G.; Mattioli, F.; Weber, D.C. A wind-oriented sticky trap for evaluating the behavioural response of the leaf-beetle *Diabrotica speciosa* to cucurbit extracts. *Int. J. Pest Manag.* **2014**, *60*, 46–51. [CrossRef]

149. Daoust, R.A.; Pereira, R.M. Stability of entomopathogenic fungi *Beauveria bassiana* and *Metarhizium anisopliae* on beetle-attracting tubers and cowpea foliage in Brazil. *Environ. Entomol.* **1986**, *15*, 1237–1243. [CrossRef]

150. Metcalf, R.L.; Ferguson, J.E.; Lampman, R.L.; Andersen, J.F. Dry cucurbitacin-containing baits for controlling diabroticite beetles (Coleoptera: Chrysomelidae). *J. Econ. Entomol.* **1987**, *80*, 870–875. [CrossRef]

151. Lance, D.R.; Sutter, G.R. Field-cage and laboratory evaluations of semiochemical-based baits for managing western corn rootworm (Coleoptera: Chrysomelidae). *J. Econ. Entomol.* **1990**, *83*, 1085–1090. [CrossRef]

152. Barbercheck, M.E.; Herbert, D.A., Jr.; Warrick, W.C., Jr. Evaluation of semiochemical baits for management of southern corn rootworm (Coleoptera: Chrysomelidae) in peanuts. *J. Econ. Entomol.* **1995**, *88*, 1754–1763. [CrossRef]

153. Schroder, R.F.W.; Martin, P.A.W.; Athanas, M.M. Effect of a phloxine B-cucurbitacin bait on Diabroticite beetles (Coleoptera: Chrysomelidae). *J. Econ. Entomol.* **2001**, *94*, 892–897. [CrossRef] [PubMed]

154. Pedersen, A.B.; Godfrey, L.D. Evaluation of cucurbitacins-based gustatory stimulant to facilitate cucumber beetle (Coleoptera: Chrysomelidae) management with foliar insecticides in melons. *J. Econ. Entomol.* **2011**, *104*, 1294–1300. [CrossRef] [PubMed]

155. Cabrera Walsh, G.; Mattioli, F.; Weber, D.C. Differential response of male and female *Diabrotica speciosa* (Coleoptera: Chrysomelidae) to bitter cucurbit-based toxic baits in relation to the treated area size. *Int. J. Pest Manag.* **2014**, *60*, 128–135. [CrossRef]

156. Tallamy, D.T.; Halaweish, F.T. Effects of age, reproductive activity, sex and prior exposure on sensitivity to cucurbitacins in southern corn rootworm (Coleoptera: Chrysomelidae). *Environ. Entomol.* **1993**, *22*, 925–932. [CrossRef]

157. Ventura, M.U.; Resta, C.C.M.; Nunes, D.H.; Fujimoto, F. Trap attributes influencing capture of *Diabrotica speciosa* (Coleoptera: Chrysomelidae) on common bean fields. *Sci. Agric.* **2005**, *62*, 351–356. [CrossRef]

158. Chandler, L.D. Corn rootworm areawide management program: United States Department of Agriculture-Agricultural Research Service. *Pest Manag. Sci.* **2003**, *59*, 605–608. [CrossRef] [PubMed]

 *insects*

Article

Automatic Field Detection of Western Corn Rootworm (*Diabrotica virgifera virgifera*; Coleoptera: Chrysomelidae) with a New Probe

Zsolt Tóth [1,*], Miklós Tóth [2], Júlia Katalin Jósvai [2], Franciska Tóth [1], Norbert Flórián [1], Veronika Gergócs [1] and Miklós Dombos [1]

[1]   Institute for Soil Sciences and Agricultural Chemistry, Centre for Agricultural Research, Herman Ottó út 15, H-1022 Budapest, Hungary; franciska.toth19@gmail.com (F.T.); florian.norbert@agrar.mta.hu (N.F.); gergocs.veronika@agrar.mta.hu (V.G.); dombos.miklos@agrar.mta.hu (M.D.)

[2]   Plant Protection Institute, Centre for Agricultural Research, Herman Ottó út 15, H-1022 Budapest, Hungary; toth.miklos@agrar.mta.hu (M.T.); josvai.julia@agrar.mta.hu (J.K.J.)

*   Correspondence: toth.zsolt@agrar.mta.hu

Received: 12 June 2020; Accepted: 28 July 2020; Published: 1 August 2020

Abstract: The Western corn rootworm (WCR), *Diabrotica virgifera virgifera* LeConte (Coleoptera: Chrysomelidae), is a significant invasive pest of maize plantations in Europe. Integrated pest management demands an adequate monitoring system which detects the activity of insects with high accuracy in real-time. In this study, we show and test a new electronic device (ZooLog KLP), which was developed to detect WCR in the field. The ZooLog KLP consists of a trapping element that attracts insects with its color and species-specific sex pheromone. The other part is an opto-electronic sensor-ring which detects the specimens when they fall into the trap. At detection, the time of catch is recorded and sent to a web interface. In this study, we followed WCR flight patterns for six weeks in two locations, using ZooLog KLP probes. We investigated sensor precision by comparing the number of catches to the number of detections. The tool reached high accuracy (95.84%) in recording WCR. We found a peak in flight activity in August and a bimodal daily pattern. This method may be beneficial in detecting the WCR during their activity, and this new device may serve as a prototype for real-time monitoring systems and improve the management of this pest.

Keywords: agricultural monitoring; infrared sensor; invasive species; maize; pest management

---

1. Introduction

Management of arthropod pests has long been a major challenge in food production for farmers throughout the world. Integrated pest management (IPM) programs [1] can achieve long term pest control with effective actions, such as monitoring the population size, estimating economic threshold, and integrating currently known chemical, biological, and physical control methods [2,3]. Therefore, the detection and identification of invertebrates are prerequisites necessity for IPM to optimize plant protection actions in agricultural fields.

Historically, different trap types have been applied for the detection and monitoring of pest species. However, most of them suffer from a number of shortcomings. For instance, during monitoring, traps should be checked at least weekly, with a visual count of the captures. Manual checking is a time-consuming and laborious task, especially in large agricultural fields. Moreover, data are delayed in time. Many initiatives approach the problem of manual checking by automatically enumerating the insects [4]. The accurate forecasting of potential pest outbreaks requires real-time detection of the insects in the field. Based on this information, pest control practices, such as selective spraying, can be arranged at the right time and location.

Insects **2020**, *11*, 486; doi:10.3390/insects11080486   www.mdpi.com/journal/insects

The Western corn rootworm (WCR), *Diabrotica virgifera virgifera* LeConte (Coleoptera: Chrysomelidae), ranks among the most critical invasive pests in Europe. It first appeared in Europe in the early 1990s [5], and currently infests most of the EU countries [6]. Larvae of WCR cause the primary damage to a maize plant by chewing and boring through the rootstock in the soil. However, beetles can also generate severe loss by feeding on the reproductive parts of the plant. The economic costs resulting from yield losses and pest management annually exceed 1 billion USD in the United States [7]. In the EU-27, potential damage to maize was estimated to be as high as 580 million USD per year, should all maize production areas be infested with WCR [8].

Protection against WCR mainly consists of cultural and chemical control. The cultural control methods, crop rotation, and delayed planting have been tested and modeled [9,10]. Annual rotation seems to be only partly effective, as the beetles have adapted to the rotation cycle (e.g., soybean *Gylcine max* L. in North America) [11,12]. As for the most common chemical control strategies, soil-applied insecticides and insecticide seed dressing are used against the larvae at planting [13]. Soil-applied insecticides showed a relatively higher efficacy to avoid root damage; however, insecticide application needs to be based on an effective monitoring program, as has been proposed for decades in many IPM programs (e.g., in the EU: Directive, 2009/128/EC). The abundance of adults on traps is used to identify population economic thresholds (e.g., 5–6 WCR adults/sticky trap/day in several growing areas).. The continuous and accurate counts of WCR over the season could be achieved more efficiently by automatic devices.

Foliage insecticide treatments are sometimes applied for WCR adults, to reduce population sizes and subsequent oviposition by females, or to avoid silk-clipping damage to maize silk. For these applications, it is necessary to know the current WCR population size; real-time, automatic measurement can also help. Annual flight patterns of WCR greatly vary according to locality, climate, and even years [14,15]. Adults generally occur from mid-July to mid-September, with a peak from the end of July to August [16,17]. The daily flight of WCR has a bimodal pattern, having two periods of maximal activity: some hours after sunrise and before sunset [18–20].

In Europe, two main types of traps are used to monitor WCR adults in the field, both containing female sex pheromone: yellow sticky traps and transparent sticky "cloak" traps called PAL (sticky sheet is shaped like a cloak, named after the Hungarian vocalist "Palást") [21]. As beetles are attracted to yellow color, yellow sticky traps (e.g., Pherocon AM) are widely used, especially in North America. However, this trap type is more effective in areas where WCR populations are well established [22]. At low-population sites, PAL traps (CSALOMON® PAL, Plant Protection Institute HAS, Budapest, Hungary) are more sensitive to detect the immigrant WCR population [23,24]. One issue with sticky traps is that non-target insects can quickly saturate them. To solve the problem of saturation, Tóth et al. [25] developed a non-sticky funnel-trap (so-called KLP "hat" trap) to catch WCR. This trap type has much higher catch capacity (5–6000 beetles compared to the 3–400 beetles in sticky traps), and is more user friendly than sticky traps.

For automatic measurement, sticky traps equipped with cameras recognize and also count dead or stuck insects in real-time, or with a relative time-lapse [26,27]. This type of machine vision technique yields an exact number of captured insects [28]; however, machine vision is still restricted only to sticky traps.

Although traps have been developed for the real-time, automatic detection of crawling or flying insects [see in 4], an automated detection technique for WCR has not yet been developed. In the case of the KLP trap, using photography is hardly possible; therefore, flying or falling insects have to be detected. For flying insects, opto-electronic devices such as those using laser beam [29] and infrared light [30,31] are common. In the case of these sensors, although good results have been achieved in distinguishing different species flying into the traps, for example, based on wingbeat frequency, e.g., [32], the problem is usually that most of the sensors are not species-specific.

Our team developed opto-electronic sensors, which automatically detect and count arthropods of different sizes [33]. These sensors effectively detect arthropods living in the soil or on the surface [34].

For agricultural pest forecasts, we equipped different pheromone traps with our sensors to achieve higher species specificity. The funnel traps (KLP trap) mentioned above are modified to automatically detect and count WCR and provide online access to those data. In this study, we provide a detailed technical description of our new probe. We present the precision and reliability of the new method by comparing the number of WCR individuals caught and detected by the ZooLog KLP probes under field conditions.

2. Materials and Methods

2.1. Description of the New ZooLog KLP Probe

The new probe was developed to catch and automatically count WCR specimens and forward those data to a central database. It comprises three main parts: a trap, a sensor, and a data communication system.

2.1.1. Description of the Trapping Element

The trapping element of our new probe is based on the CSALOMON® KLP+ "hat" trap (Figure 1), produced by the Plant Protection Institute, Centre for Agricultural Research, Budapest, Hungary [25]. This trap type is typically used for quantitative sampling of WCR in Europe and can be baited with lures containing a species-specific sex pheromone or a floral-based attractant [35]. In our test, traps baited with the commercially available sex pheromone (CSALOMON® group, Plant Protection Institute CAR, Budapest, Hungary) were used. The trap consists of a yellow, non-sticky plastic sheet panel, which attracts WCR with its color and an attached pheromone bait. The panel guides climbing WCR individuals through a funnel and a plastic tube towards a transparent plastic (PET) bottle (30 cm high with 8 cm diameter). The catch container should be transparent, because this species is attracted to light (positive phototaxis) [36]. The infrared (IR) sensor-ring (see next section) is mounted to the mouth of the PET bottle. The beetles spend approx. thirty minutes in the plastic bottle when first trapped, then eventually fall through the mouth of the bottle to the sensing unit. We used insecticidal strips (VaporTape, Hercon Environmental, Emigsville, PA, USA) to kill captured WCR adults, making manual counting of captured insects more convenient. The sensor-ring detects this event, and a sample container positioned in the lowermost part of the probe catches the specimens.

2.1.2. IR Sensor-Ring

We used the uniform IR sensor-ring, which was developed by our team and has been optimized for the detection of insects with different body sizes [33]. The sensor part contains 16 pieces of IR sensors shifted in two rows for better coverage of the detection area and focused on the receiver. This opto-electronic sensor-ring records those events that interrupt the path between the receiver and the emitter. We constructed this sensor to detect arthropods falling or flying through the sensor field, and tested the precision and accuracy of its detection under laboratory conditions for several insect species. In the case of WCR, the detection accuracy was 100% with dead individuals (see Table 1 in [33]]). According to this result, the sensor itself accurately detects WCR under stable conditions. However, the situation is entirely different under field conditions. In a noisy environment, raw detection data should be filtered to enhance precision and accuracy. For this procedure, we used artificial neural network (ANN) algorithms (see Section 2.3.1).

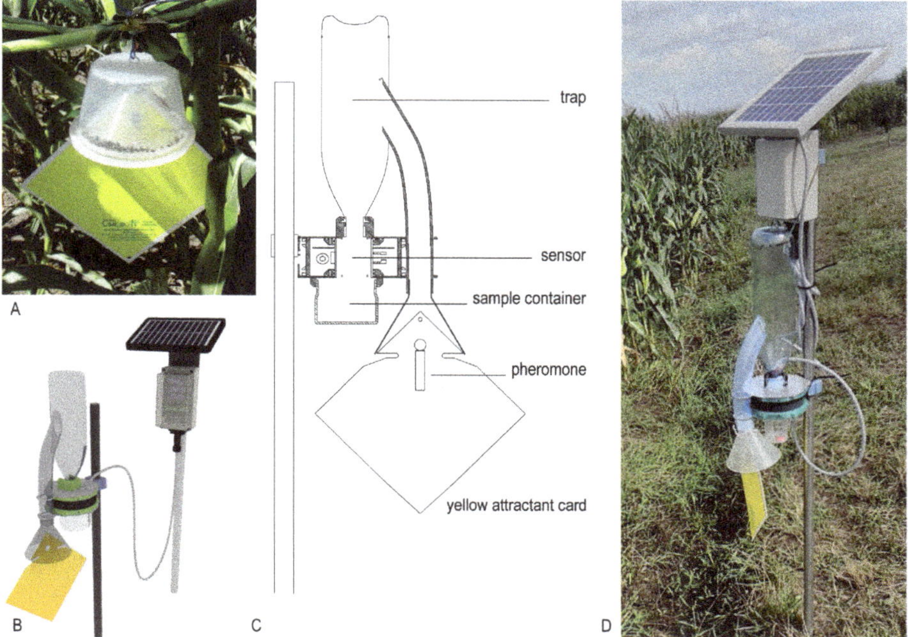

Figure 1. Construction of the new ZooLog KLP probe. (**A**): the original CSALOMON® KLP "hat" trap. Western corn rootworm (WCR) individuals are attracted by pheromone towards an inverted funnel, from where they cannot escape. (**B,C**): the design and cross section of the new probe. The yellow attractant card and the inverted funnel are connected to a PET bottle, from where WCR individuals fall through the sensor-ring into the sampling container. Detection data are transferred to the central database via the internet. A solar panel provides power for the electronics, enabling online monitoring throughout the season. (**D**): The new probe deployed adjacent to the test cornfield.

2.1.3. Data Collection

The ZooLog Monitoring system has been designed for online monitoring, including loggers, a central database, and a web interface. Data are transmitted to the central database each day (settings). Detection data can be downloaded from or directly managed on the ZooLog Online Web Interface (Figure 2). Obtained data can be reported according to the measurement series.

2.2. Field Observations

To estimate the precision and accuracy of the detection of the proposed sensor system, we conducted field surveys over six weeks, from the middle of August 2018 to the beginning of October 2018. A total of five probes were used; three probes were installed in cropland at Érd-Elvira Major Experiment Station of Research Institute for Fruit growing and Ornamentals, and two probes were placed in a privately owned cornfield near Tordas village in Hungary (Table 1). While the probes captured WCR, they continuously detected these individuals. We counted the individuals at least twice a week, found in the sample container of the probes. These datasets were the basis of data validation; we simply compared the number of detections to the number of catches obtained for a given time interval from the same probe.

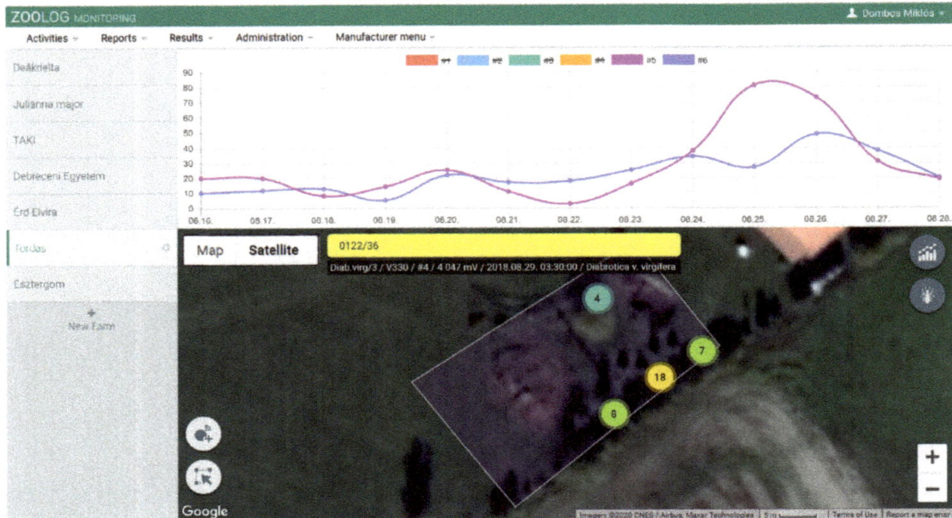

Figure 2. The ZooLog Online Web Interface: The monitoring system provides daily data via the internet. The probes and their status are visualized on a map. This interface is also used for setting the probes and data filtering. Results of the detection data can be downloaded and automatically visualized as graphs, typically the number of detected insects per day.

Table 1. The geographical location of the test sites and ZooLog KLP probes.

Location	Description	Probe	Latitude	Longitude
Tordas	cornfield in a privately owned farm	V310	47°20′54.15″ N	18°43′58.65″ E
		V330	47°20′54.43″ N	18°43′59.22″ E
		V344	47°20′54.31″ N	18°43′58.95″ E
Érd Elvira	cornfield adjacent to an orchard	V323	47°20′23.89″ N	18°51′53.44″ E
		V353	47°20′24.91″ N	18°51′52.71″ E

2.3. Filtering Detection Data and Data Analysis

2.3.1. Data Filtering

The IR sensor-ring provides a dataset at each detection, see [33]. Each detection consists of eight values generated by the sixteen IR sensors in the sensor-ring. The patterns of these raw output values are characteristic of WCR. WCR individuals usually triggered only one or two signals with higher figures due to their intermediate body size; therefore, we used these patterns to identify positive signals for WCR and filtered out noise data, e.g., signals generated by other non-target insects than WCR. For this procedure, as the main objective of field testing, we built up an interface in our database. In this operation table (Figure 3), one row corresponds to one detection, and the signal patterns appear in the column called 'Sensor data'. During field tests, we checked and noted catches of each probe, at first hourly, and later daily. For the same periods, we listed the detection table on the interface. We tried to find the positive detections (caused by WCR) by eye, based on the characteristic signal patterns, and based on the numbers we caught in a given period. For example, if we find ten individuals in a period, we cannot be entirely sure exactly which ten signals belong to these individuals. We selected ten similar signals according to the characteristic signal patterns. This procedure formed the basis of the learning dataset, which was used in the machine learning process to allow automatic identification of positive signals. Deep-learning data analysis was performed with a large-scale distributed machine learning platform named TensorFlow [37], on Github [38]. The binary classification was built into the deep learning procedure (positive class: WCR individual), and the eight sensor figures were used as features for modeling. Ten percent of the dataset was labeled as examples in the learning procedure.

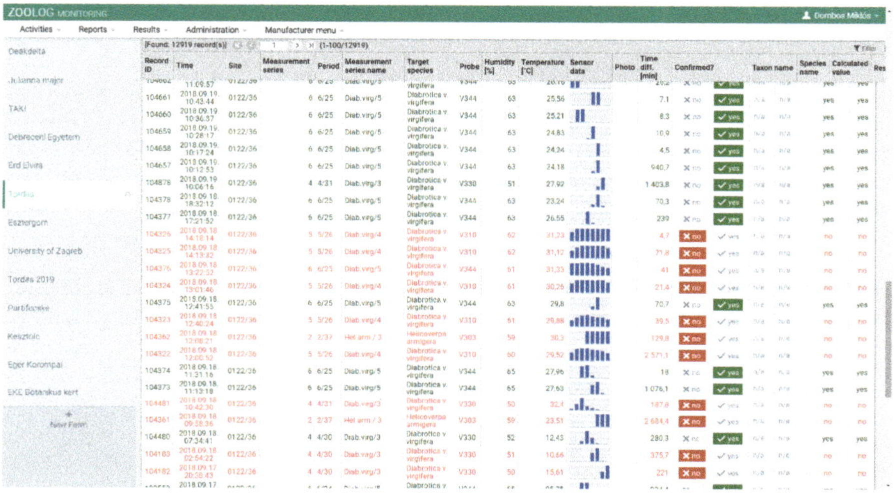

Figure 3. Filtering procedure for, and marking the records captured by, the ZooLog KLP probes. Each row corresponds to one detection. The sensor signal patterns are shown in the 'Sensor data' column. When only one or two signs were higher, we confirmed it manually in column 'Confirmed?' as "yes" (green colors) or "no" (red colors). We confirmed precisely the same number of detections that were captured in a given period. The patterns of positive detections differed significantly from the negative ones caused by other non-target insects. This procedure provided the learning dataset for deep-learning analysis used in filtering false detection data.

2.3.2. Statistical Analysis

For the analyses, automatically detected and manually captured data were summed for each day per probe at both investigation sites (altogether five measurement series). For detection data, machine learning filtered data were used. Since both the distributions of captured and detected individuals per probe were strongly right-skewed (range: 1–421, median: 14), both the daily summed number of captures and detected individuals were ln $(x + 1)$ transformed prior to the analysis.

The relationship between the automatically detected (and filtered) and the manually captured number of individuals was analyzed with a linear model to evaluate the accuracy, reliability, and overestimation rates of the sensor system. Daily data of the five probes were analyzed together. The intercept and the slope of the model were tested against 0 and 1, respectively, using the 'multcomp' package [39].

Several metrics (accuracy, precision, recall, and F1 score) were calculated to estimate the performance of the filtering procedure. In binary classification, according to the contingency matrix built up from the dataset, accuracy means true positives plus true negatives over the total number of examples. Precision is defined as the number of true positives over the number of true positives plus the number of false positives. Recall is defined as the number of true positives over the number of true positives plus the number of false negatives. These quantities are also related to the F1 score, which is defined as the harmonic mean of precision and recall. High precision is linked with a low false-positive rate, and high recall relates to a low false-negative rate.

Based on data derived from the automatic detection, some examples were presented to estimate and visualize the local daily and temporal activity of the western corn rootworm. Local polynomial regression fitting (locally estimated scatterplot smoothing, LOESS) was applied for smoothing time-series data.

All statistical analyses were carried out with R 3.6.2 software [40].

3. Results

3.1. Accuracy and Performance of the Sensor System

To achieve higher detection accuracy, we filtered out false detections. The filtering-out results of automatic detections gained by deep-learning analysis are summarized in Table 2. For the five probes together, there were 3266 true positives, 3611 true negatives, 184 false positives, and 313 false negatives out of 7374 cases. The machine learning reached a very high precision (0.95), due to relatively few false positives and negatives. The recall and F1 score for the adults of *D. v. virgifera* reached 0.91 and 0.93, respectively. The accuracy reached 0.93 for the whole data clearing procedure.

Table 2. Contingency table showing the performance summary of the sensor system tested at 5 field sites in Tordas and Érd in 2018. ANN = artificial neural network, TP = true positive, TN = true negative, FP = false positive, FN = false negative.

		Automatically Detected (ANN Predicted)		
		1 (Positive)	0 (Negative)	Sum
	1	3266 (TP)	313 (FN)	3579
Manually Counted	0	184 (FP)	3611 (TN)	3795
	Sum	3450	3924	7374

Then, we compared the filtered detection data to the number of manually caught WCR specimens. Linear regression analysis showed a high correlation between automatic and manual counting after ln (x + 1) transformation (Figure 4). The average accuracy was 95.84%, with very high reliability ($R^2 = 0.948$). The slope of the regression line was significantly lower than 1 (y = 0.96 ± 0.02, $p < 0.001$). However, there was no tendency for under- or overestimation of the number of individuals (intercept = −0.01 ± 0.07, $p = 0.967$).

3.2. Automatic Detection of WCR Over Time

The inclusion of the sensor system in the trap allows us to follow the temporal activity of the target pest. Since our field tests were launched in August, the first half of the WCR flight activity was not captured. Preliminary results are presented to provide examples from monitoring WCR during the study period. Figure 5 illustrates the number of detected individuals obtained from probes placed at Érd Elvira (2) and Tordas (3) research sites. The temporal dynamics of the two data sets followed similar general trends, even if the curves of measurement series (probes) slightly varied. The automatic counting caught a small peak at the end of August/beginning of September (Figure 5).

Employing the proposed sensor system, we estimated the local daily activity (circadian rhythm) of the Western corn rootworm. Combining all recorded events from the two research sites, WCR had periods of high activity twice a day. The daily peak periods were during the morning (between 8:00–9:00) and in the late afternoon (between 16:00–17:00) (Figure 6). The temporal pattern of data sets was similar in both sites, but the above-mentioned diurnal dynamic was more profound at the Érd Elvira site.

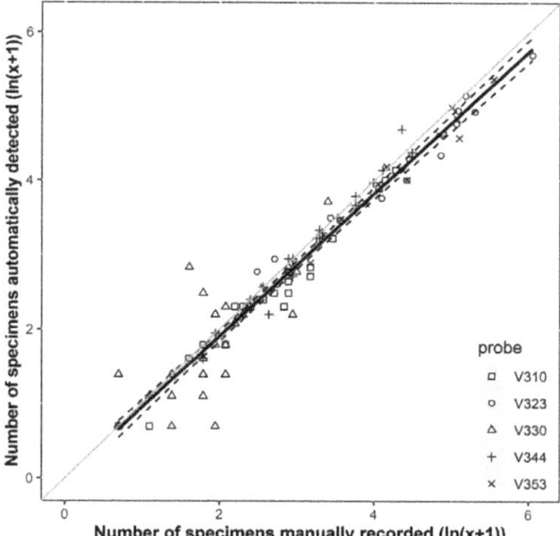

Figure 4. The relationship between the number of detections and captures of *D. v. virgifera* at Érd Elvira and Tordas sites using ZooLog KLP probes. One point represents the number of catches and the corresponding number of detections in a given period for a given probe. The solid line shows the predicted values from the linear model, and the broken lines show the 95% prediction interval. The solid grey line indicates the equality between the two variables.

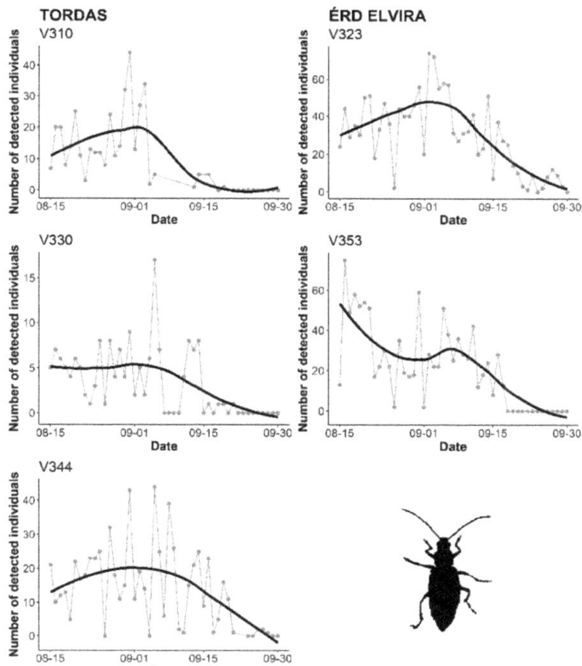

Figure 5. Continuous, automatic monitoring of *D. v. virgifera* by ZooLog KLP probes at Érd Elvira and Tordas research sites. For smooth curves, local polynomial regression fitting (locally estimated scatterplot smoothing, LOESS) was applied.

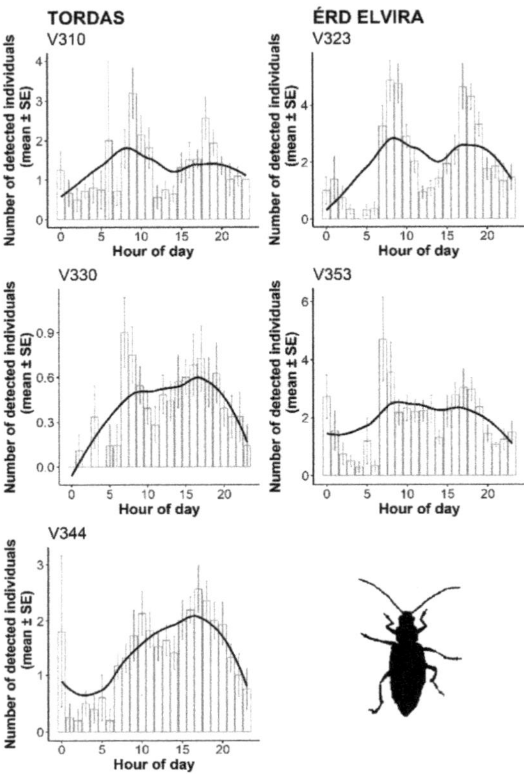

Figure 6. The activity circadian rhythms of *D. v. virgifera* based on probe recordings during six weeks from 15 August to 1 October 2018. For smooth curves, local polynomial regression fitting (locally estimated scatterplot smoothing, LOESS) was applied.

4. Discussion

The results of the present study clearly show that our new opto-electronic device (ZooLog KLP) can record WCR with high accuracy. In our previous laboratory tests with low environmental noise, the same sensor accurately detected (with 100% precision) several Coleoptera species larger than 1.40 mm, such as *Meligethes aeneus, Diabrotica virgifera, Agriotes ustulatus, Epicometis hirta, Cetonia aurata* [33]. Under semi-natural conditions, the sensor also detects small microarthopods (0.47–2.47 mm) with 88–100% accuracy, depending on species [41]. Compared to that, we reached 95.84% accuracy on average for WCR under field conditions.

For agricultural pest forecasting systems, species specificity of traps is essential. Several species-specific methods are used in the field to detect and recognize insect species automatically. Camera traps with machine vision [42–45] or using color sensors [46] have high species-specificity. However, these are useful only for sticky traps in which individuals caught can be easily photographed; photographing insects flying into the trap remains a technical challenge. Other devices recognize species according to their sound or wingbeat frequency using laser light [32] or stereo-recording [47]. Opto-electronic sensors are not able to reach species specificity; however, by using our infrared sensor-ring, we could select the positive signals of the target species (WCR) according to the body size. We were able to filter out the noise detections, which occurred in the same amount as the true-positive signals. False identifications, i.e., where the system identified WCR when the detection was not caused by WCR or vice versa, were rare; therefore, both the precision, accuracy and recall were very high, each of them was higher than 0.93. Sex pheromones are highly species-specific, and they are widely used in pest monitoring [48]. We equipped different types of the CSALOMON trap family with our sensors to detect flying insects (mostly moth species), click beetles [49], and also WCR. The ZooLog KLP trap combines the advantages of the attractiveness of the sex pheromone and the vertical climbing behavior of WCR [25], thus excluding most non-target species.

Traps baited with sex pheromone lures have the inherent disadvantage that they attract and catch only one sex (in the case of WCR, the males). Possible bias due to this in the interpretation of captures is widely discussed [48]. In the case of WCR, it is also possible to apply a synthetic floral lure [50], which acts as a feeding attractant and will attract both females and males [25]. By using the floral lure, the sex bias could be overcome. Our new trap device could be used with this lure; however, since a floral lure would attract other species more intensively than sex pheromones, the validation of the machine learning algorithm would likely need to be updated.

The advantage of our probe compared to sticky traps equipped with a camera is that it cannot easily be saturated. Since the system has low energy consumption that can be supplied by a small solar panel, our device can operate in remote sites for a long period without human intervention. Finally, cellular data transmission allows users to see online what is happening in the field.

To calculate action thresholds for *D. v. virgifera*, various characteristics have been used, such as the average and cumulative numbers of adults per plants [51]. Beyond qualitative monitoring and detection, pheromone traps can be best exploited in IPM when quantitative aspects of populations can be correlated to trap capture data. One approach is the establishment of threshold catch values connected with population densities or damage levels. Such a correlation study is very time consuming and should involve parallel experiments in several fields throughout the years. We believe that our new probe, due to its more efficient data acquisition and evaluation, will assist plant protection experts in performing such studies in the future. At present, to the best of our knowledge, no thresholds have been reliably established for any trap designs baited with the pheromone of WCR. Action thresholds calculated from different monitoring methods are not the same (e.g., for yellow sticky traps, they are lower than for pheromone traps) [24]. Moreover, in connection with WCR, even crop type affects the action threshold (e.g., the silage action threshold is lower than that for seed production or sweet corn) [13]. Future studies are needed, for example, to compare the performance of commercially used WCR traps with ZooLog KLP, to estimate the action threshold for this new device.

ZooLog KLP provides a new way of studying WCR population dynamics in the field. Flight of WCR starts in July, and adults are usually present until early October [15–17,52]. Since it gives the precise moment of each capture, creating a fine-scale data set compared to manual counting, circadian rhythms of WCR can be precisely investigated. All the studies investigating the daily flight activity of WCR, whether in the laboratory [19,53] or the field [18,20,54], and found that WCR has two flight peaks a day: in the morning and the evening. Similarly, in our study, captures also had maximal numbers between 8:00–9:00 and between 16:00–17:00. These monthly and daily patterns are usually correlated with climatic properties of the area [14,54]. Therefore, the detailed and accurate data gained by the ZooLog KLP may help to reveal the coherence between WCR flight timing and other variables. Unfortunately, our test was conducted only from the second half of the WCR flight period; therefore, we were not able to trace the whole seasonal pattern of WCR. More experimentation is needed on the entire beetle activity period to be able to evaluate trap efficacy and seasonal dynamics. However, ZooLog KLP detected the data in real-time. Most of the studies using trapping methods obtain data weekly, providing less sensitive analyses compared to the real-time approach. This detailed information may be beneficial when climatic and other variables are compared with activity data of pest insects.

The cost of the ZooLog KLP probe can only be roughly estimated as it is still in the prototype stage of development. To build the prototype, we used electronic and mechanical parts, which cost 288 Euros. The electronic board containing the infrared sensor-ring was custom designed and manufactured that led to a higher price (159 EUR). The very cheap looking infrared technology might provide an opportunity for a favorable cost-benefit ratio since, in serial production, the cost of such a board should not be higher than 25 EUR. The housing was made by 3D printing, had a fee of 65 Euros per probe. The output sensor datasets to be transferred via the internet are very small; therefore, considering the expense of online monitoring for one probe, the smallest prepaid data SIM card was enough for the whole year. Financial benefits for the growers could be the decrease in the number of on-site inspections of WCR. Since we can expect that real-time pest monitoring could contribute to reducing environmental impact, in the long run, the use of a more environmental-friendly technology could also lead to a business advantage.

5. Conclusions

Accurate and automatic monitoring methods are needed to achieve precise and effective pest control. This study is the first documented report of the ZooLog KLP as a continuous, automatic, real-time detection device to monitor WCR flight in the field. The ability to continuously and automatically measure WCR flight with high precision allows for more precise and, therefore, effective pest management of this species.

Author Contributions: Conceptualization and methodology: M.D., and M.T.; formal analysis, Z.T., V.G.; investigation, M.T., N.F., and M.D.; writing—original draft preparation, F.T., M.T., V.G., J.K.J., Z.T., and M.D.; writing—review and editing: M.T., M.D., N.F., V.G., J.K.J., and Z.T.; visualization Z.T.; supervision M.T., M.D.; funding acquisition: M.D. All authors have read and agreed to the published version of the manuscript.

Funding: This research was funded by the European Union's LIFE project (LIFE13 ENV/HU/001092), the Hungarian TALAJBIOM project (GINOP-2.3.2-15-2016-00056) and the Premium Postdoctoral Scholarship of the Hungarian Academy of Sciences (PPD2018-001/2018) (Veronika Gergócs).

Acknowledgments: We would like to express our very great appreciation to Béla Tichy (Deakdelta Ltd.) for the design and mechanical construction and Péter Liszli (Deakdelta Ltd.) for the design of the electronic board and the microcontroller programming. Deák Delta Ltd., as a project partner company, was responsible for electrotechnical developments, while Helion Ltd. was responsible for database programming. Deep-learning AI analyses were conducted by Robert Zawiasa (Z-Gen Kibernetika Ltd.), their work was indispensable; many thanks for it. We thank the landowner of Tordas site, Dénes Besenyői, for permission to work on his field. We acknowledge the three anonymous reviewers and the editors for their detailed comments, which substantially improved the manuscript.

Conflicts of Interest: The authors declare no conflict of interest.

References

1. Kogan, M. Integrated pest management: Historical perspectives and contemporary developments. *Annu. Rev. Entomol.* **1998**, *43*, 243–270. [CrossRef]
2. Nishimatsu, T.; Jackson, J.J. Interaction of insecticides, entomopathogenic nematodes, and larvae of the western corn root worm (Coleoptera: Chrysomelidae). *J. Econ. Entomol.* **1998**, *91*, 410–418. [CrossRef]
3. Toepfer, S.; Kuhlmann, U. Survey for natural enemies of the invasive alien chrysomelid, Diabrotica virgifera virgifera, in Central Europe. *BioControl* **2004**, *49*, 385–395. [CrossRef]
4. Liu, H.; Lee, S.-H.; Chahl, J.S. A review of recent sensing technologies to detect invertebrates on crops. *Precis. Agric.* **2017**, *18*, 635–666. [CrossRef]
5. Čamprag, D.; Bača, F. Diabrotica virgifera (Coleoptera, Chrysomelidae); a new pest of maize in Yugoslavia. *Pestic. Sci.* **1995**, *45*, 291–292. [CrossRef]
6. CABI. Diabrotica Virgifera Virgifera (Western Corn Rootworm). Invasive Species Compendium Datasheet. Available online: https://www.cabi.org/isc/datasheet/ (accessed on 5 June 2020).
7. Dun, Z.; Mitchell, P.; Agosti, M. Estimating Diabrotica virgifera virgifera damage functions with field trial data: Applying an unbalanced nested error component model. *J. Appl. Entomol.* **2010**, *134*, 409–419. [CrossRef]
8. Wesseler, J.; Fall, E.H. Potential damage costs of Diabrotica virgifera virgifera infestation in Europe—The 'no control' scenario. *J. Appl. Entomol.* **2010**, *134*, 385–394. [CrossRef]

9. Musick, G.; Chiang, H.; Luckmann, W.; Mayo, Z.; Turpin, F. Impact of planting dates of field corn on beetle emergence and damage by the western and the northern corn rootworms in the Corn Belt. *Ann. Entomol. Soc. Am.* **1980**, *73*, 207–215. [CrossRef]

10. Szalai, M.; Kiss, J.; Kövér, S.; Toepfer, S. Simulating crop rotation strategies with a spatiotemporal lattice model to improve legislation for the management of the maize pest Diabrotica virgifera virgifera. *Agric. Syst.* **2014**, *124*, 39–50. [CrossRef]

11. Kiss, J. Monitoring of Western Corn Rootworm (Diabrotica virgifera virgifera LeConte) in Europe 1992–2003. In *Western Corn Rootworm: Ecology and Management*; Vidal, S., Kuhlmann, U., Edwards, C.R., Eds.; CABI Publishing: Wallingford, UK, 2005.

12. Gray, M.E.; Sappington, T.W.; Miller, N.J.; Moeser, J.; Bohn, M.O. Adaptation and invasiveness of western corn rootworm: Intensifying research on a worsening pest. *Annu. Rev. Entomol.* **2009**, *54*, 303–321. [CrossRef]

13. Boriani, M.; Agosti, M.; Kiss, J.; Edwards, C. Sustainable management of the western corn rootworm, Diabrotica virgifera virgifera LeConte (Coleoptera: Chrysomelidae), in infested areas: Experiences in Italy, Hungary and the USA. *Eppo. Bull.* **2006**, *36*, 531–537. [CrossRef]

14. Lemic, D.; Mikac, K.M.; Kozina, A.; Benitez, H.A.; McLean, C.M.; Bažok, R. Monitoring techniques of the western corn rootworm are the precursor to effective IPM strategies. *Pest. Manag. Sci.* **2016**, *72*, 405–417. [CrossRef] [PubMed]

15. Levine, E.; Spencer, J.L.; Isard, S.A.; Onstad, D.W.; Gray, M.E. Adaptation of the western corn rootworm to crop rotation: Evolution of a new strain in response to a management practice. *Am. Entomol.* **2002**, *48*, 94–107. [CrossRef]

16. Meinke, L.J.; Sappington, T.W.; Onstad, D.W.; Guillemaud, T.; Miller, N.J.; Komáromi, J.; Levay, N.; Furlan, L.; Kiss, J.; Toth, F. Western corn rootworm (Diabrotica virgifera virgifera LeConte) population dynamics. *Agric. For. Entomol.* **2009**, *11*, 29–46. [CrossRef]

17. Toshova, T.B.; Velchev, D.I.; Abaev, V.D.; Subchev, M.A.; Atanasova, D.Y.; Tóth, M. Detection and monitoring of Diabrotica virgifera virgifera LeConte, 1868 (Coleoptera: Chrysomelidae) by KLP+ traps with dual (pheromone and floral) lures in Bulgaria. *Acta Zool. Bulg.* **2017**, *9*, 247–254.

18. Witkowski, J.; Owens, J.; Tollefson, J. Diel activity and vertical flight distribution of adult western corn rootworms in Iowa cornfields. *J. Econ. Entomol.* **1975**, *68*, 351–352. [CrossRef]

19. Naranjo, S.E. Comparative flight behavior of Diabrotica virgifera virgifera and Diabrotica barberi in the laboratory. *Entomol. Exp. et Appl.* **1990**, *55*, 79–90. [CrossRef]

20. Tóth, M.; Törőcsik, G.; Imrei, Z.; Vörörs, G. Diel rhythmicity of field responses to synthetic pheromonal or floral lures in the western corn rootworm Diabrotica v. virgifera. *Acta Phytopathol. Entomol. Hung.* **2010**, *45*, 323–328. [CrossRef]

21. Tóth, M. In Detection and monitoring devices for the invading western corn rootworm Diabrotica v. virgifera: A summary. In Proceedings of the Interntioanl Conference Alien Arthropods in South East Europe–Crossroad of Three Continents, Sofia, Bulgaria, 19–21 September 2007; p. 95.

22. Komáromi, J.; Kiss, J.; Tuska, T.; Pai, B. Comparison of western corn rootworm (Diabrotica virgifera virgifera LeConte) adult captures on pheromone-baited and visual traps during population build up. *Acta Phytopathol. et Entomol. Hung.* **2006**, *41*, 305–315. [CrossRef]

23. Tóth, M. Trap types for capturing Diabrotica virgifera virgifera (Coleoptera, Chrysomelidae) developed by the Plant Protection Institute, HAS,(Budapest, Hungary): Performance characteristics. *IOBC/WPRS Bull.* **2005**, *28*, 147–154.

24. Tóth, M.; Sivcev, I.; Ujváry, I.; Tomasek, I.; Imrei, Z.; Horváth, P.; Szarukán, I. Development of trapping tools for detection and monitoring of Diabrotica v. virgifera in Europe. *Acta Phytopathol. Entomol. Hung.* **2003**, *38*, 307–322. [CrossRef]

25. Tóth, M.; Csonka, É.; Szarukán, I.; Vörös, G.; Furlan, L.; Imrei, Z.; Vuts, J. The KLP+ ("hat") trap, a non-sticky, attractant baited trap of novel design for catching the western corn rootworm (Diabrotiea v. virgifera) and cabbage flea beetles (Phyllotreta spp.)(Coleoptera: Chrysomelidae). *Int. J. Hortic. Sci.* **2006**, *12*, 57–62.

26. Zhong, Y.; Gao, J.; Lei, Q.; Zhou, Y. A vision-based counting and recognition system for flying insects in intelligent agriculture. *Sensors* **2018**, *18*, 1489. [CrossRef] [PubMed]

27. Rustia, D.J.A.; Lin, C.E.; Chung, J.-Y.; Zhuang, Y.-J.; Hsu, J.-C.; Lin, T.-T. Application of an image and environmental sensor network for automated greenhouse insect pest monitoring. *J. Asia-Pac. Entomol.* **2020**, *23*, 17–28. [CrossRef]

28. Solis-Sánchez, L.O.; Castañeda-Miranda, R.; García-Escalante, J.J.; Torres-Pacheco, I.; Guevara-González, R.G.; Castañeda-Miranda, C.L.; Alaniz-Lumbreras, P.D. Scale invariant feature approach for insect monitoring. *Comput. Electron. Agric.* **2011**, *75*, 92–99. [CrossRef]

29. Batista, G.E.; Hao, Y.; Keogh, E.; Mafra-Neto, A. In Towards automatic classification on flying insects using inexpensive sensors. In Proceedings of the 10th International Conference on Machine Learning and Applications and Workshops, Honolulu, HI, USA, 18–21 December 2011; pp. 364–369.

30. Potamitis, I.; Rigakis, I.; Fysarakis, K. The electronic McPhail trap. *Sensors* **2014**, *14*, 22285–22299. [CrossRef] [PubMed]

31. Rigakis, I.; Potamitis, I.; Tatlas, N.-A.; Livadaras, I.; Ntalampiras, S. A Multispectral Backscattered Light Recorder of Insects' Wingbeats. *Electronics* **2019**, *8*, 277. [CrossRef]

32. Silva, D.F.; Souza, V.M.; Ellis, D.P.; Keogh, E.J.; Batista, G.E. Exploring low cost laser sensors to identify flying insect species. *J. Intell. Robot. Syst.* **2015**, *80*, 313–330. [CrossRef]

33. Balla, E.; Flórián, N.; Gergócs, V.; Gránicz, L.; Tóth, F.; Németh, T.; Dombos, M. An Opto-electronic Sensor-ring to Detect Arthropods of Significantly Different Body Sizes. *Sensors* **2020**, *20*, 982. [CrossRef]

34. Dombos, M.; Kosztolányi, A.; Szlávecz, K.; Gedeon, C.; Flórián, N.; Groó, Z.; Dudás, P.; Bánszegi, O. EDAPHOLOG monitoring system: Automatic, real-time detection of soil microarthropods. *Methods Ecol. Evol.* **2017**, *8*, 313–321. [CrossRef]

35. Csonka, É.; Tóth, M. Comparison of KPL+("hat") and VARL+(funnel) trap types baited with allyl isothiocyanate for capture of cabbage flea beetles (Phyllotreta spp.) (Coleoptera, Chrysomelidae). *Növényvédelem* **2006**, *42*, 597–6041.

36. Ball, H.J. Spectral response of the adult Western Corn Rootworm (Coleoptera: Chrysomelidae) to selected wavelengths. *J. Econ. Entomol.* **1982**, *75*, 932–933. [CrossRef]

37. Abadi, M.; Agarwal, A.; Barham, P.; Brevdo, E.; Chen, Z.; Citro, C.; Corrado, G.S.; Davis, A.; Dean, J.; Devin, M.; et al. Tensorflow: Large-scale machine learning on heterogeneous distributed systems. *arXiv* **2016**, arXiv:1603.04467.

38. Github. The TensorBoard Repository on GitHub. Available online: http://github.com/tensorflow/tensorboard (accessed on 15 June 2017).

39. Hothorn, T.; Bretz, F.; Westfall, P. Simultaneous inference in general parametric models. *Biom. J. Math. Methods Biosci.* **2008**, *50*, 346–363. [CrossRef] [PubMed]

40. R Development Core Team. R: A Language and Environment for Statistical Computing. R Foundation for Statistical Computing, Vienna, Austria. Available online: http://www.R-project.org (accessed on 30 July 2020).

41. Flórián, N.; Gránicz, L.; Gergócs, V.; Tóth, F.; Dombos, M. Detecting Soil Microarthropods with a Camera-Supported Trap. *Insects* **2020**, *11*, 244. [CrossRef] [PubMed]

42. Alhady, S.; Kai, X.Y. Butterfly Species Recognition Using Artificial Neural Network. In *Intelligent Manufacturing & Mechatronics*; Springer: Singapore, 2018; pp. 449–457.

43. Kaya, Y.; Kayci, L.; Uyar, M. Automatic identification of butterfly species based on local binary patterns and artificial neural network. *Appl. Soft Comput.* **2015**, *28*, 132–137. [CrossRef]

44. Li, Y.; Wang, H.; Dang, L.M.; Sadeghi-Niaraki, A.; Moon, H. Crop pest recognition in natural scenes using convolutional neural networks. *Comput. Electron. Agric.* **2020**, *169*, 105174. [CrossRef]

45. Wang, J.; Lin, C.; Ji, L.; Liang, A. A new automatic identification system of insect images at the order level. *Knowl. Based Syst.* **2012**, *33*, 102–110. [CrossRef]

46. Potamitis, I.; Rigakis, I.; Tatlas, N.-A. In On Fresnel-Based Single and Multi Spectral Sensors for Insects' Wingbeat Recording. In Proceedings of the 20th International Conference on Solid-State Sensors, Actuators and Microsystems & Eurosensors XXXIII (TRANSDUCERS & EUROSENSORS XXXIII), Berlin, Germany, 23–27 June 2019; pp. 1901–1904.

47. Potamitis, I.; Rigakis, I.; Vidakis, N.; Petousis, M.; Weber, M. Affordable bimodal optical sensors to spread the use of automated insect monitoring. *J. Sens.* **2018**, *2018*. [CrossRef]

48. Witzgall, P.; Kirsch, P.; Cork, A. Sex pheromones and their impact on pest management. *J. Chem. Ecol.* **2010**, *36*, 80–100. [CrossRef]

49. Dombos, M. Innovative Real-Time Monitoring and Pest Control for Insects, ZooLog Sensor System. LIFE Layman Report 2017. Available online: http://zoolog.hu/insectlife/wp-content/uploads/2019/04/layman_eng_min3.pdf (accessed on 30 July 2020).

50. Metcalf, R.L. Chemical Ecology of Diabroticites. In *Novel Aspects of the Biology of Chrysomelidae*; Springer Science+Business Media: Dordrecht, The Netherlands, 1994; pp. 153–169.

51. Gerber, C.K.; Edwards, C.R.; Bledsoe, L.W.; Obermeyer, J.L.; Barna, G.; Foster, R.E. Sampling devices and decision rule development for western corn rootworm (Diabrotica virgifera virgifera LeConte) adults in soybean to predict subsequent damage to maize in Indiana. *West. Corn Rootworm Ecol. Manag.* **2005**, *2005*, 169–187.

52. Spencer, J.L.; Isard, S.A.; Levine, E. Western corn rootworms on the move: Monitoring beetles in corn and soybeans. In Proceedings of the 1998 Illinois Agricultural Pesticides Conference, Cooperative Extension Service, University of Illinois, Urbana-Champaign, IL, USA, 6–8 January 1998; pp. 10–23.

53. Coats, S.A.; Tollefson, J.J.; Mutchmor, J.A. Study of migratory flight in the western corn rootworm (Coleoptera: Chrysomelidae). *Environ. Entomol.* **1986**, *15*, 620–625. [CrossRef]

54. Isard, S.A.; Spencer, J.L.; Nasser, M.A.; Levine, E. Aerial movement of western corn rootworm (Coleoptera: Chrysomelidae): Diel periodicity of flight activity in soybean fields. *Environ. Entomol.* **2000**, *29*, 226–234. [CrossRef]

MDPI AG
Grosspeteranlage 5
4052 Basel
Switzerland
Tel.: +41 61 683 77 34

Insects Editorial Office
E-mail: insects@mdpi.com
www.mdpi.com/journal/insects